Radiosensitizers and Radiochemotherapy in the Treatment of Cancer

Series in Medical Physics and Biomedical Engineering

Series Editors: John G Webster, E Russell Ritenour, Slavik Tabakov, and Kwan-Hoong Ng

Other recent books in the series:

The Physiological Measurement Handbook
John G Webster (Ed)

Diagnostic Endoscopy
Haishan Zeng (Ed)

Medical Equipment Management
Keith Willson, Keith Ison, and Slavik Tabakov

Targeted Muscle Reinnervation: A Neural Interface for Artificial Limbs
Todd A Kuiken; Aimee E Schultz Feuser; Ann K Barlow (Eds)

Quantifying Morphology and Physiology of the Human Body Using MRI
L Tugan Muftuler (Ed)

Monte Carlo Calculations in Nuclear Medicine, Second Edition: Applications in Diagnostic Imaging
Michael Ljungberg, Sven-Erik Strand, and Michael A King (Eds)

Vibrational Spectroscopy for Tissue Analysis
Ihtesham ur Rehman, Zanyar Movasaghi, and Shazza Rehman

Webb's Physics of Medical Imaging, Second Edition
M A Flower (Ed)

Correction Techniques in Emission Tomography
Mohammad Dawood, Xiaoyi Jiang, and Klaus Schäfers (Eds)

Physiology, Biophysics, and Biomedical Engineering
Andrew Wood (Ed)

Proton Therapy Physics
Harald Paganetti (Ed)

Practical Biomedical Signal Analysis Using MATLAB®
K J Blinowska and J Żygierewicz (Ed)

Physics for Diagnostic Radiology, Third Edition
P P Dendy and B Heaton (Eds)

Series in Medical Physics and Biomedical Engineering

Radiosensitizers and Radiochemotherapy in the Treatment of Cancer

Shirley Lehnert
Department of Oncology
McGill University,
Montreal, Canada

CRC Press is an imprint of the
Taylor & Francis Group, an **informa** business

CRC Press
Taylor & Francis Group
6000 Broken Sound Parkway NW, Suite 300
Boca Raton, FL 33487-2742

First issued in paperback 2019

ISBN-13: 978-1-4398-2902-8 (hbk)
ISBN-13: 978-0-367-37802-8 (pbk)

Visit the Taylor & Francis Web site at
http://www.taylorandfrancis.com

and the CRC Press Web site at
http://www.crcpress.com

For my mother, grandmother, and aunt.
They would have been pleased.

Contents

Series Preface

The *Series in Medical Physics and Biomedical Engineering* describes the applications of physical sciences, engineering, and mathematics in medicine and clinical research.

The series seeks (but is not restricted to) publications in the following topics:

- Artificial organs
- Assistive technology
- Bioinformatics
- Bioinstrumentation
- Biomaterials
- Biomechanics
- Biomedical engineering
- Clinical engineering
- Imaging
- Implants
- Medical computing and mathematics
- Medical/surgical devices
- Patient monitoring
- Physiological measurement
- Prosthetics
- Radiation protection, health physics, and dosimetry

- Regulatory issues
- Rehabilitation engineering
- Sports medicine
- Systems physiology
- Telemedicine
- Tissue engineering
- Treatment

The *Series in Medical Physics and Biomedical Engineering* is an international series that meets the need for up-to-date texts in this rapidly developing field. Books in the series range in level from introductory graduate textbooks and practical handbooks to more advanced expositions of current research.

The *Series in Medical Physics and Biomedical Engineering* is the official book series of the International Organization for Medical Physics.

The International Organization for Medical Physics

The International Organization for Medical Physics (IOMP) represents over 18,000 medical physicists worldwide and has a membership of 80 national and 6 regional organizations, together with a number of corporate members. Individual medical physicists of all national member organizations are also automatically members.

The mission of the IOMP is to advance medical physics practice worldwide by disseminating scientific and technical information, fostering the educational and professional development of medical physics, and promoting the highest quality medical physics services for patients.

A World Congress on Medical Physics and Biomedical Engineering is held every three years in cooperation with the International Federation for Medical and Biological Engineering (IFMBE) and the International Union for Physics and Engineering Sciences in Medicine (IUPESM). A regionally based international conference, the International Congress of Medical Physics (ICMP), is held between world congresses. The IOMP also sponsors international conferences, workshops, and courses.

The IOMP has several programs to assist medical physicists in developing countries. The joint IOMP Library Programme supports 75 active libraries in 43 developing countries, and the Used Equipment Programme coordinates equipment donations. The Travel Assistance Programme provides a limited number of grants to enable physicists to attend the world congresses.

IOMP cosponsors the *Journal of Applied Clinical Medical Physics*. The IOMP publishes, twice a year, an electronic bulletin, the *Medical Physics World*. The IOMP also publishes *e-Zine*, an electronic newsletter, about six times a year. The IOMP has an agreement with Taylor & Francis for the publication of the *Medical Physics and Biomedical Engineering* series of textbooks. IOMP members receive a discount.

The IOMP collaborates with international organizations, such as the World Health Organization (WHO), the International Atomic Energy Agency (IAEA), and other international professional bodies such as the International Radiation Protection Association (IRPA) and the International Commission on Radiological Protection (ICRP), to promote the development of medical physics and the safe use of radiation and medical devices.

Guidance on education, training, and professional development of medical physicists is issued by the IOMP, which is collaborating with other professional organizations in the development of a professional certification system for medical physicists that can be implemented on a global basis.

The IOMP website (www.iomp.org) contains information on all the activities of the IOMP, policy statements 1 and 2, and the "IOMP: Review and Way Forward," which outlines all the activities of the IOMP and the plans for the future.

Preface

My original intention in writing this book was to catalog and describe the mechanism of action for the entities that have been characterized as radiosensitizers. The understanding of the action of ionizing radiation at the level of molecular biology has led to the development of the so-called molecular or targeted radiosensitizers. A book dealing with radiosensitization now has to deal with a vastly increased volume of information covering everything from molecular oxygen and high Z elements to monoclonal antibodies and complex phytochemicals.

Daunting as this prospect may be, it gets worse because it is impossible to ignore the fact that the development of radiosensitizing drugs and procedures almost always takes place in the context of their medical application, providing the motivation and usually the funding for the research and development. This necessitates the second part of the book title *Radiochemotherapy in the Treatment of Cancer*; radiosensitization cannot simply be considered the austere context of molecular biology and radiation chemistry, even though the basic scientist might wish it could be so. In fact, although there are many gaps in knowledge, the combination of radiation and chemotherapy has become the standard of care for most patients with solid tumors, on the basis of well-documented improvements in locoregional disease control and survival.

Radiosensitization or, in the best case scenario, synergy between the two modalities can usually only be reliably demonstrated using preclinical models, and clinical trials evaluating drug combinations are often stimulated by claims that the agents are synergistic in preclinical models. The term *synergy* is not well understood, and there is confusion about the evidence required to conclude that there is synergy between antitumor agents. Of greater importance when evaluating the potential clinical utility from combining agents is the concept of *therapeutic index* (or *therapeutic ratio*). This term refers to the relative toxicity of a treatment of the tumor as compared with its toxicity for critical normal tissues. Improvement in therapeutic index may occur from drugs demonstrating synergy or additivity, but only if the augmented effect of the combination is greater against the tumor than against the critical normal tissues. The link between preclinical research and clinical application is often fragile, and overuse, misuse, or willful misunderstanding of the term synergy (or radiosensitization when radiation is involved) can lead to poorly designed clinical studies. In some cases, there seems to be a disconnect between claims made on the basis of preclinical data and the clinical applicability of a certain treatment, which may be based on a misunderstanding of the terms synergy or supra-additivity and also involves a substantial element of wishful thinking.

Where it is appropriate, I have tried, to the best of my ability and using the information available in the literature, to describe the clinical application of the drug or biological agent combined with radiation. There are certain well-established drug–radiation protocols of proven clinical benefit for which this is not difficult. With respect to others being investigated in different levels of clinical trials, I am not in a position to comment on how successful the given treatment might be, and any statements to that effect are direct quotes from the authors of the reports. Similarly, terms such as radiosensitization and synergy, when they are used, either apply to a clear experimental demonstration or, again, are direct quotes. In all other cases, I have tried to avoid using these terms when they are not directly supported.

The inclusion of data from clinical trials in this book is not intended to present an up-to-date and comprehensive picture of the current clinical activity but simply to summarize historical development based on the results of trials that have been published and to give an impression of the focus and extent of current interest usually on the basis of information obtained from the invaluable website http://www.clinicaltrials.gov. Of necessity, this can give only an incomplete and fragmented picture; nevertheless, I hope it will be useful as an overview for researchers and clinicians.

Acknowledgments

I WANT TO THANK ALL the people who have contributed to the preparation of this book, for showing great generosity in sharing their time and expertise. This is also a good opportunity to acknowledge fellow scientists in Canada and throughout the world who are always ready to share information, ideas, and insights. It is a privilege to belong to this community.

Author

Shirley Lehnert graduated from London University, United Kingdom, with a PhD in biophysics. She did postdoctoral work at the University of Rochester, Rochester, New York. When she moved to New York City, she did research in radiobiology and biophysics first at Sloane Kettering Institute and then at the Radiological Research Laboratory of Columbia University.

When she moved to Canada, she joined the faculty of McGill University, where she is currently a professor in the Department of Oncology. She has published extensively in the fields of radiation biology, tumor biology, and drug delivery.

CHAPTER 1

Radiosensitization and Chemoradiation

1.1 A BRIEF HISTORY OF CHEMORADIATION

Radiation therapy has been the mainstay of nonsurgical treatment of cancer for over a century, while the classic chemotherapy drugs such as cisplatin and 5-fluorouracil (5-FU) have been in clinical use for almost 50 years. The combination of these modalities is a more recent event and the optimal combinations and scheduling continue to evolve. Despite gaps in knowledge, the combination of radiation and chemotherapy has become the standard of care for most patients with solid tumors based on improvements in locoregional disease control and survival; an overview of the most common uses is shown in Table 1.1.

1.2 DEFINITIONS OF RADIOSENSITIZATION

One of the earliest pioneers in this field, G.E. Adams, divided radiosensitizers into five categories [1]:

- Suppression of intracellular-SH [thiols] or other endogenous radioprotective substances
- Radiation-induced formation of cytotoxic substances from the radiolysis of the sensitizer
- Inhibitors of postirradiation cellular repair processes

TABLE 1.1 Mechanisms of Drug and Radiation Interaction

Process	Mechanism	Drugs
Increased radiation damage	Incorporation of chemotherapy drug into DNA/RNA	5-FU: incorporation into DNA, increases susceptibility to radiotherapy damage Cisplatin: cross-links with DNA or RNA (intrastrand and interstrand)
Inhibition of DNA repair	Interference with repair of radiation-induced DNA damage	Halogenated pyrimidines (e.g., 5-FU, BrdUrd, IdUrd Nucleoside analogues (e.g., gemcitabine, fludarabine) Cisplatin, methotrexate, camptothecins and doxorubicin Etoposide, hydroxyurea
Perturbation of cell cycle distribution	Accumulation of cells in the radiosensitive G_2 and M phases. Elimination of radioresistant cells in the S phase	Taxanes lead to cell cycle arrest via tubulin stabilization Nucleoside analogues (gemcitabine, fludarabine), etoposide, methotrexate, hydroxyurea, 5-FU
Incorporation into DNA of pyrimidine analogues that enhance degree of damage by radiation	Cell cycle-specific incorporation of the agent into dividing cells results in decreased repair of DNA damage and increased biochemical effect of radiation through electron capture	Halogenated pyrimidines such as IdUrd and BrdU
Apoptosis	Resistance to apoptosis caused by radiation injury may enhance tumor cell survival	BCL-2 inhibitors, COX 2 inhibitors, drugs that work through ATM, p53, BAX, and epidermal growth factor receptor (EGFR) inhibitors
Inhibition of "prosurvival markers"	Targeted therapies (e.g., EGFR inhibition) block signaling pathways responsible for radioresistance and poor prognosis. Could be in any of the other categories	EGFR inhibitors—shown for anti-EGFR antibody, PKI-166 (small-molecule TKI), and EGFR antisense

- Sensitization by structural incorporation of thymine analogues into intracellular DNA
- Oxygen-mimetic sensitizers, for example, the electron affinic nitroimidazoles

Another interpretation of radiosensitization [2] stressed the importance of a differential effect between tumors and normal tissue and, in addition to this criterion, suggested that only two types of sensitizers had found practical use in clinical radiotherapy—the halogenated pyrimidines, based on the premise that tumor cells proliferate more rapidly and, therefore, incorporate more pyrimidine analogues than the surrounding normal tissues; and the hypoxic cell sensitizers, which increase the radiosensitivity of cells deficient in molecular oxygen, in this case, based on the premise that hypoxic cells occur only in tumors and not in normal tissues. Although both of these premises were sound, these two classes of drugs turned out to be less than effective in the clinic for reasons that are probably connected with pharmacokinetics and drug delivery issues. In fact, the hypoxic cell sensitizers (e.g., misonidazole), and thymidine analogues (e.g., bromodeoxyuridine) are examples of "true" radiosensitizers in that they have no inherent cytotoxicity. However, the most commonly used radiosensitizers (cisplatin, 5-FU, and taxanes) do have inherent cytotoxic activity and can increase damage to normal tissues, with a "true" benefit achieved only if the increase in antitumor effect is larger than the normal tissue damage. The result of combined treatment is at least an additive effect of the two modalities and at best some combination of radiosensitization and additive effects.

1.2.1 Combining Two Agents Together: What Is Meant by Additivity or Synergy?

A common definition of synergy is that the effect of two agents given together is more effective than would be predicted based on their individual activity. Thus, if agent 1 alone causes the surviving fraction to be reduced to 0.5, and agent 2 used alone also reduces survival to 0.5, then the effect of agents 1 and 2 combined will be to reduce survival to 0.25 if the effect is additive and to less than 0.25 if the effect is synergistic. This will only be correct if the survival curves generated by agents 1 and 2, when used separately, are both exponential. If the response curves are shouldered, more than one scenario is possible. If two agents are used, both of which have a survival curve with a shoulder followed by an exponential slope, then there are three possibilities: (1) if there is no overlap of damage and the shoulder representing the accumulation of sublethal damage is retained for the second agent; (2) if cells have accumulated maximum sublethal damage from agent 1, the shoulder of the survival curve is lost for the second agent; and (3) if the final slope of the survival curve is altered,

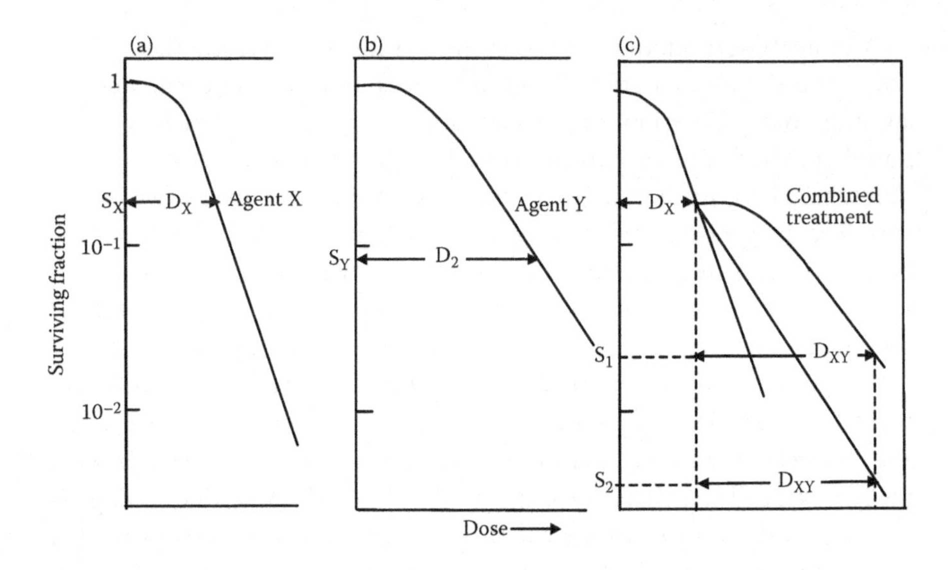

FIGURE 1.1 The effect of combining two cytotoxic agents together. Cell survival is indicated following treatment with agent X or agent Y, each of which has a survival curve with a shoulder followed by an exponential slope with increasing dose (a and b). Survival (S_{XY}) after combined use of dose D_X of X and dose D_Y of Y will be equal to $S_1 = (S_X \times S_Y)$ if there is no overlap of damage and the shoulder is retained for the second agent (c). Survival after combined treatment S_{XY} will be equal to S_2 if cells have accumulated maximum sublethal damage from the first agent X and the shoulder of the curve is lost for the second agent Y.

that is, becomes steeper, for the combined effect of two drugs (Figure 1.1). The relationships described imply that there is a range over which two agents can produce additive effects.

1.2.1.1 The Isobologram

A framework defining the interactions between combined radiation and chemotherapy was described by Steel and Peckham [3]. The isobologram is generated by plotting the dose of each agent (i.e., drug and radiation), which produces the same cytotoxic effects (an isoeffect). Two curves, named "mode 1" and "mode 2," can then be generated (Figure 1.2). The mode 1 curve results from the assumption that radiation and chemotherapy act independently, and is created by plotting a given dose of radiation against the dose of chemotherapy needed to produce an effect equal to the difference between the chosen cytotoxic effect and the effect of the current dose of radiation. The mode 2 curve assumes that radiation and chemotherapy have identical mechanisms of action. Points on this curve are

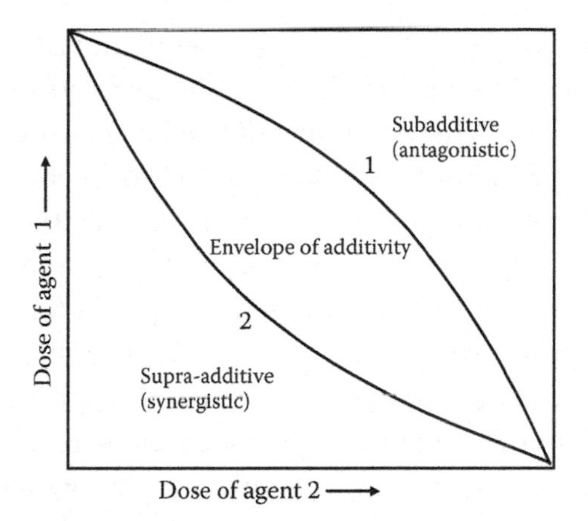

FIGURE 1.2 Schematic example of an isobologram depicting the combination of radiation and a systemic agent. The x and y axes show the isoeffective levels for radiation and drug. The thick line is the line of additivity, and the additivity envelope is based on the combined standard errors. Curves above the envelope represent antagonistic effects and curves below the envelope represent synergistic effects.

generated by plotting doses of radiation against the doses of chemotherapy needed to increase the effect of the dose of radiation to the chosen cytotoxic effect. Multiple points on both of these lines are obtained by varying the radiation dose and calculating the appropriate doses of chemotherapy. Finally, the true (empiric) survival curve is generated for radiation and of each agent needed to produce the chosen cytotoxic effect are plotted. Points that fall below the envelope between mode 1 and 2 curves indicate supra-additivity, data points occurring within the envelope (between the two curves) indicate additivity, and data points above the envelope indicate infra-additivity. The difference between the mode 1 and mode 2 curves depends on the fact the radiation and drug dose–response curves, that usually have different shapes so that their relationship changes for different parts of the response curve.

1.2.1.2 Median Effect Principle

Another approach to evaluating the degree of additivity or synergy between agents is the median effect principle. This method also depends on the availability of dose–effect relationships for each agent used alone and for both agents used together. It relies on the calculation of a combination index (CI). A computer program is available to do this [4] and a simplified explanations of the principle has been published [5]. A CI of 1 represents additivity, a CI greater than 1 represents synergy, and a CI less than 1 represents antagonism. The relationships can be portrayed graphically by a normalized isobologram in which the ratio of dose of drug needed to affect a fraction x when used in combination relative to when it is used alone is plotted (for different values of x) for drug 1 (vertical axis) against drug 2 (horizontal axis). If the combination data points fall on the curve, an additive effect is indicated. If the combination data points fall on the lower left or on the upper right, synergism or antagonism is indicated, respectively. A difference between the two methods of analysis is that the isobologram analysis of Steel and Peckham defines an envelope of additivity, but the median effect method of Chou defines only a single condition of additivity.

1.2.2 Drugs and Radiation

Mechanisms of interaction between drugs and radiation can be evaluated at the cellular level on the basis of radiation survival curves prepared with or without the drug. Drugs can influence the survival curve in three ways: the curve may be displaced downward by the amount of cell kill produced by the drug alone; the shoulder on the survival curve may be lost, suggesting that the drug is in some way preventing the repair of sublethal damage; the slope on the survival curve may be altered, indicating sensitization or protection by the drug (Figure 1.3). Most drugs influence

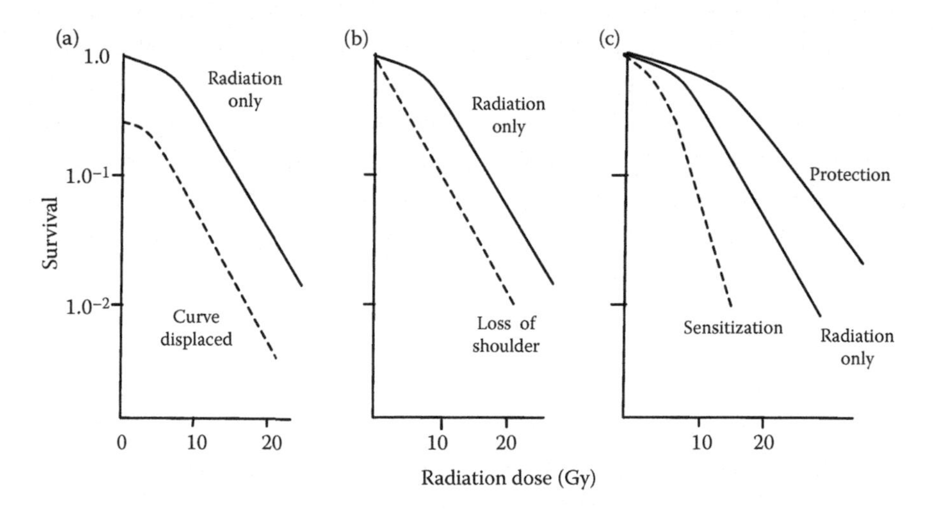

FIGURE 1.3 Possible outcomes of drug treatment on the relationship between the radiation dose and cell survival. (a) Displacement of curve; (b) loss of shoulder, indicating the effects of drug on the repair of sublethal damage; and (c) change in the slope of the curve indicating sensitization or protection.

radiation response according to the first two modes, a type of response corresponding to the limits of additivity described in Section 1.2.1. The third type of response defines drugs that are radiosensitizers or protectors.

1.2.3 Mechanisms of Benefit from Radiation–Drug Therapy

Bentzen and collaborators [6] have proposed a framework for the rational design and analysis of drug–radiation therapies from a radiation oncology perspective. This encompasses five mechanisms: spatial cooperation, cytotoxic enhancement, biological cooperation, temporal modulation, and normal tissue protection.

1.2.3.1 Cytotoxic Enhancement

The clinical outcome of treatment may reflect all of the mechanisms listed above; however, in an experimental setting, usually only one or two of them can be evaluated at one time. The end result of cytotoxic enhancement will be represented by changes in the survival curves generated when tumor cells are irradiated *in vitro*.

Table 1.2 lists the important interactions between radiation and chemotherapy, which occur at the cellular and tumor cell population level.

1.2.3.1.1 Exacerbation of DNA Damage Ionizing radiation induces DNA base damage, alkali-labile sites, single-strand breaks (SSBs), and double-strand breaks (DSBs). All this damage can be rapidly repaired except for DSBs, which, if not repaired, are considered lethal. The majority of chemotherapeutic drugs also target DNA and directly produce damage, which manifests as DNA breaks, adducts, and intercalation. In some cases, damage induced by the two agents will be simply additive whereas in others interactive processes will amplify the effects. The integration of cisplatin into DNA or RNA close to a radiation-induced SSB can act synergistically to make the defect significantly more difficult to repair. An increase in the number of radiation-induced strand breaks may occur by conversion of radiation-induced SSBs to DSBs during the repair of DNA–platinum adducts.

Compounds such as iododeoxyuridine (IdUrd) and bromodeoxyuridine (BrdUrd), when incorporated into DNA, cause no damage in the absence of radiation but enhance radiation-induced DNA damage, likely through the production of reactive uracilyl radicals and halide ions, which in turn induce DNA SSBs in neighboring DNA (Chapter 4).

TABLE 1.2 Mechanisms of Chemotherapy and Radiotherapy Interaction

Process Affected (Mode of Radiosensitization)	Mechanism	Example
Increase DNA damage (cytotoxic enhancement)	Drug incorporated into DNA/RNA	Cisplatin inter/intrastrand cross-links. 5-FU incorporation increases damage susceptibility
Inhibit DNA repair (cytotoxic enhancement)	Prevent repair of radiation damage	Halogenated pyrimidines (5-FU, I/BrUrdr), nucleoside analogues (gemcitabine), cisplatin, methotrexate, Topo I/II inhibitors (camptothecins, etoposide), hydroxyurea
Cell cycle slowing/arrest (cytotoxic enhancement)	Radiation and most chemotherapy target cycling (proliferating) cells. Accumulation of cells in radiosensitive G_2/M. Elimination of radioresistant S-phase cells	Taxanes (microtubule stabilization) nucleoside analogues, etoposide, hydroxyurea
Target radiation-induced activation of prosurvival mechanisms [promitotic, antiapoptotic] (cytotoxic enhancement)	Block receptors and signaling pathways	Drugs targeting EGFR family receptors (antibodies, small molecule TKIs)
Prevents tumor cell repopulation (biological)	Stops accelerated proliferation occurring during radiotherapy	Most chemotherapy. Antimetabolites with S-phase activity (5-FU, hydroxyurea)
Target hypoxic cells (biological)	Pro-drugs activated in reducing (hypoxic) environment. Hypoxic cell sensitizers	Tirapazamine, mitomycin selectively kill hypoxic cells. Nitroimidazoles sensitize cells to radiation

1.2.3.1.2 Inhibition of DNA Repair Chemotherapy agents have been shown to inhibit the repair of radiation damage by a number of mechanisms and, in some cases, a drug may do so by more than one means. In some cases, DNA synthesis and repair share common pathways providing a rationale for using DNA synthesis inhibitors with radiation as a means of inducing cytotoxic damage to tumor cells. Examples in this category are the nucleoside analogues, cisplatin, bleomycin, doxorubicin, and hydroxyurea. Fludarabine, for example, is a nucleoside analogue, which is incorporated into DNA and blocks DNA primase, DNA polymerase α and ε, and DNA ligase.

Many of the same agents inhibit the repair of radiation-induced DNA damage by interfering with nucleoside and nucleotide metabolism. Examples here include the fluoropyrimidines, thymidine analogues, gemcitabine, and hydroxyurea (Chapter 5).

A major problem with DNA repair inhibition as an exploitable mechanism for obtaining a therapeutic gain is the lack of evidence for a selective antitumor effect; however, there may be specific cases in which targeting the DNA repair can be effective and have an element of tumor specificity. Due to genetic instability, tumors are often defective in one aspect of DNA repair but usually have backup pathways for accomplishing repair. Attacking these backup pathways can render the tumor radiosensitive while leaving the normal tissue relatively resistant.

Inhibition of repair, or the conversion of SSBs to DSBs, has the effect of increasing the slope of the radiation survival curve and leads to an enhanced response. An important factor from the clinical standpoint is that radioenhancement, which occurs as a result of repair inhibition, will be more pronounced in fractionated schedules than for single doses.

1.2.3.1.3 Cell Cycle Effects When administered concurrently, radiotherapy and chemotherapy often target different phases of the cell cycle and may cooperate to produce an additive effect. The radiosensitivity of a cell is dependent on the phase of the cell cycle; cells in the S phase being the most radioresistant, whereas those in the G2–M phase are the most radiosensitive.

Because tumors are often defective in one of the cell cycle checkpoints (e.g., the G_1/S checkpoint), inhibiting the remaining checkpoints can shorten the cell cycle, leaving tumors with less repair time and resulting in greater cell kill than in normal tissues.

The majority of chemotherapeutic agents are inhibitors of cell division and are thus mainly active on proliferating cells; in addition, many drugs have cell cycle phase specificity. Gemcitabine, fludarabine, methotrexate, and 5-FU inhibit various enzymes involved in DNA synthesis and repair in S-phase cells. Agents such as etoposide, doxorubicin, alkylating agents, and platinum compounds induce DNA strand breaks and DNA strand cross-links in any phase of the cell cycle, but will only become potentially lethal in replicating cells. Agents such as taxol, taxotere, and *Vinca* alkaloids inhibit mitotic spindle formation and thus are mainly active during mitosis.

As a consequence of cell cycle phase-selective cytotoxicity of chemotherapeutic agents, the remaining surviving cells will be synchronized. If radiation could be delivered when these synchronized cells have reached a more radiosensitive phase of the cell cycle (e.g., G_2/mitosis), a potentiation of the radiation effect might be observed. Such a mechanism of interaction between drugs and ionizing radiation has often been reported in preclinical experimental models. However, in the clinic, because of the difficulty in assessing the appropriate timing between drug injection and radiotherapy delivery, it is unlikely that cell synchronization can be successfully exploited. Furthermore, because radiotherapy is typically delivered on a fractionated basis, it is likely that this effect will be lost between fractions because of reassortment.

1.2.3.1.4 Enhanced Apoptosis Apoptosis (or interphase cell death) is a common mechanism of cell death induced by chemotherapeutic agents. These drugs can trigger one or more of the pathways leading to apoptosis. For the antimetabolites, DNA incorporation is a necessary event to ensure a robust apoptotic response, hence the specific sensitivity of S-phase cells to these agents. Within this framework, it has been hypothesized that combining these drugs with ionizing radiation, which is very effective in inducing DNA SSBs or DSBs in every phase of the cell cycle, could facilitate their DNA incorporation and thus trigger an enhanced apoptotic reaction.

1.2.3.2 Targeted Radiosensitizers

A major part of the research effort into putatively radioenhancing drugs has been devoted, in the last few years, to agents that target either growth factor receptors or intracellular signal transduction pathways through which signaling is initiated by growth factor receptor activation. Receptor activation by mutation or overexpression, mutations in oncogenes (such as *RAS*), or tumor suppressor genes (such as *PTEN*) can lead to aberrant signaling through the PI3K-AKT, MAPK-ERK, nuclear factor-κB (NF-κB), and transforming growth factor-β (TGF-β) pathways characteristic of tumor cells. Ionizing radiation also activates the PI3K, MAPK, and NF-κB pathways through interaction with growth factor receptors and this can ultimately create radioresistance by decreasing apoptosis or increasing DNA repair (Figure 1.4). Inhibition of signaling by blocking the receptor or targeting signal transduction intermediates can reverse this mode of radioresistance, that is, act as a radiosensitizer (Chapters 10 and 11).

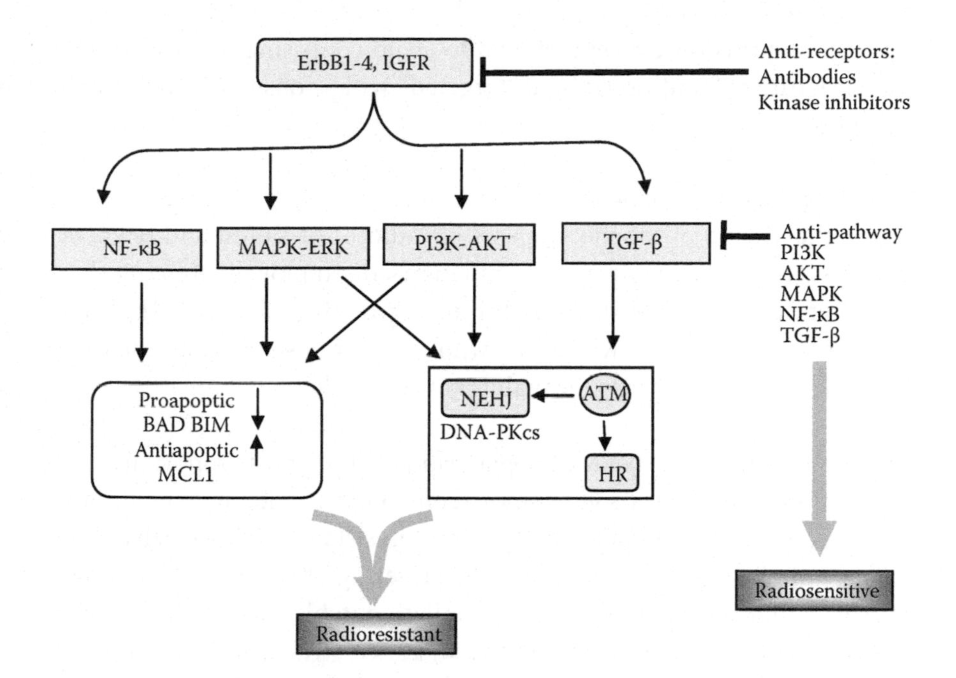

FIGURE 1.4 Activation of intracellular signaling pathways by agonists including ionizing radiation can initiate antiapoptotic and promitotic signaling effectively increasing the resistance of the irradiated cell. AKT, MAPK, and NF-κB signaling lead to phosphorylation and inactivation of the proapoptotic proteins BAD and BIM or activation of the antiapoptotic protein myeloid cell leukemia sequence 1 (MCL1). DNA–DSB repair capacity correlates with radiosensitivity. Activation of the AKT and MAPK pathways leads to the activation of the catalytic subunit of DNA-dependent protein kinase (DNA-PKcs; also known as PRKDC), a central protein in double-strand break (DSB) repair by nonhomologous end joining (NHEJ). This protein can also be activated by the receptor tyrosine kinase (RTK) EGFR after it is translocated to the nucleus. TGFβ is necessary for the full activation of ataxia telangiectasia mutated (ATM) in response to DNA damage, affecting DSB repair by NHEJ and homologous recombination (HR). IGFR, insulin-like growth factor receptor.

1.2.3.2.1 Definition of Targeted Radiosensitizers The term *targeted* is currently used to refer to a variety of drugs, such as therapeutic antibodies that target specific molecules. Examples of targeted drugs include tyrosine kinase inhibitors, PARP inhibitors, and mTOR inhibitors. Use of the word targeted can be debatable because it implies that "traditional" drugs, such as cisplatin and paclitaxel, which have well-known interactions with DNA and microtubules, respectively, are somehow "untargeted." Despite perpetuating this misconception (and to avoid confusing the reader), the nomenclature of "traditional" and "targeted" will be observed throughout this book.

1.2.3.3 Biological Cooperation

This is the second of the mechanisms of radiosensitization and refers to strategies that target distinct cell populations, or employ different mechanisms for cell killing, or delaying tumor regrowth (Table 1.2). The endpoint here is that the growth of the whole tumor is slowed and the cells targeted are not necessarily the malignant cells only.

1.2.3.3.1 Targeting the Hypoxic Subpopulation The most important distinguishing feature of a solid tumor from the radiotherapy standpoint is the presence of a subpopulation of radioresistant, hypoxic cells. One example of biological cooperation is the targeting of hypoxic tumor cells, thereby complementing the effects of radiation, which will be greater for well-oxygenated cells. A second hypoxia-related stratagem is the conversion of a pro-drug to a cytotoxic drug under reducing conditions such as those that occur in the hypoxic fraction of a solid tumor. Combining radiation with bioreductive drugs, such as tirapazamine or mitomycin C, selectively targets hypoxic tumor cells while the oxic population is targeted by radiation.

1.2.3.3.2 Targeting Nontumor Cells Another possibility is targeting nontumor cells to the overall detriment of the tumor population. An example is vascular targeting agents such as combrestatin or dimethylxanthenone acetic acid, which cause a shutdown of the tumor vasculature leading to tumor cell death via hemorrhagic necrosis. The case for combining radiation with antiangiogenic agents may seem counterintuitive because of the perception that it will create hypoxic regions. It seems, however, that antiangiogenic agents induce a normalization of tumor blood vessels which, in fact, reduces tumor hypoxia, creating potential biological cooperation between this class of drugs and radiation.

1.2.3.3.3 Targeting Drug-Resistant Cells with Radiation Genetic instability in tumors can lead to subpopulations of different sensitivities to drugs and radiation. Response to therapy may select for resistant cells, which have a survival advantage and will determine tumor response. Combined treatment with radiation and drugs may lead to improvement in therapeutic index if one modality can target a subpopulation resistant to the other. This cooperative effect requires that the mechanisms of resistance be independent. Mechanisms other than hypoxia conveying radiation resistance may include enhanced ability to repair DNA damage, increased levels of sulfhydryl compound (e.g., glutathione), or increased levels of associated enzymes (gluthione-S-transferase) that scavenge free radicals (particularly in hypoxic cells) and decreased ability to undergo apoptosis. These mechanisms may cause resistance to other drugs but many mechanisms of drug resistance do not cause resistance to radiation.

1.2.3.3.4 Targeting Repopulation Proliferation of surviving cells during fractionated radiation effectively increases the number of cells that must be killed for tumor control. If anticancer drugs (given during the course of radiation) inhibit repopulation, then combined treatment may convey a therapeutic advantage if the rate of repopulation is greater for tumor cells than for normal tissue. Greater specificity would be expected from agents that specifically inhibit tumor cell proliferation such as hormonal agents tamoxifen and antiandrogens, which are used concurrently with radiation for the treatment of breast or prostate cancer, and antiproliferative agents such as cisplatin in non-hormonal-dependent tumors. The use of molecularly targeted agents may have the same effect in some cases. Improved survival is seen for patients with head and neck cancer treated with radiation and concurrent cetuximab (an antibody-targeting EGFR described in Chapter 10) compared to those treated with radiotherapy alone. This is believed to be attributable to the inhibition of tumor cell repopulation.

1.2.3.4 Temporal Modulation

The aim of this approach is to enhance the tumor response to fractionated radiotherapy. The classic radiobiological framework for discussing dose-fractionation effects, the four R's of radiotherapy: repair, repopulation, reoxygenation, and redistribution refer to the four biological processes that take place in the time interval between radiation dose fractions: DNA damage repair, cellular repopulation or proliferation, reoxygenation of hypoxic tumor cells owing to cell killing and reduced oxygen consumption, and redistribution from more resistant (e.g., late S phase) to more sensitive phases of the cell cycle by cellular progression in the cycle. These four mechanisms are not active to the same degree in tumor and normal cells, and thus, for a given level of normal tissue damage, tumor control can be optimized by changing the distribution of dose delivered over time, that is, the dose per fraction delivered in each treatment session, the time interval between fractions and the total duration of radiotherapy from the first to the last dose fraction. Temporal modulation denotes strategies affecting one or more of these four processes, that is, the biological processes occurring from one radiation dose fraction to the next. Radioenhancing drugs in this context could function by, for instance, inhibiting repair taking place between dose fractions.

1.2.3.5 Spatial Cooperation

The term spatial cooperation is used to describe the scenario whereby radiotherapy acts locoregionally, and chemotherapy acts against distant micrometastases, without interaction between the agents. This does not involve any interaction between radiation and drugs because, in fact, the two modalities are targeting spatially separate tumor cell populations. This cooperative effect requires the agents to have nonoverlapping toxicity profiles so that both modalities can be used at effective doses without increasing normal tissue effects.

Improvement in therapeutic ratio from using drugs and radiation requires selective effects to increase damage to tumor cells compared with normal tissues. One instance of this is when radiation is used to treat the primary tumor whereas drugs are used to treat metastatic sites containing a smaller number of cells. This spatial cooperation requires no interaction between the two modalities but involves different dose-limiting toxicities.

1.2.3.6 Normal Tissue Protection

Radiotherapy is a mainstay of treatment for many forms of locally advanced cancer. Efforts to improve efficacy have focused on the identification of compounds that can sensitize targeted areas to radiotherapy while limiting bystander and systemic toxicity, thereby enhancing radiotherapy within tumor cells relative to normal cells. Drugs currently in use are nearly all standard cytotoxic agents and to date there are no ideal radiosensitizing drugs available for clinical use.

Molecularly targeted therapies may present another approach. Several drugs have been proposed to provide cytoprotection of normal cells or to modulate the cytotoxic response of normal tissue. Until relatively recently, efforts in this research area have concentrated on free-radical scavengers, but the improved understanding of the molecular pathology of radiotherapy (especially late normal-tissue effects) has highlighted several novel interventions for reducing or avoiding late toxicity.

1.2.4 Quantification of the Chemotherapy and Radiation Interaction: The Therapeutic Ratio

The five mechanisms of radiosensitization listed above all contribute to the final reckoning, the therapeutic ratio. The therapeutic ratio, or therapeutic gain, is the relative expected benefit of a combined modality treatment, integrating both the tumor and the normal tissue effects. It is defined as the ratio of dose modifying factors (DMFs) for tumor over normal tissues. This ratio is derived from sigmoid-shaped dose–response curves, calculated by plotting the response of tissues (both normal and tumor) on the ordinate axis versus the chemotherapy or radiation therapy dose on the abscissa (Figure 1.5). When chemotherapy is combined with radiation, both normal tissue and tumor control curves produced by radiation alone shift to the left because of the chemotherapy-induced sensitization of cells. Ideally, radiation sensitizers should influence the tumor response curve more than the normal tissue curve, resulting in a therapeutic ratio greater than one. This would indicate that overall, the combined modality treatment has a relatively greater effect on tumor control compared with normal tissue toxicity. Conversely, a therapeutic ratio below unity indicates that the combined treatment is relatively more toxic than beneficial. The therapeutic ratio may differ for early and late normal tissue toxicity and both should be evaluated.

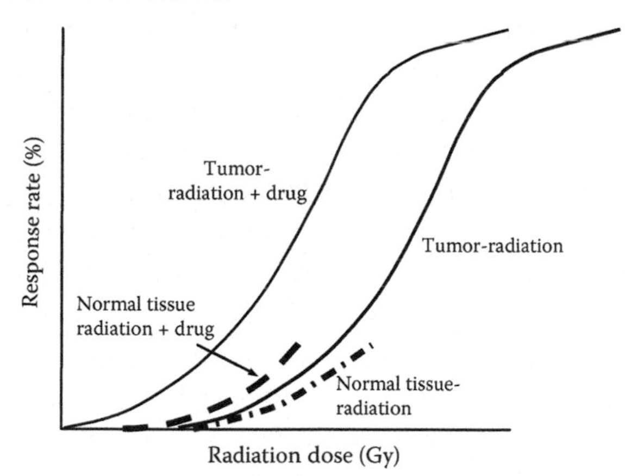

FIGURE 1.5 How the addition of a second cytotoxic agent to radiation treatment may influence the therapeutic ratio. The addition of the drug moves the response curve for both tumor and normal tissue to the left. If the drug has a greater effect on tumor control than on normal tissue morbidity, that is, the tumor curve is moved further to the left, then more effective local control of the tumor is possible for the same level of normal tissue injury.

1.2.5 Is Radiochemotherapy Preferable to Dose Escalation?

Chemoradiation in the treatment of cancer has become complex, there are many drugs and drug combinations involved particularly with the advent of targeted molecular agents. In addition, there are many different protocols believed to be specific for different types of disease. One question that has been asked is whether this is worthwhile and could the same result not be achieved by a small increment in total radiation dose while still avoiding unacceptable toxicities.

The first thing to note is that concomitant radiochemotherapy has been shown on numerous occasions to improve the outcome over radiotherapy alone. For example, prospective clinical trials have shown that the addition of concomitant chemotherapy to radiotherapy of advanced head and neck cancer increases locoregional control (LRC) and, in some trials, overall survival [7].

Several efforts have been made to actually quantitate the effect of chemotherapy in terms of equivalent radiation dose.

- In one study [8], the estimated radiotherapeutic dose equivalence of chemoradiotherapy in head and neck cancer was based on the biologically equivalent dose (BED) of radiotherapy in nine trials of standard and five trials of modified fractionated radiotherapy with or without chemotherapy as calculated using the linear-quadratic formulation. On the basis of one trial (RTOG 90-03), it was calculated that a 1% increase in BED yields a 1.1% increase in LRC. The mean BED of standard fractionated radiotherapy was 60.2 Gy_{10}, and that for modified fractionation was 66 Gy_{10}. Equivalent values for chemoradiotherapy were 71 Gy_{10} and 76 Gy_{10}. On the basis of the differences in BED, it was concluded that chemotherapy increases BED by approximately 10 Gy_{10} in standard and modified fractionated radiotherapy, equivalent to a dose escalation of 12 Gy in 2 Gy daily or 1.2 Gy twice daily (the term Gy_{10} implies that the calculations were done assuming an α/β ratio of 10 Gy).

- In a response to this article Fowler et al. [9] concluded that the estimate was too large by 40% or 50%. Criticism was made on radiobiological grounds, failure to include the start uptime for tumor repopulation, the use of an incorrect value for T_{pot} (potential doubling time), and use of an inappropriate value for the relationship between increase in biological dose and improvement in LRC (the

"S" value). In the Fowler version, the corrected BED of 8.8 Gy_{10} divided by the BED of 2.4 Gy_{10} for each 2-Gy fraction gave an average 3.6 fractions of 2 Gy for the effect of chemotherapy in head-and-neck radiotherapy. This was nearer to the informal suggestions of "approximately 3 fractions" that have been made by some radiation oncologists. The two papers were in agreement that the observed escalation in outcome could not be safely achieved by increasing radiation dose alone.

- In a study published by Jones and Sanghera [10], the biologically effective dose concept is used to estimate the α/β ratio and K (dose equivalent for tumor repopulation per day) for patients with high-grade glioma treated in a randomized fractionation trial and the equivalent radiation dose of temozolomide, chemotherapy was estimated from another randomized study. The method assumed that the radiotherapy biologically effective dose is proportional to the adjusted radiotherapy survival duration of patients with high-grade glioma. The authors used the BED linked to the radiotherapy-adjusted survival duration to initially estimate the two radiobiologic parameters, the α/β ratio and K. Using these parameters, the equivalent radiation dose of temozolomide cytotoxic chemotherapy in high-grade gliomas could be estimated from the results of another clinical trial. The median equivalent biologically effective dose of temozolomide was 11.03 $Gy_{9.3}$ (equivalent to a radiation dose of 9.1 Gy given in 2-Gy fractions). The calculated value could underestimate the dose equivalence in some patients and it was not possible to differentiate the separate contributions of the concomitant and adjuvant chemotherapy. The equivalent dose of around 9 Gy in 2-Gy fractions could, in principle, exert a marked difference in response, possibly achieving a cure in more radiosensitive tumor subtypes, although this equivalent dose would be delivered over a long overall time, inevitably "diluting" its effect because of tumor cell repopulation.

- A very recent study by Plataniotis and Dale [11] sought to estimate the radiation equivalent of chemotherapy's contribution to observed complete response rates in published results of radiochemotherapy in muscle-invasive bladder cancer. A standard logistic dose–response curve was fitted to data from radiation therapy-alone trials and then used as the platform from which to quantify the chemotherapy contribution in one-phase radiochemotherapy trials. Two possible

> mechanisms of chemotherapy effect were assumed: (1) a fixed radiation-independent contribution to local control, or (2) a fixed degree of chemotherapy-induced radiosensitization. A combination of both mechanisms was also considered. The respective best-fit values of the independent chemotherapy-induced complete response (CCR) and radiosensitization (s) coefficients were 0.40 (95% confidence interval, 0.07–0.87) and 1.30 (95% confidence interval, 0.86–1.70). The effect of independent chemotherapy was slightly favored by the analysis, and the CCR value derived was consistent with reports of pathologic complete response rates seen in neoadjuvant chemotherapy-alone treatments of muscle-invasive bladder cancer. The radiation equivalent of chemotherapy's contribution in the radiochemotherapy of bladder cancer was found to be equivalent to approximately 36.3 Gy, delivered in 2-Gy fractions, or an overall dose sensitization factor of 1.30.

The result of the last study is much larger than either of the others but somewhat different results are not surprising given that they looked at different diseases treated with different radiochemotherapy protocols and used different endpoints. Nevertheless, the common conclusion was that improvement in outcome achieved by chemoradiation equated with a significant radiation dose equivalent that could not safely be achieved by dose escalation. Whether in fact the improved response was the result of interactive cytotoxic effects of drugs and radiation, were simply additive effects of the two modalities, or some combination of the two processes, could not be determined. In fact, from the standpoint of the bottom line, which is an index of the success of treatment, it is not important how that result is achieved.

1.2.6 The Relationship between Preclinical Studies and Clinical Trials

Clinical trials evaluating drug combinations are often supported by claims that the agents are synergistic in preclinical models. The definition of the term was explored earlier in this chapter and has been the subject of rigorous analyses in several publications. Synergy implies at the very least that the combined effect of two agents is greater than would be expected from adding the effects of each agent alone. The term is not well understood and there is confusion about the evidence required to conclude that there is synergy between antitumor agents. Even when synergy is present between two or more agents, this does not necessarily mean that the combination will provide useful therapy. Of greater importance when evaluating the potential clinical utility from combining agents is the concept of therapeutic index (or therapeutic ratio). This term was also discussed earlier in this chapter and refers to the relative toxicity of a treatment of the tumor as compared with its toxicity for critical normal tissues. Improvement in therapeutic index may occur from drugs demonstrating synergy, but only if the augmented effect of the combination is greater against the tumor than against the critical normal tissues.

Overuse, misuse, or willful misunderstanding of the term synergy (or radiosensitization when radiation is involved) can lead to poorly designed clinical studies.

A recent article by Ocana et al. [5] surveyed the literature to identify reports that referred to synergy in preclinical studies to justify a drug combination being evaluated in a clinical trial. The authors identified reports of clinical trials that contained the term "synergy" or "synergistic" and reviewed the preclinical data used to justify their combination. They also evaluated the methods used to support claims for synergy, and whether the preclinical data supported potential improvement in therapeutic index. Eighty-six clinical articles met eligibility criteria and 132 preclinical articles were cited in them. Most of the clinical studies were phase I (43%) or phase II trials (56%). Appropriate methods to evaluate synergy in preclinical studies included isobologram analysis in 18 studies (13.6%) and median effect in 10 studies (7.6%). Only 26 studies using animal models (39%) attempted to evaluate therapeutic index. There was no association between the result of the clinical trial and the use of an appropriate method to evaluate synergy.

The results showed that for most clinical studies, the term *synergy* is used without an appropriate understanding of either the underlying concept or the preclinical methods that are necessary to evaluate synergy. Claims for synergy should be used appropriately when evaluating drug combinations in preclinical studies and clinical trials, but more importantly, preclinical studies should provide evidence of enhanced therapeutic index before proceeding to clinical evaluation of drug combinations. This article was concerned with drug–drug interactions but there is no doubt that the same caveats should apply to radiation–drug studies.

1.3 WHAT THIS BOOK IS ABOUT

This book set out to be about radiosensitization as precisely defined. This can usually only be reliably demonstrated using preclinical models. The many "conventional" radiosensitizing drugs and the increasing understanding and availability of conventional and targeted drugs that are being combined with radiation make this a fascinating and very complex subject, which appeals to radiation biologists, molecular biologists, and pharmacologists among others. Most often, the raison d'etre of this research is the prospect of developing drug–radiation combinations that will be effective in the clinic. Consequently, the second aspect of this book is chemoradiation, the clinical application of drugs and radiation in the treatment of cancer. This is more difficult; apart from well-established drug–radiation protocols and a much smaller number involving recently developed agents, the information accessible in the literature on completed drug trials is patchy and frequently old. There is often a disconnect between claims made on the basis of preclinical data and the clinical applicability of a certain treatment, which may be based on a misunderstanding of the terms synergy or supra-additivity and also involves a substantial element of wishful thinking.

At the end of each chapter, I have tried, to the best of my ability and using the information available in the literature, to describe the clinical application of the drug or biological agent. I am not in a position to comment on how successful the given treatment might be and any statements to that effect are direct quotes from the authors of the report. Similarly, terms such as radiosensitization and synergy, when they are used, apply either to a clear experimental demonstration or are direct quotes from the author. In all other cases, I have tried to avoid using these terms when they are not directly supported.

1.4 SUMMARY

Combinations of radiation and chemotherapy have become the standard of care for most patients with solid tumors based on improvements in locoregional disease control and survival. A theoretical framework for defining the interactions between two modalities such as radiation and chemotherapy has been developed giving precise definitions to the terms synergy and additivity on the basis of isobologram analysis and median effect principle. An evaluation of the mechanisms of interaction between drugs and radiation at the cellular level can be done on the basis of radiation survival curves prepared with or without the drug.

A framework for the rational design and analysis of drug–radiation therapies from a radiation oncology perspective has been described, and involves five mechanisms: spatial cooperation, cytotoxic enhancement, biological cooperation, temporal modulation, and normal tissue protection.

Cytotoxic enhancement can be determined at the molecular and single-cell level; the mechanisms involved are exacerbation of DNA damage, inhibition of DNA repair, cell cycle effects, and enhanced apoptosis. Biological cooperation applies at the level of populations of cells in the tumor and involves differential targeting of hypoxic cells, drug-resistant tumor cells, or nontumor cells by radiation. This can be evaluated in preclinical tumor models.

The aim of temporal modulation is to enhance the tumor response to fractionated radiotherapy. Spatial cooperation is used to describe the scenario whereby radiotherapy acts locoregionally, and chemotherapy acts against distant micrometastases, without interaction between the agents whereas normal tissue protection implies a differential effect of the drug–radiation combination on normal versus tumor tissue. These three mechanisms of radiosensitization apply at the clinical level and are difficult or impossible to evaluate in preclinical models. All five mechanisms of radiosensitization listed above contribute to the final balance sheet of treatment, the therapeutic ratio.

A number of studies based on clinical results have been done to determine the radiation dose equivalent of added chemotherapy. The results were diverse, not surprisingly, given the range of diseases, treatments, and end points involved in the study material. Nevertheless, the common conclusion was that improvement in outcome achieved by chemoradiation equated with a significant radiation dose equivalent that could not safely be achieved by dose escalation.

Finally, the interpretation of information from preclinical studies and how this is applied to the design of clinical trials is discussed.

REFERENCES

1. Adams, G. Chemical radiosensitization of hypoxic cells. *Br Med Bull* 1973; 29:48–53.
2. Hall, E. *Radiobiology for the Radiologist*. Philadelphia, PA: Lippincott, Williams & Wilkins, 2000.

3. Steel, G., and Peckham, M. Exploitable mechanisms in combined radiotherapy-chemotherapy: The concept of additivity. *Int J Radiat Oncol Biol Phys* 1979;5(1):85–91.
4. Chou, T. Drug combination studies and their synergy quantification using the Chou–Talalay method. *Cancer Res* 2010;70(2):440–446.
5. Ocana, A., Amir, E., Yeung, C., Seruga, B., and Tannock, I. How valid are claims for synergy in published clinical studies? *Ann Oncol* 2012;23:2161–2166.
6. Bentzen, S., Harari, P.M., and Bernier, J. Exploitable mechanisms for combining drugs with radiation: Concepts, achievements and future directions. *Nat Clin Pract Oncol* 2007;4:172–180.
7. Bourhis, J., Overgaard, J., Audry, H. et al. Hyperfractionated or accelerated radiotherapy in head and neck cancer: A metaanalysis. *Lancet* 2006;368:843–854.
8. Kasibhatla, M., Kirkpatrick, J., and Brizel, D. How much radiation is the chemotherapy worth in advanced head and neck cancer? *Int J Radiat Oncol Biol Phys* 2007;68:1491–1495.
9. Fowler, J. Correction to Kasibhatla et al. How much radiation is the chemotherapy worth in advanced head and neck cancer? *Int J Radiat Oncol Biol Phys* 2008;71(2):326–329.
10. Jones, B., and Sanghera, P. Estimation of radiobiologic parameters and equivalent radiation dose of cytotoxic chemotherapy in malignant glioma. *Int J Radiat Oncol Biol Phys* 2007;68(2):441–448.
11. Plataniotis, G., and Dale, R. Assessment of the radiation-equivalent of chemotherapy contributions in 1-phase radio-chemotherapy treatment of muscle-invasive bladder cancer. *Int J Radiat Oncol Biol Phys* 2014;88(4):927–932.

CHAPTER 2

Radiosensitization by Oxygen and Nitric Oxide

2.1 RADIOSENSITIZATION BY OXYGEN: THE OXYGEN FIXATION HYPOTHESIS

Oxygen is the prime mover of radiosensitization, increasing the biological effect of ionizing radiation by a factor of 2 to 3, whereas radioresistance is associated with hypoxia. It was shown that the oxygen effect only occurred if oxygen is present during irradiation or within a few milliseconds, implying that the mechanism is mediated by short-lived free radicals. The mechanism to be responsible for the enhancement of radiation damage by oxygen, is generally referred to as the oxygen-fixation hypothesis (OFH) and was developed in the late 1950s from the work of Alexander and Charlesby [1] (as reviewed by Ewing [2]).

According to the OFH, DNA sustains either direct damage or is indirectly damaged by reactive oxygen species produced by the radiolysis of water molecules. In aerobic conditions, oxygen reacts through its two unpaired electrons with DNA radicals to form DNA peroxyradicals (DNA-OO•), thus preventing DNA repair. In the absence of O_2, the DNA radical will be reduced, restoring the DNA to its original composition (DNA-H; Figure 2.1).

The OFH assumes that the addition of O_2 to the target radical fixes damage so that it cannot be repaired or restored. A development from the OFH was to consider DNA to be the target, to link sensitization by oxygen

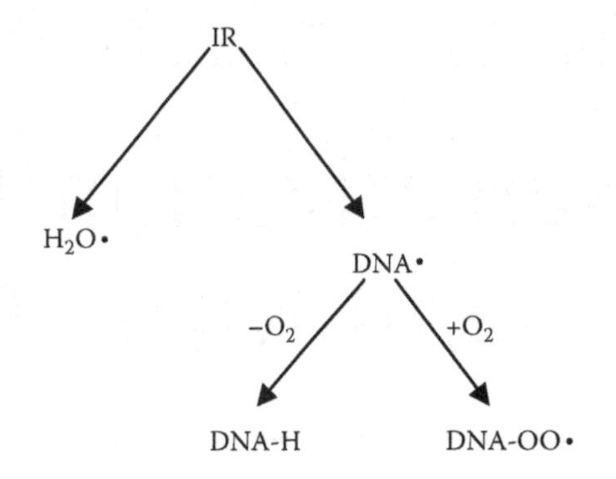

FIGURE 2.1 In OFH, the exposure of cells to ionizing radiation (IR) generates free radicals either in DNA or water molecules. Free radicals in DNA (DNA•) can react with available O_2 to generate a peroxyradical (DNA-OO•), thus chemically modifying the DNA (oxygen fixation). In the absence of O_2, the DNA radical will be reduced, restoring the DNA to its original composition (DNA-H). In addition to the direct effects on DNA, free radicals produced from H_2O (H_2O•) can further damage nearby DNA.

to the reaction of OH radicals, and to make the end product less explicit [3]. This version, which states that DNA lesions formed with the participation of O_2 are difficult or impossible to repair, is widely accepted as the best explanation of the function of oxygen. This hypothesis was, however, formulated before extensive knowledge of the DNA repair processes was accumulated and is subject to criticism on the grounds that if enzymic repair is successful, the issue of chemical restorability is not important. As it stands, the OFH is not a totally convincing explanation of the oxygen effect, which does detract from the significance of the effect itself.

In fact, the response of cells to ionizing radiation is strongly dependent on oxygen and, at high doses, the enhancement of radiation damage by oxygen is dose-modifying, that is, the radiation dose that gives a particular level of survival is reduced by the same factor at all levels of survival (Figure 2.2a), allowing the calculation of an oxygen enhancement ratio (OER)

$$\text{OER} = \frac{\text{Radiation dose in hypoxia}}{\text{Radiation dose in air}} \tag{2.1}$$

to achieve the same biological effect.

For most cells, the OER for x-rays is 2.5 to 3.0, but for some cells at radiation doses below 3.0 Gy, the OER maybe lower. The OER is also

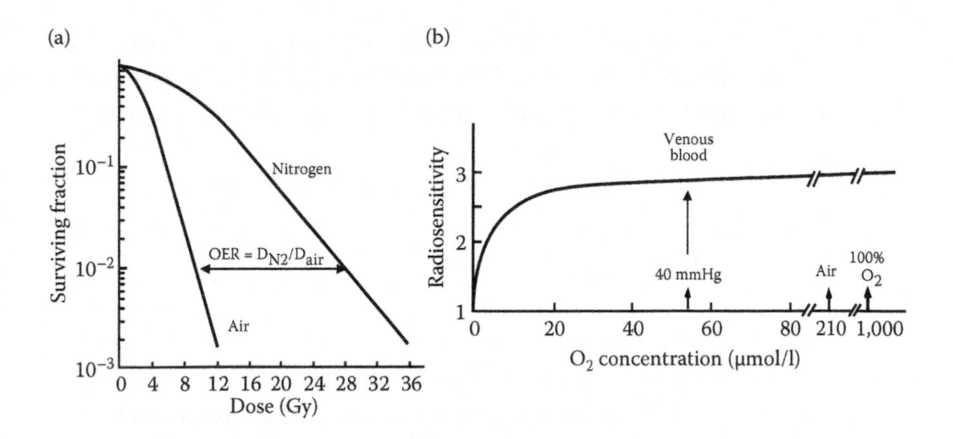

FIGURE 2.2 (a) Diagrammatic survival curve for cultured mammalian cells irradiated under oxic or hypoxic conditions illustrating the dose modifying effect of oxygen; (b) change in the OER with oxygen tension.

dependent on the type of radiation, declining to a value of 1.0 for radiation with linear energy transfer values greater than about 200 keV. The dependence of the extent of sensitization on oxygen tension is shown in Figure 2.2b. By definition, the OER under anoxic conditions is 1.0. As the oxygen level increases, there is a steep increase in radiosensitivity (and in OER) with the greatest change occurring between 0 and approximately 20 mmHg. Further increase in oxygen concentration, up to that of air (155 mmHg) or 100% oxygen (760 mmHg), produces a small but not trivial increase in radiosensitivity. From a radiobiological standpoint, most normal tissues can be considered to be well oxygenated (venous blood has a pO_2 of approximately 45 mmHg); however, moderate hypoxia can be a feature of some normal tissues such as cartilage and skin.

2.1.1 Significance of the Oxygen Effect in the Treatment of Cancer: Hypoxic Cells in Solid Tumors

Although much of the historical data is based on high radiation doses, it is clear that many common tumors include a significant fraction of cells with intracellular oxygen concentrations in the steeply rising radiosensitivity response/concentration region, such that any method to improve tumor oxygenation should result in an increase in radiosensitivity of hypoxic subpopulations. The potential gain of radiosensitizing hypoxic cells is large: severely hypoxic cells typically require two to three times higher radiation doses to kill them compared with well-oxygenated cells. This factor is independent of absolute radiosensitivity and is seen for organisms of all levels of biological complexity and at all levels of radiosensitivity.

2.1.2 Methods for Modification of Hypoxic Radioresistance

Methods that have been investigated to radosensitize hypoxic cells are summarized in Table 2.1.

2.1.2.1 Increasing the Availability of Oxygen

- *Hyperbaric oxygen* is the most direct, and one of the earliest, treatments for raising tumor oxygen levels. Well-documented clinical trials reported the most benefit associated with few/large fraction radiation regimens. The method was cumbersome and there was some increase in morbidity for some sites.

- *Oxygen carriers*, based on enhanced solubility parameters of perfluorocarbons compared with blood plasma, have been evaluated, with mixed results. One animal study [4] found benefit only when the perfluorocarbon was combined with carbogen breathing (carbogen is usually 95%–98% O_2 + 5%–2% CO_2 v/v). Newer technologies, particularly nanoparticles [5,6], are being mobilized for the development of new oxygen carriers that might have value in radiotherapy.

- *Upregulating oxygen transport.* Anemia is known to be an important prognostic factor for response to radiotherapy. Transfusion with red blood cells is usually limited to cases of severe anemia; it may be detrimental in some cases because of immunosuppression and

TABLE 2.1 Strategies to Overcome Radioresistance of Hypoxic Cells

Strategy	**Method**
Increased oxygen delivery by blood	Hyperbaric oxygen
	Normobaric oxygen, carbogen breathing
	Nicotinamide
	Blood transfusion, erythropoietin
Mimic oxygen in the radiochemical process	Nitroimidazoles
Destruction of hypoxic cells	Hypoxic cytotoxins
	Hyperthermia
Elimination of OER	High LET

is distrusted for other reasons. Derivatizing hemoglobin with polyethylene glycol to improve the biocompatibility of bovine hemoglobin was found to be useful in animal models involving fractionated radiation, particularly in carbogen-breathing mice [6].

Recombinant human erythropoietin has been evaluated clinically in some detail as a means to correct hemoglobin levels [7], but there are a number of downsides including its angiogenic, mitogenic, and antiapoptotic potential. One review, however, concluded that most studies had suggested that erythropoietic therapy either improved survival or had no negative effect on survival when used to treat anemia in patients with cancer [8].

- *Increasing the availability of molecular oxygen in tissue.* The amount of O_2 released to the tissues from hemoglobin is strongly dependent on the position of the oxyhemoglobin dissociation curve and modification of this relationship has been shown to alter the oxygenation and radiosensitivity of tumors. Some antilipidemic drugs, derived from chlorophenoxy acetic acid (clofibrate and its analogues; Figure 2.3) have the property of reducing the affinity of Hb for O_2 *in vitro* and reduction in the binding affinity of Hb for O_2 has been shown in animal models to be an effective means of increasing the oxygenation in malignant tissues. Clofibrate is an antilipidemic drug that reduces the affinity of hemoglobin for oxygen and acts as a radiosensitizer [9], but clinical studies in this context do not seem to have been carried out.

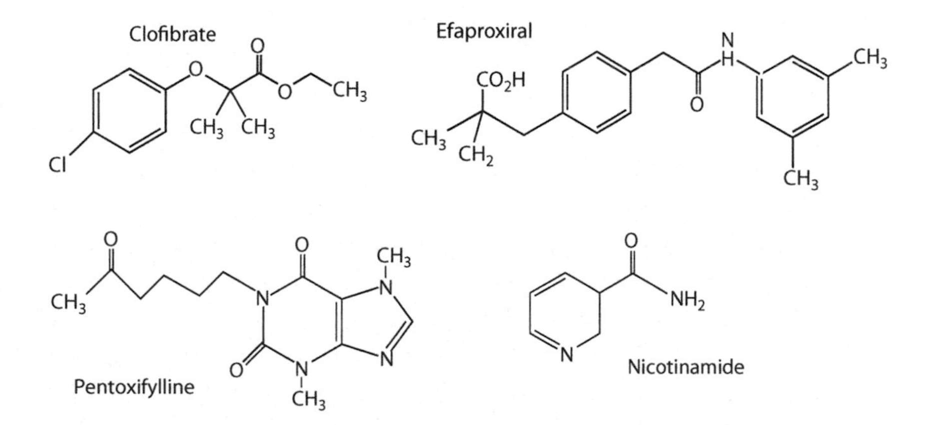

FIGURE 2.3 Drugs for improving tumor oxygenation.

Efaproxiral (RSR13) interacts with hemoglobin in a noncovalent and allosteric manner to lower the oxygen-binding affinity, and increase pO_2. Several clinical trials have shown positive results [6,10]. In one case, a phase III trial of efaproxiral, involving 515 patients treated with whole-brain radiotherapy for brain metastases, showed a significant increase in response rate for the efaproxiral arm at 3 and 6 months, with evidence that patients with breast primary tumors responded better [10]. There are currently six ongoing clinical trials assessing the effect of efaproxiral as a radiosensitizer. Four of these are concerned with brain cancer and two with non-small cell lung cancer (NSCLC) (http://clinicaltrials.gov).

Pentoxifylline (Ptx), a hemorrheologic methylxanthine derivative, is of interest in radiation oncology for several reasons. It has been reported to improve tumor oxygenation and improve radiation response if given before irradiation in several animal studies [11]. Modes of action include increased red cell deformability, reduced blood viscosity as a result of decreasing platelet aggregation, and vasodilation. Although improvement of tumor perfusion might result in better oxygenation and radiosensitivity, it is also important that the drug influences cytokine-mediated inflammation. Evidence with preclinical models suggests that Ptx improves tumor oxygenation and sensitizes p53 mutant tumors, but these findings have thus far not translated into positive clinical studies. For ongoing trials, only 1 of 18 concerned radiosensitization by Ptx (combined with radiation and hydroxyurea for the treatment of glioblastoma or astrocytoma) whereas the remainder targeted the prevention or ablation of toxicities following radiotherapy.

Nicotinamide was originally evaluated in radiobiology as an inhibitor of DNA repair via its interaction with poly(ADP ribose) polymerase (PARP) and as an analogue of known PARP inhibitors, but has found more value as a vasoactive agent. It showed good activity in combination with carbogen or normobaric oxygen with fractionated irradiation in animal models [12], where it was shown to eliminate acute hypoxia. Promising clinical studies have mainly involved the combination of nicotinamide with normobaric carbogen (to overcome diffusion-limited or chronic hypoxia), often with accelerated radiotherapy to inhibit repopulation (the "ARCON" regimen: accelerated radiotherapy with carbogen and nicotinamide). Reviewing ARCON in 2002, it was concluded that for cancers of the head and neck and bladder, the local tumor control rates were higher than in other studies. A later review confirmed this conclusion [13], although in unselected groups of patients in a phase I/II study, there were no significant differences in tumor response and local control with carbogen and nicotinamide added to conventional radiotherapy. ARCON treatment was found to reduce the prognostic significance of hemoglobin in squamous cell carcinoma of the head and neck.

A randomized phase III trial compared accelerated radiotherapy alone or with the hypoxia-modifying agents carbogen and nicotinamide (CON) in laryngeal carcinoma. This trial showed a significant improvement in 5-year regional control with CON [14,15]. Similar results have been shown in bladder cancer. The bladder carbogen nicotinamide (BCON) phase III trial showed that the addition of CON to radiotherapy significantly improved overall survival in bladder carcinoma [15].

2.1.2.2 Hypoxic Radiosensitizers

2.1.2.2.1 "Oxygen Mimetic" Radiosensitizers The concept of chemical radiosensitization of hypoxic cells was introduced in 1969 by Adams and Cooke [16], who showed that certain compounds were able to mimic oxygen and thus enhance radiation damage and that the efficiency of sensitization was directly related to the electron affinity of the compounds. It was suggested that because these agents could diffuse out of the blood vessels, they could sensitize distant hypoxic cells unlike oxygen, the concentration of which declines on a steep gradient with increasing distance from the blood vessel as it is rapidly metabolized by tumor cells. Because the drugs mimic the sensitizing effect of oxygen, they would not be expected to increase the radiation response of well-oxygenated cells in surrounding normal tissues.

The "electron-affinic" radiosensitizers were developed on the basis of comparison of the chemical properties of chemicals that radiosensitized anoxic cells with their reactivity toward radiation-produced free radicals. These compounds consist of electron-affinic nitroaromatic structures that undergo bioreductive metabolism within hypoxic cells. The reduced nitro groups thus function as oxygen mimics, which react with radiation-induced DNA free radicals and stabilize the DNA lesions under hypoxia [17,18]. There are some parallels in the reactivity of the drugs toward DNA base radicals that may justify the description as "oxygen-mimetic" radiosensitizers. Figure 2.4 shows possible mechanisms by which oxygen or nitroimidazoles and related compounds might enhance radiation-induced strand breaks [19–21].

Important experiments carried out early in the development of the nitroimidazole radiosensitizers showed that both types of radiosensitizer

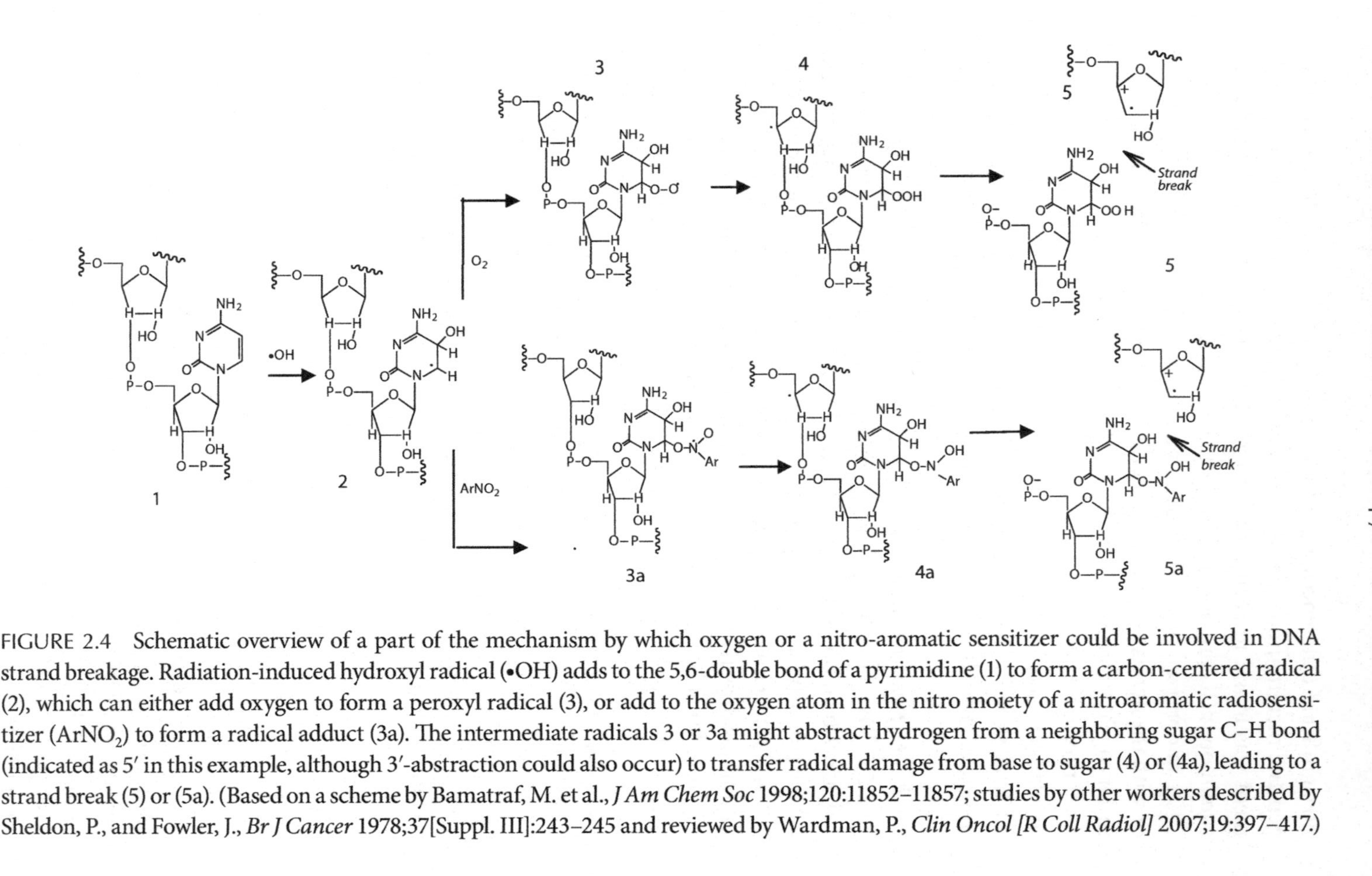

FIGURE 2.4 Schematic overview of a part of the mechanism by which oxygen or a nitro-aromatic sensitizer could be involved in DNA strand breakage. Radiation-induced hydroxyl radical (•OH) adds to the 5,6-double bond of a pyrimidine (1) to form a carbon-centered radical (2), which can either add oxygen to form a peroxyl radical (3), or add to the oxygen atom in the nitro moiety of a nitroaromatic radiosensitizer ($ArNO_2$) to form a radical adduct (3a). The intermediate radicals 3 or 3a might abstract hydrogen from a neighboring sugar C–H bond (indicated as 5′ in this example, although 3′-abstraction could also occur) to transfer radical damage from base to sugar (4) or (4a), leading to a strand break (5) or (5a). (Based on a scheme by Bamatraf, M. et al., *J Am Chem Soc* 1998;120:11852–11857; studies by other workers described by Sheldon, P., and Fowler, J., *Br J Cancer* 1978;37[Suppl. III]:243–245 and reviewed by Wardman, P., *Clin Oncol [R Coll Radiol]* 2007;19:397–417.)

(oxygen or the electron-affinic drug) had to be present at the instant of irradiation: adding either a few milliseconds after irradiation was ineffective [22]. The characteristic feature of the electron-affinic drugs is that they seem to radiosensitize hypoxic cells but do not affect well-oxygenated cells. This probably results from simple kinetic competition between oxygen and drug for reaction with DNA base radicals. The outcome of the competition is dependent on the reactivity (chemical rate constant) and the concentration of competing species. The high reactivity of oxygen ensures that it will outcompete the radiosensitizing drug except under conditions of low oxygen tension; hence, the selective hypoxic sensitization by the drug.

The earliest compounds to be investigated were nitrobenzenes, subsequently, the nitrofurans were similarly characterized and finally the nitroimidazoles. The first drug investigated, a 5-nitroimidazole, which was already in clinical use as an antibiotic (Flagyl), showed excellent activity in sensitizing mouse tumors to radiation. Shortly afterward, a more active drug (misonidazole, a 2-nitroimidazole) was developed. The structures of some of these drugs are shown in Figure 2.5.

2.1.2.2.2 Preclinical Demonstration of Hypoxic Radiosensitization Hypoxic radiosensitization by nitroimidazoles was demonstrated in a range of human and rodent cell lines [23]. One example of radiation response influenced by misonidazole is illustrated by Figure 2.6 [24], which shows that the radiation response of hypoxic cells is enhanced substantially by irradiating the cells in the presence of the optimal concentration of the drug whereas the response of the aerated cells is unaffected.

Extensive investigations were also done with preclinical tumor models. The results of more than 40 such studies with end points including cell survival, regrowth delay, and local control were summarized by Adams [25]. Sensitizer enhancement ratios (SER) of close to 2.0 were found in a variety of animal tumors when the sensitizer was administered before single-dose irradiation (Figure 2.7) [21]. When misonidazole was combined with fractionated radiation, the SER values were lower. This probably resulted from reoxygenation between radiation fractions, which reduces the therapeutic effect of hypoxia and the fact that lower drug doses had to be used with multiple administration. There is also a small but significant enhancement seen when misonidazole is given after radiation that cannot be caused by hypoxic cell radiosensitization, but probably results from the fact that misonidazole is directly toxic to hypoxic cells.

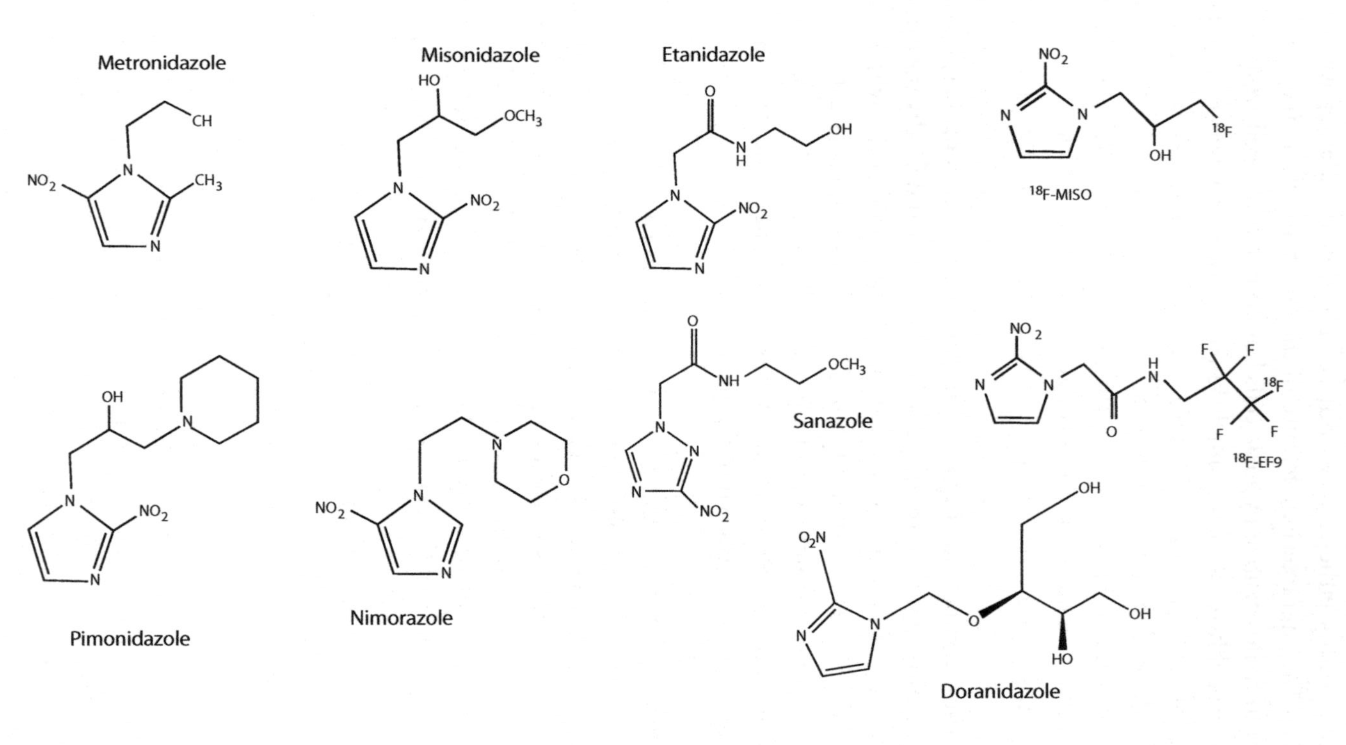

FIGURE 2.5 Hypoxic radiosensitizers.

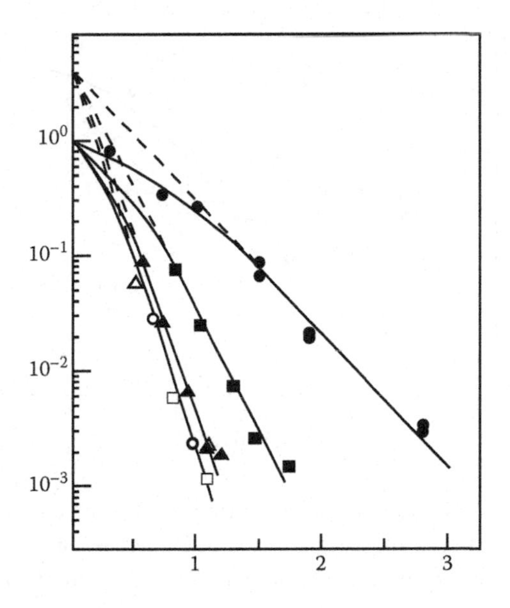

FIGURE 2.6 Survival data for aerated and hypoxic Chinese hamster cells irradiated in the presence of various concentrations of misonidazole (Ro-07-0582). At a concentration of 10 mM misonidazole, the radiation response of hypoxic cells is close to that of aerated cells. The response of aerated cells is unaffected by the drug: (○) air; (□) air + 1 mmol Ro-07-0582; (△) air + 10 mmol Ro-07-0582; (●) N_2; (▲) N_2 + 1 mmol Ro-07-0582; (■) N_2 + 10 mmol Ro-07-0582. (From Adams, G.E. et al., *Radiation Research* 67:9–20, 1976. With permission.)

2.1.2.2.3 Clinical Experience with Hypoxic Radiosensitizers The first clinical studies of radiosensitizers used metronidazole in brain tumors. These results, together with encouraging laboratory studies of misonidazole, sparked a rush of clinical trials in the late 1970s exploring the potential of misonidazole radiosensitizers [26]. However, most of the trials with misonidazole were unable to demonstrate a significant improvement in radiation response, although benefit was seen in some trials in certain subgroups of treated patients.

The dose of misonidazole was limited by a dose-dependent peripheral neuropathy and, on the whole, the results from trials using misonidazole were disappointing. The difficulty in achieving sufficiently large clinical doses of misonidazole led to a search for better (or less toxic) radiosensitizing drugs [27]. Of the many compounds synthesized and tested, two of the most promising were etanidazole and pimonidazole (Figure 2.5) [27]. Etanidazole was selected as being superior to misonidazole for two reasons. First, although the sensitizing efficiency was equivalent to that

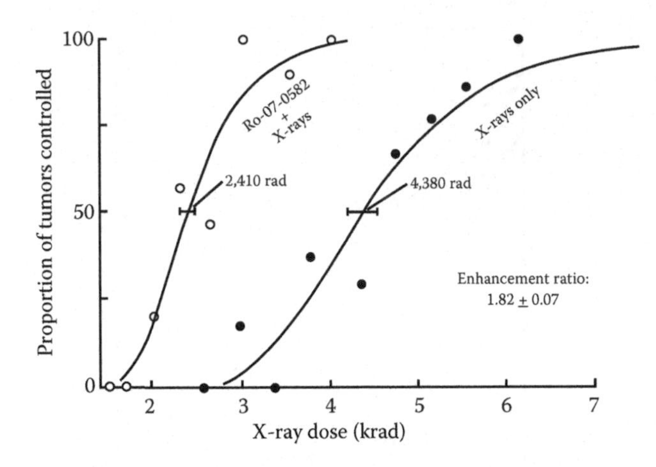

FIGURE 2.7 The proportion of mouse mammary tumors controlled at 150 days after treatment by increasing single doses of x-rays. ●—●, x-rays alone; ○—○, x-rays delivered after administration of 1 mg/g body weight misonidazole. The enhancement ratio is the ratio of x-ray doses in the absence or presence of the drug that results in 50% tumor control. (From Sheldon, P., Fowler, J., *British Journal of Cancer* 37[Suppl. III]:243–245, 1978. With permission.)

of misonidazole, it had a shorter half-life *in vivo*, which should lead to reduced toxicity. Second, it had a reduced lipophilicity (a lower octanol/water partition coefficient) and was therefore less readily taken up in neural tissue, leading to less neurotoxicity. Etanidazole was tested in two large head and neck cancer trials, one in the United States and the other in Europe. In neither case was there a significant therapeutic benefit, although in a later subgroup analysis, a positive benefit was reported [28]. Pimonidazole contains a side-chain with a weakly basic piperidine group. It is more electron-affinic than misonidazole and hence more effective as a radiosensitizer; furthermore, it is uncharged at acid pH, thus promoting its accumulation in ischemic regions of tumors. A pimonidazole trial was started in the uterine cervix, but was stopped when it became evident that those patients who received pimonidazole showed a poorer response.

An exception to the general rule of toxicity trumping efficacy, which sidelined most of the nitroimidazoles is the 5-nitroimidazole, nimorazole. Although it is a less potent sensitizer than misonidazole or pimonidazole, it was far less toxic and thus could be given in much higher doses. At a clinically relevant dose, the SER was approximately 1.3. The drug could be given in association with a conventional radiation therapy schedule, making it practical for clinical use. When given to patients with supraglottic and pharyngeal carcinomas (DAHANCA 5), a highly significant benefit in terms of improved locoregional tumor control and disease-free survival was obtained [28]. These results are shown in Figure 2.8 and were consistent with the earlier DAHANCA 2 study for misonidazole. As a consequence, nimorazole became part of the standard treatment schedule for head and neck tumors in Denmark. Unfortunately, presumably because the drug is very cheap to produce, it is not attractive for commercial development and is not widely used outside Denmark.

Although most trials with nitroimidazoles and related compounds have failed to demonstrate a significant benefit, a meta-analysis of results from more than 7,000 patients included in 50 randomized trials indicated a small but significant improvement in local control and survival, with most of the benefit attributed to an improved response in patients with head and neck cancer [29]. This suggests that the apparent lack of clinical benefit in most of the individual trials might be due to the small number

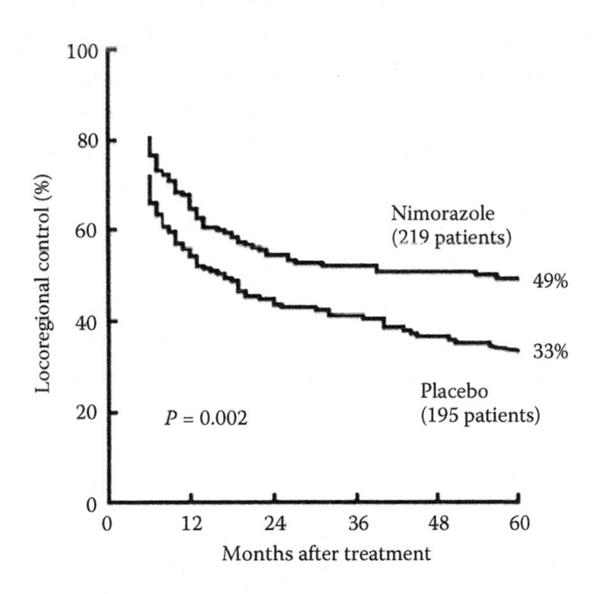

FIGURE 2.8 Actuarial estimated locoregional tumor control in patients randomized to receive nimorazole or placebo in conjunction with conventional radiotherapy for carcinoma of the pharynx and supraglottic larynx. (From Overgaard, J. et al. A randomized double-blind phase III study of nimorazole as a hypoxic radiosensitizer of primary radiotherapy in supraglottic larynx and pharynx carcinoma. Results of the Danish Head and Neck Cancer Study [DAHANCA] Protocol 5-85. *Radiotherapy and Oncology* 46:135–146, 1998. With permission.)

of patients involved rather than the lack of biological importance of tumor hypoxia. The overall conclusion is that although the nitroimidazoles have not lived up to their original promise in terms of clinical efficacy, they may be of value in the treatment of certain conditions.

2.1.2.2.4 Alternative Hypoxic Radiosensitizers A number of other nitroaromatic structures have been evaluated as alternatives to misonidazole and its nitroimidazole analogues, of which the nitrotriazole, sanazole (AK-2123), has attracted the most interest. In an International Atomic Energy Agency multicenter randomized trial [30], Sanazol was found to significantly increase local control and survival in cervical cancer following radical radiotherapy. A large (462 patients) phase III clinical trial of sanazole in the treatment of squamous cell carcinoma of the uterine cervix showed an increase in local tumor control and survival without the addition of any major toxicity [30].

Doranidazole (PR-350) is a nitroimidazole very similar to misonidazole but with greater hydrophilicity; preclinical studies demonstrated that doranidazole enhanced the local control of tumor growth when injected at a nontoxic dose before single or fractionated irradiation *in vivo* [31]. A small (48 patients) phase III trial of doranidazole combined with intraoperative radiotherapy in advanced pancreatic cancer, did not show significant gain. In contrast, long-term follow-up of a placebo-controlled phase III study in patients with unresectable pancreatic cancer revealed that doranidazole given as a systemic intravenous infusion 10 to 40 min before intraoperative radiotherapy significantly improved the 3-year survival rate as compared with placebo and radiotherapy treatment [32].

A phase I/II trial of doranidazole added to thoracic radiotherapy in patients with locally advanced NSCLC showed only minimal neurotoxicity (reviewed by Bischoff et al. [33]). The interest in these compounds seems to have waned and there are no reports of ongoing clinical trials with either doranidazole or sanazole.

Riboside derivatives of nitroimidazole with unaltered reduction potential and hydrophobicity, but with a β-nucleoside configuration to facilitate their entry into hypoxic cells through nucleoside transporters, have been described. Evidence has been presented that 1-β-D-(5-deoxy-5-iodoarabinofuranosyl)-2-nitroimidazole behaves as a superior radiosensitizer of hypoxic colorectal carcinoma cells in a clonogenic survival assay *in vitro* [34].

It has been suggested that the elevated membrane transporter-mediated uptake of glucose, associated with enhanced glycolysis in hypoxic tumor cells, might offer another approach to boost the affinity of nitroimidazoles for tumoral tissue independently of hydrophobicity. To that end, sugar-hybrid 2-nitroimidazole derivatives were synthesized and characterized, leading to the identification of TX-2224, which contains an acetylated glucose moiety, as a potent radiosensitizer [35]. When tested using a colony formation assay in mammary sarcoma cells under hypoxic conditions, TX-2224 showed a higher radiosensitizing activity than misonidazole, although the two compounds displayed similar hydrophobicity [35].

Apart from nimorazole, the nitroimidazoles mentioned above are represented on the clinical scene as probes for hypoxia. Pimonidazole, misonidazole, and the 2-nitroimidazole EF5 have all been used to visualize hypoxic tissue at the microscopic level and, when tagged with fluorine-18, for noninvasive assay by positron emission tomography (PET) scanning.

2.1.2.3 Hypoxic Cytotoxins

The hypoxic cytotoxins or bioreductive drugs are compounds that undergo intracellular reduction to form active cytotoxic species under low oxygen tensions, selectively killing hypoxic cells. Thus, they are not radiosensitizers in the true sense but when combined with radiation for tumor treatment, the result is an enhancement of radiation response. This arises because two tumor cell populations are being targeted: radiation-sensitive oxic cells and radiation-resistant hypoxic cells, which are sensitive to the hypoxic cytotoxins. The bottom line is apparent "radiosensitization" and improved tumor control.

The development of hypoxic cytotoxins followed the discovery that electron-affinic radiosensitizers not only sensitize hypoxic cells to radiation but also are preferentially toxic to them. These drugs can be divided into three major groups: quinones (mitomycin C), nitroimidazoles (RSU-1069), N-oxides (tirapazamine; Table 2.2; Figure 2.9).

2.1.2.3.1 Quinones The prototype bioreductive agent in this category is the quinone-alkylating agent mitomycin C (MMC), which was in routine clinical use long before it was realized that its bioreduction and activation were facilitated by the hypoxic tumor environment. *In vitro* measurements of the cytotoxicity of MMC in hypoxic compared with normoxic cells (the hypoxic cytotoxic ratio; HCR) showed the differential effect to be generally fivefold or less. MMC and other agents

TABLE 2.2 Classes of Bioreductive Drug and Their Metabolism

Prodrug	Cytotoxic Species	Product	Enzymes
Benzotriazines, e.g., tirapazamine	Free radical	Nontoxic	P450R; iNOS
Bifunctional and monofunctional alkylating agents (mitomycin C, porfiromycin, RB6145, RSU, EO9)	DNA adduct	Nonfunctional reduced drug	P450R; DTD
Alkyl-amino-anthraquinones (AQ4N, NLCQ-1)	Nitro ($1e^-$), nitroso ($2e^-$), and hydroxylamine ($4e^-$) metabolites	Unknown, shows weak DNA-intercalating activity	Cytochrome b5, other one-electron reductases
Phosphate esters of dinitrobenzamide mustards (PR-04 [ProActa])	Stable cytotoxin	Stable persistent cytotoxin	Phosphatases to generate alcohol, then nitro-reductases, including one-electron reductases

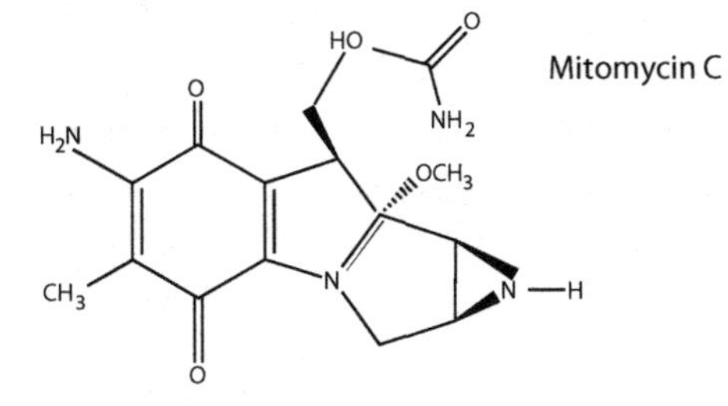

Tirapazamine
O⁻
N
N
NH2
O⁻

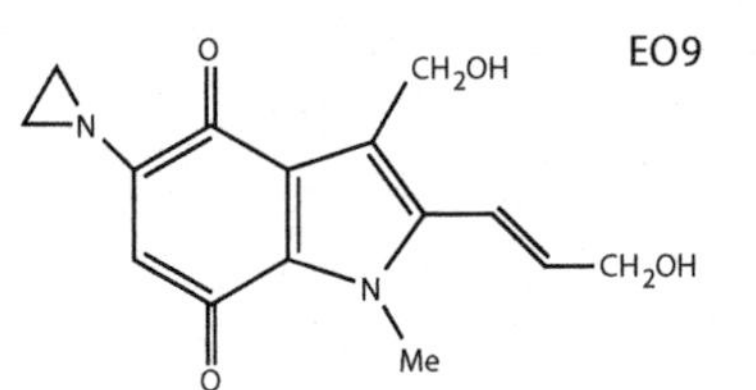

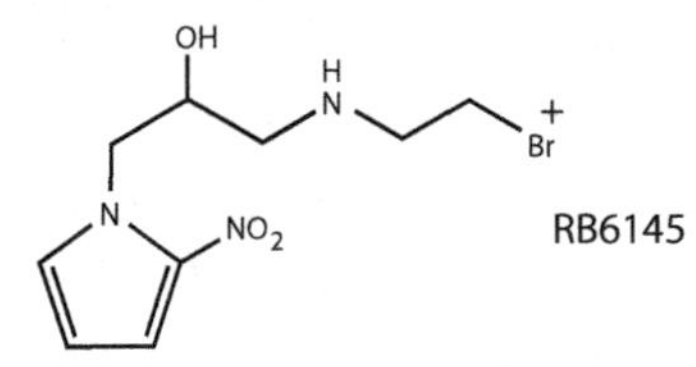

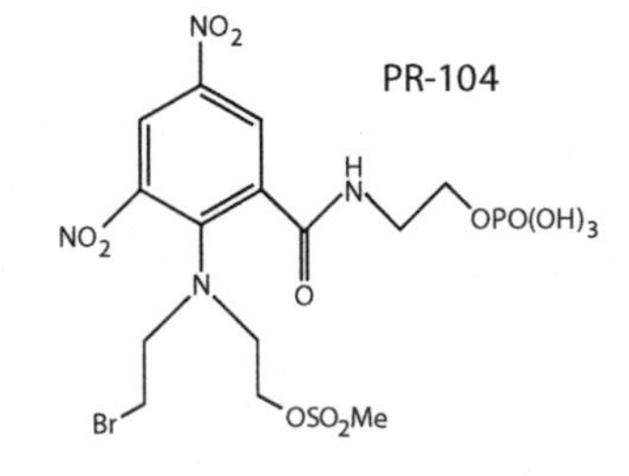

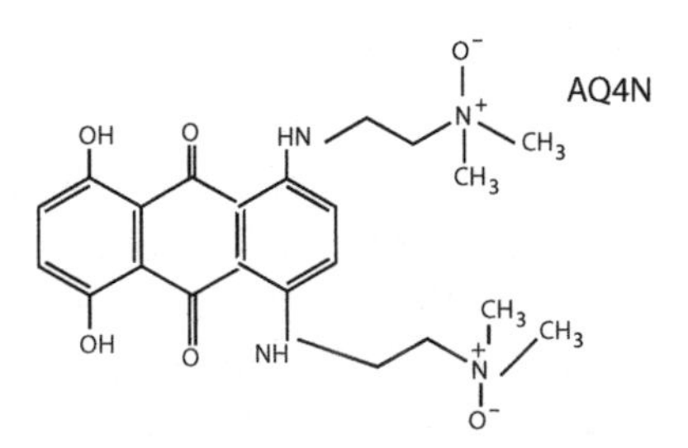

FIGURE 2.9 Bioreductive drugs.

within this class are substrates for the two-electron reducing enzyme, DT-diaphorase [DTD; NAD(P)H: quinone oxidoreductase, NQO1]. DTD is overexpressed in a number of tumor types, suggesting it as an attractive target for bioreductive drug design. However, it is an oxygen-independent reductase and bioactivation by DTD can occur in aerobic conditions, compromising hypoxic selectivity. The predominant flavoenzyme implicated in the bioactivation of MMC (and other quinone agents) under hypoxic conditions is the one-electron reductase NADPH/cytochrome P450 reductase (P450R) [36]. One-electron reduction produces the semiquinone radical anion, which covalently interacts with DNA causing cross-links as the major cytotoxic lesion. In the presence of oxygen, however, the free radical would be back-oxidized, accounting for the requirement of hypoxia for the generation of the cytotoxic species.

The identification of MMC as a potential bioreductive agent led to the search for analogues that showed a greater degree of preferential toxicity toward hypoxic cells. Porfiromycin, a methylated analogue of MMC, was found to have a much greater HCR than the parent compound [37]. Preclinical studies combining porfiromycin with radiotherapy showed an additive effect *in vitro* and a supra-additive effect *in vivo*, consistent with the idea that porfiromycin successfully targeted the radioresistant hypoxic tumor compartment. Further development resulted in the synthesis of a series of indolequinones, of which EO9 showed the most promise in preclinical studies. The HCR of EO9 is approximately 30, significantly superior to that of MMC [38]. Studies in cell-free systems showed that P450R-mediated activation of EO9 resulted in the generation of a DNA-damaging radical [38]. However, EO9 is also an excellent substrate for oxygen independent reduction by DTD compromising the hypoxia selectivity of the drug in high DTD backgrounds.

2.1.2.3.2 Nitroaromatics The finding that misonidazole was preferentially toxic toward hypoxic cells led to numerous efforts to find other nitroimidazoles that were better. To that end, RSU-1069 was developed. This compound has the classic 2-nitroimidazole radiosensitizing properties, but also an aziridine ring at the terminal end of the chain, which gave the molecule substantial potency as a hypoxic cell cytotoxin, both *in vitro* and *in vivo*. In large animal studies, it was found to cause gastrointestinal toxicity and a less toxic prodrug was therefore developed (RB-6145), which is reduced *in vivo* to RSU-1069. Although this drug was found to have potent antitumor activity in experimental systems, further animal studies revealed that it induced blindness, and thus further development was halted. However, other nitro-containing compounds including NLCQ-1, CB1954, SN23862, and PR-104 are currently under development.

Nitroaromatic prodrugs (e.g., the hypoxic radiosensitizer misonidazole) are bio-reduced via the stepwise addition of up to six electrons catalyzed by various one-electron reductases. As with the quinones, the first reduction one-electron intermediate can be readily back-oxidized in the presence of oxygen, although subsequent reductions are effectively irreversible. Misonidazole was found to have a HCR similar to, or greater than, that for MMC and could enhance radiotherapeutic outcome *in vivo* when given after radiotherapy, a result consistent with hypoxic cytotoxicity.

Further development based on the nitroimidazoles identified RSU1069 [39], a 2-nitroimidazole containing an aziridine group in the N1 side chain enabling RSU1069 to act as a bifunctional alkylating agent upon reduction. The one-electron reductase P450R was identified as an important bioactivator of this compound. *In vivo* studies demonstrated that RSU1069 could enhance radiation response when the drug was given before or after radiotherapy and that it was also effective when used in combination with fractionated radiotherapy. Clinical development of RSU 1069 and of RB6145, a supposedly less toxic prodrug of RSU1069, was terminated by the discovery of marked toxicities associated with both drugs [40].

A more recently developed nitroaromatic bioreductive agent, NLCQ-1 exhibits weak DNA-intercalating ability, and shows up to 40-fold selectivity toward hypoxic tumor cells *in vitro*. NLCQ-1 has been shown to be a potent radiosensitizer *in vitro*, demonstrating a synergistic interaction with radiotherapy *in vivo* [41]. Synergistic interactions with a wide range of chemotherapy drugs have been reported whereas efficacy studies suggest that NLCQ-1 compares favorably with the current lead bioreductive agent tirapazamine (TPZ) [40].

The dinitrobenzamide mustards show bioreductive potential but are activated only under severe hypoxia and cause a bystander effect due to the formation of relatively stable cytotoxic metabolites. They were initially developed as analogues of the weak monofunctional alkylating agent CB1954. Under the right conditions, CB1954 is reduced to the 4-hydroxylamino derivative, which undergoes further reaction with acetyl coenzyme A to produce a potent DNA interstrand cross-linking agent. The first generation dinitrobenzamide mustards (e.g., SN 23862) had poor aqueous solubility and limited hypoxic selectivity. The second generation compounds were phosphate ester analogues with good solubility and formulation characteristics, which in fact act as pre-prodrugs; the systemic phosphatase activity generates the corresponding alcohols (prodrugs), which are then activated by nitroreductases. A lead compound, PR-104 (Proacta), was identified and underwent phase I trial in solid tumors [42].

2.1.2.3.3 Aliphatic N-Oxides Another group of compounds that has shown potential as bioreductive drugs are the aliphatic N-oxides. The lead compound is the bis-N-oxide banoxantrone (AQ4N), which is reduced under hypoxic conditions to yield the cytotoxic product AQ4 [43]. AQ4, an analogue of mitoxantrone, has high DNA-binding affinity, and also acts as a topoisomerase II inhibitor. Bioreductive activation of AQ4N to AQ4 occurs through two two-electron additions via the intermediate AQ4M. The key enzymes implicated in this bioactivation are the cytochrome P450s (CYPs); in particular, CYP1A1, CYP2B6, and CYP3A4 [43]. An important feature of AQ4N is that the reduction product AQ4 is stable and persistent, thereby enhancing the potential for a bystander effect. AQ4N tested as a single agent *in vivo* had limited effect on tumor growth but, in combination with methods to increase tumor hypoxia, a substantial growth delay was seen consistent with it acting as a hypoxic selective cytotoxin.

2.1.2.3.4 Heteroaromatic N-Oxides The most successful bioreductive agent to date, TPZ, is a heteroaromatic N-oxide with an impressive HCR in the range of 50 to 100. Combined *in vivo* treatments with radiation showed positive interactions, particularly with fractionated protocols [44]. TPZ is also beneficial when combined with a wide range of chemotherapeutic agents, whereas enhancement of normal tissue toxicity was generally small. TPZ has been shown to be an excellent substrate for the one-electron reductases, cytochrome P450 and P450R, and for nitric oxide synthase (NOS). One-electron reduction causes the generation of the nitroxide radical intermediate, which can be readily back-oxidized in the presence of oxygen. In the absence of oxygen, the radical undergoes rearrangement by the loss of water to form an oxidizing radical that can cause DNA damage through the abstraction of a hydrogen atom (Figure 2.10). The short-lived nature of the one-electron products means that TPZ has no bystander effect.

TPZ behaves differently from other bioreductive agents in terms of the relationship between cytotoxicity and oxygen tension. TPZ becomes

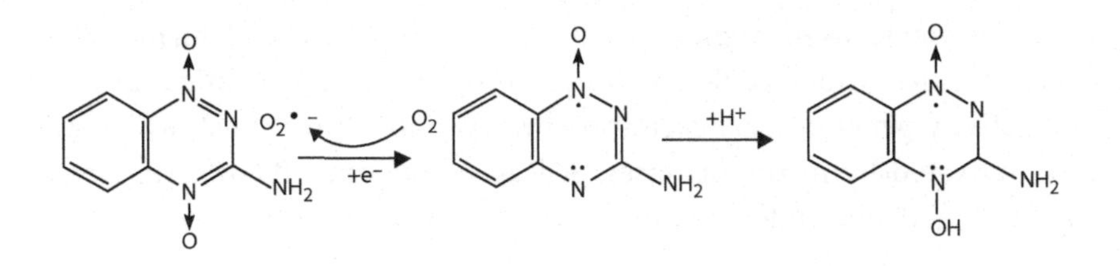

FIGURE 2.10 The mechanism of the preferential hypoxic cytotoxicity of TPZ. Under both aerobic and hypoxic conditions, the drug is reduced by an intracellular reductase to form a highly reactive radical, which can cause DNA single-strand and double-strand breaks. In the presence of oxygen, however, oxygen will take the free electron to itself, thereby back-oxidizing the radical to the inactive parent drug.

increasingly cytotoxic as oxygen levels are diminished [4] but does not require extremely low oxygen tensions to elicit bioactivation. This means that TPZ can effectively target cells at intermediate oxygen tensions that are radioresistant but not sufficiently hypoxic for targeting with other bioreductive drugs. Broad oxygen selectivity is beneficial when the reactive product is a very short-lived radical, but there is a downside—the results of some studies have suggested that if TPZ is metabolized at intermediate oxygen tensions, this would reduce its delivery to chronically hypoxic sites within the tumor.

2.1.2.3.5 Gene-Directed Enzyme Prodrug Therapy Enzyme profiling studies of human tumors compared with healthy tissues, have shown that several endogenous reductases have increased activity in tumors, notably DTD, iNOS, carbonyl reductase, and cytochrome P450s. However, the activity levels of these enzymes are heterogeneous between patients and between different tumor types. One approach to this problem is the enhancement of bioreductive drug activation using gene-directed enzyme prodrug therapy (GDEPT).

GDEPT targets the enhancement of bioreductive drug metabolism by teaming genes encoding for reductases with a specific bioreductive drug; one example being the P450R/TPZ GDEPT strategy based on the fact that P450R plays a major role in TPZ metabolism [45]. Another approach proposes achieving hypoxia-specific overexpression of P450R by the incorporation of hypoxia-responsive enhancer sequences within the promoter of the P450R expression cassette, which bind the transcription factor HIF-1 [46], resulting in HIF-dependent transcriptional activation of the P450R gene.

For several reasons, AQ4N would be an ideal prodrug for inclusion in a GDEPT strategy. Unlike TPZ, whose active metabolite is highly reactive and short-lived, the reduction product of AQ4 is very stable and can elicit a bystander effect offsetting one of the major limitations of gene therapy, the inability to deliver and express the gene in all tumor cells. Several CYP enzymes mediating the bioreduction of AQ4N have been tested in GDEPT strategies. CYP3A4 and CYP1A1 injection into murine tumors increased tumor metabolism of AQ4N and enhanced tumor control with radiation [47]. Although AQ4N is not a substrate for P450R, it is readily metabolized by iNOS. The use of iNOS gene therapy to sensitize tumors to AQ4N is particularly attractive because iNOS also catalyzes the conversion of L-arginine to citrulline with the concomitant production of nitric oxide, which has been shown to be both directly cytotoxic and a potent radiosensitizer in tumors.

2.1.2.4 Clinical Trials with Hypoxic Cytotoxins

2.1.2.4.1 Mitomycin C Several randomized clinical trials in patients with squamous cell carcinoma of the head and neck have been undertaken, specifically using MMC to counteract the effects of hypoxia. Initial studies reported an improvement in local tumor control and survival, without any enhancement of radiation reactions in normal tissues, perhaps to be expected because MMC actually has a small HCR. It seems that the protocol for MMC administration was suboptimal, so its ability to preferentially kill hypoxic cells and thus enhance radiation therapy was limited [40].

Other quinones investigated in clinical trials included porfiromycin and EO9. EO9 has gone through only preliminary phase I/II testing whereas porfiromycin was included in a prospective randomized trial in combination with radiation therapy in head and neck cancer, but was found to be no better than MMC [40].

2.1.2.4.2 Tirapazamine The first phase I trial of TPZ was reported in 1994. By 2007, results of the phase II trials involving more than 1100 patients with a range of tumor types had been published (summarized by McKeown et al. [40]; Table 2.3). The results of these trials were generally promising and indicative that progress to phase III investigations should be pursued. However, only a few randomized phase II or III studies were subsequently completed.

An early study of particular interest was one in which a screening procedure for tumor hypoxia was incorporated into the protocol. This trial compared chemoradiation with cisplatin/5-FU to chemoradiation with

TABLE 2.3 Reported Clinical Trials Using TPZ to Target Hypoxic Tumor Cells

Tumor	Phase	N	Treatment	Outcome	Remarks
Stage III/IV HNSCC	II	39	Radiotherapy (70 Gy in 7 weeks) concurrent TPZ (159 mg/m^2 3×/week for 12 doses)	1- and 2-year local control rates: 64% and 59%, respectively	33% grade 3 or 4 drug-related toxicities. No excessive RT-associated acute normal tissue reactions
Stage III/IV HNSCC	II	122	RT (70 Gy in 7 weeks). CIS, TPZ (290 mg/m^2): day 2, weeks 1, 4, and 7, TPZ alone (160 mg/m^2); days 1, 3, 5, of weeks 2 and 3, TPZ/CIS or CIS (50 mg/m^2); day 1 + infusional 5-FU	3-year failure-free survival, 55% with TPZ/CIS and 44% with chemoboost arm; 3-year locoregional failure-free rates 84% in the TPZ/CIS arm, 66% in the chemoboost arm	Both regimes feasible; significant but acceptable toxicity. Based on promising efficacy TPZ/CIS evaluated in phase III trial
Stage IV HNSCC	II	62	TPZ, CIS, 5-FU: induction CT (2×), followed by simultaneous CRT (TPZ, CIS, and 5-FU) or same regime without TPZ. RT (2 Gy fractions): 66–70 Gy to tumor	Clinicopathologic response rate in lymph nodes: standard (90%) vs. TPZ (74%; $P = 0.08$), response at the primary site and 89% and 90% ($P = 0.71$); 5-year OS, 59%; cause-specific survival rate, 68%; locoregional control rate, 77% for entire group	No difference with regard to any of the outcome parameters between the two treatment arms. Hematologic toxicity greater with TPZ
Glioblastoma multiforme	II	124	RT: 30 × 2 Gy. TPZ: 3×/week for 12 treatments during RT. 55 patients received TPZ 159 mg/m^2; 69 received 260 mg/m^2	There was no significant survival advantage compared with controls (standard treatment population)	Grade 3/4 toxicities were more frequent in the higher dose regimen

Note: RT, radiotherapy; CRT, chemoradiotherapy; n, number of patients; 5-FU, 5-fluorouracil; CIS, cisplatin. Reviewed by McKeown et al. [40].

cisplatin/TPZ. Positron emission tomography, before treatment, was used to stratify the tumors into hypoxic and nonhypoxic. Results of this study clearly showed that TPZ improved local control of hypoxic head and neck tumors. Cisplatin/5-FU showed that 8/18 tumors recurring in the hypoxic group, whereas 0/26 recurred in patients treated with cisplatin/TPZ. When the nonhypoxic tumors were analyzed, TPZ did not provide an improvement in outcome: 2/27 (5-FU) versus 3/21(TPZ) [48]. Thus, local control of hypoxic tumors is improved when a hypoxic cell cytotoxin is included in the treatment regimen.

Several recent reviews of results from different disease have generally supported this conclusion:

- Reddy and Williamson [49] reviewed all clinical trials published up to 2009 and summarized the results. They concluded that the very promising results obtained in various preclinical studies and early-phase clinical trials were not borne out by the results of several phase III trials. In reviewing the mechanism of action, toxicity, and antitumor activity of tirapazamine, and providing insights into factors that might have contributed to the disappointing results in some of the phase III trials, the need to develop dependable markers of tumor hypoxia was identified.
- Peters et al. [50] describe a study designed as a follow-up to the promising results that had been obtained in a randomized phase II trial with tirapazamine (TPZ) combined with cisplatin (CIS) and radiation.

In a phase III trial of head and neck cancer, 861 patients were accrued from 89 sites in 16 countries. Radiotherapy (70 Gy in 7 weeks) was administered concurrently with either cisplatin or cisplatin plus TPZ. The primary end point was overall survival (OS). The 2-year OS rates were 65.7% for CIS and 66.2% for TPZ/CIS. There were no significant differences in failure-free survival, time to locoregional failure, or quality of life. It was concluded that there was no evidence that the addition of TPZ to chemoradiotherapy, in patients with advanced head and neck cancer who were not selected for the presence of hypoxia, improves overall survival.

- In a 2012 review, Ghatage and Sabagh [51] noted that the addition of TPZ to conventional chemoradiation protocols in the management of cervical cancer had shown promise in initial phase I and II clinical

> trials in delaying recurrence and improving survival. This conclusion was not supported by later findings. In a review of results of all clinical trials published up to 2012, with special emphasis on cervical cancer, the authors concluded that despite the earlier promising results, it seemed that the addition of tirapazamine to chemoradiation treatment conferred no benefit on progression-free or overall survival in patients with cervical cancer.

These results suggest that addition of TPZ to chemoradiation is only of value if the tumor can be identified as having a hypoxic fraction.

2.1.2.4.3 AQ4N Apart from TPZ, the most promising hypoxic cytotoxin is AQ4N, which has been shown to be tolerated up to high doses [43].

2.2 NITRIC OXIDE: A VERSATILE SMALL MOLECULE

NO is one of the oldest and most primitive molecules, which may have evolved as a critical defense mechanism against ozone toxicity. In the biological context, nitric oxide, a diffusible multifunctional second messenger, is implicated in numerous physiological functions in mammals, ranging from immune response and potentiation of synaptic transmission, to dilation of blood vessels and muscle relaxation. Nobel Prize–winning research led to the discovery of its key roles in these areas and it was named “Molecule of the Year” in 1992.

2.2.1 Metabolism of Nitric Oxide

NO is generated endogenously by NOS in mammals from the oxidation of L-arginine to L-citrulline (Figure 2.11). NO can also be formed independently of NOS by the reduction of nitrate and nitrite catalyzed by reductive enzymes including deoxyhemoglobin and nitrate reductase.

Three NOS isoforms have been characterized: neuronal NOS (nNOS, NOS1), which is primarily found in neuronal tissue and skeletal muscle; inducible NOS (iNOS, NOS2), which was originally isolated from macrophages and later discovered in many other cells types; and endothelial NOS (eNOS, NOS3), which is present in vascular endothelial cells, cardiac myocytes, and in blood platelets. The enzymatic activity of all three isoforms is dependent on calmodulin, which binds to nNOS and eNOS at elevated intracellular calcium levels, but is constitutively associated with iNOS, even at basal calcium levels. As a result, the enzymatic activity of nNOS and eNOS is modulated by changes in intracellular calcium levels,

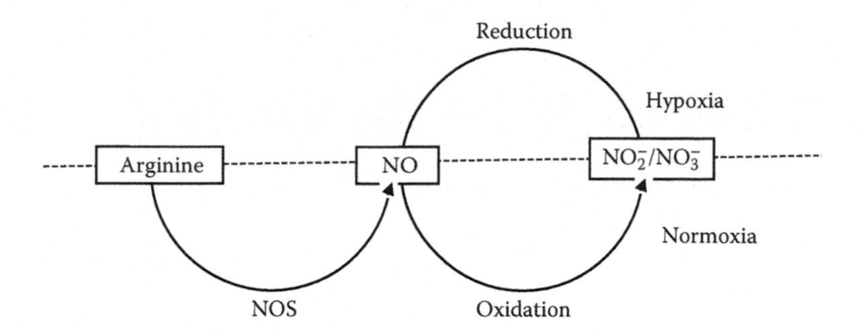

FIGURE 2.11 The nitrate-nitrite-NO pathway. NO is generated from the precursor L-arginine by the enzyme NOS under normoxic conditions. Under these conditions, NO is oxidized to nitrite and nitrate. Under hypoxia, nitrite is reduced by a variety of NOS-independent processes to form NO.

leading to transient NO production, whereas iNOS continuously releases NO independent of fluctuations in intracellular calcium and is mainly regulated at the gene expression level.

A gaseous, short-lived free radical that is uncharged and freely diffusible across tissues and cell membranes, NO is able to bind to its most sensitive known target, soluble guanylate cyclase, to stimulate the production of cyclic GMP (cGMP). cGMP, in turn, can activate cGMP-dependent protein kinase G (PKG) and can cause smooth muscles and blood vessels to relax, decrease platelet aggregation, and alter neuron function among other effects. NO acts through the stimulation of the soluble guanylate cyclase, which is a heterodimeric enzyme with subsequent formation of cGMP. Nitric oxide relaxes vascular smooth muscle by binding to the heme moiety of cytosolic guanylate cyclase, activating guanylate cyclase and increasing intracellular levels of cyclic-guanosine 3′,5′-monophosphate, which then leads to vasodilation. Signaling from the endothelium (inner lining) of blood vessels by nitric oxide causes the surrounding smooth muscle to relax, resulting in vasodilation and increasing blood flow (Figure 2.12).

2.2.2 Mechanisms of Radiosensitization by Nitric Oxide

The radiosensitizing properties of nitric oxides have been studied for almost 60 years. Nitric oxide is able to penetrate well into tumors as it is readily soluble in lipid membranes and has a high diffusion coefficient in water. It has also been associated with the radiation-induced bystander effect. Elevated radiation response by NO in hypoxic cells and tumors has

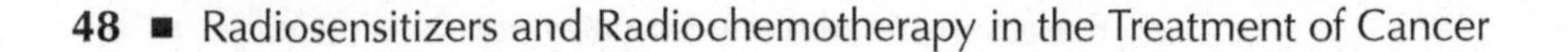

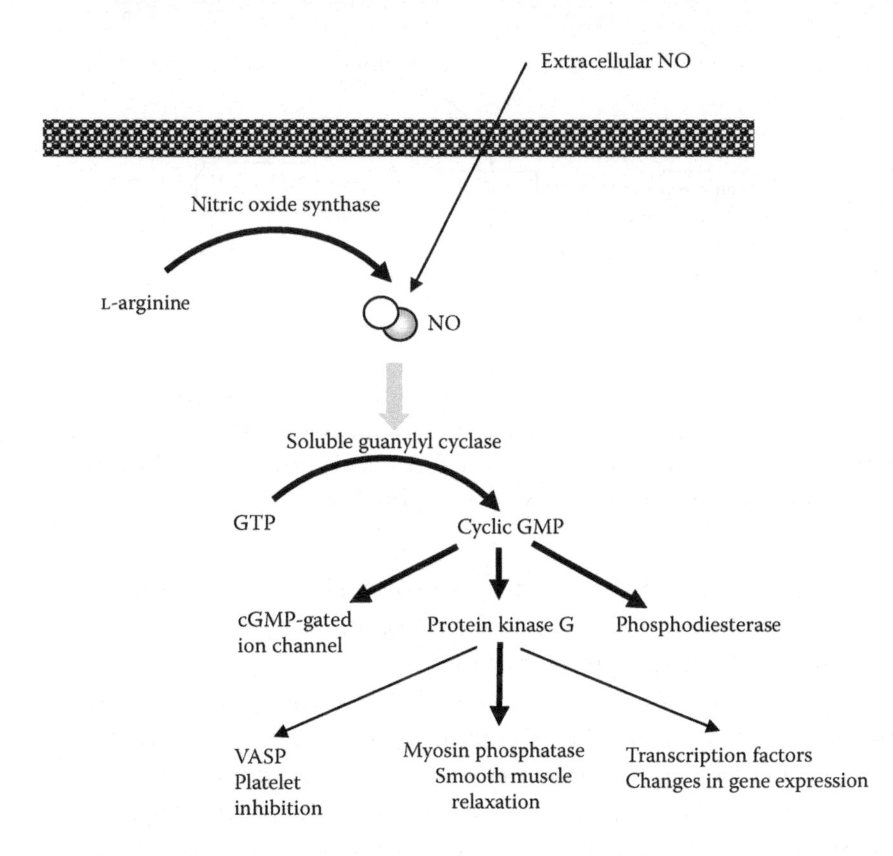

FIGURE 2.12 The effects of nitric oxide are mediated in the production of the second messenger cGMP. Nitric oxide stimulates the production of cGMP by interacting with soluble guanylate cyclase (sGC). This interaction allows sGC to convert GTP into cGMP. Many of the effects of cGMP are mediated through the activation of protein kinase G (PKG). In the case of vasodilation, activation of PKG by cGMP leads to activation of myosin phosphatase, which in turn leads to the release of calcium from intracellular stores in smooth muscle cells causing the relaxation of the smooth muscle cells.

been attributed to many factors, including an increase in tumor oxygenation through the ability of NO to act as a vasodilator, the induction of apoptosis through activation of p53, inhibition of DNA repair enzymes and enhanced radiation-induced DNA damage seen through increased formation of double-strand breaks, detected as γH2AX foci.

2.2.2.1 Direct Radiosensitization by NO

Nitric oxide is sometimes described as "oxygen mimetic" because it shares with oxygen the property of being a free radical in a "stable" form and is highly reactive toward many other free radicals. The basis for comparison is limited however because NO, like many other chemicals, reacts rapidly with hydrated electrons produced by the radiolysis of water but it has a relatively low electron affinity, lower in fact than that of metronidazole. Another important distinction is that whereas in the case of oxygen or the nitroimidazoles, the product is a DNA base peroxyl radical (as in Figure 2.4); the unpaired electron of NO pairs up with an unpaired electron in a DNA base radical to form a non-radical species. Nevertheless, nitric oxide enhances the yield of DNA double-strand breaks in irradiated cells, although by a less well-defined mechanism than that for oxygen. The repair times in the γH2AX assay were reported to be longer for cells irradiated in nitric oxide compared with those irradiated in air or anoxia, suggesting dysfunctional DNA repair [52].

A remarkable property of nitric oxide as a hypoxic cell radiosensitizer is its efficiency. The effect on cell survival *in vitro* using clinically relevant radiation doses of only 40 ppm v/v nitric oxide ~70 nM, a concentration similar to that used in inhaled nitric oxide therapy for respiratory conditions [52], is shown in Figure 2.13. In this study, at low radiation doses, nitric oxide seemed to be significantly more efficient than oxygen as a hypoxic cell radiosensitizer.

2.2.2.1.1 Mechanism of Radiosensitization in the Presence of Oxygen In aerated cells, the cytotoxic effects of NO derive from the reaction of NO with the superoxide anion to form the peroxynitrite anion $ONOO^-$ $O_2^- + \bullet NO \rightarrow ONOO^-$ and in fact the short half-life of NO may be partly explained by its rapid reaction with superoxide to form the peroxynitrite ion. Radiosensitization occurs because the level of reactive oxygen species in the cell, including the superoxide anion, is increased by interaction with ionizing radiation. Peroxynitrite causes apoptotic or necrotic cell death through nitration of tyrosine residues in proteins, lipid peroxidation, oxidation of critical thiols, DNA strand breakage, and activation of the nuclear enzyme poly(ADP-ribose) polymerase, leading to NAD+ depletion and energy failure. At physiologic pH, peroxide combines with NO to form nitrogen dioxide and hydroxyl radicals, before oxidizing to nitrate, usually considered inert, and nitrite, which can be cycled back to NO and other reactive nitrogen oxide species through hypoxia-mediated reduction (reviewed by Oronsky et al. [53]).

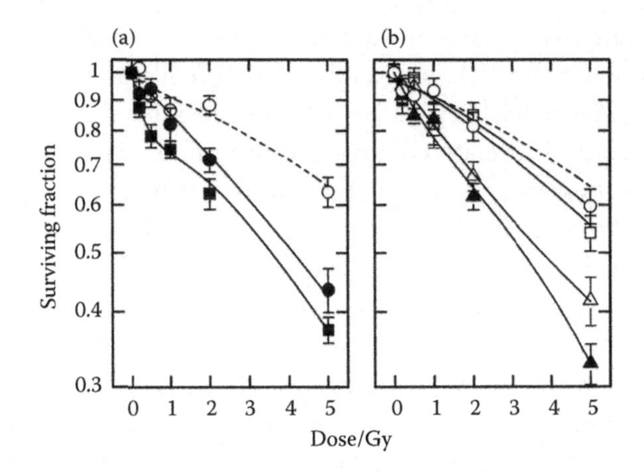

FIGURE 2.13 Clonogenic survival of V79 cells after irradiation in suspension continuously bubbled with the saturating gas. (a) (○) N_2; (●) 40 ppm v/v NO• in N_2; (■) 80 ppm NO• in N_2. (b) (○) 50 ppm O_2 in N_2; (□) 100 ppm O_2; (△) 1,000 ppm O_2; (▲) 800 ppm O_2 80 ppm NO•; dashed line is the curve for anoxic cells from (a). (From Wardman, P. et al., *Radiation Research* 167:475–484, 2007. With permission.)

2.2.2.1.2 Radiosensitization of Hypoxic Cells: NO as Substitute for Oxygen

Nitric oxide is a very effective radiosensitizer of hypoxic mammalian cells. *In vivo* NO may have effects on tumor vasculature and hence on tumor oxygenation (discussed in Section 2.2.2) but it may also interact with radiation-produced radicals to modify DNA lesions although, until recently, few publications have addressed this aspect. In a recently published study [54], it is proposed that nitric oxide (NO) is an effective radiosensitizer of hypoxic mammalian cells, which is at least as efficient as oxygen in enhancing cell death *in vitro*. Based on experimental findings, it is hypothesized that NO induces cell death through the formation of difficult-to-repair base lesions which, if they occur within complex clustered damage common to ionizing radiation, can lead to replication-induced DNA strand breaks. This conclusion was supported by a number of lines of evidence: cells irradiated in exponential growth in the presence of NO were twice as radiosensitive as those irradiated in anoxia without NO whereas confluent cells were less radiosensitive to NO. The numbers of DNA double-strand breaks observed as γH2AX staining following radiosensitization by NO was higher in exponential cells than in confluent cells. DNA damage, detected as 53BP1 foci, was also higher in cells in S and G_2

phases of the cell cycle following radiosensitization by NO. RAD51 foci are highest in V79-4 cells irradiated in the presence of NO compared with anoxic cells 24 h after radiation. It was concluded that radiosensitization of hypoxic cells by NO is in part through the formation of specific, difficult to repair DNA damage which in dividing cells may induce the formation of stalled replication forks and replication-induced DNA strand breaks which may lead to cell death.

Another study by the same authors reported on specific base modifications that result from the reaction of NO with radicals in DNA bases and in plasmid DNA after irradiation [55]. Potent radiosensitization of mammalian cell lines by low levels of nitric oxide was reported to be consistent with near diffusion-controlled reaction of nitric oxide with purine and pyrimidine radicals (observed by pulse radiolysis), with nitric oxide reacting two to three times faster than oxygen [52].

2.2.2.2 *Indirect Radiosensitization: Increased NO-Mediated Tumor Oxygenation*

A large component of the radiosensitizing effect of NO results from its effects on the tumor and systemic vasculature acting in such a way as to increase tumor oxygenation. These effects are summarized in Table 2.4 (reviewed by Oronsky et al. [53]).

TABLE 2.4 Mechanism of NO Radiosensitization Effects

Treatment	Systemic Effects	Local Effects
NO donors (tumor type-dependent)	Vasodilation and steal effect Red blood cell rheology modification	Increased oxygenation Release of NO by RBCs under hypoxic conditions Local oxygen sparing through NO-mediated mitochondrial effects
TSP-1/CD47 modulation	Increased TSP inhibits NO generation leading to vasoconstriction	Increased blood flow through antisteal effect
Insulin and electrical stimulus	Increased iNOS activity and NO	Increased tumoral blood flow
Focused low-dose radiation	None	Increased tumor NO through eNOS upregulation
Hypoxia-activated NO donors	None	Increased local NO concentration, modulation of tumor blood flow

2.2.2.2.1 Vasodilatation and the Steal Effect The effect of administration of NO donors to control tumor blood flow is not easily described. Some findings indicate improved tumor oxygenation and radiosensitization through increased blood flow, whereas others are consistent with the promotion of tumor growth and decreased tumor perfusion. These apparently contradictory results may be attributable to the peculiarities of the tumor microenvironment resulting in blood flow redistribution through steal or anti–steal effects. The "steal effect," describes a process by which systemic blood pressure reduction in response to nitric oxide can make the tumor more hypoxic. Because tumor blood vessels are fully dilated, a "steal effect" would involve a relocation of blood away from the tumor as a result of NO-mediated systemic vasodilation. An anti-steal effect, on the other hand, would increase tumor blood flow and oxygenation through preferential vessel relaxation or systemic vasoconstriction [56].

2.2.2.2.2 Red Cell Rheological Effects It is known that renitration of stored blood with exogenous NO donors can reverse the rheological effects (crenation) on RBCs due to hypoxia and acidosis and thereby improve tissue perfusion and oxygenation upon transfusion. By analogy, in the *in vivo* situation, NO should restore normal red cell shape and normalize blood viscosity in the relatively hypoxic, malformed, and tortuous tumor microvessels; thus, increasing overall flow through the tumor. These hemodynamic changes would be expected to result in a radiosensitizing effect by improved oxygen delivery through relief of hypoxia.

2.2.2.2.3 Hypoxia and Hypoxic Vasodilation The hypoxic microenvironment would be expected to influence the susceptibility of tumors to nitrovasodilation with the two types of hypoxia, chronic and acute, responding differently. Long-term or diffusion-limited hypoxia is related to the maximum distribution distance of oxygen in actively respiring tumor tissue, creating an O_2 tension gradient as the distance from the vessels increases. Acute or transient hypoxia involves a temporary restriction of blood flow to areas of the tumor by transiently constricted or blocked vessels, resulting in local ischemia.

It has been suggested that the vasculature of acutely hypoxic tumors may be more susceptible to the vasoactive and rheological properties of NO reduction through relief or prevention of temporary flow stasis. NO is known to react with a highly conserved reduced cysteine at position 93 on the β-chain of hemoglobin to form an *S*-nitrosothiol (SNO). Under normoxic conditions, NO binds to this cysteinyl residue on deoxyhemoglobin to form *S*-nitrosohemoglobin (SNOHb). Under hypoxic conditions, the allosteric transition in *S*-nitrosohemoglobin from the R (oxygenated) to the T (deoxygenated) conformation transfers NO to a cysteine in the membrane of the red cell. The red cell thus serves as the NO gatekeeper, sequestering NO equivalents in normoxia and dispensing them in hypoxia [57]. This would result in selective vasodilation occurring under transient hypoxic conditions with resultant radiosensitization through increased oxygenation.

An alternative hypothesis proposes that it is nitrite, rather than SNOHb, which is responsible for preserving NO bioactivity in the circulation. In either case, whether the carrier is nitrite or SNOHb, NO bioactivity can be exported selectively to hypoxic zones of the tumor, dilating mature blood vessels, and thereby facilitating O_2 delivery.

2.2.2.2.4 Oxygen-Sparing Effects Another possibility is that reoxygenation of tumors can be linked to decreased consumption of oxygen during metabolism. NO and its oxidative product nitrite have been reported to increase tumor oxygenation through inhibition of mitochondrial respiration by binding reversibly to cytochrome c oxidase in complex IV. Oxygen is thus redistributed away from the electron transport chain toward nonrespiratory oxygen-dependent targets like hypoxia-inducible factor 1 (HIF-1; see Chapter 12). HIF-1, consisting of two subunits, HIF-1α and HIF-1β, is the master regulator of hypoxic stress. Under well-oxygenated conditions, HIF-1α is stable being continuously synthesized and degraded. At low O_2 concentrations, HIF-1α accumulates to heterodimerize with HIF-1β and activates the expression of HIF-dependent target genes such as vascular endothelial growth factor (VEGF), contributing to radioresistance. By contrast, NO-mediated tumor reoxygenation increases tumor radiosensitivity through resumption of HIF-1α degradation. These NO-elicited events can also generate peroxynitrite, which induces oxidative stress and apoptosis [58] through different mechanisms, including DNA cross-linking.

2.2.3 NO-Based Radiosensitization Strategies

As indicated by Section 2.2.2, the effects of NO donors on tumor blood flow and radiosensitization may not be consistent or predictable. In addition, the potential for dose-limiting hypotension and vascular steal limits the clinical utility of NO donors as radiosensitizers for both the currently approved NO donors (organic nitrates including glyceryl trinitrate, isosorbide mononitrate, and dinitrate) and the direct NO donors, which includes sodium nitroprusside and SIN-1. A number of strategies to optimize the activity/toxicity profile of NO manipulation have been studied. These include hypoxia-activated NO donors, thrombospondin 1/CD-47 modulation, and indirect activation of endogenous NOS (summarized in Oronsky et al. [53]).

2.2.3.1 Hypoxia-Activated NO Donors

2.2.3.1.1 *S*-Nitroso-*N*-Acetylpenicillamine and *S*-Nitrosoglutathione The nitrosonium ion (NO+) donors, *S*-nitroso-*N*-acetylpenicillamine (SNAP) and *S*-nitrosoglutathione (GSNO), generate NO bioreductively under hypoxic conditions, leading to *in vitro* radiosensitization [59]. The mechanism is thought to include delivery and release of NO through transnitrosylation steps. However, these NO donors have also been associated with the stabilization of HIF-1α, leading to the induction of HIF-1α target genes, and radioresistant phenotypes, which would limit their clinical utility.

2.2.3.1.2 RRx-001 A nonexplosive pernitro compound RRx-001 is a small-molecule NO donor that has undergone phase I clinical trials. Preclinical studies demonstrated profound radiosensitization *in vivo* at nontoxic doses, which was accompanied by a significant increase in tumor blood flow for up to 72 h. At these doses, normal gastrointestinal epithelium was not sensitized to the effects of radiation and may even have been protected.

2.2.3.1.3 Thrombospondin 1/CD-47 Modulation In contrast with directly modulating NO levels, thrombospondin 1 (TSP-1), acting through its receptor CD-47, antagonizes the effects of NO. Antagonism of NO signaling by TSP results in acute inhibition of tissue perfusion; however, the tumor vasculature is resistant to acute vasodilation by NO and increased TSP may thus indirectly increase tumor blood flow by limiting circulation elsewhere. Some tumors induce an increase in systemic TSP-1 derived from nontumorigenic stromal cells. The systemic increase in TSP-1 limits NO-driven responses in normal tissue and thereby induces an antisteal effect by increasing tumor perfusion to the detriment of healthy tissue circulation [60]. Whereas local NO production drives tumor angiogenesis, systemic NO-mediated vasodilation preferentially enhances normal tissue perfusion at the expense of the tumor, similar to the steal effect.

2.2.3.2 Indirect Activation of Endogenous NOS

2.2.3.2.1 Insulin and Electrical Stimulus Jordan et al. [61] reported that insulin infusion and electrical stimulation of the host tissue radiosensitized experimental tumors *in vivo* through increased NO production from the endothelial isoform of NOS (eNOS). The authors speculate that both treatments resulted in tumor reoxygenation from increased blood flow and reduced oxygen consumption due to decreased mitochondrial respiration. The increase in tumor oxygenation, and the radiosensitizing effect, was completely abolished in eNOS knockout mice.

2.2.3.2.2 Lipid A Analogue, ONO-4007 A synthetic analogue of the lipid A moiety of Gram-negative bacterial lipopolysaccharide. Preclinical data have demonstrated that the compound radiosensitizes with considerably less toxicity than lipopolysaccharide in animal models of malignancy [62]. The mechanism of action is thought to involve the activation of interferon-γ pathways and the induction of NOS [52]. In a phase I trial, a maximum tolerated dose was reached, and although no objective responses were seen, disease stabilization was observed in five patients for the duration of the study (18 weeks) [63].

2.3 SUMMARY

Oxygen is a potent radiosensitizer, increasing the biological effect of ionizing radiation by a factor of 2 to 3, whereas radioresistance is associated with hypoxia. One mechanism which has been suggested to be responsible for the oxygen effect is the OFH. The OFH proposes that DNA sustains direct or indirect radiation damage which, in the presence of oxygen, is converted to DNA peroxyradicals, preventing DNA repair. Radioresistance associated with hypoxia is of particular significance in cancer therapy because it is known that many solid tumors contain a significant fraction of hypoxic cells.

A number of methods have been investigated for radosensitizing hypoxic cells, including hyperbaric oxygen, upregulation of oxygen transport, and increasing the availability of molecular oxygen by reducing the oxygen-binding affinity of hemoglobin. It was discovered in the late 1960s that certain electron-affinic compounds were able to mimic oxygen and thus enhance radiation damage. Based on this discovery, a number of compounds were investigated in preclinical models, with the most successful being nitroimidazoles, which were highly effective for acute radiation doses but less so for fractionated treatment. A number of clinical trials in the late 1970s centered on 5-nitroimidazole (misonidazole) but most of these were unable to demonstrate a significant improvement in radiation response, although benefits were seen for certain subgroups of treated patients. An important reason for this failure was that the dose of misonidazole was limited by a dose-dependent peripheral neuropathy.

Other nitroimidazoles were developed to overcome this problem; one being etanidazole but this again failed to show significant therapeutic benefit although, in a later subgroup analysis, some positive benefit was reported. The overall conclusion is that although the nitroimidazoles have not lived up to their original promise in terms of clinical efficacy, they may be of value in the treatment of certain conditions.

The hypoxic cytotoxins or bioreductive drugs are compounds that undergo intracellular reduction to form active cytotoxic species under low oxygen tensions, selectively killing hypoxic cells. Thus, they are not radiosensitizers in the true sense but when combined with radiation for tumor treatment the result is an enhancement of radiation response. These drugs can be divided into three major groups: quinones (mitomycin C), nitroimidazoles (RSU-1069), and N-oxides (tirapazamine).

The most successful bioreductive agent to date is tirapazamine (TPZ) a heteroaromatic N-oxide, which showed positive interactions in combined *in vivo* treatments with radiation, particularly with fractionated protocols. TPZ is also beneficial when combined with a wide range of chemotherapeutic agents, and enhancements in normal tissue toxicity were generally small. Most of the clinical studies to date have combined tirapazamine with chemotherapy notably cis-platinum. Trials with combined tirapazamine and radiation have had mixed results, possibly because insufficient care has been taken to target patients whose tumors contained a significant proportion of hypoxic cells.

Nitric oxide (NO) is a diffusible multifunctional second messenger, which is implicated in numerous physiological functions in mammals, ranging from immune response and potentiation of synaptic transmission, to dilation of blood vessels and muscle relaxation. NO is generated endogenously by NO synthase from the oxidation of L-arginine to L-citrulline. A gaseous short-lived free radical that is uncharged and freely diffusible across tissues and cell membranes, NO is able to bind to its most sensitive known target, soluble guanylate cyclase, to stimulate the production of cGMP, which is responsible for downstream NO-mediated signaling.

Radiosensitization by NO depends on two processes, neither of which is exclusive to the NO molecule itself but depends on the oxygen effect in one form or another. First, NO reacts with $O_2^{\dot{}}$ to form the peroxynitrite anion, $ONOO^-$, a highly cytotoxic species; and second, the physiological effects of NO involve the modulation of oxygen consumption and delivery in the tumor.

NO has a number of effects on the tumor vasculature and oxygenation: a steal effect involves NO-mediated systemic vasodilation causing a relocation of blood away from fully dilated tumor blood vessels whereas a reverse steal effect can result from systemic vasocontraction; NO donors can reverse the rheological effects (crenation) on RBCs due to hypoxia and acidosis, which results in improved oxygen delivery; hypoxic vasodilation occurs due to differential binding of NO under hypoxic and aerated conditions allowing selective vasodilation under hypoxic conditions; and oxygen sparing occurs when NO and its oxidative product nitrite act to increase tumor oxygenation through inhibition of mitochondrial respiration.

Although NO has been shown to act as a radiosensitizer in a number of studies, there are also reports that it is ineffective or may even exert a radioprotective effect. One interpretation of this mixed response is that it depends on local NO concentration, the biological milieu, and interactions with oxygen. At low concentrations, NO promotes tumor cell survival and angiogenesis, whereas at higher levels, when the homeostatic balance is perturbed, NO acts as an antitumor agent and radiosensitizer. NO-based radiosensitization strategies include the use of hypoxia-activated NO donors and methods to bring about indirect activation of endogenous NOS.

REFERENCES

1. Alexander, P., and Charlesby, A. Energy transfer in macromolecules exposed to ionizing radiations. *Nature* 1954;173:578–579.
2. Ewing, D. The oxygen fixation hypothesis: A reevaluation. *Am J Clin Oncol* 1998;21:355–361.
3. Johansen, I., and Howard-Flanders, P. Macromolecular repair and free radical scavenging in the protection of bacteria against X-rays. *Radiat Res* 1965;24:184–200.
4. Koch, C., Oprysko, P., Shuman, A., Jenkins, W., Brandt, G., and Evans, S. Radiosensitization of hypoxic tumor cells by dodecafluoropentane: A gas-phase perfluorochemical emulsion. *Cancer Res* 2002;62:3626–3629.
5. Chang, T. Evolution of artificial cells using nanobiotechnology of hemoglobin based RBC blood substitute as an example. *Artif Cells Blood Substit Immobil Biotechnol* 2006;34:551–556.
6. Rowinsky, E. Novel radiation sensitizers targeting tissue hypoxia. *Oncol Rep* 1999;13(Suppl. 5):61–70.

7. Harrison, L., Chadha, M., Hill, R., Hu, K., and Shasha, D. Impact of tumor hypoxia and anemia on radiation therapy outcomes. *Oncologist* 2007;7:492–508.
8. Aapro, M., and Vaupel, P. Erythropoietin: Effects on life expectancy in patients with cancer-related anaemia. *Curr Med Res Opin* 2006;22(Suppl. 4):5–13.
9. Calais, G., and Hirst, D.G. In situ tumour radiosensitization induced by clofibrate administration: Single dose and fractionated studies. *Radiother Oncol* 1991;22:99–103.
10. Stea, B., Suh, J., Boyd, A., Cagnoni, P., Shaw, E., and Group, R. Whole brain radiotherapy with or without efaproxiral for the treatment of brain metastases: Determinants of response and its prognostic value for subsequent survival. *Int J Radiat Oncol Biol Phys* 2006;64:1023–1030.
11. Collingridge, D., and Rockwell, S. Pentoxifylline improves the oxygenation and radiation response of BA1112 rat rhabdomyosarcomas and EMT6 mouse mammary carcinomas. *Int J Cancer* 2000;90:256–264.
12. Kjellen, E., Joiner, M., Collier, J., Johns, H., and Rojas, A. Therapeutic benefit from combining normobaric carbogen or oxygen with nicotinamide in fractionated X-ray treatments. *Radiother Oncol* 1991;22:81–89.
13. Kaanders, J., Bussink, J., and van der Kogel, A. Clinical studies of hypoxia modification in radiotherapy. *Semin Radiat Oncol* 2004;14:233–240.
14. Janssens, G., Rademakers, S., Terhaard, C. et al. Accelerated radiotherapy with carbogen and nicotinamide for laryngeal cancer: Results of a phase III randomized trial. *J Clin Oncol* 2012;30:1777–1783.
15. Hoskin, P., Rojas, A., Bentzen, S., and Saunders, M. Radiotherapy with concurrent carbogen and nicotinamide in bladder carcinoma. *J Clin Oncol* 2010;28:4912–4918.
16. Adams, G., and Cooke, M. Electron-affinic sensitization. I. A structural basis for chemical radiosensitizers in bacteria. *Int J Radiat Biol* 1969;15:457–471.
17. Brown, J. The hypoxic cell: A target for selective cancer therapy—Eighteenth Bruce F. Cain Memorial Award Lecture. *Cancer Res* 1999;59:5863–5870.
18. Overgaard, J. Hypoxic radiosensitization. Adored and ignored. *J Clin Oncol* 2007;25:4066–4074.
19. Bamatraf, M., O'Neill, P., and Rao, B. Redox dependence of the rate of interaction of hydroxyl radical adducts of DNA nucleobases with oxidants: Consequences for DNA strand breakage. *J Am Chem Soc* 1998;120:11852–11857.
20. Wardman, P. Chemical radiosensitizers for use in radiotherapy. *Clin Oncol (R Coll Radiol)* 2007;19:397–417.
21. Sheldon, P., and Fowler, J. Radiosensitization by misonidazole of fractionated X-rays in a murine tumor. *Br J Cancer* 1978;37(Suppl. III):243–245.
22. Shenoy, M., Asquith, J., Adams, G., Michael, B., and Watts, M. Time-resolved oxygen effects in irradiated bacteria and mammalian cells: A rapid-mix study. *Radiat Res* 1975;62:498–512.
23. Adams, G., Fowler, J., and Wardman, P., eds. Hypoxic cell sensitizers in radiobiology and radiotherapy. *Br J Cancer* 1978;37(Suppl. III).

24. Adams, G.E., Flockhart, I.R., Smithen, C.E., Stratford, I.J., Wardman, P., and Watts, M.E. VII. A correlation between structures, one-electron reduction potentials and efficiences of nitroimidazoles as hypoxic cell radiosensitizers. *Radiat Res* 1976;67:9–20.
25. Adams, G. Hypoxia-mediated drugs for radiation and chemotherapy. *Cancer* 1981;48:696–707.
26. Overgaard, J. Sensitization of hypoxic tumour cells—Clinical experience. *Int J Radiat Biol* 1989;56:801–811.
27. Overgaard, J. Clinical evaluation of nitroimidazoles as modifiers of hypoxia in solid tumors. *Oncol Res* 1994;6:509–518.
28. Overgaard, J., Hansen, H., Overgaard, M. et al. A randomized double-blind phase III study of nimorazole as a hypoxic radiosensitizer of primary radiotherapy in supraglottic larynx and pharynx carcinoma. Results of the Danish Head and Neck Cancer Study (DAHANCA) Protocol 5-85. *Radiother Oncol* 1998;46:135–146.
29. Overgaard, J., and Horsman, M.R. Modification of hypoxia-induced radioresistance in tumors by the use of oxygen and sensitizers. *Semin Radiat Oncol* 1996;6:10–21.
30. Dobrowsky, W., Huigol, N., Jayatilake, R. et al. AK-2123 (Sanazol) as a radiation sensitizer in the treatment of stage III cervical cancer. Results of an IAEA multicentre randomised trial. *Radiother Oncol* 2007;82:24–29.
31. Murata, R., Tsujitani, M., and Horsman, M. Enhanced local tumour control after single or fractionated radiation treatment using the hypoxic cell radiosensitizer doranidazole. *Radiother Oncol* 2008;87:331–338.
32. Karasawa, K., Sunamura, M., and Okamoto, A. Efficacy of novel hypoxic cell sensitiser doranidazole in the treatment of locally advanced pancreatic cancer: Long-term results of a placebo-controlled randomised study. *Radiother Oncol* 2008;87:326–330.
33. Bischoff, P., Altmeyer, A., and Dumont, F. Radiosensitising agents for the radiotherapy of cancer: Advances in traditional and hypoxia targeted radiosensitisers. *Expert Opin Ther Pat* 2009;19(5):643–662.
34. Emami, S., Kumar, P., Yang, J. et al. Synthesis, transportability and hypoxia selective binding of 1-beta-D-(5-deoxy-5-fluororibofuranosyl)-2-nitroimidazole (beta-5-FAZR), a configurational isomer of the clinical hypoxia marker, FAZA. *J Pharm Pharm Sci* 2007;10:237–245.
35. Nakae, T., Uto, Y., Tanaka, M. et al. Design, synthesis, and radiosensitizing activities of sugar-hybrid hypoxic cell radiosensitizers. *Bioorg Med Chem Lett* 2008;16:675–682.
36. Jaffar, M., Williams, K., and Stratford, I. Bioreductive and gene therapy approaches to hypoxic diseases. *Adv Drug Deliv Rev* 2001;53:217–228.
37. Fracasso, P., and Sartorelli, A. Cytotoxicity and DNA lesions produced by mitomycin C and porfiromycin in hypoxic and aerobic EMT6 and Chinese hamster ovary cells. *Cancer Res* 1986;46:3939–3944.

38. Saunders, M., Jaffar, M., Patterson, A. et al. The relative importance of NADPH: Cytochrome c (P450) reductase for determining the sensitivity of human tumour cells to the indolequinone EO9 and related analogues lacking functionality at the C-2 and C-3 positions. *Biochem Pharmacol* 2000;59:993–996.
39. Stratford, I., Adams, G., Godden, J., and Howells, N. Induction of tumour hypoxia post-irradiation: A method for increasing the sensitizing efficiency of misonidazole and RSU 1069 *in vivo*. *Int J Radiat Biol* 1989;55:411–422.
40. McKeown, S., Cowen, R., and Williams, J. Bioreductive drugs: From concept to clinic. *Clin Oncol* 2007;19:427–442.
41. Papadopoulou, M., and Bloomer, W. NLCQ-1 (NSC 709257): Exploiting hypoxia with a weak DNA-intercalating bioreductive drug. *Clin Cancer Res* 2003;9:5714–5720.
42. Wilson, W., Pullen, S., Degenkolbe, A. et al. Water-soluble dinitrobenzamide mustard phosphate pre-prodrugs as hypoxic cytotoxins. *Eur J Cancer* 2004;Suppl. 2:151.
43. Patterson, L., and McKeown, S. AQ4N: A new approach to hypoxia-activated cancer chemotherapy. *Br J Cancer* 2000;83:1589–1593.
44. Zeman, E.M., Brown, J.M., Lemmon, M.J., Hirst, V.K., and Lee, W.W. SR-4233: A new bioreductive agent with high selective toxicity for hypoxic mammalian cells. *Int J Radiat Oncol Biol Phys* 1986;12:1239–1242.
45. Cowen, R., Williams, K., Chinje, E. et al. Hypoxia targeted gene therapy to increase the efficacy of tirapazamine as an adjuvant to radiotherapy: Reversing tumor radioresistance and effecting cure. *Cancer Res* 2004;64:1396–1402.
46. Semenza, G., and Wang, G. A nuclear factor induced by hypoxia via de novo protein synthesis binds to the human erythropoietin gene enhancer at a site required for transcriptional activation. *Mol Cell Biol* 1992;12:5447–5454.
47. McCarthy, H., Yakkundi, A., McErlane, V. et al. Bioreductive GDEPT using cytochrome P450 3A4 in combination with AQ4N. *Cancer Gene Ther* 2003;10:40–48.
48. Rischin, D., Hicks, R.J, Fisher, R. et al. Prognostic significance of [18F]-misonidazole positron emission tomography-detected tumor hypoxia in patients with advanced head and neck cancer randomly assigned to chemoradiation with or without tirapazamine: A substudy of Trans-Tasman Radiation Oncology Group Study 98.02. *J Clin Oncol* 2006;24:2098–2104.
49. Reddy, S., and Williamson, S. Tirapazamine: A novel agent targeting hypoxic tumor cells. *Expert Opin Investig Drugs* 2009;18(1):77–87.
50. Peters, L., O'Sullivan, B., Giralt, J. et al. Tirapazamine, cisplatin, and radiation versus cisplatin and radiation for advanced squamous cell carcinoma of the head and neck (TROG 02.02, HeadSTART): A phase III trial of the Trans-Tasman Radiation Oncology Group. *J Clin Oncol* 2010;28(18):2996–3001.
51. Ghatage, P., and Sabagh, H. Is there a role for tirapazamine in the treatment of cervical cancer? *Expert Opin Drug Metab Toxicol* 2012;8(12):1589–1597.
52. Wardman, P., Rothkamm, K., Folkes, L., Woodcock, M., and Johnston, P. Radiosensitization by nitric oxide at low radiation doses. *Radiat Res* 2007;167:475–484.

53. Oronsky, B., Knox, S., and Scicinski, J. Is nitric oxide (NO) the last word in radiosensitization? A review. *Transl Oncol* 2012;5(2):66–71.
54. Folkes, L., and O'Neill, P. Modification of DNA damage mechanisms by nitric oxide during ionizing radiation. *Free Radic Biol Med* 2013;58:14–25.
55. Folkes, L., and O'Neill, P. DNA damage induced by nitric oxide during ionizing radiation is enhanced at replication. *Nitric Oxide* 2013;34:47–55.
56. Trotter, M., Chaplin, D., and Olive, P. Effect of angiotensin II on intermittent tumour blood flow and acute hypoxia in the murine SCCVII carcinoma. *Eur J Cancer* 1991;27:887–893.
57. Sonveaux, P., Lobysheva, I., Feron, O., and McMahon, T. Transport and peripheral bioactivities of nitrogen oxides carried by red blood cell hemoglobin: Role in oxygen delivery. *Physiology (Bethesda)* 2007;22:97–112.
58. Brown, G. Nitric oxide and mitochondria. *Front Biosci* 2007;12:1024–1033.
59. Janssens, M., Verovski, V., Van den Berge, D., Monsaert, C., and Storme, G. Radiosensitization of hypoxic tumour cells by S-nitroso-*N*-acetylpenicillamine implicates a bioreductive mechanism of nitric oxide generation. *Br J Cancer* 1999;79:1085–1089.
60. Isenberg, J., Martin-Manso, G., Maxhimer, J., and Roberts, D. Regulation of nitric oxide signalling by thrombospondin 1: Implications for antiangiogenic therapies. *Nat Rev Cancer* 2009;9:182–191.
61. Jordan, B., Sonveaux, P., Feron, O. et al. Nitric oxide as a radiosensitizer: Evidence for an intrinsic role in addition to its effect on oxygen delivery and consumption. *Int J Cancer* 2004;109:768–773.
62. Ridder, M.D., Verellen, D., Verovski, V., and Storme, G. Hypoxic tumor cell radiosensitization through nitric oxide. *Nitric Oxide* 2008;19:164–169.
63. deBono, J., Dalgleish, A., Carmichael, J. et al. Phase I study of ONO-4007, a synthetic analogue of the lipid A moiety of bacterial lipopolysaccharide. *Clin Cancer Res* 2000;6:397–405.

CHAPTER 3

Radioenhancement by Targeting Cellular Redox Pathways and/or by Incorporation of High-Z Materials into the Target

3.1 INTRODUCTION

This chapter is concerned with what seems to be very disparate approaches, using physical or biological approaches to dose enhancement. In fact, although this is true, it became apparent while researching this topic that certain compounds can be active as radiosensitizers based on either or both mechanisms.

3.2 DOSE ENHANCEMENT BY COMPOUNDS WITH HIGH ATOMIC NUMBER

The rationale for using metal-containing compounds or complexes as radiosensitizers is based on their ability to increase the dose deposited in the target volume due to differences in their mass energy absorption coefficient when compared with soft tissue.

3.2.1 Mechanism of Dose Absorption

In soft tissue, the primary mechanism by which photons lose energy is the Compton effect, in which an energetic photon is scattered by a weakly bound electron leading to a transfer of energy from the photon to the electron, followed typically by ejection of the electron from the atom (Figure 3.1). Photons retain the majority of their energy following these collisions and, in soft tissue, photons with energies of more than a few kiloelectron volts slow gradually over a long range, leading to very sparse distributions of ionizing events.

The photoelectric effect is the competing process at lower energies. In this process, a photon is wholly absorbed by a bound electron, which is then ejected from the atom. In contrast to the Compton effect, which can occur for free electrons as well as those in atoms, the complete absorption of a photon in the photoelectric effect can only occur in the presence of an atomic nucleus to allow for conservation of momentum. As a result, the cross-section of the photoelectric effect depends on Z, the atomic number

(a)

(b)

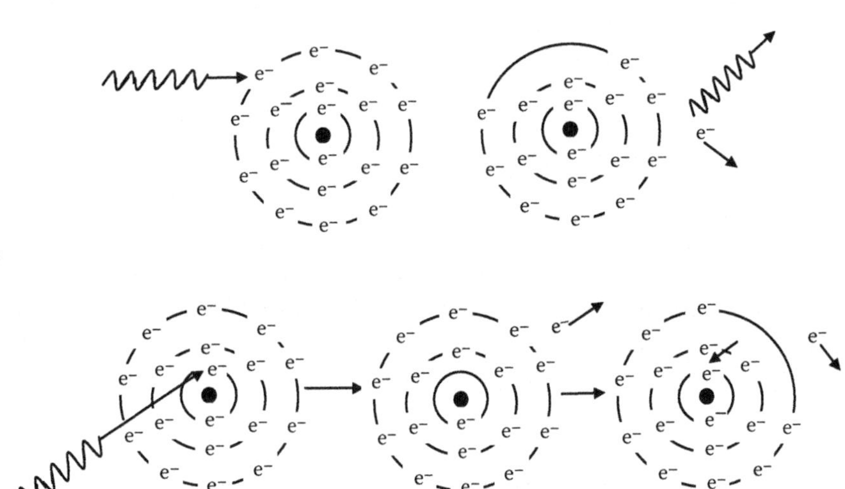

FIGURE 3.1 Schematic illustration of ionizations caused by the Compton and photoelectric effects. In the Compton effect (a), an incident photon is scattered by an electron losing some energy, which is transferred to the electron ejecting it from the atom. In the photoelectric effect (b), an incident photon is wholly absorbed by an electron also ejecting it from the atom. Because this effect preferentially occurs in inner atomic orbitals, further electrons can be ejected as outer-shell electrons fall to fill inner-shell vacancies.

of the atom and is roughly proportional to Z^4. The photoelectric effect is also very strongly dependent on the relationship between the energy of the photon and the strength of the electron binding in the atom and has a very strong maximum cross-section when the photon has just enough energy to liberate the electron and falls off steeply with increasing energy (approximately as E^{-3}, where E is the photon energy). This means that at higher energies, the photoelectric effect preferentially liberates electrons from inner shells due to their higher binding energies. This process is apparent as sharp jumps in the mass energy absorption curve of gold, as shown in Figure 3.2, which can be attributed to photons being able to liberate electrons from different orbitals, with the inner-shell electrons in gold being bound with 79 keV and in subsequent shells with energies around 13 and 3 keV [1].

Inner-shell electrons in soft tissue, on the other hand, tend to have binding energies on the order of 1 keV or less that, coupled with the much lower average atomic number of organic materials, means the photoelectric effect is a relatively small contribution to the absorption in soft tissue whereas it is the dominant component of gold ionization events at energies

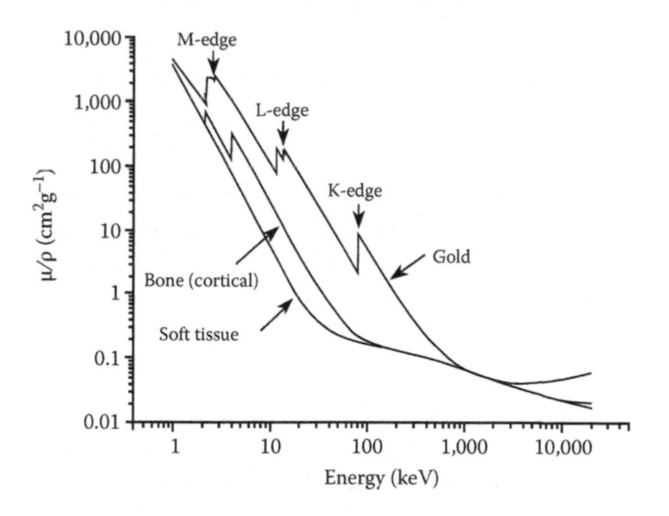

FIGURE 3.2 Plot of gold, soft tissue, and bone attenuation versus x-ray energy. For gold, absorption edges occur at: K (80.7 keV), L (L1 14.4 keV; L2 13.7 keV; L3 11.9 keV), and M (2.2–3.4 keV); the K-edge is the energy where the incident energy is just enough to eject an electron from the inner K shell. (From Hainfeld, J. et al., *J Pharm Pharmacol* 60:977–985, 2008. With permission.)

up to hundreds of kiloelectron volts. Taking these effects together, the significantly increased photoelectric cross-section means high-*Z* metals absorb substantially more energy per unit mass than do soft tissue, translating to a significant increase in local dose when even a small amount of metal is present in the tumor [2]. For certain photon beam qualities, this also results in a change of photon interaction cross-section within the tumor. This fact has been exploited by the development of phototherapy, whereby an orthovoltage (140–250 kVp) x-ray beam is used for treatment. This low photon beam quality (i.e., low in comparison to the normally used megavoltage in radiation therapy) results in a predominance of the photoelectric absorption interaction in the CM-containing tumor. Because the atomic cross-section for photoelectric absorption exhibits a Z^4 to $Z^{4.8}$ dependence, photon absorption is augmented specifically in the tumor volume and the absorbed dose is enhanced relative to surrounding tissues.

3.2.2 Radiosensitization by Contrast Media

It has been known for a long time that the dose was increased when a high-*Z* material was in the targeted zone [3]. Observations of high-*Z* material dose enhancement were first made clinically in patients with reconstructive metal implants receiving radiotherapy for mandibular and head and neck cancers [4]. Following on the understanding of the effect of high-*Z* material on dose, agents that were already in use as contrast media were reassessed in terms of their capacity to increase photoelectric absorption in proximity to the tumor. This is an attractive option because injected contrast media are formulated to localize in tumors. High-*Z* materials such as iodine ($Z = 53$) and gadolinium ($Z = 64$) were already in use as contrast medium improving the contrast between target and surrounding normal tissue [5].

3.2.2.1 Iodine

Iodine-based contrast media used to enhance contrast in computed tomography (CT) imaging, is incorporated preferentially into a variety of tumors. It is particularly effective in tumors of the central nervous system, cranial nerves, and dura due to mechanisms that include hypervascularity and damage to the blood–brain barrier caused by invasive tumor growth.

One of the earliest biological reports of radiosensitization by iodine was of chromosome damage noticed in circulating lymphocytes from patients undergoing iodine contrast angiography [6].

A number of studies have demonstrated radiosensitization in rodent tumor models:

- Remission was seen in 80% of radioresistant tumors in mice directly injected with iodine contrast media followed by 100 kVp x-rays [7].
- Stereotactic delivery of 50 keV monochromatic x-ray beams was used to irradiate rat F98 brain gliomas. Before the irradiation, the tumors were loaded with 1% iodine by intracarotid injection with mannitol. The tumor concentration was 1.0% iodine and that of the surrounding tissue 0.3%. Following a dose of 15 Gy, life spans were increased by 169%. A dose of 25 Gy, however, conferred no additional benefit over iodine controls apparently due to the excessive damage to normal tissue [8].
- A modified CT scanner was used to deliver tomographic orthovoltage (140 kVp) x-rays to spontaneous canine brain tumors after intravenous injection with iodine contrast medium resulting in 53% longer survival [9].
- A similar modified CT device was used in a phase I human trial of brain tumors using intravenous injection of iodine contrast media [10]. This trial proved the method to be safe and potentially beneficial, although further work would have been required to establish statistically significant efficacy.

3.2.2.2 Gadolinium

Gadolinium-based contrast media variously called gadolinium(III) texaphyrin (Gd-Tx), or motexafin gadolinium (MGd; Figure 3.3) are used in magnetic resonance imaging (MRI) to provide selective image enhancement of the tumor relative to the surrounding anatomy. One theoretical study examined the magnitude of tumor dose enhancement achieved by injection of gadolinium or iodine contrast media and treatment using modified x-ray photon spectra from linear accelerators. Monte Carlo modeling of the linear accelerator and patient geometry was used to explore the effect of removing the flattening filter for various beam qualities and the resultant effect on dose enhancement and it was concluded that, under certain circumstances, significant dose enhancement could be achieved with gadolinium using megavoltage x-rays [11].

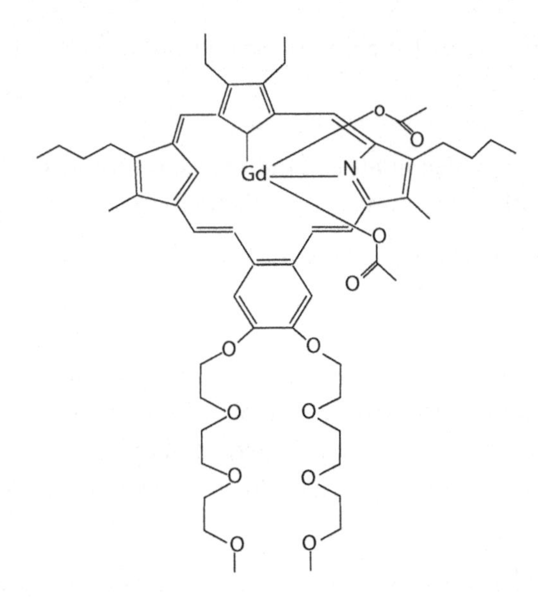

FIGURE 3.3 Motafexin gadolinium.

In fact, it has been determined that the mechanism by which MGd acts as a radiosensitizer probably depends more on its role as a redox-active drug than on its mass energy absorption coefficient. This will be discussed further in Section 3.3.

3.2.2.3 *Gold*

The use of gold as a radiosensitizer-based high-mass energy absorption coefficient seems more promising compared with earlier trials with contrast media such as iodine. Gold has a higher *Z* number than iodine and greater biocompatibility.

Theoretical and experimental studies of high-*Z* dose enhancement over both the megavolt and orthovoltage range have been reported. In one study, cells grown on a gold foil were irradiated (40–120 kVp) and a dose enhancement factor (DEF) of more than 100 was achieved with secondary electrons travelling over a range of up to 10 nm [12]. In *in vivo* experiments, 1.5 to 3.0 nm gold particles were injected directly into the tumor followed by irradiation. Excised cells had reduced plating efficiency, but histological examination showed gold particles only at injection sites. Greater success with this system was not achievable because the large particles did not diffuse far enough to provide adequate tumor coverage [13].

These experiments provided proof-of-principle for the use of gold as a radiosensitizer based on the difference in mass energy absorption coefficient of gold and soft tissue but failed to provide a practical method for delivery of the gold to the tumor target.

3.2.2.3.1 Gold Nanoparticles The need for smaller particles was met by gold nanoparticles (GNPs), which are sufficiently small to penetrate the tumor mass and have been shown to accumulate preferentially in tumors *in vivo* and, in addition, are reported to have high biocompatibility. These characteristics would suggest that GNPs are ideal candidates for use as high-*Z* radiosensitizers.

Nanoparticles with dimensions up to 100 nm can traverse the cell membrane and may accumulate preferentially in cancer cells [14,15]. Such nanoparticles (1–100 nm) are smaller than the typical cutoff size of the pores in tumor vasculature (e.g., up to 400 nm), so they may access cells in tumors [16]. According to a theoretical study, GNPs between 2 and 10 nm in size or larger particles with micrometer dimensions would be expected to have a very low cell uptake, and this has been confirmed by *in vitro* studies, which demonstrated that cellular uptake of nanoparticles to be dependent on their size with nanoparticles 50 nm in diameter showing the highest uptake [16–18].

Simple predictions based on the ratio of the mass energy absorption coefficients of gold and soft tissue suggest that the addition of 1% of gold by mass to the tumor would result in approximately a doubling of the amount of energy deposited by a kilovoltage x-ray source. These predictions have been confirmed on a theoretical basis in several reports investigating the dose-modifying properties of GNPs for a variety of combinations of nanoparticle concentration, radiation source, and target geometry [11,14,15,19].

3.2.2.3.2 Experimental Evidence of Radiosensitization by GNPs Among the earliest experimental evidence supporting the use of GNPs as radiosensitizers was an *in vivo* study of the effect of treatment on tumor regression. One-year survival was significantly increased from 20% for treatment with 250 kVp x-rays alone to 86% for x-rays in combination with GNPs (Figure 3.4) [20–22].

A review by Butterworth and coworkers [22] summarized *in vitro* studies characterizing the effect of GNPs in combination with ionizing radiation for therapeutic advantage. The experimental conditions used were very diverse in terms of the size, synthesis, and surface functionalization of the GNPs as well as differences in cell model, incubation time, and GNP concentration that were investigated. In addition, the radiation sources

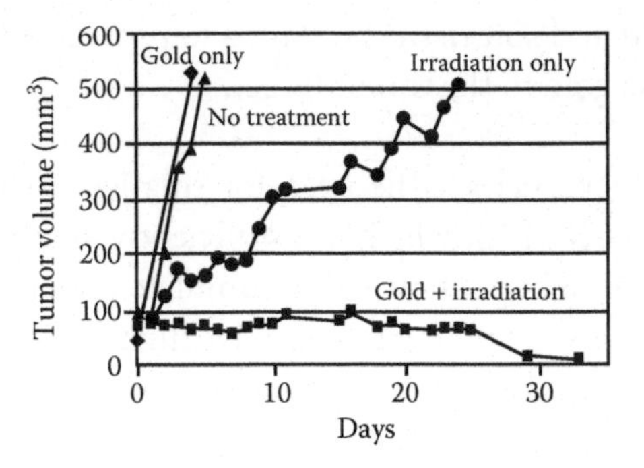

FIGURE 3.4 Average tumor volume after: no treatment (▲, $n = 12$); gold only (◆, $n = 4$); irradiation only (30 Gy, 250 kVp; ●, $n = 11$); intravenous gold injection (1.35 g Au kg^{-1}) followed by irradiation (■, $n = 10$). (From Hainfeld, J. et al., *Phys Med Biol* 49:N309–N315, 2004. With permission.)

of different energies were used. Despite these variations, all 11 studies reported some degree of radiosensitizing effect with a wide variety of cell lines and nanoparticle preparations.

There were two important deviations from the physical predictions of dose modification apparent in these reports. First, many studies reported significant radiosensitizing effects when GNPs were used at concentrations much lower than the 0.1% to 1% by mass that is typically associated with theoretical predictions of significant dose increases. Second, although enhancements have widely been observed with kilovoltage x-rays, as predicted, other studies have investigated megavoltage x-rays or electrons and in many cases reported significant radiosensitization where little or no increase in overall dose deposition would be expected.

These findings raised questions concerning the mechanism of the cellular radiobiological response, which is apparently not primarily driven by increasing the total dose delivered to the cells. When the various x-ray dose modification results for the data were plotted against the degree of enhancement that would be predicted based on the gold concentrations and x-ray energies used, the observed enhancement was much greater than the overall physical dose increase. In fact, the correlation between the predicted dose enhancement and observed radiosensitization was weak with the variation with physical dose being much smaller than that seen between different nanoparticle preparations and cell lines. The conclusion was that although GNPs are a viable radiosensitizer, their sensitizing properties are not solely due to the absorption of higher levels of dose and there is a significant biological component to the observed radiosensitization.

An important implication is that GNPs might be useful clinical radiosensitizers when used in combination with current conventional radiotherapy protocols offering sensitization to megavoltage radiation at GNP concentrations achievable *in vivo* in a clinical setting.

3.2.2.3.3 Mechanisms of GNP Radiosensitization: Physical Aspects As already noted, numerous theoretical studies have investigated dose localization through the introduction of GNPs, as well as how the energy profile of radiation sources could be used to maximize this effect. Typically, it was concluded that GNP concentrations on the order of 0.1% to 1% by mass would significantly increase dose when combined with kilovoltage x-rays. However, due to the lack of contrast between tissue and gold, these models predict little benefit at the megavoltage energies typically used in therapy.

In vitro experimental studies of GNP radiosensitization report radiation-sensitizing effects substantially greater than the additional dose due to the presence of GNPs [2,23,24] with sensitization observed at concentrations as low as 0.05% by mass. Of particular relevance to the application of GNPs as a therapeutic agent, several studies have reported significant radiosensitizations with clinical megavoltage x-ray sources. A recently published model [25] presented an alternative view of GNP radiosensitization. In this model, instead of simply calculating the average dose deposited throughout a volume, energy deposits were scored on the nanoscale in the vicinity of individual nanoparticles. This showed that the presence of GNPs led to highly inhomogeneous dose distributions, as ionizations in gold were typically followed by the production of large numbers of low-energy secondary electrons, which deposited their energy very densely in the vicinity of the GNP. This inhomogeneous dose distribution was then analyzed with the local effect model (LEM) [26] to predict the radiosensitizing effects of GNPs. The LEM was originally developed to describe the effects of the highly localized doses found in charged particle radiotherapy, but it should be equally applicable to other sources of inhomogeneous radiation doses.

This approach was used to generate predictions for the radiosensitizing effects of 0.05% by mass of 1.9 nm GNPs on MDA-MB-231 breast cancer cells exposed to 160 kVp x-rays. The predictions were found to agree well with experimental observations [24,25], suggesting that this approach may provide a physical basis for understanding the radiosensitizing effects of GNPs. This work extended the model to megavoltage x-ray sources, to determine if similar subcellular dose localization could be responsible for the experimentally observed ability of GNPs to sensitize cells to megavoltage radiation.

3.2.2.3.4 Biological Mechanisms of GNP Radiosensitization

3.2.2.3.4.1 Oxidative Stress In the absence of radiation, metallic nanoparticles are known to induce the production of reactive oxygen species (ROS) leading to oxidative stress [18]. Specifically, GNPs have been reported to induce significant levels of ROS causing oxidative DNA damage at concentrations that do not alter mitochondrial activity. Other studies have shown increased levels of ROS and DNA damage in cells exposed to 1.9 nm GNPs [2].

There are only a few experimental reports of elevated levels of oxidative stress induced by GNPs in combination with ionizing radiation; however, in one study, a high level of intracellular ROS leading to higher levels of oxidative stress and apoptosis was shown in ovarian cancer cells treated with 14 nm particles irradiated with kilovoltage and megavoltage x-rays [27].

3.2.2.3.4.2 Cell Cycle Effects Radiosensitization with cell cycle arrest in the G_2/M phase has been observed for other metal nanomaterials [28], and there are several reports that GNPs cause alterations in cell cycle distributions:

- Acceleration in the G_0/G_1 phase and accumulation of cells in G_2/M was observed in cells exposed to 10.8 nm glucose-capped GNPs irradiated using a Cs-137 source. These changes were accompanied by decreased expression of p53 and cyclin A, and increased expression of cyclin B1 and cyclin E [29].
- Similar findings were reported by Geng et al. [27], who observed an increase in the G_2/M cell population for cells irradiated with 6 MV x-rays following exposure to 14 nm glucose-capped GNPs.
- Nuclear-targeted GNPs were shown to cause the accumulation of cells in the G_1 phase and disruption of the G_1/S transition inducing apoptosis in cancer cells [30].

These reports would suggest that changes in cell cycle kinetics induced by GNPs either induced additive toxicity as a result of an increase in the sub-G_1 population or there is accumulation in G_2/M, either or both of these could be the mechanism of GNP radiosensitization.

3.3 TARGETING CELLULAR REDOX PATHWAYS

3.3.1 The Cellular Antioxidant System

Ionizing radiation exerts biological effects through the production of highly reactive free radical species. This takes place against a background of the normal cellular processes (oxygen metabolism, immune-mediated response to pathogens, and signal transduction/gene expression pathways), which generate free radicals and ROS. There are many ways by which cells and tissues protect themselves against free radicals and ROS, under certain circumstances, however, these protective mechanisms can be either overwhelmed or inefficient in handling radicals/ROS resulting in "oxidative stress" (Figure 3.5).

The frontline intracellular defenses against ROS include superoxide dismutase, catalase, glutathione (GSH) and related enzymes, protein thiols, and a variety of intracellular redox couples. Collectively, they constitute a complex intracellular "redox buffer" network of molecules that function to maintain a slightly reducing environment.

There are three main intracellular redox couples: (1) NAD(P)H/NAD(P)_, (2) GSH/GSSG, and (3) Trx(SH)2/TrxSS (thioredoxin system). All these are linked both kinetically and thermodynamically and, for most cells, the GSH/GSSG couple is the dominant redox buffer. The thioredoxin- and glutathione-dependent systems each reduce hydrogen peroxide to two molecules of water. Glutathione peroxidase consumes two molecules of

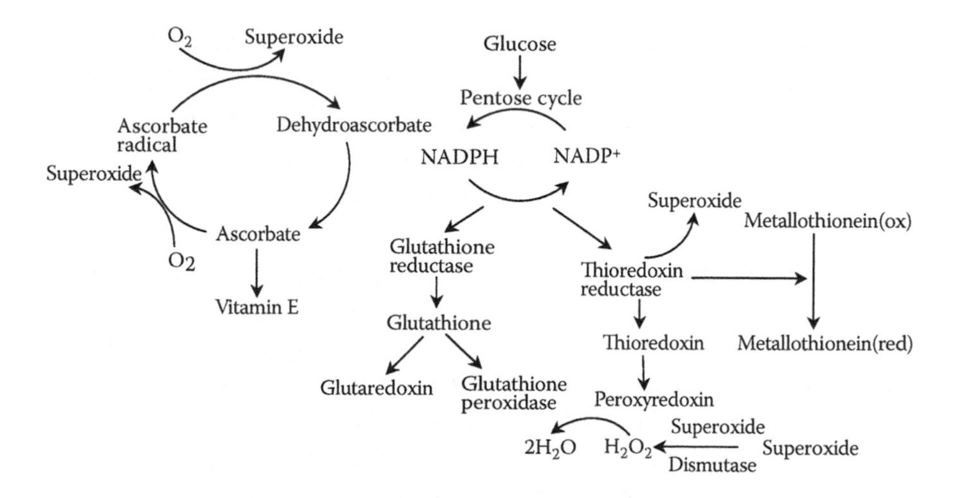

FIGURE 3.5 Cellular anti-oxidant system showing relationship of ascorbate, glutathione, and NADPH.

reduced glutathione (GSH) to form one molecule of oxidized glutathione (GSSG) and the disulfide linkage in the latter species is reduced by the enzyme glutathione reductase, with the concomitant oxidation of one molecule of NADPH cofactor. In the thioredoxin system, peroxiredoxin is consumed by reaction with hydrogen peroxide and is replenished through the stepwise transfer of reducing equivalents from NADPH, thioredoxin reductase, and thioredoxin. NADPH involved in these reactions is primarily generated from NADP+ by the action of the oxidative pentose cycle, a series of reactions that consume glucose.

Another pool of molecules involved in the cellular redox buffer are protein sulfhydryls (PSH), which are substantially higher in concentration than GSH. Several forms exist, including protein thiols, disulfides, or mixed disulfides. In addition, there are a number of small intracellular molecule antioxidants including glutathione, ascorbate, and vitamin E.

When pro-oxidant production begins to exceed the ability of the cellular antioxidant capacity to maintain the normal steady state redox potential, as would be the case if the cell were exposed to ionizing radiation, a condition of oxidative stress exists threatening cell survival. This can be demonstrated experimentally when the manipulation of antioxidants (i.e., thiols, superoxide dismutases, hydroxyl radical scavengers, and hydroperoxide-metabolizing enzyme systems) at the time of irradiation alters the reactions of free radicals and ROS, leading to alterations in oxidative damage and to expression of the biological effects of radiation.

3.3.2 Radiosensitization by Targeting the Thioredoxin System

Thioredoxin reductases (TxnRds) are essential mammalian selenocysteine-containing flavoenzymes that act in homodimeric form to catalyze NADPH-dependent reduction of thioredoxin and small molecular weight oxidants including ROS [31,32]. TxnRd1 and TxnRd2 (the cytosolic/nuclear and mitochondrial forms, respectively) are two ubiquitously expressed isoforms of this enzyme family. They play a key role in maintaining redox-regulated cellular functions, including transcription, DNA damage recognition and repair, proliferation, and apoptosis [31]. In response to oxidative stress, TxnRds sustain signaling pathways that regulate transcription of genes to protect the cell from oxidative damage [33–35]. Cytosolic TxnRd1 expression is often upregulated in human cancers, where it is associated with aggressive tumor growth and poor prognosis. TxnRd1 has been shown to confer protection against the lethal effects of ionizing radiation (IR) in tumor cells [33], and agents that selectively target TxnRd1 have shown promising results as

TABLE 3.1 Trx System Inhibitors

Drug	Class of Drug	Target	Clinical Application	Radiosensitizer
SAHA	HDAC inhibitor	Trx1	Cutaneous T cell lymphoma, clinical trials, other cancers	Described in Chapter 8
Cisplatin	Platinum compound	TrxR	FDA approved (1978), often combined with other chemotherapy	Described in Chapter 6
MGd	Texaphyrins	TrxR	Clinical trials, multiple cancers	This chapter
Curcumin, flavonoids	Polyphenols	TrxR	Clinical trials, various cancers	Described in Chapter 13
Auranofin	Gold-containing compound	TrxR	Approved as an antiarthritic drug	Demonstrated experimentally for gold compounds. This chapter
ATO Darinoparsin	Arsenic-containing compounds	TrxR	ATO approved for treatment of acute myelocytic leukemia	ATO demonstrated preclinical models. Darinoparsin active against solid tumors, radiosensitizes, and improves therapeutic ratio. Mechanism(s) not identified [36]

anticancer drugs in preclinical and clinical studies [37,38] when used alone or when combined with IR [33]. In fact, it is becoming apparent that targeting the thioredoxin system is common to a diverse group of radiosensitizers, some of which are described elsewhere in this book (Table 3.1).

3.3.2.1 Trnx System Inhibitors

Targeting the Trnx system will result in alterations of the intracellular redox state and will adversely affect a number of intracellular systems. Either Trnx or TrnxR can be targeted by a chemotherapeutic reagent to achieve the same effect because a nonfunctional TrxR will result in lower levels of reduced Trx in a cell, thus preventing the redox function of Trx. Targeting TrxR also has some additional effects because TrxR has its own substrates, which it reduces directly without the need for Trx. Some drugs are also capable of converting TrxR into a ROS generating system, in direct contrast with its usual role as a defensive enzyme. Figure 3.6 summarizes the effects of inhibition of TrxR and its possible physiological consequences.

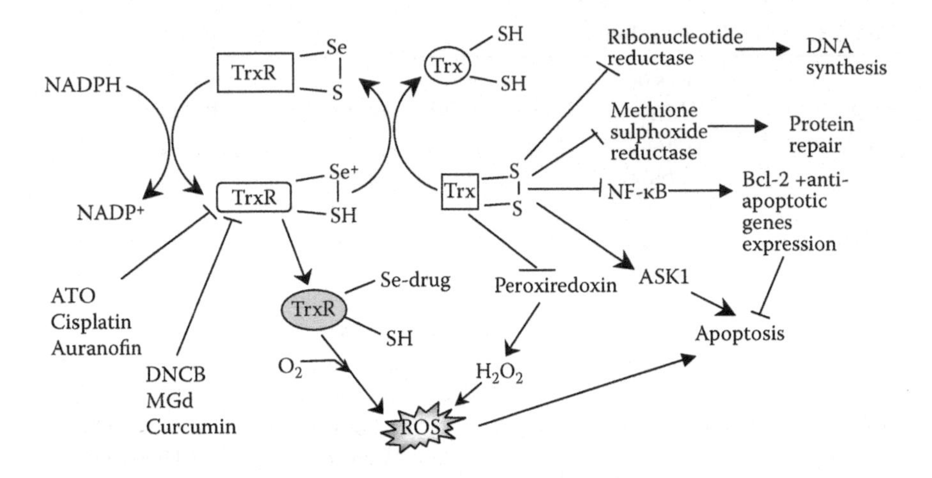

FIGURE 3.6 Inhibition of TrxR results in enhanced levels of oxidized Trx and a reduction in the level of reduced Trx. Lack of reduced Trx results in an inability to activate listed substrates, and causes an inhibition of downstream cell functions including DNA synthesis, protein repair, transcription factor activity, and peroxidase function. An accumulation of hydrogen peroxide and free radicals subsequently occurs. This leads to oxidative stress conditions, which promote apoptosis. Another consequence is that oxidized Trx can no longer repress ASK1, allowing it to initiate apoptosis. In addition, some TrxR inhibitors modify TrxR such that it acquires an oxidase function, which causes an accumulation of ROS.

Four of the compounds listed in Table 3.1 have demonstrated radiosensitization capability. HDAC inhibitors (SAHA) and the polyphenols (curcumin) are described elsewhere in this book. Motexafin texaphyrin and gold-containing compounds will be discussed in this chapter.

3.3.2.2 Motexafin Texaphyrin

Gadolinium is a transition metal and under certain circumstances participates in redox interactions; however, motexafin texaphyrin (MGd) is a texaphyrin coordinated to a non-redox-active gadolinium(III) cation. It is the aromatic texaphyrin ring system of MGd which is easily reduced and in the presence of oxygen, this can result in redox cycling, oxidative stress, and disruption of the cellular redox homeostasis. The texaphyrins were first identified as potential radiation sensitizers in 2000 [39]. Texaphyrins are Schiff base macrocycles that resemble porphyrins and other naturally occurring tetrapyrrolic prosthetic groups of which MGd (Figure 3.3) is a prototypical complex. In contrast to the porphyrins, texaphyrins have five, rather than four, coordinating nitrogen atoms in their central core.

This slightly larger core allows the formation of stable, nonlabile 1:1 complexes with a range of larger metal cations, resulting in ligand oxidation and a very high barrier for disassociation. In addition, texaphyrin complexes are more readily reducible than typical metalloporphyrins. These features and the belief that texaphyrins would show the tumor selectivity similar to that of porphyrins led to the investigation of MGd as a radiation sensitizing agent. As a bonus feature, MGd can be detected by MRI so that tissue localization and clearance can be assessed noninvasively.

3.3.2.2.1 MGd: Mechanism of Action Reducing metabolites are maintained in their reduced state through the generation of reducing equivalents via the pentose oxidative pathway, which reduces NADPH. These reducing equivalents are required for a variety of redox reactions and for the maintenance of the redox balance of the cell. MGd exhibits high electron affinity and is easily reduced, ensuring that intermediates such as ascorbate, glutathione, and NADPH preferentially reduce MGd in the presence of oxygen, generating superoxide and ROS [40,41]. In effect, MGd acts to "sponge up" electrons formed as the result of the interactions of x-rays with water, effectively depleting the pool of substrates available to repair the radiation-induced oxidative damage. This mechanism is known as futile redox cycling and is the basis for a number of specific processes contributing to radiosensitization by MGd.

The cytosolic TrxR1 is one of the target molecules of MGd [42], and MGd inhibits TrxR1 disulfide reductase activity and induces an NADPH oxidase activity in TrxR1, generating ROS such as superoxide and hydrogen peroxide. Manipulation of the Trx redox system together with an increased production of ROS induces a pro-oxidant state in the cell triggering injury and induction of apoptotic events in tumor cells by several mechanisms. Inhibition of ribonucleotide reductase blocks DNA synthesis, repair, and cellular growth. The indirect action of MGd on oxidized Trx may induce apoptosis via ASK-mediated cell death [42].

Other studies have indicated an involvement of MGd with alterations of zinc metabolism. Microarray analyses conducted in human cancer cell lines (A549 lung cancer cells and Ramos B-cell lymphoma line) have shown that-MGd treatment causes an upregulation in the expression of genes controlling free zinc levels [43]. Increased intracellular free zinc levels may have detrimental effects on the redox active SH groups of enzymes such as TrxR or ribonucleotide reductase. Inhibition of TrxR activity by zinc has been previously reported [43], suggesting an additional indirect action of MGd on TrxR [41].

3.3.2.2.2 Radiosensitization and Chemosensitization by MGd The action of MGd in targeting several aspects of redox cycling activity is likely to be significant in conditions where oxidative stress is high, as in cells exposed to ionizing radiation, hypoxia and inflammation, tumors, or any other circumstance where the cellular antioxidant system is compromised. In cancer cells, as described above, MGd targets oxidative stress proteins such as metallothioneins and thioredoxin reductase. Targeting these proteins leads to oxidative damage, impaired metabolism, and altered metal ion homeostasis, which can make the cells more vulnerable to apoptotic cell death.

3.3.2.2.3 *In Vitro* Demonstration of Radiosensitization by MGd *In vitro* studies with MGd have been complicated by the fact that the texaphyrins are highly reactive with the reducing metabolites present in cell culture media, and with other factors that may influence the effect of the compound or the redox status of the cells. Factors that are not generally considered important in the design of cell culture studies may be critical determinants of the activity of MGd. In one case, it was shown that the radiation dose–response curves for EMT6 cells were altered by ascorbate when cells were irradiated under either normoxic or hypoxic conditions with MGd [44] because MGd radiation enhancement required the presence of ascorbate in the culture medium used in the clonogenic assay system. *In vitro* studies demonstrating radiosensitization by MGd include:

- Experiments with the radiation-sensitive, LYAS, and the radiation-resistant, LYAR, B-lymphoid murine cell lines demonstrated enhanced radiation sensitivity with MGd in the LYAS cell lines but not the LYAR cells by clonogenic assay [45]. It was shown by DNA microarray analysis that LYAS cells overexpress apoptosis-inducing genes and underexpress various antioxidant genes.
- MGd radiation enhancement in the MES-SA, human ovarian cancer cell line and A549 human lung cancer cell line requires concomitant treatment with BSO, which decreased glutathione levels to approximately 30% of controls [45].
- The CHO cell line variant, E89, is deficient in the pentose cycle enzyme glucose-6-phosphate dehydrogenase and consequently could not produce normal levels of NADPH [46]. Incubation with MGd led to greater radiation enhancement in this cell line, as compared with the wild-type cell line, K1. Measurement of protein and nonprotein

thiols in the E89 cells confirmed that protein and nonprotein thiols were depleted by 11% and 31%, respectively, presumably as a consequence of MGd redox cycling and hydrogen peroxide production.

3.3.2.2.4 Tumor Radiosensitization by MGd

- A549 tumors growing in nude mice were irradiated (15 Gy) and tumors were excised for analysis of clonogenic survival. There was an improvement in clonogenic survival when irradiated tumor cells were allowed to incubate in the stationary phase before plating in clonogenic assays and improvement in tumor cell survival or recovery, was attributed to repair of potentially lethal radiation damage. Tumor cell recovery was inhibited when mice were treated with MGd [43].
- Intramuscular or subcutaneous EMT 6, SMT-F and MCa murine tumors were injected i.v. with MGd and other metallotexaphyrins before acute or fractionated radiation. MGd, in combination with radiation, produced significant tumor growth delay compared with irradiated control animals in both single and multifraction studies [47]. Only MGd produced radiation enhancement whereas texaphyrin ligands containing the metal ions lutetium, europium, yttrium, or cadmium did not (Figure 3.7).

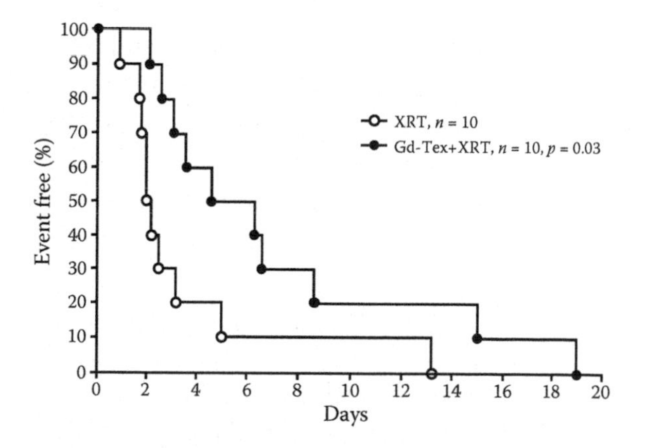

FIGURE 3.7 Gd-Tex radiation enhancement in the MCa tumor model. Animals were given 40 μmol/kg of Gd-Tex, 2 h before treatment with 32 Gy (●) of radiation. Radiation control animals are shown (○). Gd-Tex radiation enhancement was observed ($p = 0.03$, log rank). (From Miller, R. et al., *Int J Radiat Oncol Biol Phys* 45(4):981–989, 1999. With permission.)

3.3.2.2.5 Clinical Findings with MGd A randomized controlled study in 401 patients with brain metastases of various histologies, comparing whole-brain radiotherapy (WBRT) alone versus WBRT with motexafin gadolinium failed to show any significant difference in median survival or tumor response [48]. However, the median time to neurologic progression was increased by 0.5 months ($p = 0.018$) for the group that received MGd. When patients were stratified by histology (lung, breast, or other), a subset analysis revealed that the time to neurological progression favored the MGd and WBRT arm for patients with lung cancer (median, 5.5 months for MGd vs. 3.7 months for WBRT alone; $p = 0.025$), but no difference was seen in the other strata. Furthermore, a companion study of neurocognitive function suggested that MGd may preserve memory and executive function and prolong time to neurocognitive and neurologic progression in patients with brain metastases from lung cancer [49].

In an international, phase III trial study, patients with brain metastases from non-small cell lung cancer (NSCLC) were randomized to WBRT with or without MGd. The primary endpoint was the interval to neurologic progression. The trial was conducted in a number of centers and there were differences in the time after diagnosis that WBRT was delivered. Of 554 patients, 275 were randomized to WBRT and 279 to WBRT + MGd. MGd improved the interval to neurologic progression compared with WBRT alone and the interval to neurocognitive progression, but the differences were not significant. For patients treated in locations where treatment was more prompt, a statistically significant prolongation of the interval to neurologic progression, from 8.8 months for WBRT to 24.2 months, was observed. The overall conclusion was that MGd exhibited a favorable trend in neurologic outcomes, the interval to neurologic progression in patients with NSCLC with brain metastases receiving prompt WBRT was significantly prolonged and the toxicity was acceptable [50]. It was later concluded in a review of the results of the two large trials described above that MGd did not improve the results of therapy of brain metastases [51]. It was suggested, however, that the results were sufficiently encouraging to merit further clinical investigation.

A number of clinical trials involving radiation and MGd are listed as being ongoing or recently completed. These are largely phase I and phase II trials of various types of brain tumor and CNS sites (11 of 15). Additional chemotherapy includes rituximab, cytarabine, methotrexate, procarbazine, vincristine, and temozolomide. There is one phase III trial of brain metastases from NSCLC with additional chemotherapy (carboplatin and paclitaxel), one phase I trial for head and neck cancer (5-FU, cisplatinum), and one phase I trial for gallbladder and pancreatic cancer (www.clinicaltrials.gov).

3.3.2.3 Gold-Containing Compounds

It has been demonstrated that various gold compounds with known therapeutic benefit in cancer, rheumatoid arthritis, and inflammatory diseases are efficient inhibitors of mammalian TrxR. In particular, phosphine gold(I) complexes such as auranofin [Au(I)(PEt3) (2,3,4,6-tetra-*O*-acetyl-1-thio-D-thioglucose-S)] inhibit TrxR even at nanomolar concentrations [52].

A series of gold(I) complexes were synthesized with the intent of developing TrxR inhibitors [53]. One of these [Au(SCN)(PEt3)] had selective and efficient TrxR-inhibiting characteristics and was further tested as a potential radiosensitizing agent on human lung cancer cells. Human radioresistant lung cancer cells were subjected to a combination of single fractions of radiation at clinically relevant doses and nontoxic levels of [Au(SCN)(PEt3)]. The combination of the TrxR inhibitor and ionizing radiation reduced the surviving fraction and impaired the ability of the U1810 cells to repopulate by approximately 50%. In addition, inhibition of thioredoxin reductase caused changes in the cell cycle distribution, suggesting a disturbance of the mitotic process. Global gene expression analysis also revealed clustered genetic expression changes connected to several major cellular pathways including cell cycle, cellular response to stress, and DNA damage. Specific TrxR inhibition as a factor behind the achieved results was confirmed by correlation of gene expression patterns between gold and siRNA treatment. These results implicated TrxR as an important factor conferring resistance to radiation and identified [Au(SCN)(PEt3)] as a radiosensitizing agent.

3.4 RADIOSENSITIZATION BY TARGETING THE GSH/GSSG SYSTEM

The overall cellular redox state is regulated by three systems, two of which, the reduced glutathione (GSH)/oxidized glutathione (GSSG) system and the glutaredoxin (Grx) system, are dependent on glutathione. The GSH/GSSG couple is the dominant redox buffer and homeostatic control of intracellular redox status is largely exerted by reduced GSH to protect the cell from oxidative stress. In the cytoplasm, the ratio of reduced GSH to GSSG may be as high as 100:1 whereas in the endoplasmic reticulum, the ratio is lower at approximately 3:1.

3.4.1 Functions of GSH

GSH along with superoxide dismutase, catalase, and thioredoxin reductase, a variety of intracellular redox couples function in concert to detoxify ROS and rescue cells from oxidative damage, thus, providing a first-line defense against ROS. GSH can directly scavenge free radicals and peroxides that accumulate in cells during oxidative stress via mixed disulfide formation or upon oxidation to GSSG. In addition, GSH is a cofactor or a substrate for various enzymes, and together with these GSH-associated enzymes, it provides a second line of defense. These redox reactions are one of the most important functions of GSH, and are catalyzed by GSH peroxidases (GPx) and GSSG reductase (GR). Glutathione peroxidase (GPx) detoxifies hydroperoxide substrates, such as, H_2O_2 fatty acid hydroperoxides and phospholipid hydroperoxides with GSH acting as an electron donor in the reduction reaction, GSH is oxidized to GSSG, in these GPx-mediated reactions. It has been shown that GSSG is capable of thiol-exchange reactions with protein sulfydryls, leading to the formation of protein–glutathione mixed disulfides (Figure 3.8).

Maintaining GSH/GSSG equilibrium in the cell is critical to survival and extreme levels of oxidative or nitrosative stress can quickly and substantially diminish GSH levels in favor of GSSG. To maintain a constant cellular GSH pool, the NADPH-dependent enzyme GR is required to catalyze the reduction of GSSG to GSH [54], GSH (once synthesized) can undergo transport across biological membranes, notably across the plasma membrane, to be part of an inter-organ transport network [55].

3.4.2 GSH Biosynthesis

(This section is based on a review by Singh et al. [56] and the articles referenced therein). GSH is synthesized in a two-step ATP-dependent pathway. In the first step, enzyme glutamate-cysteine ligase (GCL; formerly called γ-glutamyl cysteine synthase), catalyzes the formation of γ cysteine

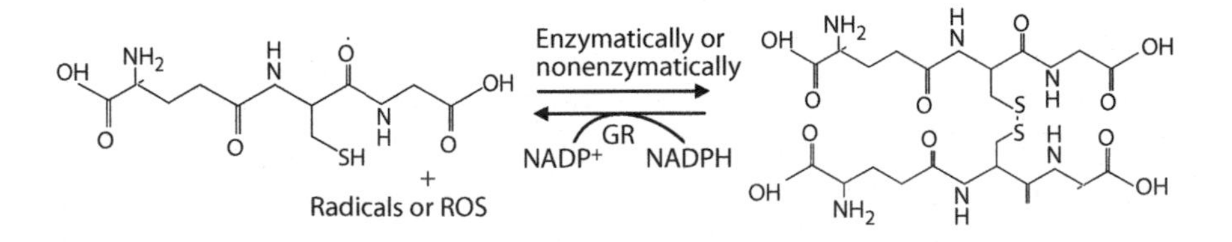

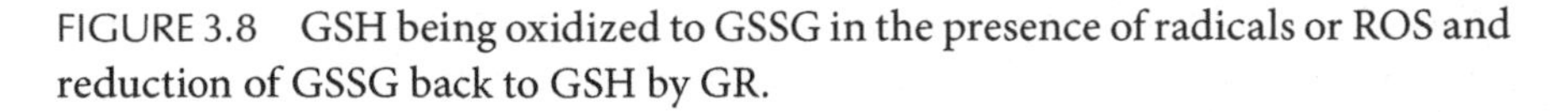

FIGURE 3.8 GSH being oxidized to GSSG in the presence of radicals or ROS and reduction of GSSG back to GSH by GR.

from L-glutamine and L-cysteine. In the second reaction, catalyzed by the enzyme glutathione synthase (GS; also called glutathione synthetase), L-glycine is added to form GSH (Figure 3.9). *De novo* synthesis of GSH is influenced by at least three factors: (i) the cellular levels of GCL; (ii) the availability of L-cysteine; and (iii) feedback competitive inhibition by

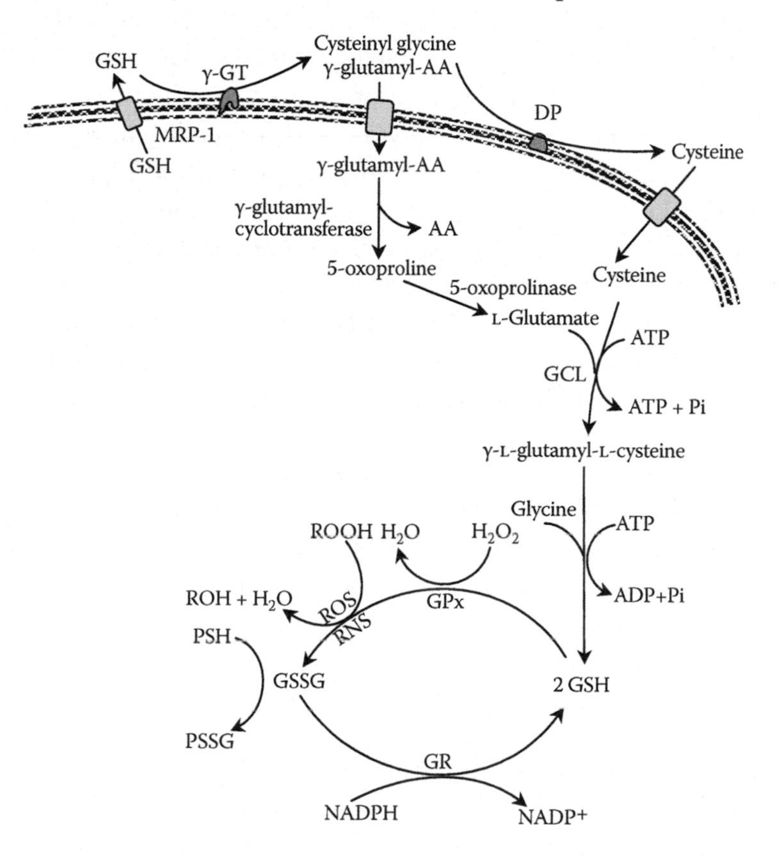

FIGURE 3.9 Biosynthesis and metabolism of glutathione outside the cell. GSH is partially degraded by γ-glutamyl transpeptidase (γ-GT) to yield the γ-glutamyl moiety, which couples to an amino acid (AA) and cysteinyl glycine, which in turn is broken down by dipeptidase (DP). AA transporters return the γ-glutamyl-AA complex into the cytosol, where γ-glutamyl moiety is separated from the AA and converted to glutamate. Now inside the cell, glutamate and cysteine are combined by the enzyme glutamate-cysteine ligase (GCL) to form glutamyl-cysteine, which is then combined with glycine by glutathione synthase (GS) at the expense of two ATP molecules to yield GSH. GSH acts as a reductant in peroxidase reactions by glutathione peroxidases (GPx) with various hydroperoxides (ROOH) to form the corresponding alcohol (ROH) and glutathione disulfide (GSSG), which can be reduced by glutathione reductase (GR) at the expense of NADPH to regenerate GSH.

GSH. Expression of GCL and GS is ubiquitous, but it may vary from tissue to tissue with the liver and kidney being the major producer and exporter of GSH.

3.4.3 Targeting GSH as a Therapeutic Strategy

GSH, the most prevalent intracellular nonprotein thiol, functions directly or indirectly in many important biological phenomena, including the synthesis of protein and DNA, transport, enzyme activity, metabolism, and cell protection. More than 40 years ago, low endogenous GSH was observed to increase the radiosensitivity of bacteria and diploid human cells to low LET radiation and subsequently the radiobiological effects of GSH have been studied in many cell lines. Investigations have shown that GSH is an important intracellular radioprotective agent under the hypoxic condition and that GSH depletion by thiol modifiers results in the enhancement of cellular radiosensitivity. There are a number of strategies that have been shown to effectively deplete GSH.

3.4.3.1 Agents That Oxidize or Derivatize GSH

One class of compounds that oxidize GSH to GSSG are the diazenecarboxylic acid derivatives. One example, sodium tetrathionate, has been shown to deplete GSH stores by 15% to 60% within 2 to 5 h in kidney, liver, and erythrocytes in rats [57]. These compounds are nephrotoxic, precluding any clinical application.

3.4.3.2 Inhibition of GSH Biosynthesis

Increasing the rate of GSH utilization or inhibiting its biosynthesis will result in decreased GSH levels. In a search for a specific inhibitor of γ-GCLC, different methionine sulfoximine analogues were synthesized and evaluated for their potency *in vitro* and *in vivo*. The most specific inhibitor was found to be L-buthionine-SR-sulfoximine (L-SR-BSO), which is an amino acid analogue of cysteine [58]. BSO is a mixture of L,S- and L,R-diastereomers; however, only the R-stereomer of BSO is a specific and potent inhibitor of γ-GCL.

3.4.3.2.1 Experimental Demonstration of Radiosensitization by BSO Attempts to reduce tumor GSH concentrations with BSO produced widely varying results. These studies mostly failed to take into consideration the biological half-life of GSH in the respective tumors, as well as the variation in the rate of recovery of GSH between BSO doses.

Greater success was achieved with an intracranial interstitial model in athymic nude mice bearing intracranial D-54 MG human glioma xenografts used for evaluation of the therapeutics benefit of BSO-mediated depletion of tumor GSH levels. Administration of L-BSO [2.5 mmol/kg intraperitoneal injection × four doses plus concomitant availability in acidified water (pH 3.0) at a concentration of 20 mM] resulted in depletion of tumor glutathione levels to 0.15 μmol/g wet weight (7.9% of control). The therapeutic activity of intracranial interstitial radiotherapy with an ^{125}I seed was enhanced after L-BSO with increases in median survival of 13.4% to 30.5% over that seen with ^{125}I alone. It was suggested that these results demonstrated a potential role for BSO in enhancing the therapeutic activity of interstitial radiotherapy [36].

3.4.3.2.2 Potentiation of Effect of Hypoxic Radiosensitizers by Thiol Depletion
Radiosensitization of hypoxic cells by nitroimidazoles is described in the next chapter. The results of some preclinical studies support the use of agents such as BSO to deplete GSH before administering nitroimidazole radiosensitizers. In one study, it was demonstrated that BSO effectively reduced tumor GSH concentrations to levels that enhanced the radiosensitizing efficacy of SR-2508, and that this effect was observed at clinically relevant doses of radiation (i.e., 2.5–3.0 Gy). The BSO dosing schedule adopted in this case produced an 80% to 95% decrease in tumor GSH. A later study showed that tumor GSH levels can be maintained by prolonged BSO administration at 10% to 15% of control values for several days. Under these conditions, SER values of 1.4 and 1.6 were obtained in RIF and MMCa tumor models, respectively, when chronically GSH-depleted tumors were treated with SR-2508 and fractionated radiation treatment [59].

3.4.3.3 Inhibition of Glutathione Reductase

Depletion of the reduced form of glutathione (GSH) has been studied for its effect on sensitizing cancer to radiation. However, little attention has been paid to the effects of thiol oxidative stress created through an increase in glutathione disulfide (GSSG) on cancer sensitivity to radiation. This was addressed in a study in which an increase in GSSG was effectively created using 2-acetylamino-3-[4-(2-acetylamino-2 carboxyethylsulfanylthiocarbonylamino) phenylthiocarbamoylsulfanyl]propionic acid (2-AAPA), an irreversible glutathione reductase (GR) inhibitor. It was shown that the GSSG increase significantly enhanced cancer sensitivity to x-ray irradiation in four human cancer cell lines (A431, MCF7, NCI-H226, and OVCAR-3). When cells were pretreated with 2-AAPA followed by x-ray irradiation, the

IC_{50} values were all reduced. The synergistic effects observed from the combination of x-rays plus 2-AAPA were comparable to those from the combination of x-rays plus BSO. The synergistic effect was correlated with an increase in cell thiol oxidative stress, which was reflected by a fivefold to sixfold increase in GSSG and a 25% increase in total disulfides. No change in GSH or total thiols was observed as a result of GR inhibition [60].

There are several other approaches to targeting GSH metabolism but to date the only methods shown to be effective in terms of modulating radiation response are inhibition of γ-GCL (by BSO) and inhibition of GR (by 2-AAPA). Some other methods have been shown to have clinical potential as chemotherapy, but interaction with radiation has not been shown. All these methods are summarized in Table 3.2.

TABLE 3.2 GSH-Based Approaches to Chemosensitization and Radiosensitization

Mode of Action	**Agent**	**Radiosensitization/ Chemosensitization**
Deplete GSH by targeting GCL	BSO and analogues	Radiosensitizer
	GCL antisense, c-Jun antisense	
Target GST	TLK 286 (Canofosfamide)	GSTPl-specific GSH analogue linked to a cytotoxic agent chemosensitizer, phase I, II clinical trials [59,61]
	Ethacrynic acid	
	Ezathiostat (TLK 199)	Pro-drug, bioactivated to TLKl 17, binds to GSTPl and activates JNK, leading to growth and differentiation of normal cells and apoptosis of cancer cells [36]; chemosensitizer
Oxidize GSH	Imexon	
	Disulfiram	
Inhibition of S-glutathiolation	NOV-002	Contains oxidized GSH, alters GSH/GSSG ratio by increasing GSSG levels, induces 5-glutathionylation; inhibitory effects tumor invasion, proliferation, and survival. In phase II trials [60]
Increase in glutathione disulfide (GSSG)	2-AAPA	Increase in radiosensitivity

3.5 SUMMARY

Compounds or complexes containing high atomic number elements act as radiosensitizers based on their ability to increase the dose deposited in the target volume. This is due to differences in their mass energy absorption coefficient when compared with soft tissue. Understanding the effect of high-*Z* material on dose led to a reassessment of the agents that were already in use as contrast media and which were formulated to localize in tumors in terms of their capacity to increase photoelectric absorption in proximity to the tumor. A number of experiments demonstrated radiosensitization in rodent tumor models when animals were injected with iodine contrast media and treated with orthovoltage radiation and this approach reached the level of a phase 1 clinical trial.

Gadolinium-based contrast media, which are used in MRI to provide selective image enhancement, were also investigated. Theoretical dose modeling indicated that Gd ($Z = 64$) would be at least as effective as iodine ($Z = 53$) in terms of dose enhancement; however, it became apparent that the mechanism by which MGd acts as a radiosensitizer probably depends more on its role as a redox-active drug than on its mass energy absorption coefficient.

The greatest interest in a radiosensitizer-based high-mass energy absorption coefficient is now concentrated on gold, which has a number of advantages including higher *Z* number than iodine and greater biocompatibility. Several experiments provided proof-of-principle for the use of gold as a radiosensitizer based on the difference in mass energy absorption coefficient of gold and soft tissue. A practical method for delivery to the tumor target was provided by GNPs, which are sufficiently small to penetrate the tumor mass, are reported to have high biocompatibility, and accumulate preferentially in tumors *in vivo*. There is extensive evidence that GNPs are effective radiosensitizers, but experimental findings deviate from the physical predictions of dose modification. Significant radiosensitization was seen when GNPs are used at concentrations much lower than that predicted to give a significant dose increase, and enhancement was observed not only with kilovoltage x-rays, as predicted, but also with megavoltage x-rays or electrons for which little or no increase in overall dose deposition would be expected. Understanding of the mechanism of sensitization of cells to megavoltage radiation by GNPs has been achieved by a method of modeling dose at the subcellular level. Production of ROS leading to oxidative stress by metallic nanoparticles may also be a factor.

Cells and tissues protect themselves against free radicals and ROS generated by intracellular metabolism and by special circumstances such as ionizing radiation. The thioredoxin system is part of the cellular defense against oxidative stress, targeting thioredoxin reductase is a characteristic shared by a diverse group of radiosensitizers including thioredoxin, the phytochemicals curcumin motatexafin gadolinium (MGd), and some gold compounds.

MGd, an expanded porphyrin that contains Gd^{3+} in the central cavity of the macrocycle, has a strong affinity for electrons and has been shown to selectively localize to tumor cells and to enhance the effects of radiation therapy. MGd reacts with electrons formed as the result of the interactions of x-rays with water, effectively depleting the pool of substrates available to repair the radiation-induced oxidative damage; a mechanism known as futile redox cycling and the basis for a number of processes contributing to radiosensitization by MGd. Studies with MGd *in vitro* have demonstrated enhancement of tumor cell cytotoxicity with ionizing radiation and several chemotherapeutic agents. Manipulation of the Trx redox system together with an increased production of ROS induces a pro-oxidant state in the cell triggering injury and induction of apoptotic events in tumor cells by several mechanisms.

A review of the results of the two large clinical trials concluded that MGd did not improve the therapy of brain metastases but it was nevertheless suggested that the results were sufficiently encouraging to merit further clinical investigation. A number of clinical trials involving radiation and MGd are in fact listed as ongoing or recently completed. These are largely phase I and phase II trials of various types of brain tumor and CNS sites.

In addition to Trxn, the cellular redox state is regulated by two other systems, the reduced glutathione (GSH)/oxidized glutathione (GSSG) system and the glutaredoxin (Grx) system, which are dependent on glutathione. The GSH/GSSG couple is the dominant redox buffer and homeostatic control of intracellular redox status is largely exerted by reduced GSH to protect the cell from oxidative stress. GSH, the most prevalent intracellular nonprotein thiol, functions directly or indirectly in many important biological phenomena including the synthesis of protein and DNA, transport, enzyme activity, metabolism, and cell protection. Investigations have shown that GSH is an important intracellular radioprotective agent under hypoxic conditions and that GSH depletion by thiol modifiers results in the enhancement of cellular radiosensitivity. There are a number of strategies that have been shown to deplete GSH, including inhibiting its biosynthesis. The most specific inhibitor of γ-GCLC found thus far is L-buthionine-SR-sulfoximine (L-SR-BSO), which is an amino acid analogue of cysteine. When adequate GSH depletion is achieved, significant changes in radiation response are observed including potentiation of the effects of hypoxic radiosensitizers.

REFERENCES

1. Hainfeld, J., Slatkin, D., Dilmanian, F., and Smilowitz, H. Radiotherapy enhancement with gold nanoparticles. *J Pharm Pharmacol* 2008;60:977–985.
2. Butterworth, K., Coulter, J.A., Jain, S. et al. Evaluation of cytotoxicity and radiation enhancement using 1.9 nm gold particles: Potential application for cancer therapy. *Nanotechnology* 2010;21:295101.
3. Spiers, F. The influence of energy absorption and electron range on dosage in irradiated bone. *Br J Radiol* 1949;22:521–533.
4. Niroomand-Rad, A., Razavi, R., Thobejane, S., and Harter, K.W. Radiation dose perturbation at tissue–titanium dental interfaces in head and neck cancer patients. *Int J Radiat Oncol Biol Phys* 1996;34:475–480.
5. Mesa, A.V., Norman, A., Solberg, T., Demarco, J., and Smathers, J. Dose distributions using kilovoltage x-rays and dose enhancement from iodine contrast agents. *Phys Med Biol* 1999;44:1955–1968.
6. Callisen, H., Norman, A., and Adams, F. Absorbed dose in the presence of contrast agents during pediatric cardiac catheterization. *Med Phys* 1979;6(6):504–509.
7. Santos Mello, R., Callisen, H., Winter, J., Kagan, A., and Norman, A. Radiation dose enhancement in tumors with iodine. *Med Phys* 1983;10(1):75–78.
8. Adam, J., Joubert, A., Biston, M. et al. Prolonged survival of Fischer rats bearing F98 glioma after iodine-enhanced synchrotron stereotactic radiotherapy. *Int J Radiat Oncol Biol Phys* 2006;64(2):603–611.
9. Norman, A., Ingram, M., Skillen, R., Freshwater, D., Iwamoto, K., and Solberg, T. X-ray phototherapy for canine brain masses. *Radiat Oncol Invest* 1997;5:8–14.
10. Rose, J., Norman, A., and Ingram, M. First experience with radiation therapy of small brain tumors delivered by a computerized tomography scanner. *Int J Radiat Oncol Biol Phys* 1994;30:24–25.
11. Robar, J. Generation and modelling of megavoltage photon beams for contrast-enhanced radiation therapy. *Phys Med Biol* 2006;51(21):5487–5504.
12. Regulla, D., Hieber, L., and Seidenbusch, M. Physical and biological interface dose effects in tissue due to x-ray induced release of secondary radiation from metallic gold surfaces. *Radiat Res* 1998;150:92–100.
13. Herold, D., Das, I., Stobbe, C., Iyer, R., and Chapman, J. Gold microspheres: A selective technique for producing biologically effective dose enhancement. *Int J Radiat Biol* 2000;76(10):1357–1364.

14. Cho, S. Estimation of tumour dose enhancement due to gold nanoparticles during typical radiation treatments: A preliminary Monte Carlo study. *Phys Med Biol* 2005;50:N163–N173.
15. Robar, J., Riccio, S., and Martin, M. Tumour dose enhancement using modified megavoltage photon beams and contrast media. *Phys Med Biol* 2002;47:2433–2449.
16. Wallach-Dayan, S., Izbicki, G., Cohen, P., Gerstl-Golan, R., Fine, A., and Breuer, R. Bleomycin initiates apoptosis of lung epithelial cells by ROS but not by Fas/FasL pathway. *Am J Physiol Lung Cell Mol Physiol* 2006;290(4):L790–L796.
17. Ott, M., Gogvadze, V., Orrenius, S., and Zhivotovsky, B. Mitochondria, oxidative stress and cell death. *Apoptosis* 2007;12:913–922.
18. Jia, H., Liu, Y., Zhang, X. et al. Potential oxidative stress of gold nanoparticles by induced-NO releasing in serum. *J Am Chem Soc* 2009;131(1):40–41.
19. McMahon, S., Mendenhall, M., Jain, S., and Currell, F. Radiotherapy in the presence of contrast agents: A general figure of merit and its application to gold nanoparticles. *Phys Med Biol* 2008;53(20):5635–5651.
20. Hainfeld, J., Slatkin, D., and Smilowitz, H. The use of gold nanoparticles to enhance radiotherapy in mice. *Phys Med Biol* 2004;49:N309–N315.
21. Hainfeld, J., Dilmanian, F., Zhong, Z., Slatkin, D., Kalef-Ezra, J., and Smilowitz, H. Gold nanoparticles enhance the radiation therapy of a murine squamous cell carcinoma. *Phys Med Biol* 2010;55:3045–3059.
22. Butterworth, K., McMahon, S., Currell, F., and Prise, K. Physical basis and biological mechanisms of gold nanoparticle radiosensitization. *Nanoscale* 2012;4:4830.
23. Liu, C.-J., Wang, C.-H., Chien, C.-C. et al. Enhanced X-ray irradiation-induced cancer cell damage by gold nanoparticles treated by a new synthesis method of polyethylene glycol modification. *Nanotechnology* 2008;9:295104.
24. Jain, S., Coulter, J., Hounsell, A. et al. Cell-specific radiosensitization by gold nanoparticles at megavoltage radiation energies. *Int J Radiat Oncol Biol Phys* 2011;79(2):531–539.
25. McMahon, S., Hyland, W., Muir, M. et al. Biological consequences of nanoscale energy deposition near irradiated heavy atom nanoparticles. *Sci Rep* 2011;1:18.
26. Elsässer, T., and Scholz, M. Cluster effects within the local effect model. *Radiat Res* 2007;167:319–329.
27. Geng, F., Song, K., Xing, J. et al. Thio-glucose bound gold nanoparticles enhance radio-cytotoxic targeting of ovarian cancer. *Nanotechnology* 2011;22:28.
28. Turner, J., Koumenis, C., Kute, T. et al. Tachpyridine, a metal chelator, induces G2 cell-cycle arrest, activates checkpoint kinases, and sensitizes cells to ionizing radiation. *Blood* 2005;106(9):3191–3199.
29. Roa, W., Zhang, X., Guo, L. et al. Gold nanoparticle sensitize radiotherapy of prostate cancer cells by regulation of the cell cycle. *Nanotechnology* 2009;20:37.
30. Kang, B., Mackey, M., and El-Sayed, M. Nuclear targeting of gold nanoparticles in cancer cells induces DNA damage, causing cytokinesis arrest and apoptosis. *J Am Chem Soc* 2010;132(5):1517–1519.

31. Tanaka, H., Arakawa, H., Yamaguchi, T. et al. A ribonucleotide reductase gene involved in a p53-dependent cell-cycle checkpoint for DNA damage. *Nature* 2000;404(6773):42–49.
32. Kaufmann, S., and Vaux, D. Alterations in the apoptotic machinery and their potential role in anticancer drug resistance. *Oncogene* 2003;22:7414–7430.
33. Marcu, M., Jung, Y., Lee, S. et al. Curcumin is an inhibitor of p300 histone acetylatransferase. *Med Chem* 2006;2:169–174.
34. Ghosh, S., and Hayden, M. New regulators of NF-kB in inflammation. *Nat Rev Immunol* 2008;8:837–848.
35. Sandur, S., Deorukhkar, A., Pandey, M. et al. Curcumin modulates the radiosensitivity of colorectal cancer cells by suppressing constitutive and inducible NF-kB activity. *Int J Radiat Oncol Biol Phys* 2009;75:534–542.
36. Lippitz, B., Halperin, E., Griffith, O. et al. L-buthionine-sulfoximine-mediated radiosensitization in experimental interstitial radiotherapy of intracerebral D-54 MG glioma xenografts in athymic mice. *Neurosurgery* 1990;26(2):255–260.
37. Mendonca, M., Chin-Sinex, H., Gomez-Millan, J. et al. Parthenolide sensitizes cells to X-ray-induced cell killing through inhibition of NF-kappaB and split-dose repair. *Radiat Res* 2007;168:689–697.
38. Cheng, J., Chou, C., Kuo, M., and Hsieh, C. Radiation-enhanced hepatocellular carcinoma cell invasion with MMP-9 expression through PI3K/Akt/NF-kB signal transduction pathway. *Oncogene* 2006;25:7009–7018.
39. Sessler, J., and Miller, R. Texaphyrins: New drugs with diverse clinical applications in radiation and photodynamic therapy. *Biochem Pharmacol* 2000;59(7):733–739.
40. Khuntia, D., and Mehta, M. Motexafin gadolinium: A clinical review of a novel radioenhancer for brain tumors. *Expert Rev Anticancer Ther* 2004;4(6):981–989.
41. Magda, D., and Miller, R. Motexafin gadolinium: A novel redox active drug for cancer therapy. *Semin Cancer Biol* 2006;16:466–476.
42. Hashemy, S., Ungerstedt, J., Avval, A.Z., and Holmgren, A. Motexafin gadolinium, a tumor-selective drug targeting thioredoxin reductase and ribonucleotide reductase. *J Biol Chem* 2006;281:10691–10697.
43. Magda, D., Lecane, P., Miller, R. et al. Motexafin gadolinium disrupts zinc metabolism in human cancer cell lines. *Cancer Res* 2005;65(9):3837–3845.
44. Rockwell, S., Donnelly, E., Liu, Y., and Tang, L. Preliminary studies of the effects of gadolinium texaphyrin on the growth and radiosensitivity of EMT6 cells *in vitro*. *Int J Radiat Oncol Biol Phys* 2002;54(2):536–541.
45. Magda, D., Lepp, C., Gerasimchuk, N. et al. Redox cycling by motexafin gadolinium enhances cellular response to ionizing radiation by forming reactive oxygen species. *Int J Radiat Oncol Biol Phys* 2001;51(4):1025–1036.
46. Biaglow, J., Ayene, I., Koch, C., Donahue, J., Stamato, T., and Tuttle, S. G6PD deficient cells and the bioreduction of disulfides: Effects of DHEA, GSH depletion and phenylarsine oxide. *Biochem Biophys Res Commun* 2000;273(3):846–852.
47. Miller, R., Woodburn, K., Fan, Q., Renschler, M., Sessler, J., and Koutcher, J. *In vivo* animal studies with gadolinium (III) texaphyrin as a radiation enhancer. *Int J Radiat Oncol Biol Phys* 1999;45(4):981–989.

48. Mehta, M., Rodrigus, P., Terhaard, C. et al. Survival and neurologic outcomes in a randomized trial of motexafin gadolinium and whole-brain radiation therapy in brain metastases. *J Clin Oncol* 2003;21(13):2529–2536.
49. Meyers, C., Smith, J., Bezjak, A. et al. Neurocognitive function and progression in patients with brain metastases treated with whole-brain radiation and motexafin gadolinium: Results of a randomized phase III trial. *J Clin Oncol* 2004;22(1):157–165.
50. Mehta, M., Shapiro, W., Phan, S. et al. Motexafin gadolinium combined with prompt whole brain radiotherapy prolongs time to neurologic progression in non-small-cell lung cancer patients with brain metastases: Results of a phase III trial. *Int J Radiat Oncol Biol Phys* 2009;73(4):1069–1076.
51. Olson, J., Paleologos, N., Gaspar, L. et al. The role of emerging and investigational therapies for metastatic brain tumors: A systematic review and evidence-based clinical practice guideline of selected topics. *J Neurooncol* 2010;96:115–142.
52. Marzano, C., Gandin, V., Folda, A., Scutari, G., Bindoli, A., and Rigobello, M.P. Inhibition of thioredoxin reductase by auranofin induces apoptosis in cisplatin-resistant human ovarian cancer cells. *Free Radic Biol Med* 2007;42:872–881.
53. Gandin, V., Fernandes, A., Rigobello, M. et al. Cancer cell death induced by phosphine gold(I) compounds targeting thioredoxin reductase. *Biochem Pharmacol* 2008;79:90–101.
54. Anderson, M. Glulathione: An overview of biosynthesis and modulation. *Chem Biol Interact* 1998;111–112:1–14.
55. Dalle-Donne, I., Rossi, R., Colombo, G., Giustarini, D., and Milzani, A. Protein 5-glutathionylation: A regulatory device from bacteria to humans. *Trends Biochem Sci* 2009;34:85–96.
56. Singh, S., Khan, A., and Gupta, A. Role of glutathione in cancer pathophysiology and therapeutic interventions. *J Exp Therapeut Oncol* 2012;9:303–316.
57. Richardson, R., and Murphy, S. Effect of glutathione depletion on tissue deposition of methylmercury in rats. *Toxicol Appl Pharmacol* 1975;31:505–519.
58. Griffith, O. Mechanism of action, metabolism, and toxicity of buthionine sulfoximine and its higher homologs, potent inhibitors of glutathione synthesis. *J Biol Chem* 1982;257:13704–13712.
59. Kramer, R., Soble, M., Howes, A., and Montoya, V. The effect of glutathione GSH depletion *in vivo* by buthione sulfoxomide (BSO) on the radiosensitization of SR2508. *Int J Radiat Oncol Biol Phys* 1989;16(5):1325–1329.
60. Zhao, Y., Seefeldt, T., Chen, W. et al. Increase in thiol oxidative stress via glutathione reductase inhibition as a novel approach to enhance cancer sensitivity to X-ray irradiation. *Free Radic Biol Med* 2009;47:176–183.
61. Dote, H., Burgan, W., Camphausen, K., and Tofilon, P. Inhibition of Hsp90 compromises the DNA damage response to radiation. *Cancer Res* 2006;66:9211–9220.

CHAPTER 4

Radiosensitization by Halogenated Pyrimidines

4.1 HALOGENATED PYRIMIDINES

The halogenated thymidine (TdR) analogues, bromodeoxyuridine (BrdUrd) and iododeoxyuridine (IdUrd; Figure 4.1), are a class of pyrimidine analogues that have been recognized as potential radiosensitizing agents since the early 1960s. A great deal has been discovered about their mode of action but the interest in their clinical applications has abated, although it still persists on the basis of novel analogues and new delivery systems.

The combining size of a halogen (chlorine, bromine, or iodine) atom is close enough to that of the methyl (CH_3) group to allow the halogenated thymine to be metabolized in the same way as TdR, and the similarity is so close that they are incorporated into DNA in place of TdR. The substitution makes the polynucleotide chain more sensitive to damage by ionizing and UV radiation for reasons that will be discussed in the following sections. Halogenated pyrimidines sensitize cells to a degree dependent on the amount of the analogue incorporated (Figure 4.2) and a differential effect between tumor and normal tissue depends on the fact that tumor cells generally proliferate more rapidly compared with cells of normal tissue.

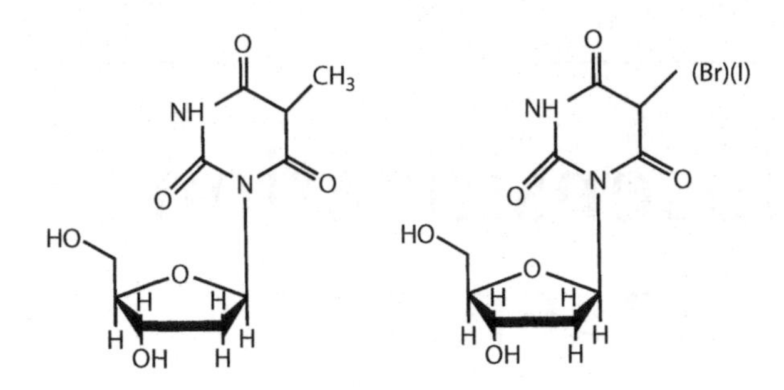

FIGURE 4.1 Thymidine and bromo(iodo)deoxyurine.

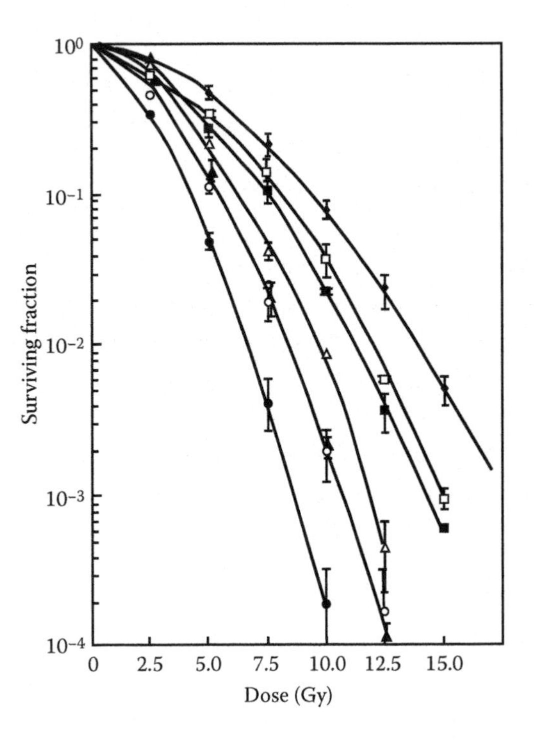

FIGURE 4.2 Radiation survival curves as a function of BrdUrd labeling (concentration and time). The results are means ± SD of two or three experiments. Concentrations of BrdUrd and times of exposure: (◆) control; (○) 10^{-5} M, 12 h; (●) 10^{-5} M, 24 h; (Δ) 5×10^{-6} M, 12 h; (▲) 5×10^{-6} M, 24 h; (□) 10^{-6} M, 12 h; (■) 10^{-6} M, 24 h. (From Ling, L., and Ward, J., *Radiation Research* 121:76–83, 1990. With permission.)

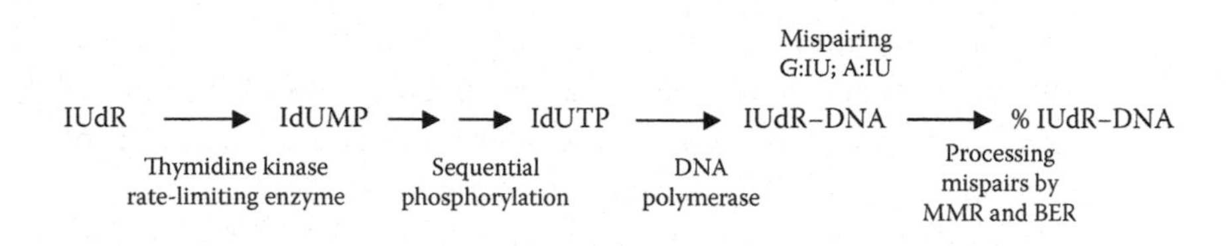

FIGURE 4.3 Intracellular processing of halogenated TdR analogues.

4.1.1 Pharmacology

Cellular uptake and metabolism are dependent on the TdR salvage pathway, where they undergo initial intracellular phosphorylation to the monophosphate derivative (BdUMP, IdUMP) by the rate-limiting enzyme, thymidine kinase (Figure 4.3). Sequential phosphorylation to triphosphates (BdUTP, IdUTP) results in the use of the modified analogues during scheduled (S-phase) and unscheduled DNA synthesis, in competition with deoxythymidine triphosphase (dTTP) by DNA polymerase. DNA incorporation is a prerequisite for radiosensitization in human tumors as well as in normal cells, and the extent of radiosensitization correlates directly with the percentage of TdR replacement in DNA [1].

Although there is little to choose from between the bromine and iodine analogues in terms of radiosensitivity, BrdUrd is a more efficient sensitizer for fluorescent light. This affects the clinical application because a side effect of BUdR may be a rash resulting from phototoxicity, which is much less of a problem for IdUrd, which would thus be the choice for clinical application. Much of the experimental work, however, has been done with BrdUrd.

4.2 MECHANISMS OF RADIOSENSITIZATION

BrdUrd incorporation increases DNA damage induced by ionizing radiation (IR), particularly single-strand and double-strand breaks [2] and may influence the rate of DNA repair of sublethal damage or potentially lethal damage [3]. A proposed mechanism of strand breakage by bromouracil involves hydrated electrons (e^-_{aq}), produced by the radiolysis of water, which interact with pyrimidine and halogenated bases [4,5]. The pyrimidines, thymine, and particularly cytosine react with e^-_{aq} to form a thermally stable anion radical that can be potentially converted back to the initial base or can be stabilized by transfer of the hydrogen involved in the interaction with the complementary base [5]. The halogenated pyrimidines (I,Br)dUrd, on the other hand, are irreversibly destroyed by dissociative attachment of a thermal or solvated electron, yielding a reactive uridinyl radical [6]. It has been suggested that in double-stranded DNA, base pairing can stabilize bromine in bromouracil, thus inhibiting this crucial step in the DNA radiosensitization process. Some theoretical work has suggested that an increased activation barrier for bromine loss in double-stranded DNA may account for its resistance to strand breakage compared with single-strand breaks in DNA [7].

4.2.1 The Role of DNA Repair Pathways

Mismatch repair (MMR) and base excision repair (BER) are central pathways in DNA damage processing of certain types of cancer therapeutics, including the alkylating and antimetabolite classes of chemotherapeutic drugs and ionizing radiation. Thymidine (TdR) replacement in DNA by halogenated analogues can result in DNA mismatches and it has been shown that specific DNA mismatches (particularly G:IU mispairs) are recognized by both DNA MMR and BER [8,9]. As a consequence, the cellular status of these two DNA repair pathways in human tumors may affect the extent of (I)BrdUrd–DNA incorporation and, subsequently, the extent of TdR analogue-mediated IR radiosensitization. It has been suggested that human tumors known to be deficient in DNA MMR (i.e., microsatellite instability–high cancers; MSI-H) or those deficient in BER could be targeted for IR radiosensitization by the halogenated TdR analogues.

There is an important therapeutically relevant distinction between these two DNA repair pathways; MMR processing is required for the cytotoxicity of these drug types, whereas BER processing sometimes leads to reduced drug-related cytotoxicity. A similar distinction can be made for ionizing radiation damage processing by MMR versus BER although, in this case, nonhomologous end-joining and homologous recombination are believed to be the major ionizing radiation repair pathways for double-strand breaks.

4.2.1.1 Base Excision Repair

DNA-SSBs are one of the most frequent and lethal lesions occurring in cellular DNA, either spontaneously or as intermediates of enzymatic repair of base damage during BER. In this repair pathway, after the removal of a damaged base by a DNA glycosylase, the resulting AP site can be processed by AP endonuclease cleavage, leaving a 5′-deoxyribose-phosphate, and by an AP lyase activity leaving a 3′ β-elimination product. The subsequent removal of these AP sites by DNA polymerase β (Pol β) or by a proliferating cell nuclear antigen (PCNA)-dependent polymerase allows the repair synthesis to fill a single nucleotide (for Pol β) or a longer repair patch (for Pol δ/ε), which is then religated. Although the two pathways use different subsets of enzymes, there is cooperation and compensation between the short-patch and long-patch pathways. It is believed that short-patch BER accounts for most BER activity after chemotherapeutic or ionizing radiation treatments [8].

4.2.1.2 Role of BER in Radiosensitization by TdR Analogues

Dillehay et al. [10] first suggested a possible role for BER in the cytotoxicity of halogenated TdR analogues. More recently, computational and mathematical modeling approaches have also been used to better understand BER processing of alkylating and antimetabolite chemotherapeutic damage as well as ionizing radiation damage [11]. These data support a model in which BER protein levels are directly controlled by the type and extent of DNA damage. After repair is accomplished, the key BER proteins may be recycled or degraded by a proteasomal-dependent pathway if the BER protein levels exceed the levels of DNA damage. Unlike MMR processing of these types of DNA damage, BER processing sometimes leads to chemotherapeutic and ionizing radiation resistance. It is recognized that BER removes a wide spectrum of DNA adducts such as those caused by alkylating drugs (particularly N7-methylguanine and N3-methyladenine), as well as some adducts caused by ionizing radiation and radiomimetic drugs (e.g., bleomycin, particularly 8-oxoguanine). Other BER glycosylases such as thymine–DNA glycosylase, uracil–DNA glycosylase, and methyl-binding domain protein 4 (MBD4, also known as MED1) can efficiently process antimetabolite–DNA base damage, leading to antimetabolite drug resistance [1].

The role of BER in the radiosensitization by TdR analogues has been demonstrated in a number of experimental situations including the following:

- Experiments with EM9 cells, which are derived from the AA8 CHO cell line and characterized by hypersensitivity to agents that induce DNA base damage, specifically simple alkylating agents including methylmethane sulfonate and ethylmethane sulfonate. EM9 cells also show a 10-fold increased frequency of spontaneous sister chromatid exchanges and are affected by a deficiency in rejoining

DNA-SSB induced by exposure to alkylating agents and ionizing radiation. The repair defect has been defined as a frameshift mutation in the endogenous *XRCC1* gene resulting in a truncated polypeptide lacking approximately two-thirds of the gene sequence (reviewed by Thompson and West [12]), and this repair defect of EM9 cells is fully corrected by human gene *XRCC1*. Recent studies have shown the role of XRCC1 protein to be that of a "scaffolding" protein, which binds tightly to at least three other factors involved in BER and DNA-SSB repair mechanisms: DNA ligase III, DNA polymerase β, and PARP [13]. In addition, it has been shown that the human polynucleotide kinase enzyme also binds XRCC1, and that this interaction stimulates SSB repair reactions *in vitro* [14].

EM9 cells were found to be significantly more sensitive than parental AA8 cells to IdUrd alone and to IdUrd plus ionizing radiation. The EM9 cells showed increased DNA damage after IdUrd treatment as evaluated by pulsed-field gel electrophoresis and single-cell gel electrophoresis (Comet Assay). In contrast, BER-competent EM9 cells, which were stably transfected with a cosmid vector carrying the human *XRCC1* gene showed a response to IdUrd similar to that of AA8 cells.

Incorporation of IdUrd into DNA of XRCC1-mutant cells is equivalent to that observed in wild-type CHO cells. This suggests that the differences in cytotoxicity and DNA damage evident in these cell lines after IdUrd treatment results from a defect in DNA repair rather than differential incorporation of IdUrd in DNA.

- Other studies were done with methoxyamine (MX), an alkoxyamine derivative that blocks short-patch BER by reacting with an aldehyde-sugar group of the AP site, causing a methoxyamine-AP–stable intermediate adduct [15]. This adduct blocks the endonuclease activity of AP endonuclease 1. Human colorectal carcinoma, RKO cells were exposed to IdUrd or methoxyamine (or both) for 48 h before ionizing radiation (5 Gy). Pretreatment of cells with IdUrd/methoxyamine before ionizing radiation alters cell cycle kinetics, particularly at the G_1–S transition, causing a prolonged G1 cell cycle arrest. The treatment reduced ionizing radiation-induced apoptotic cell death while promoting stress-induced premature senescence. Overall, the effects of ionizing radiation were enhanced by IdUrd/methoxyamine pretreatment.

- Experiments with methoxyamine (MX) were also done with the human colorectal cancer cell model HCT116, this time in two versions, with and without intact MMR. HCT116/3-6 cells (MMR$^-$) have a hemizygous nonsense mutation in the MMR *hMLH1* gene located on chromosome 3 and are MMR deficient, lacking hMLH1 protein expression.

In HCT116 cells, methoxyamine not only increased IdUrd cytotoxicity but also increased the incorporation of IdUrd into DNA cells, leading to greater radiosensitization. The increase in the DNA incorporation with IdUrd was explained on the basis that that MX-adducted AP sites, although refractory to the action of the single-nucleotide BER pathway, can still be processed by the long-patch BER pathway. Consequently, DNA synthesis associated with long-patch BER events results in a continuous accumulation of IdUrd in DNA, engaging the cell in a "suicide" cycle of excision and incorporation of IdUrd with increased cytotoxicity because of the halogenated analogue alone and the potential greater radiosensitization. Evidence was also presented that the increased IdUrd cytotoxicity observed in cells lacking functional single-nucleotide BER can be explained by the increased number of DNA breaks left unrepaired after the removal of IdUrd. Linear-quadratic modeling was used to analyze the radiosensitization data, which led to the conclusion that combined treatment with IdUrd and MX resulted in a greater than additive effect on the IR-induced cell killing in both HCT116 and HCT116/3-6 cells [8].

It was concluded, on the basis of all these results, that cells with the capacity to carry out BER are protected to a certain extent against IdUrd-mediated radiation damage, and that DNA repair deficiencies associated with XRCC1 polymorphisms may predict a more favorable outcome of anticancer therapies using halogenated pyrimidines and radiation.

4.2.1.3 Mismatch Repair

DNA MMR (Figure 4.4) helps safeguard the genome from errors that arise as a result of normal cellular processes including DNA repair and from chemically induced DNA damage. MMR of naturally occurring errors in base-pairing prevents point mutations, transversion mutations, and instability throughout the genome at microsatellite sequences. MMR is crucial for correcting single base-pair mismatches occurring during DNA replication that are not corrected by the proof-reading activity of DNA polymerase, by mispairs resulting from recombination, or from the deamination of

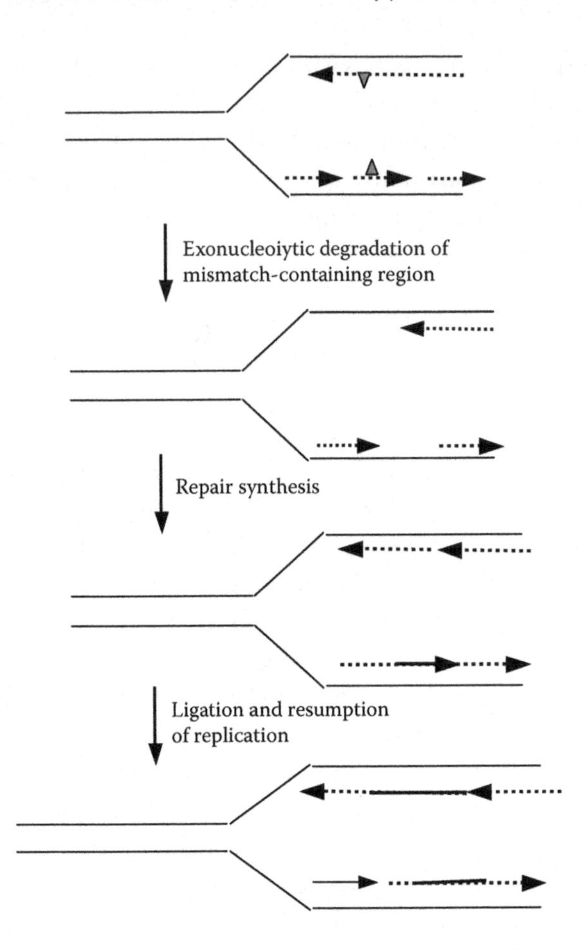

FIGURE 4.4 In mammalian cells, the degradation of the mismatch-containing leading strand (dashed line) could begin at the 3′ terminus of the primer strand. Once the mispair is removed, the polymerase resynthesizes the degraded region. Mismatch repair (MMR) in the lagging strand could remove an entire Okazaki fragment, with degradation commencing at either end. Extension of the fragment closest to the replication fork would replace the degraded one. Removal of the RNA termini of the Okazaki fragments, followed by ligation, would give rise to a continuous, error-free, lagging strand.

5-methylcytosine in DNA, as well as insertion/deletion loops (IDLs) resulting from slippage of the DNA polymerase as it replicates simple repetitive sequences in DNA (reviewed by Kunkel and Erie [16] and Jiricny [17]).

At least six different MMR proteins are required to fulfill the primary function of the MMR system (Table 4.1). For mismatch recognition, the MSH2 protein forms a heterodimer with either MSH6 or MSH3, depending on the type of lesion to be repaired (MSH6 is required for

TABLE 4.1 Human MutS and MutL Homologue Complexes That Are Involved in Mismatch Repair

Complex	Components	Functions
MutSα	MSH2, MSH6	Recognition of base–base mismatches and small IDLs
MutSβ	MSH2, MSH3	Recognition of IDLs
MutLα	MLH1, PMS2	Forms a ternary complex with mismatch DNA and MutSα increases discrimination between heteroduplexes and homoduplexes also functions in meiotic recombination
MutLβ	MLH1, PMS1	
MutLγ	MLH1, MLH3	Primary function in meiotic recombination; backup for MutLα in the repair of base–base mismatches and small IDLs

the correction of single-base mispairs, whereas both MSH3 and MSH6 may contribute to the correction of insertion/deletion loops). A heterodimer of MLH1 and PMS2 coordinates the interplay between the mismatch recognition complex and other proteins necessary for MMR. These additional proteins include at least one nuclease, exonuclease 1 (EXO1), PCNA, single-stranded DNA-binding protein (RPA), and DNA polymerases δ and ε. In addition to PMS2, MLH1 may heterodimerize with two additional proteins, MLH3 and PMS1. Recent observations indicate that PMS2 is required for the correction of single-base mismatches, and PMS2 and MLH3 both contribute to the correction of insertion/deletion loops [8,18].

In eukaryotes, MMR is initiated when complexes of MutS homologues, either MSH2–MSH6 (MutSα) or MSH2–MSH3 (MutSβ), bind to a mismatch. MutSα is primarily responsible for repairing single base-base and IDL mismatches whereas MutSβ is primarily responsible for repairing IDL mismatches containing up to 16 extra nucleotides in one strand. The two complexes can share responsibility for repairing some IDL mismatches, especially those with one extra base. The MutSα and MutSβ complexes interact with PCNA, which contributes to several steps in MMR. Eukaryotes also encode multiple MutL homologues that form different heterodimers. MutLα (MLH1–PMS2) is involved in repairing a wide variety of mismatches. Several eukaryotic exonucleases are implicated in MMR, and several other proteins are also required. DNA resynthesis is catalyzed by an aphidicolin-sensitive polymerase, the most probable being DNA polymerase δ.

The postreplicative MMR process is believed to be coupled to DNA replication and has been shown to be directed to the daughter strand of DNA, using the parental strand as a template to ensure that genomic integrity is maintained. During the repair process, a "patch" of DNA surrounding the mispair or IDL is removed in a reaction requiring a multiprotein complex. In eukaryotes, one of two dimers provides recognition specificity. The hMutSα dimer recognizes single base-pair mismatches and small IDLs, whereas the hMutSβ dimer, specifically binds larger IDLs. After recognition by one of the MutS complexes, the hMLH1/hPMS2 dimer hMutSα is recruited to DNA. Subsequent steps in the reaction include incision of the newly synthesized strand, unwinding of the two strands, and excision of the daughter strand from the point of incision to just past the mismatch. This is followed by resynthesis of DNA and, finally, ligation.

The PCNA is thought to physically couple MMR to the replication machinery, thus providing the means for targeting MMR to the daughter strand in eukaryotes. Other proteins that may be required include DNA polymerase δ, responsible for resynthesis of DNA during repair, and the proofreading exonucleases of DNA polymerases δ and ε.

In addition to correcting single base-pair mismatches and IDLs, MMR also plays an important role in cell cycle arrest and the cellular cytotoxic response to many cancer chemotherapy agents. The same proteins seem to be involved in this response, as MutSα has been shown to recognize many of the DNA adducts resulting from treatment with cisplatin, 6-TG, and MNNG, and both MSH6 and MLH1 have been demonstrated to be important for G_2/M cell cycle arrest after radiation treatment [19].

4.2.1.4 Role of MMR in Radiosensitization by TdR Analogues

The results from a number of investigations have demonstrated that human tumor and murine embryonic cell lines mutated in either *MLH1* or *MSH2* genes are increasingly radiosensitized by TdR analogues compared with genetically matched MMR-proficient cells [8,18]. It has also been shown that differences exist in the DNA levels of both IdUrd and BrdUrd, consistent with MMR status of the cell. MMR-deficient cells accumulate more BrdUrd and IdUrd than do MMR competent cells of both human and rodent origin, and the DNA levels of the halogenated TdR analogues are related to both MLH1 and MSH2 status. It has also been shown that there are no consistent or significant differences in the cellular metabolic pathways for the halogenated thymidine analogues and, in fact, differential DNA levels of IdUrd and BrdUrd correlate consistently with MMR status but not with intrinsic dNTP pools or TS and TK enzyme activity.

Typically comparable levels of IUdR–DNA incorporation were seen immediately after short drug exposures (one cell cycle), but in MMR-deficient cells, up to two to three times higher levels persist after drug removal suggesting that IdUrd–DNA is more efficiently repaired (removed) in MMR-proficient cells. To explain this, it has been suggested that MMR removes IdUrd–DNA (particularly IdUrd-G mispairs) in the daughter strand after replication in a manner similar to the repair of other single base pair mismatches. As has already been described, radiosensitization by IdUrd and BrdUrd, after incorporation into DNA, correlates directly with the amount of analogue in DNA suggesting an explanation as to why, in MMR-deficient cells, failure to remove analogues accounts for the greater degree of radiosensitization seen in these cells. This interpretation is supported by a number of experimental findings:

- Using human colon cancer cells with truncating mutations in both alleles of the *hMLH1* gene and a derived cell line corrected for MMR deficiency by transfer of a human chromosome 3 containing a wild-type copy of *hMLH1*, it was shown that IUdR and BUdR DNA levels are much lower in MMR-proficient cells [20].
- A similar effect was seen in murine cells from *Mlh1* knockout and wild-type mice and in a second colon cancer cell system in which untreated human colon cancer (RKO) cells, which lack hMLH1 expression as a result of promoter hypermethylation, were compared with RKO colon cancer cells treated with 5-azadeoxycytidine, a demethylating agent that causes re-expression of MLH1. In these cell systems, MLH1-deficient cells displayed twofold or threefold higher levels of TdR analogue in DNA compared with matched MLH1-proficient cell lines [18].
- Results of other experiments provide proof that both MMR proteins are involved in mediating TdR analogue levels in DNA. The human endometrial cancer cell line HEC59, which is MMR-deficient due to mutations in both alleles of the *hMSH2* gene, has twofold higher levels of both IdUrd and BrdUrd in DNA than do the MMR-proficient HEC59/2-4 cells, which contain a wild-type *hMSH2* gene via transfer of a human chromosome 2. Data from E1A-transformed embryonic stem cells derived from *Msh2* knockout mice or their wild-type siblings showed the same trend, with higher levels of TdR analogue in *Msh2–/–* cells than in *Msh2+/+* cells [18].
- The differences between MMR-proficient and MMR-deficient cells become apparent within the same population doubling because

incorporation of the analogue continues to increase in MMR-deficient cells over time, whereas it remains low in MMR-proficient cells. This is consistent with a postreplicational MMR process that can remove the TdR analogues themselves from DNA, similar to mismatched bases overlooked by the proofreading activity of DNA polymerase.

- There are only very modest differences in the cytotoxicity of IdUrd in MMR^+ versus MMR^- isogenic cell systems; however, MMR^- human colorectal and endometrial tumor cells/xenografts show persistently higher IdUrd DNA levels compared with MMR^+ (normal) cells after IdUrd treatment because the resulting G:IU mispairs are efficiently repaired in MMR^+ cells but are not repaired in MMR^- cells [21].

4.2.2 Cell Cycle Regulation and Cell Death Signaling in Cells Irradiated after Pretreatment with TdR Analogues

It had been demonstrated that cells deficient in MutL homologue-1 (MLH1) expression had a reduced and shorter G_2 arrest after high-dose rate ionizing radiation (IR), suggesting that the MMR system mediates this cell cycle checkpoint. This was confirmed in several studies with human cell lines. Differences in IR-induced G_2 arrest between MMR-proficient and MMR-deficient cells were found regardless of whether synchronized cells were irradiated in G_0/G_1 or S phase, indicating that MMR indeed dramatically affects the G_2/M checkpoint arrest. However, no significant difference in the clonogenic survival of MMR-deficient cells compared with MMR-proficient cells was observed after high-dose rate IR.

The results of other studies, however, have made it clear that radiosensitization with TdR analogues influences cell signaling pathways and influence cell cycle regulation and the mode of cell death. This was apparent from the results described in Section 4.2.1.2 in which IdUrd (±MX) and IR resulted in a decrease in IR-induced apoptosis, no effect on IR-induced necrosis or autophagy, but a clear enhancement of IR-induced senescence in human colorectal cells. Thus, in this case, cell death signaling pathways were changed by combined preradiation treatment using the halogenated TdR analogues.

4.2.3 Structure of DNA as a Factor in Radiosensitization by ThdR Analogues

It is established that the radiosensitization properties of BrdUrd result primarily from the electrophilic nature of bromine, making it a good leaving group and leading to the irreversible formation of a uridinyl radical (dUrd•) or uridinyl anion ($dUrd^-$) upon the addition of an electron. In one study, it was reported that the radiolytic loss of the bromine atom is greatly suppressed in double-stranded compared with single-stranded DNA, and this conclusion is supported by results from experiments with model DNA systems.

Using a model DNA containing a bulge formed by five mismatched bases, a linear dose–response was observed for the formation of strand breaks on the single-stranded regions of both the brominated strand and the opposite nonbrominated strand. The formation of interstrand cross-links occurred exclusively in the mismatched region. In the same study, the creation of γ-radiation induced interstrand cross-links was reported. These appear exclusively in the mismatched single-stranded brominated region of a double-stranded oligonucleotide. The authors suggested that the interstrand cross-links might contribute to the radiosensitization effects of BRdU, and that both DNA breaks and cross-links resulting from the presence of BRdU in the DNA of γ-irradiated cells should form primarily in regions of DNA that were single-stranded at the time of irradiation. On this basis, it was proposed that the radiosensitization effects of BrdUrd *in vivo* would be limited to single-strand regions such as those found in transcription bubbles, replication forks, mismatched DNA, and possibly the loop region of telomeres [22].

DNA conformation has also been implicated as a factor governing the ability of 5-bromodeoxyuridine (BrdUrd) to radiosensitize DNA. In one study, DNA conformation was altered by gradually rehydrating lyophilized DNA samples, which induces an A to B form transition, whereas in other experiments, DNA was irradiated in solution, in the presence or absence of 80% ethanol to induce an A or B form, respectively. Alkali-labile DNA lesions were revealed using hot piperidine to transform both base and sugar lesions into strand breaks. Strand breaks were reported to be specific for B-form DNA, whereas A-DNA only undergoes formation of piperidine-sensitive DNA lesions. Interstrand cross-links were only found in semicomplementary B-DNA.

The location of damage as a function of DNA structure was also analyzed: piperidine-sensitive lesions were observed exclusively at the site of BrdU substitution, whereas strand breaks were able to migrate along the DNA strand, with a clear preference for the adenine 5′ of the BUdR. Thus, the hybridization state and the DNA conformation affected the degree of sensitization by BUdR by influencing the amount and type of damage produced [23].

4.2.4 Optimizing Access of the TdR Analogue Radiosensitizer to the Tumor Cell

Useful radiosensitization with thymidine analogues depends on achieving significant uptake into the DNA of cancer cells and minimizing the number of cells that have no thymidine replacement. This can be facilitated by intervention at several levels: by pharmacological stratagems to bring the drug close to the tumor cell, by manipulating nucleotide metabolism to facilitate incorporation of the analogue, and by modifying the molecule itself to facilitate delivery.

4.2.4.1 Pharmacological Stratagems for Drug Delivery

Various pharmacologic approaches have been tried experimentally and clinically to improve the therapeutic gain of TdR analogue radiosensitization in poorly radioresponsive human tumors.

4.2.4.1.1 Intra-arterial Infusion The use of selective intra-arterial infusions to increase tumor bed drug concentrations has been used clinically for primary brain tumors and hepatic metastases with a modest improvement in the therapeutic gain. The downside of this approach is that prolonged infusion of drugs at a dose level that might approach maximal sensitization may be accompanied by an unacceptable level of systemic toxicity. Even if this is not the case, variability in tumor cell kinetics may be the limiting factor in TdR incorporation.

4.2.4.1.2 Intratumoral Delivery of TdR Analogue Another approach is to increase the BrdUrd concentration selectively in the tumor by intralesional drug delivery, which can enhance both cytotoxicity and sensitization of the tumor cells while reducing normal tissue toxicity. Several implantable intratumorally controlled drug release systems have been developed in addition to biodegradable microspheres and other devices. The involvement of several drug delivery devices as vehicles for radiosensitizing molecules will be discussed at more length in Chapter 14, and some results relevant to this chapter will be mentioned here:

- One report described radiosensitization of a subcutaneous RIF-1 tumor by release of BrdU from an intratumoral polymer implant (Figure 4.5) [24].

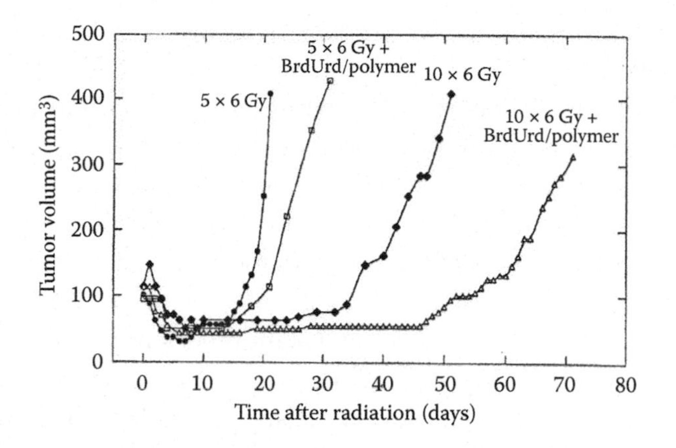

FIGURE 4.5 Tumor volume after treatment with fractionated radiation: (●) 5 × 6 Gy; (□) 5 × 6 Gy + BrdUrd/polymer; (◆) 10 × 6 Gy; (Δ) 10 × 6 Gy polymer/BrdUrd. Polymers were implanted 3 days before the first radiation fraction. Experimental groups consisted of five to eight mice. SDs of the mean volumes, which were calculated for each time interval, did not exceed 10% of the mean. Error bars are omitted for clarity. (From Doiron, A. et al., *Cancer Research* 59:3677–3681, 1999. With permission.)

- The use of the PCCP/SA polymer for the delivery of IdUrd to human U251 glioblastoma xenografts has been reported [25].
- Another study reported the optimization of radiosensitization of an intracranial glioma using an intratumoral slow-release device for combined delivery BUdR and a biomodulator (Figure 4.6) [26].

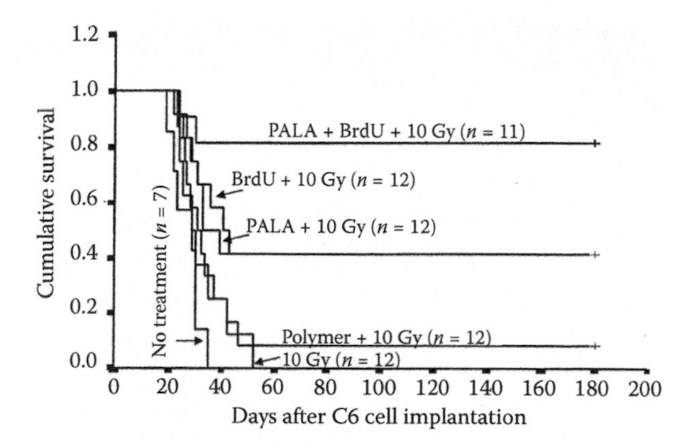

FIGURE 4.6 Survival of rats implanted intracranially with BrdUrd/polymer with and without RT. (From Li, Y. et al., *International Journal of Radiation Oncology – Biology – Physics* 58:519–527, 2004. With permission.)

4.2.4.2 Manipulation of Nucleotide Metabolism: To Promote TdR Analogue Uptake

Br(I)dUrd is a potent radiosensitizer when incorporated into the DNA of target cells; the degree of radiosensitization depending on the extent of thymidine replacement by the analogue and on the number of cells labeled as already emphasized. Given that rapid uncontrolled proliferation is a hallmark of cancer cells, the problem of persuading those cells to incorporate high levels of TdR analogue would seem to be a no-brainer. In fact, this is not the case as the clinical findings indicate and many other studies have confirmed. In fact, the failure of proliferating cells to take up thymidine analogues such as BrdUrd is attributable to the low availability of the drug to the tumor cell or to the analogue being diluted to below useful levels by the availability of endogenous nucleotide precursors. The latter results from the activity of enzymes in the *de novo* synthesis pathways or the enzymes in the alternative pathways for pyrimidine or purine biosynthesis, the so-called salvage pathways (Figure 4.7). A marked increase in the activities of enzymes in both categories has been observed in cancer cells in logarithmic growth.

- The first category of antimetabolites that have been evaluated in terms of the extent to which they enhance radiosensitization by TdR analogues are those that block TS to enhance TdR analogue incorporation: 5-FU, fluorodeoxyuridylate, and folinic acid (leucovorin) (Figure 4.8).

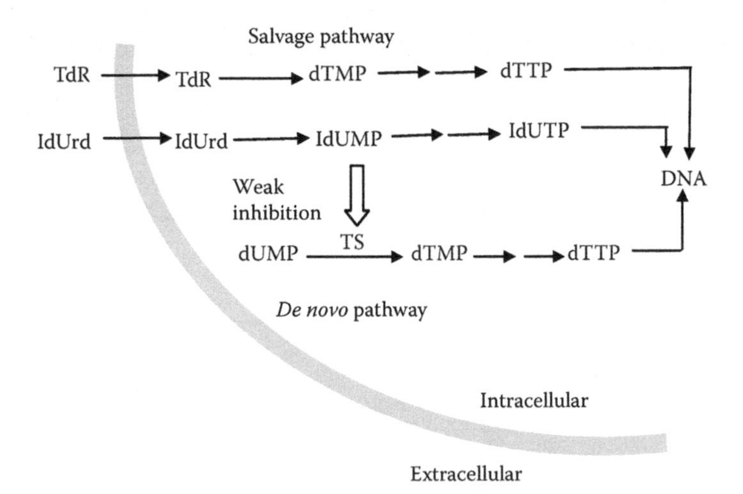

FIGURE 4.7 A biochemical pathway of activation of IdUrd (through TK) occurs in the thymidine salvage pathway. This pathway is favored after inhibition of the *de novo* pathway by IdUMP.

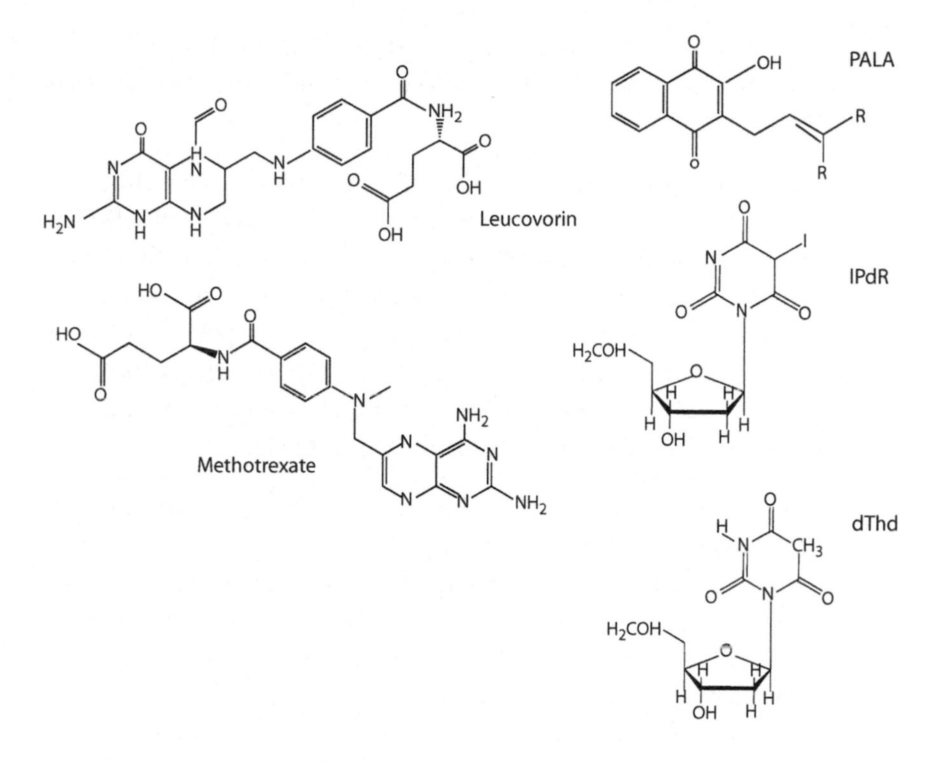

FIGURE 4.8 Structures of compounds used to facilitate BrdUrd incorporation into DNA.

5-Fluorouracil (5-FU), is converted to fluorodeoxyuridylate and decreases thymidylate pools by inhibiting thymidylate synthase (see Chapter 5). In addition, 5-FU acts as a radiosensitizer by causing cell cycle redistribution and it has been shown to increase the incorporation of IdUrd into human bladder cells. Clinical phase I trials using concomitant continuous infusions of IUdR with either fluorodeoxyuridine or folinic acid (leucovorin) showed significant improvements in the therapeutic gain, which were not seen with IUdR infusions alone [27].

- Methotrexate. Methotrexate (MTX) is a folate antagonist that blocks the regeneration of the dihydrofolate required for the synthesis of purines and the conversion of deoxyuridine monophosphate to deoxythymidine monophosphate. Blocking the regeneration of dihydrofolates results in two specific sites in the *de novo* pathway for purine synthesis being blocked. It also inhibits thymidylate synthetase in the pyrimidine biosynthesis pathway. Thus, MTX depletes three nucleotide pools: guanosine triphosphate, adenosine triphosphate, and thymidine triphosphate. MTX has been shown to modulate the uptake of BrdUrd into tumor and normal tissue DNA [28].

- PALA. Another drug that can be used to achieve the same effect but by different means is *N*-(phosphonacetyl)-L-aspartic acid (PALA), a stable inhibitor of L-aspartate transcarbamylase [29]. By tightly binding to L-aspartate transcarbamylase, the second enzyme in the pyrimidine biosynthetic pathway, PALA acts as a potent inhibitor of pyrimidine biosynthesis and is cytotoxic. Short-term exposure to PALA has been used experimentally and clinically to enhance the incorporation of halogenated pyrimidines into DNA.
- In a tissue culture study using a human ovarian carcinoma in which 72-h exposure to 25 μM of PALA produced an approximately 90% decrease in intracellular uridine triphosphate and cytidine triphosphate levels and promoted a fivefold increase in the incorporation of BrdUrd into DNA when BrdUrd was added for the final 24 h of PALA treatment. In addition, PALA alone was cytotoxic and was a radiosensitizer independent of the presence of BrdUrd [30].
- In another study, a glioma model, the C6 astrocytoma, proved to be extremely refractory to radiosensitization by BrdUrd *in vitro*, and the situation was not greatly improved by the incorporation of 5-FU or MTX into the treatment scheme. The addition of PALA was more effective, and treatment with PALA and BUdR achieved some degree of radiosensitization *in vitro* and was even more effective *in vivo* [26].

4.2.4.3 IpDR, a Novel Oral Radiosensitizer

5-Iodo-2-pyrimidinone-2′-deoxyribose (IPdR) is an orally available, halogenated thymidine (TdR) analogue (Figure 4.8). IPdR is a prodrug that is efficiently converted to 5-iodo-2′-deoxyuridine (IUdR) by a hepatic aldehyde oxidase, resulting in high IPdR and IUdR plasma levels in preclinical experimental models. Conversion takes place primarily in the liver. Pharmacologic studies with IPdR demonstrate that IUdR–DNA incorporation into tumor tissue is significantly higher after oral administration of IPdR than after intravenous administration of IdUrd. Athymic mice tolerated oral IPdR at up to 1500 mg/kg per day given three times per day for 6 to 14 days without significant systemic toxicities. The results of several preclinical studies support the conclusion that oral IPdR is a superior radiosensitizer compared with continuous intravenous infusions of IdUrd in terms of safety and efficacy, with a significantly lower toxicity profile, including gastrointestinal and hematologic side effects [31]. In addition, toxicology studies have demonstrated that the maximum tolerated dose (MTD) of oral IPdR is considerably lower than the MTD for IdUrd. These data indicate that IPdR used as a prodrug for IUdR can substantially decrease toxicity, increase IUdR–DNA incorporation, and thereby increase tumor radiosensitization.

4.3 CLINICAL RESULTS

Halogenated thymidine (TdR) analogues, such as 5-iododeoxyuridine (IUdR) and 5-bromodeoxyuridine (BUdR) have been recognized as potential radiosensitizers since the early 1960s. The earliest reports of the clinical application of TdR radiosensitization go back almost as long (reviewed by Phuphanich et al. [32]). These early studies involved small numbers of patients and the mode of administration, intracarotid or intravertebral arterial infusion, caused unacceptable levels of morbidity in the patients.

More recent phase I and II clinical trials have used prolonged continuous or repeated intermittent intravenous infusions of BrdUrd or IdUrd before and during radiation therapy, principally in patients with high-grade brain tumors. These clinically radioresistant tumors are reported to have a rapid proliferation rate (potential tumor doubling times of 5–15 days) and are surrounded by nonproliferating normal brain tissues in adults, where little or no DNA incorporation of the TdR analogues occurs. Thus, high-grade brain tumors are ideal targets for TdR analogue-mediated radiosensitization and the results of these phase I/II trials were deemed to be encouraging. Similarly, the combination of IdUrd and radiation therapy seemed effective in the treatment of high-grade sarcomas, another clinically radioresistant cancer [33].

On the whole, however, the results of clinical studies of brain tumors combining halogenated pyrimidines with fractionated radiotherapy (RT) have been varied. The results published in 1991 by the Northern California Oncology Group for patients with glioblastoma multiforme and anaplastic astrocytoma indicated that glioblastoma multiforme patients who tolerated large cumulative doses of BUdR (the major toxicity was skin rash) had improved progression-free survival compared with those receiving reduced doses, in line with laboratory findings in which increased thymidine replacement improves cell killing by radiation [34]. The results reported for patients with anaplastic astrocytoma in this study showed an estimated 4-year survival rate of 46% after combined multiagent chemotherapy [35], concurrent with RT, and intravenous BrdUrd.

In contrast, the authors of a retrospective analysis of data from 2,000 patients treated with BUdR in four separate studies with considerable heterogeneity of treatment groups concluded that they were not able to provide a definitive answer as to whether BrdUrd given during RT improved survival in patients with malignant glioma. Although a treatment effect did seem to exist in the glioblastoma group, this was not consistent among all the strata studied, and it was concluded that this approach, as it had been applied, was not overly promising [36]. An intergroup study of the efficacy of BUdR radiosensitization in the treatment of anaplastic astrocytoma also yielded disappointing results. A treatment arm of RT plus BUdR followed by procarbazine, chloroethylcyclohexylnitosourea (CCNU or lomustine), and vincristine showed no survival advantage over the control arm of RT followed by the same chemotherapy regimen [37]. Clinical trials involving gastrointestinal and head-and-neck tumors gave more encouraging results in terms of local control [33,38], and these results suggested that significant levels of RT enhancement with tolerable drug levels could be achievable, given the right conditions.

A 2004 publication described a phase III trial that had been based on earlier phase II and III trials, with the hypothesis that BrdUrd radiosensitization would improve the survival outcome in patients with anaplastic glioma. The conclusion was that there was no survival advantage by adding BUdR to external beam radiotherapy and PCV (procarbazine, cyclohexylchloroethylnitrosurea, and vincristine) chemotherapy in this patient population. On the basis of an interim analysis and the decision of the Radiation Therapy Oncology Group (RTOG) Data Monitoring Committee, accrual was suspended and then ultimately discontinued before the full-anticipated accrual [39].

Currently, there are no clinical trials listed involving TdR analogues as radiosensitizers. Uptake of these compounds is a highly effective indicator of cell proliferation and thus they may sometimes have a diagnostic or predictive role in clinical studies.

Very recently, a phase 0 trial of 5-iodo-2-pyrimidinone-20-deoxyribose (IPdR), an oral prodrug of IUdR, was conducted in patients with advanced malignancies to assess whether the oral route was a feasible alternative to continuous intravenous infusion before embarking on large-scale clinical trials. Plasma concentrations of IPdR, IUdR, and other metabolites were measured after a single oral dose of IPdR. There were no drug-related adverse events. Plasma concentrations of IUdR generally increased as the dose of IPdR escalated from 150 to 2,400 mg. All patients at the 2,400 mg dose achieved peak IUdR levels in approximately 1.5 h after IPdR administration. The authors felt that the adequate plasma levels of IUdR achieved justified proceeding with a phase I trial of IPdR in combination with radiation (Kummar). Possibly, this signals the start of another chapter in the history of TdR analogues as radiosensitizers.

4.4 SUMMARY

The halogenated thymidine (TdR) analogues, BrdUrd and IdUrd, have been recognized as potential radiosensitizing agents for more than 40 years. The structure of the TdR allows it to be metabolized and incorporated into DNA in place of TdR making the polynucleotide chain more sensitive to damage by ionizing and UV radiation. TdR analogues sensitize cells to a degree dependent on the amount of the analogue incorporated and a differential effect between tumor and normal tissue depends on the fact that generally tumor cells proliferate more rapidly compared with cells of normal tissue.

BrdUrd incorporation increases DNA damage induced by ionizing radiation, particularly single-strand and double-strand breaks, and may influence the rate of DNA repair of sublethal damage or potentially lethal damage. Thymidine (TdR) replacement in DNA by halogenated analogues can result in DNA mismatches and it has been shown that specific DNA mismatches are recognized by both DNA MMR and BER. As a consequence, the cellular status of these two DNA repair pathways in human tumors may affect the extent of TdR analogue incorporation and, subsequently, the extent of TdR analogue-mediated radiosensitization. The important distinction between the two DNA repair pathways is that whereas MMR processing is required for the cytotoxicity of TdR analogues, BER processing may in fact protect against cytotoxicity.

The conformation and structure of DNA has also been implicated as a factor in radiosensitization by TdR analogues. Useful radiosensitization with thymidine analogues depends on achieving significant uptake into the DNA of cancer cells and minimizing the number of cells that have no thymidine replacement. Failure to optimize incorporation results in poor levels of radiosensitization. Incorporation can be facilitated by pharmacological stratagems to bring the drug close to the tumor cell: selective intra-arterial infusions or intratumoral implant of a drug delivery device; manipulation of nucleotide metabolism by blocking thymidylate synthase (FU, fluorodeoxyuridylate, leucovorin) or other means (methotrexate, PALA) to facilitate incorporation of TdR analogue; or modifying the molecule itself to facilitate delivery. IPdR is a prodrug that is efficiently converted to IdUrd, resulting in high IdUrd levels in plasma. IdUrd–DNA incorporation into tumor tissue is significantly higher after oral administration of IPdR than after intravenous administration of IdUrd.

The clinical investigation of TdR analogues as radiosensitization was spread over a number of years starting in the 1980s. Phase I and II clinical trials used prolonged continuous or repeated intermittent intravenous infusions of BrdUrd or IdUrd before and during radiation therapy, principally in patients with high-grade brain tumors. These clinically radioresistant tumors have a rapid proliferation rate and are surrounded by nonproliferating normal brain tissues in adults making them ideal targets for TdR analogue-mediated radiosensitization. Results of these phase I/II trials were deemed to be encouraging and the combination of IdUdR and radiation therapy seemed effective in the treatment of high-grade sarcomas, another clinically radioresistant cancer. Subsequently, a number of phase II and III trials involving cumulatively large numbers of patients were conducted. Analysis of the results indicated that there was no survival advantage to adding BrdUrd to radiotherapy and chemotherapy in this patient population, and clinical trials involving TdR analogues were discontinued. Interest may be revived by encouraging results with the orally available pro-drug IPdR.

REFERENCES

1. McGinn, C., Shewach, D., and Lawrence, T. Radiosensitizing nucleosides. *J Natl Cancer Inst* 1996;88:1193–1203.
2. Ling, L., and Ward, J. Radiosensitization of Chinese hamster V79 cells by bromodeoxyuridine substitution of thymidine: Enhancement of radiation-induced toxicity and DNA strand break production by monofilar and bifilar substitution. *Radiat Res* 1990;121:76–83.
3. Iliakis, G., Pantelias, G., and Kurtzman, S. Mechanism of radiosensitization by halogenated pyrimidines: Effect of BrdUrd on cell killing and interphase chromosome breakage in radiation-sensitive cells. *Radiat Res* 1991;125:56–64.
4. Gilbert, E., Volkert, O., and Schulte-Frohlinde, D. Radiochemistry of aqueous, oxygen-containing solutions of 5-bromouracil (I). Identification of the radiolysis products. *Z Naturforsch B* 1967;22:477–480.
5. Steenken, S. Electron-transfer-induced acidity/basicity and reactivity changes of purine and pyrimidine bases. Consequences of redox processes for DNA base pairs. *Free Radic Res Commun* 1992;16:349–379.

6. Neta, P. Electron spin resonance study of radicals produced in irradiated aqueous solutions of 5-halouracils. *J Phys Chem* 1972;76:2399–2402.
7. Li, X., Sevilla, M., and Sanche, L. DFT investigation of dehalogenation of adenine-halouracil base pairs upon low energy electron attachment. *J Am Chem Soc* 2003;125:8916–8920.
8. Taverna, P., Hwang, H.S., Schupp, J.E. et al. Inhibition of base excision repair potentiates iododeoxyuridine-induced cytotoxicity and radiosensitization. *Cancer Res* 2003;63:838–846.
9. Seo, Y., Yan, T., Schupp, J., Colussi, V., Taylor, K., and Kinsella, T. Differential radiosensitization in DNA mismatch repair-proficient and -deficient human colon cancer xenografts with 5-iodo-2-pyrimidinone-2′-deoxyribose. *Clin Cancer Res* 2004;10:7520–7528.
10. Dillehay, L., Thompson, L., and Carrano, A. DNA-strand breaks associated with halogenated pyrimidine incorporation. *Mutat Res* 1984;131:129–136.
11. Hwang, H., Davis, T., Houghton, J., and Kinsella, T. Radiosensitivity of thymidylate synthase-deficient human tumor cells is affected by progression through the G1 restriction point into S-phase: Implications for fluoropyrimidine radiosensitization. *Cancer Res* 2000;60:92–100.
12. Thompson, L., and West, M. XRCC1 keeps DNA from getting stranded. *Mutat Res* 2000;459:1–18.
13. Caldecott, K., McKeown, C., Tucker, J., Ljungquist, S., and Thompson, L. An interaction between the mammalian DNA repair protein XRCC1 and DNA ligase III. *Mol Cell Biol* 1994;14:68–76.
14. Whitehouse, C., Taylor, R., Thistlethwaite, A. et al. XRCC1 stimulates human polynucleotide kinase activity at damaged DNA termini and accelerates DNA single-strand break repair. *Cell* 2001;104:107–117.
15. Liu, L., Taverna, P., Whitacre, C., Chatterjee, S., and Gerson, S.L. Pharmacologic disruption of base excision repair sensitizes mismatch repair-deficient and -proficient colon cancer cells to methylating agents. *Clin Cancer Res* 1999;5:2908–2917.
16. Kunkel, T., and Erie, D. DNA mismatch repair. *Annu Rev Biochem* 2005;74:681–710.
17. Jiricny, J. The multifaceted mismatch-repair system. *Nat Rev Mol Cell Biol* 2006;7(5):335–346.
18. Berry, S., Davis, T., Schupp, J. et al. Selective radiosensitization of drug-resistant, MSH2 mismatch repair-deficient cells by halogenated thymidine (dThd) analogs; Msh2 mediates dThd analog DNA levels, and the differential cytotoxicity and cell cycle effects of the dThd analogs and 6-TG. *Cancer Res* 2000;60:5773–5780.
19. O'Brien, V., and Brown, R. Signalling cell cycle arrest and cell death through the MMR system. *Carcinogenesis* 2006;27:682–692.
20. Berry, S., Garces, C., Hwang, H. et al. The mismatch repair protein, hMLH1, mediates 5-substituted halogenated thymidine analogue cytotoxicity, DNA incorporation, and radiosensitization in human colon cancer cells. *Cancer Res* 1999;59:1840–1845.
21. Berry, S., and Kinsella, T. Targeting DNA mismatch repair for radiosensitization. *Semin Radiat Oncol* 2001;11:300–315.

22. Cecchini, S., Girouard, S., Huels, M., Sanche, L., and Hunting, D. Single-strand-specific radiosensitization of DNA by bromodeoxyuridine. *Radiat Res* 2004;162(6):604–615.
23. Dextraze, M., Wagner, J., and Hunting, D. 5-bromodeoxyuridine radiosensitization: Conformation-dependent DNA damage. *Biochemistry* 2007;46: 9089–9097.
24. Doiron, A., Yapp, D., Olivares, M., Zhu, J., and Lehnert, S. Tumor radiosensitization by sustained intratumoral release of bromodeoxyuridine. *Cancer Res* 1999;59:3677–3681.
25. Williams, J.A., Yuan, X., Dillehay, L.E., Shastri, V., Brem, H., and Williams, J.R. Synthetic implantable polymers for local delivery of IUDR to experimental malignant human glioma. *Int J Radiat Oncol Biol Phys* 1998;42:631–639.
26. Li, Y., Owusu, A., and Lehnert, S. Treatment of intracranial rat glioma model with implant of radiosensitizer and biomodulator drug combined with external beam radiotherapy. *Int J Radiat Oncol Biol Phys* 2004;58:519–527.
27. Speth, P., Kinsella, T., Belanger, K. et al. Fluorodeoxyuridine modulation of the incorporation of iododeoxyuridine into DNA of granulocytes: A phase I and clinical pharmacological study. *Cancer Res* 1988;48:2933–2937.
28. Kassis, A., Kirichian, A., Wang, K., Semnani, E., and Adelstein, S. Therapeutic potential of 5-[125I]iodo-2′-deoxyuridine and methotrexate in the treatment of advance neoplastic meningitis. *Int J Radiat Biol* 2004;80:941–946.
29. Collins, K., and Stark, G. Aspartate transcarbamylase. Interaction with the transition state analogue *N*-(phosphonacetyl)-L-aspartate. *J Biol Chem* 1971;246:6599–6605.
30. Yang, J., Fernandes, D., and Wheeler, K. PALA enhancement of bromodeoxyuridine incorporation into DNA increases radiation cytotoxicity to human ovarian adenocarcinoma cells. *Int J Radiat Oncol Biol Phys* 1996;34:1073–1079.
31. Saif, M., Berk, G., Cheng, Y., and Kinsella, T. IPdR: A novel oral radiosensitizer. *Expert Opin Investig Drugs* 2007;16(9):1415–1424.
32. Phuphanich, S., Levin, E., and Levin, V. A phase-1 study of intravenous bromodeoxyuridine used concomitantly with radiation therapy in patients with primary malignant brain tumors. *Int J Radiat Oncol Biol Phys* 1984;10:1769–1772.
33. Robertson, J., Sondak, V., Weiss, S., Sussman, J., Chang, A., and Lawrence, T.S. Pre-operative radiation therapy and iododeoxyuridine for large retroperitoneal sarcomas. *Int J Radiat Oncol Biol Phys* 1995;31:87–92.
34. Phillips, T., Levin, V., Ahn, D. et al. Evaluation of bromodeoxyuridine in glioblastoma multiforme: A Northern California Cancer Center phase II study. *Int J Radiat Oncol Biol Phys* 1991;21:709–714.
35. Levin, V., Prados, M., Wara, W. et al. Radiation therapy and bromodeoxyuridine chemotherapy followed by procarbazine, lomustine and vincristine for the treatment of anaplastic gliomas. *Int J Radiat Oncol Biol Phys* 1995;32:75–83.
36. Kassis, A., Dahman, B., and Adelstein, S. *In vivo* therapy of neoplastic meningitis with methotrexate and 5-[125I]iodo-2-deoxyuridine. *Acta Oncol* 2000;39:731–737.

37. Prados, M., Scott, C., Sandler, H. et al. A phase III randomized study of radiotherapy plus procarbazine, CCNU and vincristine (PCV) with or without BUdR for the treatment of anaplastic astrocytoma: A preliminary report of RTOG 9404. *Int J Radiat Oncol Biol Phys* 1999;45:1109–1115.
38. Cook, J., Glass, J., Lebovics, R. et al. Measurement of thymidine replacement in patients with high-grade gliomas, head and neck tumors, and high-grade sarcomas after continuous intravenous of 5-iododeoxyuridine. *Cancer Res* 1992;52:719–725.
39. Prados, M., Seiferheld, W., Sandler, H.M. et al. Phase III randomized study of radiotherapy plus procarbazine, lomustine, and vincristine with or without BUdR for treatment of anaplastic astrocytoma: Final report of RTOG 9404. *Int J Radiat Oncol Biol Phys* 2004;58(4):1147–1152.

CHAPTER 5

Radiosensitization by Antimetabolites

5.1 ANTIMETABOLITES

Fluropyrimidine-based drugs were developed in the 1950s following the observation that rat hepatomas used the pyrimidine uracil more rapidly than normal tissues, suggesting that the uracil metabolism was a target for antimetabolite chemotherapy. The first drug designed to exploit this aspect of tumor metabolism was fluorouracil (FU), the synthesis of which was reported in 1957. In fact, FU can be considered one of the first molecules to selectively target tumor cells based on the observation that uracil is preferentially used by cancer cells, and that a fluorinated analogue of this base might selectively alter cancer cell metabolism. Chemoradiotherapy using fluorouracil has been in use since the 1960s in radical and adjuvant treatment programs. To date, many millions of patients have been treated worldwide with FU and a large percentage of those also received FU-based chemoradiotherapy.

In addition to FU, several other antimetabolites have been shown to have radiosensitizing properties *in vitro* and to have clinical application, including 5-fluoro-2′-deoxyuridine (FdUrd) and hydroxyurea (HU). The first antimetabolites to be shown to specifically enhance cell killing by ionizing radiation were the halogenated thymidine analogues, 5-bromo-2′-deoxyuridine

(BrdUrd) and 5-iodo-2′-deoxyuridine (IdUrd). These TdR analogues were discussed in Chapter 4.

5.2 ANTIMETABOLITES: MODE OF ACTION

Antimetabolites can be classified on the basis of their mode of action as either drugs that inhibit the biosynthesis of deoxyribonucleotides for DNA replication, the thymidylate synthase (TS) inhibitors, such as FU, or the ribonucleotide reductase (RR) inhibitors, such as HU or, the third possibility, as drugs that become fraudulent substrates for DNA polymerases (nucleoside/nucleobase analogues, such as gemcitabine; Figure 5.1). Some of the antimetabolite radiosensitizers have more than one of these actions, and it may be difficult to determine whether one or more actions contribute to radiosensitization. Moreover, the primary mechanism of cytotoxicity may not be the primary mechanism for radiosensitization (Table 5.1).

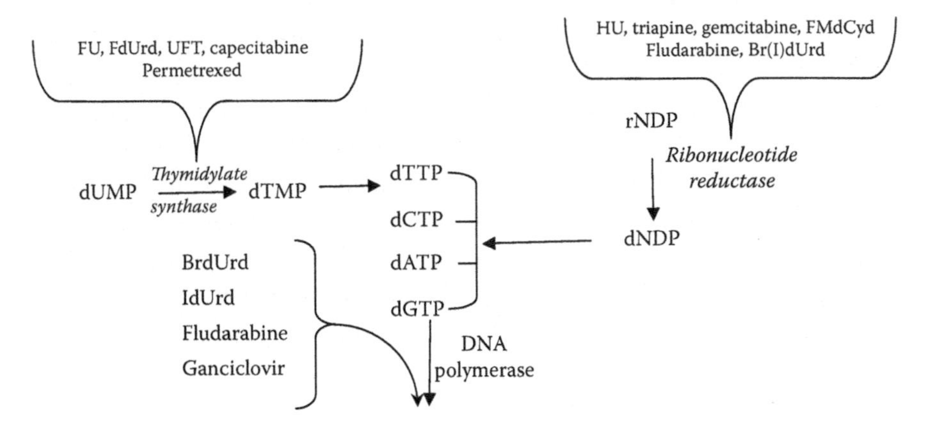

FIGURE 5.1 Targets for antimetabolite radiosensitizers. Antimetabolites are grouped according to the mechanisms of radiosensitization. Abbreviations: BrdUrd, 5-bromo-2′deoxyuridine; dATP, deoxyadenosine triphosphate; dCTP, deoxycytidine triphosphate; dGTP, deoxyguanosine triphosphate; dNDP, deoxyribonucleoside diphosphate; dTMP, thymdine monophosphate; dTTP, deoxythymidine triphosphate; dUMP, deoxyuridine monophosphate; dUrd, 2′-deoxyridine; FdUrd, 5-fluoro-2′-deoxyuridine; FMdCyd, 2′-fluoromethylene-2′-deoxycytidine; FU, fluorouracil; HU, hydroxyurea; IdUrd, 5-iodo-2′-deoxyuridine; rNDP, ribonucleoside diphosphate; and UFT, uracil/Ftorafur.

TABLE 5.1 Mechanisms of Cytotoxicity and Radiosensitization for Antimetabolites

	Cytotoxicity				Radiosensitization	
Drug	**Inhibit TS**	**Inhibit RR**	**Incorporated into DNA**	**Other Effects**	**Increase DNA dsb**	**Decrease DNA Repair**
FU	+	–	+	Disrupt RNA synthesis	–	+
FdUrd	+	–	+	Elevate dUTP	–	+
HU	–	+	–		–	+
Gemcitabine	–	+	–	Inhibit dCMP deaminase	–	+ deficient HRR, MMR
Pemetrexed	+	–	–		not known	
Ganciclovir	–	–	+	Inhibitor of DHFR, GARFT, AICARFT	not known	

Note: AICARFT, aminoimidazole carboxamide ribonucleotide formyltransferase; dCMP, deoxycytidine monophosphate; DHFR, dihydrofolate reductase; dUTP, deoxyuridine triphosphate; FdUrd, 5-fluoro-2′-deoxyuridine; FU, fluorouracil; GARFT, glycinamide ribonucleotide formyltransferase; and HU, hydroxyurea.

5.3 INHIBITORS OF THYMIDYLATE SYNTHASE: 5-FU AND FDURD

(This section and Section 5.4 are based on reviews by Rich et al. [1], Longley et al. [2], and Shewach and Lawrence [3] and the articles referenced therein.)

5-FU is an analogue of uracil with a fluorine atom at the C5 position in place of hydrogen, the corresponding nucleoside is FdUrd (Figure 5.2). The substitution of a fluorine atom for the hydrogen on the 5-position of the uracil base allows FU and FdUrd to be activated along the same molecular pathways as the endogenous compounds uracil and deoxyuridine. FU rapidly enters the cell using the same facilitated transport mechanism as uracil and is converted intracellularly to several active metabolites: fluorodeoxyuridine monophosphate (FdUMP), fluorodeoxyuridine triphosphate (FdUTP), and fluorouridine triphosphate (FUTP; Figure 5.3).

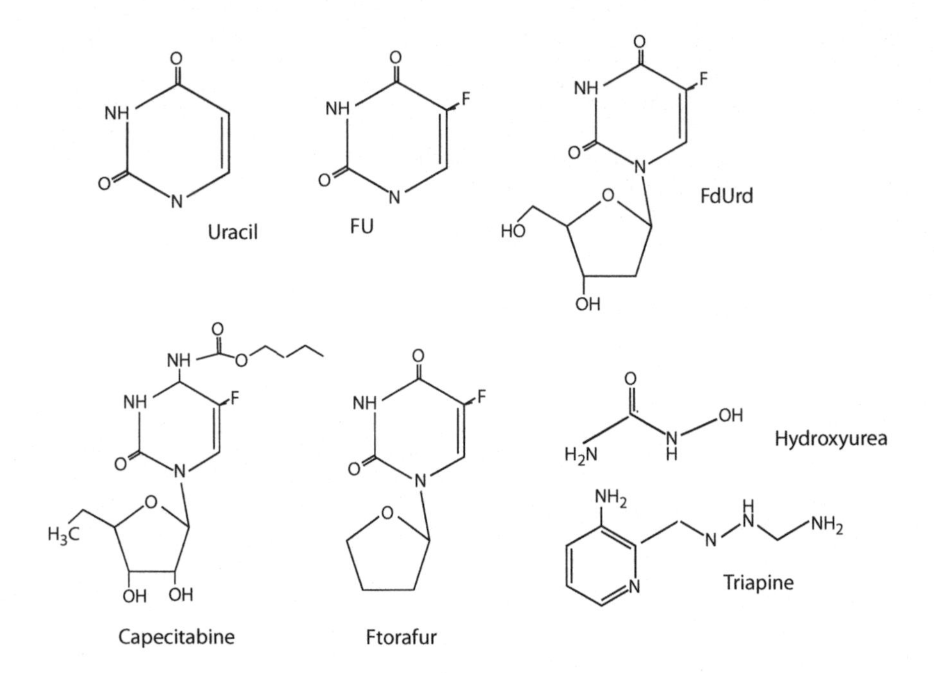

FIGURE 5.2 Structures of inhibitors of TS: FU, FdUrd, capecitabine, Ftorafur (UFT) and inhibitors of ribonucleotide reductase: hydroxyurea and triapine.

The metabolite FdUMP, a potent inhibitor of TS, is believed to be primarily responsible for the cytotoxic and radiosensitizing effects of FU. The rate-limiting enzyme in 5-FU catabolism is dihydropyrimidine dehydrogenase (DPD), which converts 5-FU to dihydrofluorouracil (DHFU) and, under normal circumstances, more than 80% of administered 5-FU will be catabolized, primarily in the liver, where DPD is highly expressed. Conversion of 5-FU to the active metabolite FdUMP requires a combination of phosphoribosylation (via orotic acid phosphoribosyltransferase) followed by reduction at the diphosphate level (by RR). FdUrd is phosphorylated by endogenous thymidine kinase (Figure 5.3).

5.3.1 Cytotoxicity Resulting from TS Inhibition

TS catalyzes the reductive methylation of deoxyuridine monophosphate (dUMP) to deoxythymidine monophosphate (dTMP), with reduced folate 5,10-methylenetetrahydrofolate (CH_2THF) as the methyl donor. This reaction provides the sole *de novo* source of thymidylate, which is necessary for DNA replication and repair. The 36-kDa TS protein is a dimer, both subunits of which contain a nucleotide-binding site and a binding site for

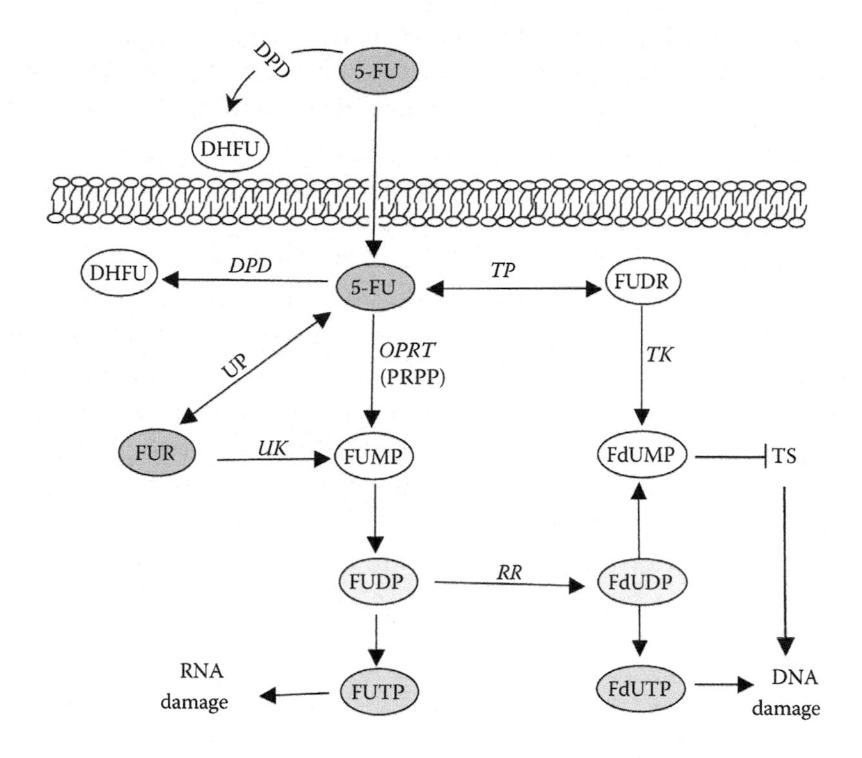

FIGURE 5.3 The main mechanism of 5-FU activation is conversion to fluorouridine monophosphate (FUMP). This may occur directly by the action of orotate phosphoribosyltransferase (OPRT) with phosphoribosyl pyrophosphate (PRPP) as the cofactor, or indirectly via fluorouridine (FUR) by the sequential action of uridine phosphorylase (UP) and uridine kinase (UK). FUMP is phosphorylated to fluorouridine diphosphate (FUDP), which can be either further phosphorylated to the active metabolite fluorouridine triphosphate (FUTP) or converted to fluorodeoxyuridine diphosphate (FdUDP) by ribonucleotide reductase (RR). FdUDP can then be either phosphorylated or dephosphorylated to produce the active metabolites FdUTP and FdUMP. An alternative activation pathway goes through the thymidine phosphorylase catalyzed conversion of 5-FU to fluorodeoxyuridine (FUDR), which is then phosphorylated by thymidine kinase (TK) to FdUMP. Dihydropyrimidine dehydrogenase (DPD)-mediated conversion of 5-FU to dihydrofluorouracil (DHFU) is the rate-limiting step of 5-FU catabolism in normal and tumor cells, and up to 80% of administered 5-FU is broken down by DPD in the liver.

CH_2THF. The 5-FU metabolite FdUMP binds to the nucleotide-binding site of TS, forming a stable ternary with the enzyme and CH_2THF, thereby blocking binding of the normal substrate dUMP and inhibiting dTMP synthesis.

Depletion of dTMP results in subsequent depletion of deoxythymidine triphosphate (dTTP), which causes perturbations in the levels of the other deoxynucleotides (dATP, dGTP, and dCTP) through various feedback mechanisms. The deoxynucleotide pool imbalances (in particular, the dATP/dTTP ratio) are thought to severely disrupt DNA synthesis and repair, resulting in lethal DNA damage. In addition, TS inhibition results in the accumulation of dUMP, which could lead to increased levels of deoxyuridine triphosphate (dUTP). Both dUTP and another 5-FU metabolite FdUTP can be misincorporated into DNA (Figure 5.4).

Several repair pathways have been implicated in the cytotoxicity of FU. Base excision repair (BER) can recognize uracil and fluorouracil bases

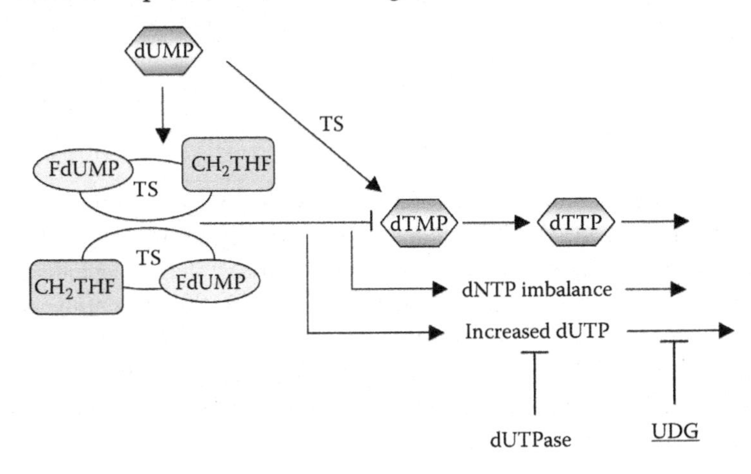

FIGURE 5.4 Thymidylate synthase (TS) inhibition by 5-fluorouracil. TS catalyzes the conversion of deoxyuridine monophosphate (dUMP) to deoxythymidine monophosphate (dTMP) with 5,10-methylene tetrahydrofolate (CH_2THF) as the methyl donor. Fluorodeoxyuridine monophosphate (FdUMP), a metabolite of 5-FU, binds to the nucleotide-binding site of TS and forms a stable ternary complex with TS and CH_2THF. This blocks access of dUMP to the nucleotide-binding site and inhibits dTMP synthesis. This causes imbalances in the deoxynucleotide (dNTP) pool i and increases levels of deoxyuridine triphosphate (dUTP), both of which cause DNA damage. The extent of DNA damage caused by dUTP is dependent on the levels of dUTPase and uracil-DNA glycosylase (UDG). dTMP can be salvaged from thymidine through the action of thymidine kinase (TK).

in DNA and excise them to decrease cytotoxicity. However, attempts to repair uracil and 5-FU-containing DNA mediated by the nucleotide excision repair enzyme uracil-DNA-glycosylase (UDG) are ineffective in the presence of high (F)dUTP/dTTP ratios and result in further false nucleotide incorporation. Futile cycles of misincorporation, excision, and repair eventually lead to DNA strand breaks and cell death. DNA damage due to dUTP misincorporation is highly dependent on the levels of the pyrophosphatase dUTPase, which limits the intracellular accumulation of dUTP.

A role for DNA mismatch repair (MMR) has been postulated based on findings that cells deficient in MMR, due to a lack of either MLH1 or MSH2, were significantly less sensitive to FU- or FdUrd-mediated cell killing. Compared with MMR-proficient cells, MMR-deficient cells treated with FdUrd contain more fluoropyrimidine in DNA and do not exhibit a G_2 arrest, leading to the conclusion that the lack of MMR prevented the excision of misincorporated FdUTP in DNA. A model has been proposed in which MMR and BER excise the fluorinated pyrimidine in DNA and, in so doing, generate DNA double-strand breaks leading to cell death [2].

Thymidylate can be salvaged from thymidine through the action of thymidine kinase, alleviating some of the effects of TS deficiency, and this salvage pathway also represents a potential mechanism of resistance to 5-FU.

5.3.2 Radiosensitization due to TS Inhibition

As already described, cells treated with FU or FdUrd exhibit a decrease in deoxythymidine triphosphate (dTTP; due to TS inhibition) and an increase in deoxyadenosine triphosphate (dATP) resulting in inhibition of DNA synthesis and accumulation of cells in early S phase. A number of experimental findings support the hypothesis that this cell cycle effect is crucial for radiosensitization and, in fact, that radiosensitization by 5-FU occurs in those cells that have made inappropriate progression through S phase in the presence of drug (i.e., from a disordered S phase checkpoint). This conclusion is supported by the results of many experimental studies including:

- Radiosensitization by fluorodeoxyuridine (FdUrd) was found to occur in HT29 human colon cancer cells, which express activated G_1/S cyclins in the presence of drug but not under the same drug treatment conditions in SW620 cells, which did not have activated cyclins [4].

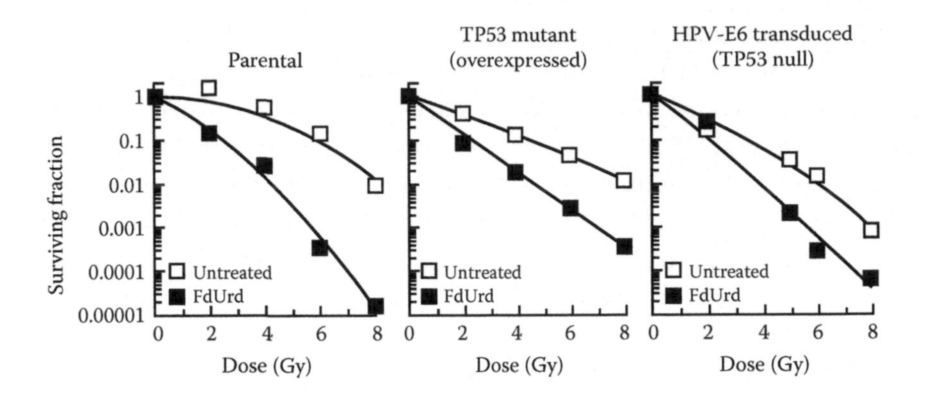

FIGURE 5.5 FdUrd-mediated radiosensitization in RKO cells. Cells were irradiated at the end of a 14-h exposure to 100 nM FdUrd and processed for determination of clonogenicity. The results of one of three similar experiments are shown. The data are shown as mean ± SEM. The error bars are contained within the size of the symbol. (From Lawrence, T. et al., *Radiat Res* 2000;154:140–144. With permission.)

- Blockade of S phase entry by producing G_1 arrest [5] or by inhibition of progression into S by treatment with the DNA α-polymerase aphidicolin can block radiosensitization.
- SW620 cells, which were shown to be resistant to FdUrd-mediated sensitization when arrested at the G_1/S boundary, were sensitized by FdUrd when they were transduced with the viral protein HPV E6 [6]. HPV E6 inactivates the retinoblastoma protein, releasing E2F and other S phase transcription factors and driving cells through S phase. The events producing sensitization seem to occur after the classic G_1 checkpoint in the cell cycle, which is consistent with the lack of dependence on p53 (Figure 5.5) [7].

5.3.3 FU as a Radiosensitizer: Clinical Application

Table 5.2 (based on information summarized by Seiwert et al. [8]) lists the cancers for which treatment with FU and concurrent radiation is most frequently used. In particular, the combination of FU and radiation is a mainstay of treatment for gastrointestinal tumors, where it has a proven role in improving locoregional control and survival [1,3,9].

TABLE 5.2 Summary of Diseases and Disease Entities for Which 5-FU Is Used with Concurrent Radiation

Disease	Indication and Treatment	Frequently Used Agents	Benefit
Head and neck cancer	Locally advanced HNC, primary or adjuvant treatment	Cisplatin, 5-FU FHX, cetuximab	Improved organ preservation and survival compared with radiation alone
Esophageal cancer	Locally advanced disease	Cisplatin/5-FU	Survival benefits, increased cure rate
Anal cancer	Mainstay of curative treatment	5-FU	Improved sphincter preservation, decrease in local and distal failures
Gastric cancer	Adjuvant?	Cisplatin, 5-FU	Some data indicate survival benefit
Pancreatic cancer	Adjuvant, unresectable locoregionally advanced tumors	5-FU	Improved locoregional control, possibly a survival benefit
Cholangio-carcinoma	Adjuvant, unresectable locoregionally advanced tumors	5-FU	Some data indicate a survival benefit
Cervical cancer	Primary modality	Cisplatin, 5-FU, hydoxyurea	Improved local and distal control, organ preservation

Source: Based on information reviewed by Seiwert, T. et al., *Nat Clin Pract Oncol* 2007;4(2):86–100.

5-FU was introduced almost 60 years ago and, for some types of disease, it is an established treatment modality; nevertheless, research is continuing into the biochemistry and optimal clinical application of this deceptively simple molecule. This is apparent from the fact that 164 ongoing or recently completed clinical trials are listed in which 5-FU is involved in combination with other drugs and concurrent radiotherapy (clinicaltrials.gov).

The timing of drug administration with respect to radiation exposure is critical for the optimization of radiosensitization. When used as a radiosensitizer, 5-FU is typically given by continuous venous infusion rather than as bolus therapy because of its short plasma half-life. Protracted infusion may be difficult to administer and this problem has been addressed by the development of oral forms of FU including Ftorafur (UFT) and capecitabine.

- Capecitabine (Xeloda) is an orally administered fluoropyrimidine that is preferentially converted to 5-FU in tumor tissue through a three-step enzymatic pathway involving two intermediary metabolites, 5′-deoxy-5-fluorocytidine (5′-DFCR) and 5′-deoxy-5-fluorouridine (5′-DFUR), to form 5-fluorouracil [10]. Capecitabine is now widely used as an alternative to 5-FU for the treatment of gastrointestinal cancers. Randomized trials have shown that capecitabine gives at least equivalent outcomes as 5-FU and leucovorin for the treatment of metastatic colorectal cancer, as well as for adjuvant treatment of colon cancer [11].

As was the case with 5-FU, optimization of the clinical application of capecitabine is being vigorously investigated. There are 69 ongoing or recently concluded clinical trials involving capecitabine and concurrent radiation (clinicaltrials.gov).

5.4 INHIBITORS OF RIBONUCLEOTIDE REDUCTASE

Ribonucleotide reductase (RR) is the enzyme responsible for the conversion of ribonucleotides to deoxyribonucleotides, the prerequisites of DNA synthesis and DNA repair. These molecules are produced by an enzymatically difficult radical-induced reduction of ribonucleotides, a multistep chemical process catalyzed by RR.

Ribonucleotide Reductase consists of a constitutively expressed subunit containing regulatory and substrate-binding sites (R1), and a non-heme iron-containing subunit (R2) expressed only during S phase that is responsible for catalysis [12]. The drugs that target ribonucleotide reductase inhibit the biosynthesis of deoxyribonucleotides but by a different mechanism from that of the TS inhibitors, such as FU.

5.4.1 Mechanism of Cytotoxicity

Inhibition of RR by the free radical scavenger, hydroxyurea (HU; Figure 5.2) targets the tyrosyl free radical that is necessary to initiate the reduction of ribonucleotides bound to the catalytic site. Inhibition of RR results in a decrease in deoxynucleoside triphosphate in cells with subsequent inhibition of DNA synthesis.

Other inhibitors of RR include the thiosemicarbazone triapine and the nucleoside analogue 2-fluoromethylene-2-deoxycytidine, both of which have been shown to sensitize cells to radiation in preclinical studies [13]. Nucleoside analogue radiosensitizers such as bromo-deoxyuridine can also inhibit RR and this presumably contributes to their radiosensitizing capability.

5.4.2 Radiochemotherapy with RR Inhibitors

Hydroxyurea was one of the first recognized radiosensitizers, and had some initial clinical success, which encouraged the evaluation of other RR inhibitors as radiosensitizers. Early studies had demonstrated that, at cytotoxic concentrations, HU could inhibit repair of DNA strand breaks generated by ionizing radiation [14], enhancing radiosensitization by nucleoside analogues presumably by increasing their incorporation into DNA. Clinically, radiosensitization with HU has been most successful in cancer of the cervix or head and neck, where it has been shown to improve control of local disease [15]. More recent studies of concomitant radiotherapy and HU have included other radiosensitizers such as FU and paclitaxel [16]. This approach, although it improved local control, was associated with increased toxicity, particularly leukopenia and mucositis in head and neck cancer. There are currently 16 trials ongoing or recently completed involving HU, plus other drugs with concurrent radiation (clinicaltrials.gov).

5.5 DNA POLYMERASE INHIBITORS/SUBSTRATES

DNA polymerase inhibitors include the nucleotide/nucleoside analogues, which are incorporated into DNA. There are a number of compounds in this group, but not all of them are effective as radiosensitizers. The action of fludarabine and gemcitabine will be described in this chapter (structures are shown in Figure 5.6).

5.5.1 Fludarabine

Fludarabine phosphate (fludarabine) is a single phosphorylated and fluoridated adenine nucleoside derivate (9-β-D-arabinofuranosyl-2-fluoroadenine-50-monophosphate), which is established in the treatment of chronic lymphatic leukemia.

5.5.1.1 Cytotoxicity

The pro-drug fludarabine phosphate is rapidly dephosphorylated *in vivo* to 2-F-ara-A, which is then actively transported into the cell where it undergoes three successive phosphorylation steps to become the active form of the drug—fludarabinetriphosphate (2-F-ara-ATP). 2-F-ara-ATP inhibits enzymes that are involved in DNA synthesis and DNA repair, including DNA polymerase α and ε, DNA primase and ligase, and ribonucleotide

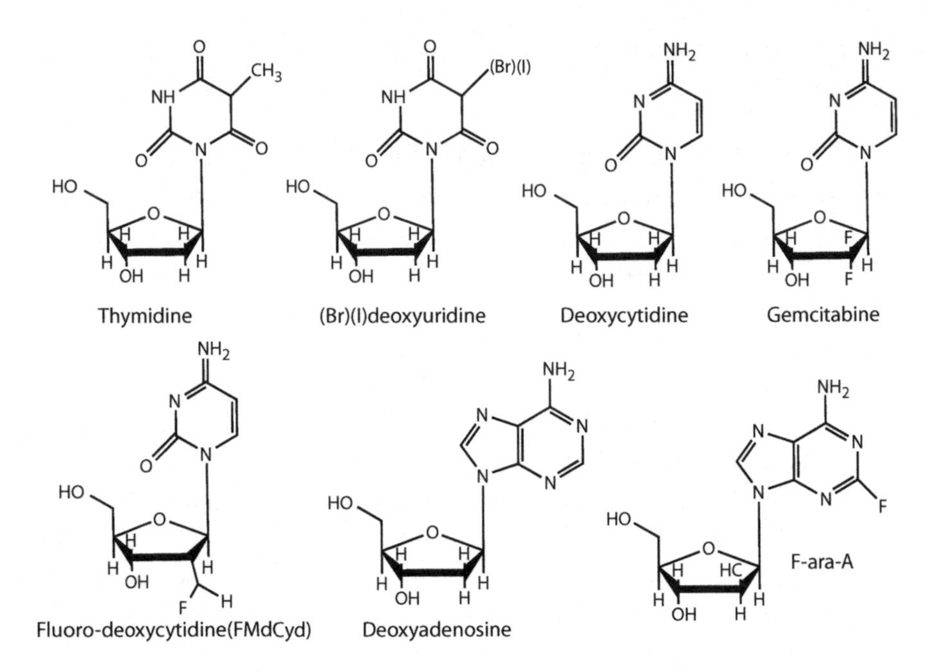

FIGURE 5.6 Structures of nucleoside analogues. Thymidine analogues: thymidine, (Br)(I)deoxyuridine. Deoxycytidine analogues: deoxycytidine, gemcitabine, fluoromethylene deoxycytidine (FMdCyd). Deoxyadenosine analogues: deoxyadenosine, F-araA, fludarabine.

reductase. If the analogue is incorporated into the DNA, it induces a termination of chain elongation.

5.5.1.2 Radiosensitization

Like many other drugs that inhibit DNA repair, fludarabine enhances the cytotoxic effects of radiation [17]. Preclinical studies with fludarabine in different cell types of cancer showed continuous exposure at high concentrations to have notable cytotoxic activity against acute leukemia and non-Hodgkin's lymphomas but no effect was seen against the majority of cell lines from solid tumors. Later, fludarabine was reported to increase radiation-induced clonogenic cell death in several mouse sarcoma cell lines *in vitro* to an extent greater than would be expected from simple additivity [18]. The potential of fludarabine as a radiosensitizer was also demonstrated by the results of animal experiments in which it was shown that fludarabine *in vivo* acts as a radiosensitizer in several mouse tumor models after single and fractionated irradiation [19,20].

5.5.1.3 Clinical Application

The antitumor activity of fludarabine against major tumor types has been investigated in clinical phase II trials. The results were generally disappointing and only in head and neck cancer and breast cancer did a small proportion of patients have objective remissions [21,22]. Although the effectiveness of fludarabine in combination with radiation on solid tumor cells still has to be proven, phase I studies demonstrated that it could be safely administered concurrently with radiation [21].

Despite some positive preclinical findings, fludarabine has shown little promise clinically in the treatment of solid tumors as part of a chemotherapy or radiochemotherapy regimen. The proven clinical role for fludarabine is in the treatment of chronic lymphocytic leukemia, indolent non-Hodgkin lymphomas, and acute myeloid leukemia where it is used in various combinations with other chemotherapy.

5.5.2 Gemcitabine

Difluoro-2-deoxycytidine (gemcitabine) is a deoxycytidine analogue with structural similarities to cytarabine. The antitumor activity of gemcitabine depends on transport into the cell followed by a series of phosphorylations. Nucleoside analogues require three successive phosphorylations to become 5-triphosphate substrates for DNA incorporation. The rate-limiting step in this process is the initial phosphorylation of the nucleoside 2′2′ difluoro-2′-deoxycytidine by deoxycytidine kinase, which converts gemcitabine to the monophosphorylated metabolite, dFdCMP. Subsequent phosphorylations lead to the accumulation of gemcitabine diphosphate and triphosphate (dFdCDP and dFdCTP), both of which are active metabolites.

5.5.2.1 Cytotoxicity

The products of phosphorylation of gemcitabine are cytotoxic for two reasons. First, fraudulent nucleotides can inhibit DNA polymerases directly interfering with DNA synthesis and this is the case with dFdCTP, which competes with endogenous deoxycytidine triphosphate (dCTP) for incorporation into DNA. DNA misincorporation is the lesion believed to be primarily responsible for cytotoxicity of gemcitabine.

Second, in addition to their ability to be incorporated into DNA, all these analogues can inhibit deoxynucleotide production required for DNA synthesis. The 5′-diphosphate of gemcitabine is an inhibitor of RR that causes a decrease primarily in dATP and deoxyguanosine triphosphate (dGTP) in solid tumor cells, whereas similar treatment of leukemia cells causes a decrease in levels of dCTP and dTTP (reviewed by Morgan et al. [23]).

5.5.2.2 Radiosensitization

The potent radiosensitizing effect of gemcitabine has been demonstrated *in vitro* and *in vivo*. Radiosensitization has correlated with the inhibition of RR, primarily resulting in the depletion of dATP at low concentrations of gemcitabine, whereas incorporation of gemcitabine into DNA does not seem to affect radiosensitization. The contribution of the different inputs of gemcitabine to overall radiosensitization were analyzed in experiments that showed: first, cells transduced with the active subunit of ribonucleotide reductase become relatively resistant to gemcitabine-mediated radiosensitization [24]; and second, intracellular concentrations of dFdCTP do not correlate with radiosensitization [25], suggesting that dATP pool depletion and not incorporation of dFdCMP into DNA underlies radiosensitization. Support for this interpretation came from the results of a clinical trial with gemcitabine and concurrent irradiation in patients with head and neck cancer, which showed that the levels of phosphorylated gemcitabine measured in tumor biopsy specimens were sufficient to inhibit RR.

In fact, gemcitabine-induced dATP pool depletion is necessary, but not sufficient for radiosensitization—the ability of gemcitabine to cause the redistribution of cells into S phase is also required [26]. Although high concentrations of gemcitabine cause near-complete dATP pool depletion within just a few hours, cells irradiated at this time are minimally radiosensitized. Maximum sensitization requires both dATP pool depletion and sufficient time to permit the redistribution of cells into early S phase [27]. Sensitization is maximized *in vivo* by a continuous fixed dose-rate exposure to gemcitabine, compared with a bolus administration [28], presumably due to the continuous exposure facilitating the production of more intracellular metabolites.

The S phase specificity for gemcitabine radiosensitization is consistent with the dependence on homologous recombination repair, which occurs during S and G_2, but is not consistent with nonhomologous end-joining, which occurs primarily in G_1. Although a deficiency in MMR is not required, radiosensitization with gemcitabine is enhanced in the absence of this repair process [29]. It has been suggested that the depletion of dATP results in mismatches in DNA that, if not repaired before irradiation, will augment cell killing by ionizing radiation. Most tumor cells that are radiosensitized by gemcitabine are capable of carrying out MMR to some degree, but it has been shown that at high enough concentrations of gemcitabine, mutations are produced in DNA that persist after irradiation [30].

5.5.2.3 Clinical Response to Radiochemotherapy with Gemcitabine

The combination of gemcitabine with concurrent irradiation has demonstrated promising activity in limited clinical trials, although toxicity was significant in patients with head and neck cancer or non-small cell lung cancer (NSCLC). Results of early phase I and II clinical trials were reviewed by Pauwels et al. [31]. These trials were predominantly concerned with pancreatic cancer (10 out of 18 trials listed) as well as non-small cell lung cancer (3), head and neck cancer (2), gastrointestinal cancer (1), glioblastoma (1), and cervical cancer (1). A slightly later review of clinical trials involving gemcitabine and radiotherapy again reported studies largely with pancreatic cancer but also head and neck cancer, breast cancer, bladder cancer, and lung cancer (reviewed by Spalding and Lawrence [9]).

Early clinical trials with pancreatic cancer investigated low-dose gemcitabine concurrent with standard radiation protocols (50.4 Gy in 1.8-Gy fractions) in patients with locally advanced pancreatic cancer and determined the maximum tolerated dose of gemcitabine to be 40 mg/m^2 given twice a week. In later trials, patients treated with much higher weekly doses of gemcitabine and with 30 to 33 Gy in 3-Gy fractions [32] experienced unacceptable toxicities. It was suggested that the relatively large standard radiation fields including clinically uninvolved regional lymph nodes increased the toxicity of the combination therapy.

Results from another trial supported this conclusion. In this case, a standard dose of gemcitabine (1,000 mg/m^2) designed to maximize systemic control, with dose-escalated three-dimensional conformal radiotherapy to the gross disease only, and with the exclusion of clinically uninvolved regional lymph nodes [33], was found to be tolerable and to produce a favorable objective response rate (10 of 33 patients). In subsequent preclinical and clinical trials, cisplatin or oxaliplatin were added to gemcitabine. Neither of these significantly prolonged survival compared with gemcitabine alone in the treatment of metastatic disease [34], suggesting that these combinations could only modestly improve the treatment of locally advanced, nonmetastatic disease. Similarly likewise, adding capecitabine to gemcitabine marginally improved the survival of patients with metastatic disease in one study but not in another [35,36].

The conclusion drawn from the early trials was that the combination of gemcitabine and radiation was associated with a high level of acute toxicity; the severity of which was strongly related to the dose and schedule of gemcitabine administration as well as to the radiation field size. Despite the mixed results, there is continued belief that radiochemotherapy with gemcitabine could be a promising approach provided that strategies could be devised to reduce the incidence of severe acute side effects particularly by developing techniques of radiation delivery that would decrease normal tissue toxicity and allow radiation dose escalation.

Clinical trials carried out within the last 2 to 3 years have gone some way to achieving this aim, with most of the interest in this area continuing to focus on pancreatic cancer.

- An exception was a phase II trial to evaluate the activity of gemcitabine as radiosensitizer for newly diagnosed glioblastoma multiforme. Patients were treated using standard cranial irradiation plus concomitant gemcitabine given intravenously for 6 weeks. After chemoradiotherapy, irrespective of tumor response, patients went on to receive oral temozolomide. Four patients responded to treatment (17.5%), with an additional 14 patients (61%) experiencing stable disease for an overall disease control rate of 78.5%. Median progression-free and overall survival were 6.8 and 10.1 months, respectively. The treatment was well tolerated and severe adverse events were rare. It was concluded that concomitant radiotherapy–gemcitabine is active and well tolerated in newly diagnosed glioblastoma multiforme [36].
- Phase I dose escalation study using two cycles of full-dose gemcitabine, escalating-doses of oxaliplatin, and 27 Gy of concurrent radiation therapy in pancreatic adenocarcinoma. Patients with unresectable or low-volume metastatic disease proceeded with additional, planned therapy to a total of four cycles of chemotherapy and 54 Gy of concurrent radiation therapy with concurrent full-dose gemcitabine and oxaliplatin. This regime was well tolerated and resulted in favorable rates of local tumor response and 1-year freedom from local progression. Low rates of toxicity and high rates of tumor response and 1-year freedom from local progression were shown [37].
- A phase I/II trial in 50 patients with unresectable pancreatic cancer to determine the maximum tolerated dose delivered by

intensity modulated radiation (IMRT) with fixed dose-rate gemcitabine (FFLP), and radiation dose of 55 Gy in 25 fractions was well tolerated. The 2-year freedom from local progression (FFLP) and overall survival (OS) were 59% and 30%, respectively. Twelve patients who underwent resection survived a median of 32 months. This trial established that high-dose radiation therapy could be administered safely with concurrent full-dose, fixed dose-rate gemcitabine (FDR-G), with IMRT delivered during breath hold. For an equivalent toxicity, it was possible to escalate the biologically effective dose by approximately 60% compared with a previously used protocol.

- Patients with nonmetastatic unresectable adenocarcinoma of the pancreas were randomly assigned to receive gemcitabine alone (1,000 mg/m^2 per week for 6 weeks, followed by 1 week rest, then five more cycles of 1,000 mg/m^2 for 3 out of 4 weeks) or gemcitabine (600 mg/m^2 per week) concurrently with three-dimensional conformal radiotherapy (50.4 Gy in 28 fractions) followed by additional gemcitabine (five cycles of 1,000 mg/m^2 for 3 out of 4 weeks). The study was closed early owing to poor accrual but, in the 74 patients enrolled, median survival improved from 9.2 months to 11.1 months ($P = 0.017$). This came at a cost of increased frequency of grade 4 toxic effects (although combined grade 3 or 4 toxic effects were the same in each arm) [38].

Discussing the results of the last trial Ben-Josef and Lawrence [39] observed that because a significant proportion of patients with pancreatic cancer die of complications of uncontrolled growth, the demonstration that improved local control and survival in patients could be achieved with high-dose radiotherapy gemcitabine supported the use of this protocol as a standard of care for these patients.

5.5.2.4 Combining Gemcitabine with Targeted Molecular Radiosensitizers

Efforts to improve therapy for pancreatic cancer by using multimodality therapies including gemcitabine and radiation with cisplatin or oxaliplatin have been investigated, but the use of multiple cytotoxic agents produces additional toxicity. Another approach to pancreatic cancer therapy would integrate molecularly targeted therapies with standard chemoradiation regimens. This approach was reviewed by Morgan et al. [23], who identified epidermal growth factor receptor (EGFR) and checkpoint kinase 1 (CHK1) as promising molecular targets. Clinically proven antibody and small molecule inhibitors of EGFR are available and have been evaluated in combination with gemcitabine and radiation.

5.5.2.4.1 Targeting EGFR EGFR is a transmembrane receptor tyrosine kinase that is activated in response to binding of ligands such as EGF, transforming growth factor-α, or amphiregulin. Ligand binding results in receptor dimerization and activation of a number of downstream pathways (STAT, AKT, extracellular signal-regulated kinase, and protein kinase C [PKC]), which promotes survival, angiogenesis, cell cycle progression, and transformation. EGFR is also phosphorylated in response to a variety of other cytotoxic agents (see Chapter 10), and it is hypothesized that this phosphorylation may promote survival through stimulation of stress/survival response pathways. Gemcitabine has been shown to stimulate the phosphorylation of EGFR in head and neck and pancreatic cancer cells [23], providing a rationale for the targeting of EGFR in conjunction with gemcitabine radiotherapy/chemotherapy.

EGFR inhibitors including the small molecule tyrosine kinase inhibitors, erlotinib and gefititinib, and the anti-EGFR antibody, cetuximab have been used in conjunction with gemcitabine and concurrent radiotherapy. Studies in head and neck cancer xenografts showed that gefitinib, which blocked gemcitabine-mediated EGFR phosphorylation, enhanced gemcitabine-mediated tumor growth delay [40]. In other studies, both cetuximab and erlotinib were found to enhance pancreas tumor growth delay when combined with gemcitabine and radiation [41].

As already described, the cytotoxic and radiosensitizing actions of gemcitabine involve a number of processes, and interaction with molecular targeted sensitizers is similarly complex affecting cell cycle, DNA repair, and cell survival functions. First, sensitization to gemcitabine by EGFR inhibitors is sequence-dependent. This schedule-dependent cell killing may be attributable to the cell cycle effects of EGFR inhibitors which upregulate the cyclin-dependent kinase inhibitors, p27 [40,42] and p21 [43], and bring about G_1 cell cycle arrest. Second, EGFR plays a role in the repair of DNA damage produced by ionizing radiation and chemotherapy. In response to radiation, EGFR translocates to the nucleus, where it is associated with increased DNA-dependent protein kinase activity. Inhibition of EGFR activation by cetuximab blocks nuclear EGFR import, DNA-dependent protein kinase activity, radiation-induced DNA damage repair, and induces radiosensitization. Finally, activation of EGFR in response to gemcitabine can also result in activation of antiapoptotic survival signaling through AKT [40].

5.5.2.4.2 Clinical Application of Gemcitabine and Targeted Radiosensitizers
Several clinical trials of pancreatic cancer with gemcitabine and erlotinib or cetuximab have had mixed results ranging from no benefit to statistically significant but clinically modest improvement in survival (reviewed by Chakravarthy et al. [44]). Seven phase I/phase II clinical trials with concurrent radiation, gemcitabine, and EGFR inhibitor (cetuximab, erlotinib) are listed as ongoing or recently concluded (clinicaltrials.gov).

5.6 NEW GENERATION ANTIMETABOLITES

Antimetabolite radiosensitizers developed recently include pemetrexed, a multitargeted folate analogue, and ganciclovir, an analogue of deoxyguanosine (Figure 5.7).

5.6.1 Pemetrexed

The antifolate pemetrexed is an antimetabolite with antitumor activity against a broad range of human malignancies (reviewed by Wouters et al. [45]). Pemetrexed acts as a multitargeted antifolate by inhibiting multiple

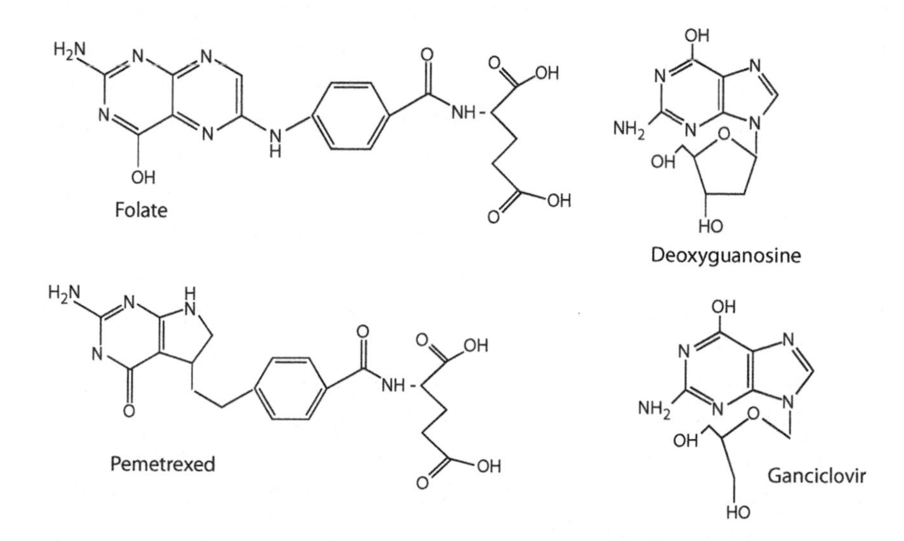

FIGURE 5.7 Structures of more recently developed radiosensitizers. Pemetrexed, a multitargeted folate analogue, and ganciclovir, an analogue of deoxyguanosine.

key enzymes involved in both pyrimidine and purine synthesis, its primary targets being TS, dihydrofolate reductase, and glycinamide ribonucleotide formaldehyde transferase (GARFT). Pemetrexed enters the cell mainly by a reduced folate carrier system. Once inside the cell, pemetrexed is an excellent substrate for the enzyme folylpolyglutamate synthase (FPGS), and is rapidly converted to its active polyglutamate derivatives that have a substantially higher potency for inhibition of TS and to GARFT. It is believed that polyglutamation of pemetrexed plays an important role in determining both the selectivity and the antitumor activity of this agent. The ability of pemetrexed to deplete cellular nucleotide pools, to modulate the cell cycle, and to induce apoptosis suggests the drug as a cytotoxic agent that should be combined with radiotherapy. In preclinical studies, radiosensitization by pemetrexed was observed in human colon, breast, cervix, and lung carcinoma cells [46]. *In vivo*, the combination of pemetrexed with fractionated radiotherapy produced additive to greater than additive antitumor activity in murine and human tumor xenografts [47].

5.6.1.1 Clinical Applications of Pemetrexed

Pemetrexed was approved by the Food and Drug Administration (FDA) for first-line treatment of inoperable malignant mesothelioma in combination with cisplatin and was also FDA-approved as second-line therapy in patients with previously chemotherapy-treated advanced NSCLC, as first-line therapy, in combination with cisplatin, for patients with chemotherapy-naive NSCLC, and most recently, for maintenance treatment of patients with locally advanced or metastatic nonsquamous NSCLC whose disease had not progressed after four cycles of platinum-based first-line chemotherapy [45]. The results of several trials of pemetrexed with concurrent radiotherapy have been reported, while others are ongoing: a phase I trial evaluated chest radiotherapy with pemetrexed or pemetrexed and carboplatin in patients with thoracic malignancies. Results indicated that that pemetrexed could be administered at systemically active doses in combination with radiotherapy [48].

In another phase I trial, the combination of pemetrexed and cetuximab, with radiotherapy in poor prognosis head and neck cancer was studied. It was concluded that the addition of pemetrexed (500 mg/m^2) to cetuximab and radiotherapy is to be recommended for further study in patients not previously irradiated [49].

Twenty clinical trials (phases I–III) involving pemetrexed and concurrent radiation are listed as ongoing or recently completed (clinicaltrials.gov).

Non-small cell lung cancer is the disease most frequently studied (14 of 20 trials); also investigated are small cell lung cancer (2/20), brain metastases from NSCLC (1/20), head and neck cancer (2/20), and pancreatic cancer (1/20). In four trials, pemetrexed and radiation were investigated whereas in others additional chemotherapy was given including cisplatin (4), carboplatin (3), cetuximab (2), and erlotinib (1). In other cases, additional chemotherapy involved two or more agents, carboplatin and afatinib or paclitaxel, cisplatin with doxecetal or etoposide and soy isoflavones, and in one trial, pemetrexed, etoposide, and vinelrelbine.

5.6.2 Ganciclovir

Suicide gene therapy involves the transfer and expression of nonmammalian genes encoding enzymes that convert nontoxic pro-drugs into toxic antimetabolites. The antiviral ganciclovir (Figure 5.7) is used in one form of suicide gene therapy in which the activating enzyme, herpes simplex virus thymidine kinase, is administered to tumor cells to sensitize them to the cytotoxic effects of ganciclovir. *In vitro* and *in vivo* data have demonstrated the radiosensitization of cells that can activate ganciclovir to its 5′-triphosphate, which is incorporated into DNA [50].

This approach has been combined with another suicide gene therapy stratagem, cytosine deaminase, to activate the antifungal 5-flucytosine to the radiosensitizer FU, and the combined effects of ganciclovir and 5-flucytosine or FU plus radiotherapy have been evaluated [51]. There was increased radiosensitization both *in vitro* and *in vivo* with ganciclovir and FU administered before irradiation, and strong preclinical data encouraged clinical trials. Although single suicide gene therapy with ganciclovir and radiation did not affect outcome in patients with glioblastoma [52], results from a phase II trial were suggestive of antitumor activity using two suicide genes to activate ganciclovir and 5-flucytosine in patients with prostate cancer [53].

5.7 SUMMARY

5-Fluoruracil (5-FU) is used widely in the treatment of solid tumors and belongs to a class of anticancer agents termed antimetabolites. It is an analogue of uracil with a fluorine atom at the C5 position in place of hydrogen. The mechanism of cytotoxicity of 5-FU has been ascribed to the misincorporation of fluoronucleotides into RNA and DNA, and to the inhibition of the nucleotide synthetic enzyme TS. 5-FU rapidly enters the cell using the same facilitated transport mechanism as uracil, and is converted intracellularly to several active metabolites that disrupt RNA synthesis and the action of TS. TS represents a critical enzyme for the conversion of dUMP to dTMP, a precursor of thymidylate, which is necessary for DNA replication and repair. The 5-FU metabolite fluorodeoxyuridine monophosphate competitively inhibits dUMP from binding to TS, thereby abrogating dTMP synthesis. Another metabolite of 5-FU, fluoridine triphosphate (FUTP), is extensively incorporated into RNA, disrupting normal RNA processing and function at several levels. In addition to the effect on cellular RNA, 5-FU may also be misincorporated into cellular DNA. This is accomplished through another one of its active metabolites, fluorodeoxyuridine triphosphate (FdUTP). Based on this mechanism, 5-FU has particular activity in cells in S phase through incorporation into DNA, leading to both single and double-strand DNA breaks. There are a number of mechanisms by which 5-FU could increase radiation sensitivity at the cellular level. The toxicity of 5-FU seems to be S phase-specific and increased radiation sensitivity occurs in cells that have inappropriate progression though S phase in the presence of drug, suggesting a disordered S phase checkpoint response.

Preclinical studies indicated that protracted infusion of 5-FU during a course of fractionated radiotherapy would be required to achieve optimal radiosensitization, a finding resulting from the mechanism of 5-FU inhibition of TS and the short half-life of 5-FU and its metabolites. Protracted venous infusion of 5-FU is a standard therapy in several tumor types and the introduction of oral form of 5-FU, the prodrug capecitabine, has facilitated protracted treatment. The cytidine analogue gemcitabine is a potent radiosensitizer, both in the laboratory and in the clinic, with activity against a variety of cancers. Gemcitabine treatment results in the depletion of dATP pools and arrests cells at early S phase after 24 h of treatment. Radiosensitization by gemcitabine also depends on intact homologous recombination and mismatch repair processes. Because the effects of gemcitabine (i.e., dATP depletion and cell cycle arrest) do not materialize immediately, no synergy is seen when irradiation is administered prior to the full 24-h period of drug exposure. On the basis of results from early trials, it seems that the combination of gemcitabine and radiation was associated with a high level of acute toxicity; the severity of which was strongly related to the dose and schedule of gemcitabine administration, as well as to the radiation field size. More recent trials have gone some way to reversing the initial negative impression with most of the interest in this area focused on pancreatic cancer.

REFERENCES

1. Rich, T., Shepard, R., and Mosley, S. Four decades of continuing innovation with fluorouracil: Current and future approaches to fluorouracil chemoradiation therapy. *J Clin Oncol* 2004;22:2214–2232.
2. Longley, D., Harkin, D., and Johnston, P. 5-fluorouracil: Mechanisms of action and clinical strategies. *Nat Rev Cancer* 2003;3:330–338.
3. Shewach, D., and Lawrence, T. Antimetabolite radiosensitizers. *J Clin Oncol* 2007;25:4043–4050.
4. Malet-Martino, M., and Martino, R. Clinical studies of three oral prodrugs of 5-fluorouracil (capecitabine, UFT, S-1): A review. *Oncologist* 2002;7:288–323.
5. Lamont, E., and Schilsky, R. The oral fluoropyrimidines in cancer chemotherapy. *Clin Cancer Res* 1999;5:2289–2296.
6. Miwa, M., Ura, M., Nishida, M. et al. Design of a novel oral fluoropyrimidine carbamate, capecitabine, which generates 5-fluorouracil selectively in tumours by enzymes concentrated in human liver and cancer tissue. *Eur J Cancer* 1998;34:(8)1274–1281.
7. Lawrence, T., Burke, R., Davis, M., and Wygoda, M. Lack of effect of TP53 status on fluorodeoxyuridine-mediated radiosensitization. *Radiat Res* 2000; 154:140–144.
8. Seiwert, T., Salama, J., and Vokes, E. The concurrent chemoradiation paradigm—General principles. *Nat Clin Pract Oncol* 2007;4(2):86–100.
9. Spalding, A., and Lawrence, T. New and emerging radiosensitizers and radioprotectors. *Cancer Invest* 2006;24:444–456.
10. Miwa, M., Ura, M., Nishida, M. et al. Design of a novel oral fluoropyrimidine carbamate, capecitabine, which generates 5-fluorouracil selectively in tumours by enzymes concentrated in human liver and cancer tissue. *Eur J Cancer* 1998;34(8):1274–1281.
11. Das, P., Wolff, R., Abbruzzese, J. et al. Concurrent capecitabine and upper abdominal radiation therapy is well tolerated. *Radiat Oncol* 2006;1:41–44.
12. Nordlund, P., and Reichard, P. Ribonucleotide reductases. *Annu Rev Biochem* 2006;75:681–706.
13. Ostruszka, L., and Shewach, D. The role of DNA synthesis inhibition in the cytotoxicity of 2,2-difluoro-2-deoxycytidine. *Cancer Chemother Pharmacol* 2003;52:325–332.
14. Sinclair, W. The combined effect of hydroxyurea and x-rays on Chinese hamster cells *in vitro*. *Cancer Res* 1968;28:198–206.
15. Beitler, J., Anderson, P., Haynes, H. et al. Phase I clinical trial of parenteral hydroxyurea in combination with pelvic and para-aortic external radiation and brachytherapy for patients with advanced squamous cell cancer of the uterine cervix. *Int J Radiat Oncol Biol Phys* 2002;52:637–642.
16. Rosen, F., Haraf, D.J., Kies, M. et al. Multicenter randomized phase II study of paclitaxel (1-hour infusion), fluorouracil, hydroxyurea, and concomitant twice daily radiation with or without erythropoietin for advanced head and neck cancer. *Clin Cancer Res* 2003;9:1689–1697.

17. Dahlberg, W., and Little, J. Differential sensitization of human tumor cells by Ara-A to X-irradiation and its relationship to inherent radioresponse. *Radiat Res* 1992;130:303–308.
18. Grégoire, V., Ruifrok, A., Price, R. et al. Effect of intra-peritoneal fludarabine on rat spinal cord tolerance to fractionated irradiation. *Radiother Oncol* 1995;36:50–55.
19. Milas, L., Fuji, T., Hunter, N. et al. Enhancement of tumor radioresponse *in vivo* by gemcitabine. *Cancer Res* 1999;59:107–114.
20. Grégoire, V., Hittelmann, W., Rosier, J., and Milas, L. Chemo-radiotherapy: Radiosensitizing nucleoside analogues (Review). *Oncol Rep* 1999;6:949–957.
21. Grégoire, V., Ang, K., Rosier, J. et al. A phase I study of fludarabine combined with radiotherapy in patients with intermediate to locally advanced head and neck squamos cell carcinoma. *Radiother Oncol* 2002;63(2):187–193.
22. Mittelmann, A., Ashikari, R., Ahmed, T., Savona, S., Arnold, P., and Arlin, Z. Phase II trial of fludarabine phosphate (f-ara-AMP) in patients with advanced breast cancer. *Cancer Chemother Pharmacol* 1988;22:63–64.
23. Morgan, M., Parsels, L., Maybaum, J., and Lawrence, T. Improving gemcitabine-mediated radiosensitization using molecularly targeted therapy: A review. *Clin Cancer Res* 2008;14(21):6744–6750.
24. Lawrence, T., Blackstock, A., and McGinn, C. The mechanism of action of radiosensitization of conventional chemotherapeutic agents. *Semin Radiat Oncol* 2003;13(1):13–21.
25. Robinson, B., Im, M., Ljungman, M., Praz, F., and Shewach, D. Enhanced radiosensitization with gemcitabine in mismatch repair-deficient HCT116 cells. *Cancer Res* 2003;63:6935–6941.
26. Pauwels, B., Korst, A., Pattyn, G. et al. Cell cycle effect of gemcitabine and its role in the radiosensitizing mechanism *in vitro*. *Int J Radiat Oncol Biol Phys* 2003;57:1075–1083.
27. Ostruszka, L., and Shewach, D. The role of cell cycleprogression in radiosensitization by 2′,2′-difluoro-2′-deoxycytidine. *Cancer Res* 2000;60:6080–6088.
28. Morgan, M., Shaikh, M.E., Abu-Isa, E., Davis, M., and Lawrence, T. Radiosensitization by gemcitabine fixed-dose-rate infusion versus bolus injection in a pancreatic cancer model. *Transl Oncol* 2008;1:44–49.
29. van Bree, C., Rodermond, H., deVos, J. et al. Mismatch repair proficiency is not required for radioenhancement by gemcitabine. *Int J Radiat Oncol Biol Phys* 2005;62:1504–1509.
30. Flanagan, S., Robinson, B., Krokosky, C., and Shewach, D. Mismatched nucleotides as the lesions responsible for radiosensitization with gemcitabine: A new paradigm for antimetabolite radiosensitizers. *Mol Cancer Ther* 2007;6:1858–1868.
31. Pauwels, B., Korst, A., Lardon, F., and Vermorken, J. Combined modality therapy of gemcitabine and radiation. *Oncologist* 2005;10:34–51.
32. Wolff, R., Evans, D., Gravel, D. et al. Phase I trial of gemcitabine combined with radiation for the treatment of locally advanced pancreatic adenocarcinoma. *Clin Cancer Res* 2001;7:2246–2253.

33. McGinn, C., Zalupski, M., Shureiqi, I. et al. Phase I trial of radiation dose escalation with concurrent weekly full-dose gemcitabine in patients with advanced pancreatic cancer. *J Clin Oncol* 2001;19:4202–4208.
34. Heinemann, V., Quietzsch, D., Gieseler, F. et al. Randomized phase III trial of gemcitabine plus cisplatin compared with gemcitabine alone in advanced pancreatic cancer. *J Clin Oncol* 2006;24:3946–3952.
35. Cunningham, D., Chau, I., Stocken, D. et al. Phase III randomised comparison of gemcitabine (GEM) versus gemcitabine plus capecitabine (GEM-CAP) in patients with advanced pancreatic cancer. *Eur J Cancer* 2005;3.
36. Metro, G., Fabi, A., Mirri, M. et al. Phase II study of Wxed dose rate gemcitabine as radiosensitizer for newly diagnosed glioblastoma multiforme. *Cancer Chemother Pharmacol* 2010;65:391–397.
37. Hunter, K., Feng, F., Griffith, K. et al. Radiation therapy with full-dose gemcitabine and oxaliplatin for unresectable pancreatic cancer. *Int J Radiat Oncol Biol Phys* 2012;83(3):921–926.
38. Loehrer, P., Feng, Y., Cardenes, H. et al. Gemcitabine alone versus gemcitabine plus radiotherapy in patients with locally advanced pancreatic cancer: An Eastern Cooperative Oncology Group trial. *J Clin Oncol* 2011;29:4105–4112.
39. Ben-Josef, E., and Lawrence, T. The importance of local control in pancreatic cancer. *Nat Rev Clin Oncol* 2012;9:9–10.
40. Chun, P., Feng, F., Scheurer, A., Davis, M., Lawrence, T., and Nyati, M. Synergistic effects of gemcitabine and gefitinib in the treatment of head and neck carcinoma. *Cancer Res* 2006;66:1–8.
41. Buchsbaum, D., Bonner, J., Grizzle, W. et al. Treatment of pancreatic cancer xenografts with Erbitux (IMC-C225) anti-EGFR antibody, gemcitabine, and radiation. *Int J Radiat Oncol Biol Phys* 2002;54:1180–1193.
42. Ling, Y., Li, T., Yuan, Z., Haigentz, M., Weber, T., and Perez-Soler, R. Erlotinib, an effective epidermal growth factor receptor tyrosine kinase inhibitor, induces p27KIP1 up-regulation and nuclear translocation in association with cell growth inhibition and G1/S phase arrest in human non-small-cell lung cancer cell lines. *Mol Pharmacol* 2007;72:248–258.
43. DiGennaro, E., Barbarino, M., Bruzzese, F. et al. Critical role of both p27KIP1 and p21CIP1/WAF1 in the antiproliferative effect of ZD1839 ('Iressa'), an epidermal growth factor receptor tyrosine kinase inhibitor, in head and neck squamous carcinoma cells. *J Cell Physiol* 2003;195(1):139–150.
44. Chakravarthy, A., Tsai, C., O'Brien, N. et al. A phase I study of cetuximab in combination with gemcitabine and radiation for locally advanced pancreatic cancer. *Gastrointest Cancer Res* 2012;5(4):112–118.
45. Wouters, A., Pauwels, B., Lardon, F. et al. *In vitro* study on the schedule-dependency of the interaction between pemetrexed, gemcitabine and irradiation in non-small cell lung cancer and head and neck cancer cells. *BMC Cancer* 2010;10:441.
46. Bischof, M., Weber, K., Blatter, J., Wannenmacher, M., and Latz, D. Interaction of pemetrexed disodium (Alimta, multitargeted antifolate) and irradiation *in vitro*. *Int J Radiat Oncol Biol Phys* 2002;52:1381–1388.

47. Teicher, B., Chen, V., Shih, C. et al. Treatment regimens including the multi-targeted antifolate LY231514 in human tumor xenografts. *Clin Cancer Res* 2000;6:1016–1023.
48. Seiwert, T., Connell, P., Mauer, A. et al. A phase I study of pemetrexed, carboplatin, and concurrent radiotherapy in patients with locally advanced or metastatic non-small cell lung or esophageal cancer. *Clin Cancer Res* 2007;13:515–522.
49. Argiris, A., Karamouzis, M., Smith, R. et al. Phase I trial of pemetrexed in combination with cetuximab and concurrent radiotherapy in patients with head and neck cancer. *Ann Oncol* 2011;22:2482–2488.
50. Desaknai, S., Lumniczky, K., Esik, O., Hamada, H., and Safrany, G. Local tumour irradiation enhances the anti-tumour effect of a double-suicide gene therapy system in a murine glioma model. *J Gene Med* 2003;5:377–385.
51. Wu, D., Liu, L., and Chen, L. Antitumor effects and radiosensitization of cytosine deaminase and thymidine kinase fusion suicide gene on colorectal carcinoma cells. *World J Gastroenterol* 2005;11:3051–3055.
52. Rainov, N. A phase III clinical evaluation of herpes simplex virus type 1 thymidine kinase and ganciclovir gene therapy as an adjuvant to surgical resection and radiation in adults with previously untreated glioblastoma multiforme. *Hum Gene Ther* 2000;11:2389–2401.
53. Freytag, S., Stricker, H., and Pegg, J. Phase I study of replication-competent adenovirus mediated double-suicide gene therapy in combination with conventional-dose three-dimensional conformal radiation therapy for the treatment of newly diagnosed, intermediate- to high-risk prostate cancer. *Cancer Res* 2003;63:7497–7506.

CHAPTER 6

Radiosensitization by Platinum Drugs and Alkylating Agents

6.1 THE PLATINUM DRUGS

After the accidental discovery of the anticancer properties of cisplatin, the drug was introduced clinically in the 1970s. Carboplatin, a second-generation analogue that is safer but shows a similar spectrum of activity to cisplatin, was introduced in the 1980s. The platinum-based drugs comprising cisplatin, carboplatin, and oxaliplatin are among the most widely utilized classes of cancer therapeutics.

6.1.1 Cytotoxicity of Cisplatin

Cisplatin is a neutral inorganic, water-soluble, coplanar complex (Figure 6.1). The drug enters the cell partly by diffusion and is converted to its active form through aquation reactions replacing its chloride ions with water or hydroxyl groups. The mono-aquated active metabolite reacts with cellular DNA to form interstrand and intrastrand cross-links. These distort DNA structure and block nucleotide replication and transcription leading to DNA breaks and miscoding (Figure 6.2). These damages may be repaired but, if not repaired, will be mutagenic or lethal. In the last case, the lesion activates an irreversible sequence of events culminating in apoptosis. The formation of the active species is, however, subject to

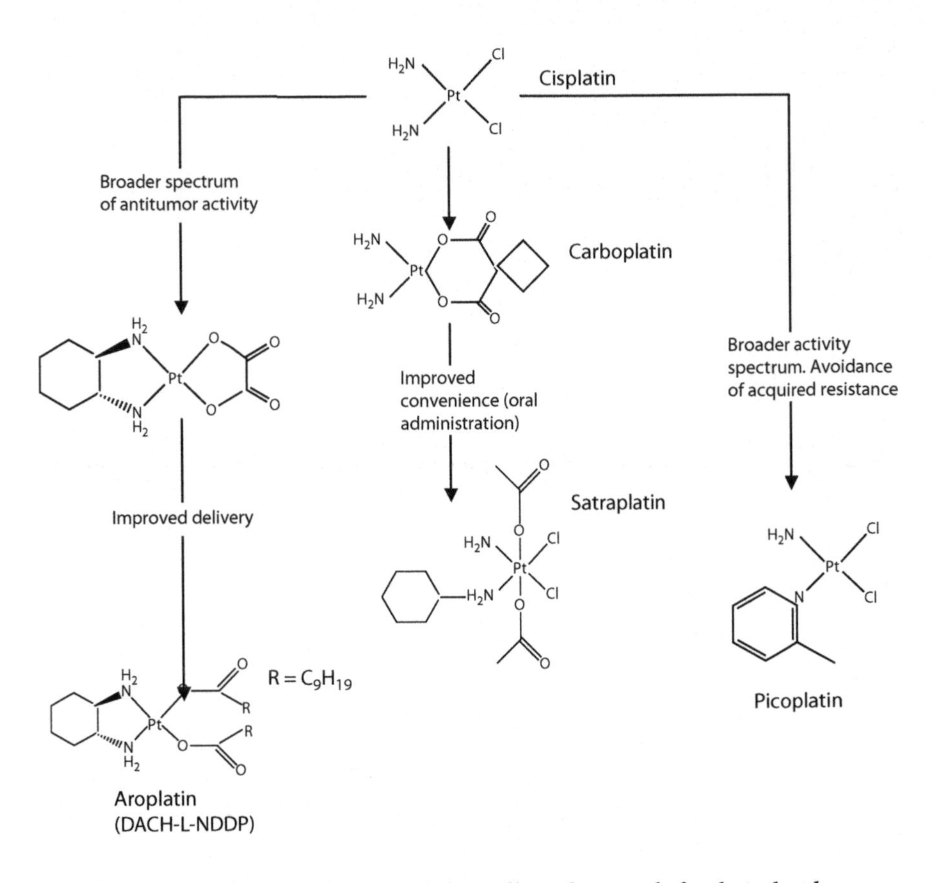

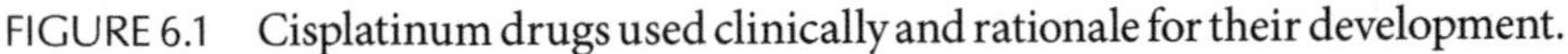
FIGURE 6.1 Cisplatinum drugs used clinically and rationale for their development.

inactivation by endogenous nucleophiles such as glutathione, methionine, and metallothionein.

The cytotoxicity of cisplatin is primarily ascribed to its interaction with nucleophilic N_7 sites of purine bases in DNA [1], to form both DNA–protein and DNA–DNA interstrand and intrastrand cross-links (Figure 6.2). The intrastrand *cis*-Pt(NH_3)2-d(GpG) and *cis*-Pt(NH_3)2-d(ApG) cross-links represent approximately 65% and 25%, respectively, of the total lesions in DNA. Interstrand cross-links are far less common, but also play a role in the cytotoxicity of the drug [2].

Cisplatin–DNA lesions trigger apoptosis by blocking replication and transcription and causing replication-mediated DSBs. Several pathways of apoptosis have been implicated including the extrinsic pathway mediated by death receptors and the intrinsic pathway centered on mitochondria.

In the extrinsic pathway, cisplatin-induced DNA lesions provoke activation of SAPK (stress-activated protein kinase), also known as c-Jun

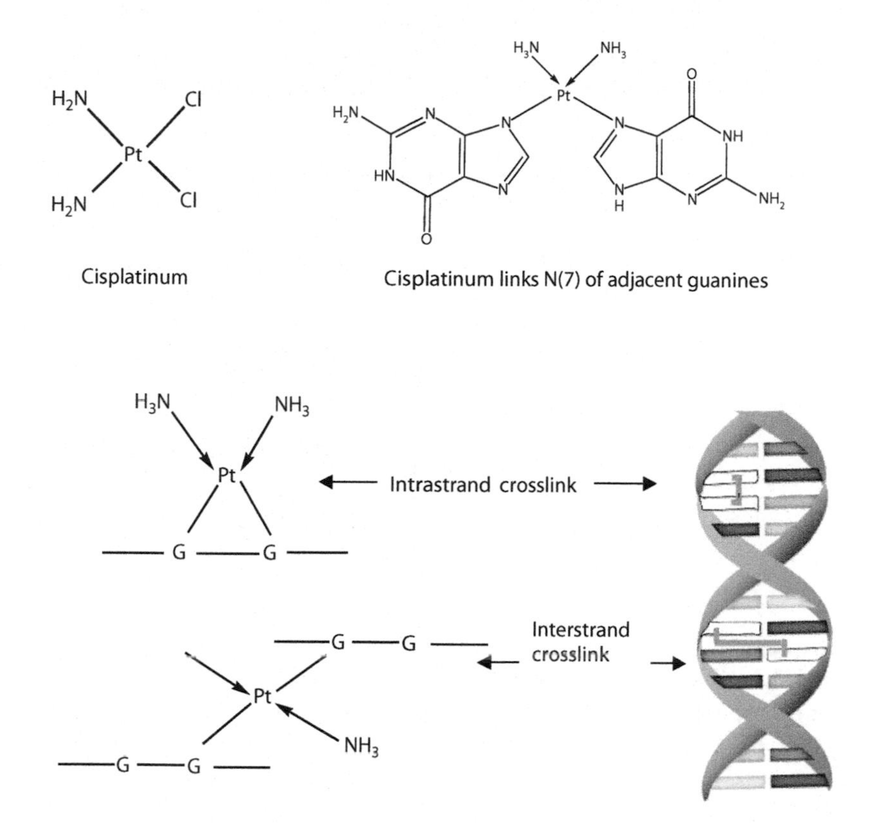

FIGURE 6.2 Activated aqua platinum species of cisplatin or carboplatin enters the nucleus, preferential covalent binding to the nitrogen on position 7 of guanine occurs. The major covalent bis-adduct that is formed involves adjacent guanines on the same strand of DNA (the intrastrand cross-link); a minor adduct involves binding to guanines on opposite DNA strands (the interstrand cross-link).

N-terminal kinase (JNK), and p38 kinase. JNK and p38 kinase activation occurs quickly after treatment of cells, but the induction is not transient but long-lasting, persisting for several days. This sustained activation of JNK and p38 kinase by cisplatin treatment, is accompanied by prolonged up-regulation of AP-1 (activator protein-1), a transcription factor. AP-1 transcribes the death receptor ligand FAS-L continuously leading to low-level death receptor activation and accumulation of activated caspase-8 initiating apoptosis over time (Figure 6.3).

In the intrinsic pathway, cellular stress leads to the activation of the proapoptotic Bcl-2 family proteins Bax and Bak, which form porous defects on the outer membrane of mitochondria, resulting in the release of apoptogenic factors from the mitochondria including cytochrome c and

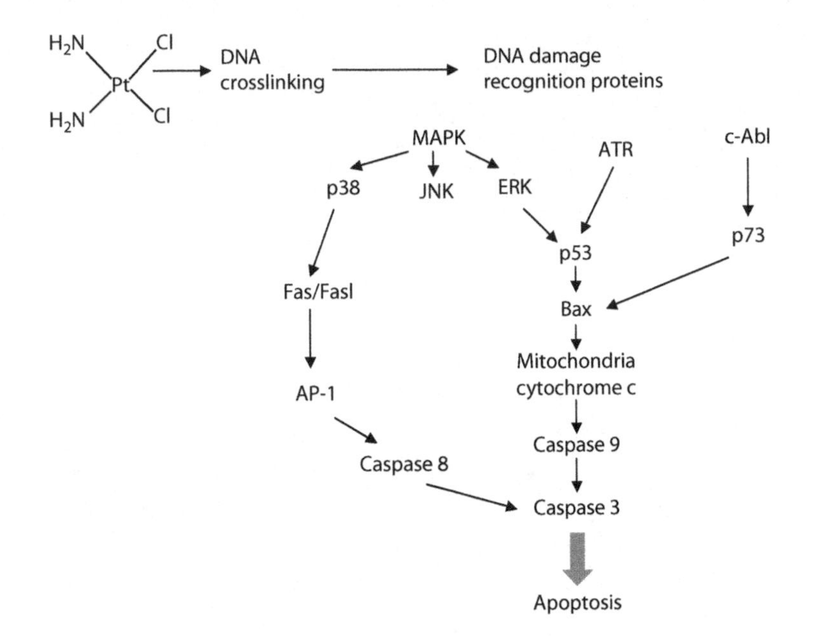

FIGURE 6.3 Cisplatin–DNA lesions trigger apoptosis by blocking replication and transcription and causing replication-mediated DSBs. Several pathways of apoptosis have been implicated including the extrinsic pathway mediated by death receptors and the intrinsic pathway centered on mitochondria.

AIF (apoptosis-inducing factor). Cytochrome c, after being released into the cytosol, binds to and induces conformational changes in the adaptor protein Apaf-1, leading to the recruitment and activation of caspase-9, which in turn after proteolytic processing activates downstream caspases for caspase-dependent apoptosis. In contrast, AIF, after being released from mitochondria, accumulates in the nucleus to induce apoptosis in a caspase-independent manner. p53-independent DNA damage-induced apoptosis can also occur via the p53 homologs p63 and p73. DNA damage activates ATM and/or ATR and initiates a sequence of events culminating in stimulation of p73 transcription and up-regulation of the level of p73 protein. p73 is proapoptotic in the absence of p53 mediating BAX mitochondrial translocation and cytochrome c release [3,4].

6.1.2 Cisplatin Resistance

This section is based on reviews by Kelland [5], Galluzzi et al. [6] and Siddik [7] and the articles referenced therein. Not long after the first promising clinical trial data with cisplatin, and later with carboplatin, it became apparent that some tumors were intrinsically resistant to cisplatin whereas, for others, resistance was acquired during the course of therapy. Studies of cisplatin and carboplatin drug resistance using cell lines demonstrated that resistance might be mediated through two broad mechanisms: first, failure of a sufficient amount of platinum to reach the target DNA, and second, failure to achieve cell death after platinum–DNA adduct formation. In fact, many resistant cells show a pleomorphic phenotype involving drug uptake, DNA-damage recognition and repair, and apoptosis. Intrinsic or acquired resistance to cisplatin diminishes its interaction with radiation, creating a need for alternative agents.

6.1.2.1 Resistance Mediated through Insufficient Cisplatin Binding to DNA

Cisplatin is highly polar and enters cells relatively slowly in comparison to other classes of small-molecule cancer drugs. The uptake of cisplatin is influenced by factors such as sodium and potassium ion concentrations, pH, and the presence of reducing agents; and a role for transporters or gated channels has been postulated in addition to passive diffusion. The major plasma membrane transporter involved in copper homeostasis, copper transporter-1 (CTR1), has been shown to have a substantial role in cisplatin influx. It is generally decreased uptake rather than increased efflux that predominates in platinum-drug resistance.

The second reason for the nonavailability of cisplatin for binding to DNA is related to increased levels of cytoplasmic thiol-containing species such as the tripeptide glutathione and metallothioneins. These species are rich in the sulfur-containing amino acids, cysteine and methionine, and platinum avidly binds to sulfur resulting in detoxification. The conjugation of cisplatin with glutathione may be catalyzed by glutathione *S*-transferases (GSTs), which makes the compound more anionic and more readily exported from cells by the ATP-dependent glutathione *S*-conjugate export (GS-X) pump.

6.1.2.2 Resistance Mediated after DNA Binding

After platinum–DNA adducts have been formed, cellular survival (and therefore, tumor drug resistance) can occur either by DNA repair or removal of these adducts, or by tolerance mechanisms. Many cisplatin-resistant cell lines derived from tumors have been shown to have increased DNA-repair capacity in comparison to sensitive counterparts. Of the four major DNA-repair pathways—nucleotide-excision repair (NER), base-excision repair (BER), mismatch repair (MMR), and double-strand break (DSB) repair—NER is the major pathway known to remove cisplatin lesions from DNA.

Particular attention, in both cell lines and clinical biopsy specimens, has focused on the NER endonuclease protein ERCC1 (excision repair cross-complementing-1) and resistance to platinum drugs. ERCC1 forms a heterodimer with xeroderma pigmentosum (XP) complementation group F (XPF) and acts to make a 5′ incision into the DNA strand, relative to the site of platinated DNA. It has been shown in tissue culture that increased NER in cisplatin-resistant ovarian cancer cells was associated with increased expression of ERCC1 and XPF (predominantly ERCC1), and that knockdown of ERCC1 by small interfering RNAs enhanced cellular sensitivity to cisplatin and decreased NER of cisplatin-induced DNA lesions. Clinical studies in patients with ovarian cancer have found a correlation between increased *ERCC1* mRNA levels and resistance to platinum-based chemotherapy.

Increased tolerance to platinum-induced DNA damage can also occur through loss of function of the MMR pathway. Loss of this repair pathway leads to low-level resistance to cisplatin and carboplatin (but not to oxaliplatin). During MMR, cisplatin-induced DNA adducts are recognized by the MMR proteins MSH2, MSH3, and MSH6 and it is believed that cells then undergo several unsuccessful repair cycles, finally triggering an apoptotic response. Loss of MMR surveillance with respect to cisplatin–DNA adducts thus results in reduced apoptosis and greater drug resistance. The clinical relevance of the loss of MMR to platinum drug resistance has been particularly studied in patients with ovarian cancer and although some data indicate a role in acquired drug resistance, other results show no correlation with intrinsic resistance.

The final tolerance mechanism involved in response to DNA adducts is the enhancement of replicative bypass where DNA polymerases β and η can bypass cisplatin–DNA adducts by translesion synthesis. At the cellular level, polymerase η has been shown to have a role in cellular tolerance to cisplatin and carboplatin.

The bottom line in terms of cisplatin toxicity is the activation of the DNA damage response by cisplatin-generated DNA lesions leading to the induction of mitochondrial apoptosis. What might be termed post-target resistance to cisplatin can result from alterations that interfere with the generation of apoptosis including defects in the signal transduction pathways that normally elicit apoptosis in response to DNA damage as well as deficiencies with the cell death executioner machinery itself.

Finally, there is evidence to suggest that the cisplatin-resistant phenotype can be sustained (if not entirely generated) by alterations in signaling pathways that are not directly engaged by cisplatin, yet compensate for (and hence interrupt) cisplatin-induced lethal signals. These include inhibition of ERBB2 (HER-2) protooncogene, PI3-Akt1, and Ras signaling and activation of PTEN. Inhibition of autophagy and of the chaperone heat shock proteins (HSP) have also been implicated.

6.1.3 Strategies to Improve the Performance of Platinum Drugs

The use of platinum-based drugs is limited not only by intrinsic or acquired drug resistance but also by severe dose-limiting side effects. A number of strategies have been proposed to circumvent platinum-drug resistance in patients with cancer. The most important of these are the development of new, improved platinum drugs, improved drug delivery, and the combination of platinum drugs with molecularly targeted agents.

6.1.3.1 Improved Platin Drugs

Over the past 30 years, more than 20 other platinum-based drugs have entered clinical trials with only two (carboplatin and oxaliplatin) of these gaining international marketing approval, and another three (nedaplatin, lobaplatin, and heptaplatin) gaining approval in individual jurisdictions (reviewed by Wheate et al. [8]). The development of many more was halted during clinical trials.

- *Oxaliplatin.* (cis-[(1R,2R)-1,2-cyclohexanediamine-*N,N*=] [oxalato (2-)-*O,O′*] platinum) is a third-generation platinum compound with a 1,2-diaminocyclohexane carrier ligand, which has been shown to have a wide spectrum of anticancer activity *in vitro* and has displayed preclinical and clinical activity in a variety of tumors. It became the third platin to be approved by the United States Food and Drug Administration in 2002. The drug has consistently shown activity in cell lines with acquired cisplatin resistance and in cisplatin-resistant human tumors. Contrary to what might be expected, oxaliplatin forms fewer DNA adducts than cisplatin at equimolar or isoeffective concentrations, and plasmid reactivation studies have shown that oxaliplatin lesions are no more difficult to repair than cisplatin lesions using the excision repair system. The disparity between the two platinum compounds can be partly accounted for by a difference in input from the MMR system. Cisplatin adducts are recognized by the MMR complex and loss of expression of the MMR proteins hMLH1 and hMSH2 has been implicated in resistance to cisplatin. In contrast, the MMR system does not seem to efficiently detect oxaliplatin-DNA adducts. These data suggest that defects in MMR contribute to increased net replicative bypass of cisplatin adducts and therefore to drug resistance by preventing futile cycles of translesion synthesis and mismatch correction.

- *Other platins of clinical significance.* Based on drugs currently entered in clinical trials, some interest is being generated by satraplatin (18 trials) and picoplatin (8 trials). No new small molecule platinum drug has entered clinical trials since 1999; possibly indicative of a shift in emphasis away from drug design and toward drug delivery and targeting. Liposomal preparations of cisplatin-like molecules have been prepared, and one, SPI-77, has been tested clinically [9,10]. The current lead platinum liposomal drug is based on the DACH stable ligand found in oxaliplatin (DACH-L-NDDP; Aroplatin; Figure 6.1). Clinical application of this approach is currently generating modest interest (Aroplatin is being investigated in six clinical trials, and Lipoplatin, another liposomal platin, is being investigated in one).

6.1.3.2 Combined Treatment with Targeted Molecular Drugs

Most contemporary cancer drug discovery and development involves the targeting of specific molecular abnormalities that are characteristic of cancer. This strategy has been successful in a number of cases such as glivec, trastuzumab, bevacizumab, and many others (Chapters 8–12 of this book describe radiosensitization by molecularly targeted drugs). An important clinical observation is that, in some cases, these agents do not have outstanding activity as monotherapy, but can be used optimally in combination with existing cytotoxic drugs including the platinum drugs. This approach again has the advantage of circumventing resistance to cisplatin while improving the targeting of the drug. Radiochemotherapy protocols with platinum drugs combined with molecularly targeted agents will be described later in this chapter.

6.1.4 Radiosensitization by Cisplatin

Early in the development of cisplatin as a clinical agent, it was found to have radiosensitizing properties.

6.1.4.1 Preclinical Studies

A number of preclinical experiments, over a period of years, demonstrated that cisplatin (and other Pt analogues) induced greater than additive cell killing when used in combination with radiation both *in vitro* and *in vivo* (reviewed by Gorodetsky et al. [10] and Yapp et al. [11]).

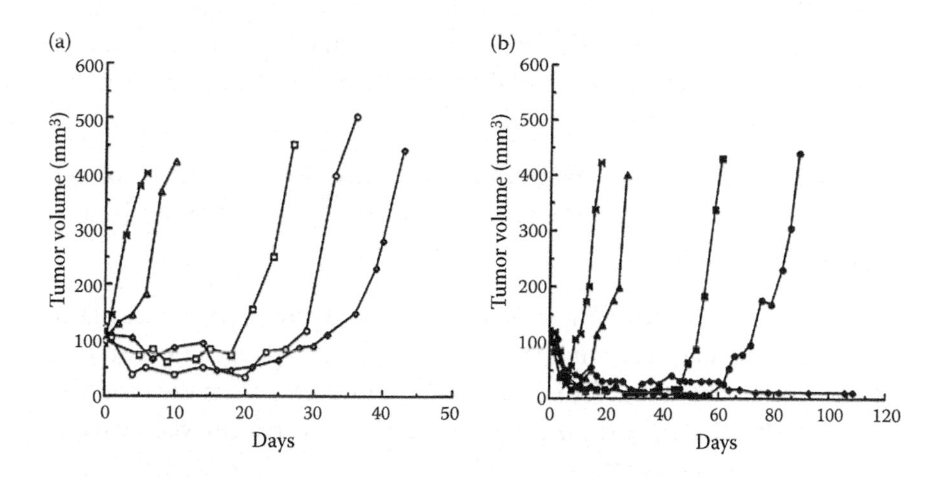

FIGURE 6.4 Tumor regrowth curves after treatment with acute or fractionated radiation with or without cisplatin delivered intratumorally by biodegradable polymer implant. (a) Tumors treated with radiation only: ×—×, no treatment; △—△ 10 Gy; □—□, 5 × 9 Gy; ○—○, 8 × 9 Gy; ◇—◇ 12 × 9 Gy. (b) Tumors treated with cisplatin polymer implant: ×—×, implant only; ▲—▲, implant + 10 Gy, ■—■, implant + 5 × 9 Gy; ●—●, implant + 8 × 9 Gy; ◆—◆, implant + 12 × 9 Gy. (From Yapp, D. et al., *International Journal of Radiation Oncology - Biology - Physics* 39:497–504, 1997. With permission.)

The potentiation of radiation effects by cisplatin has been shown in animal tumor models. For single doses of cisplatin used in combination with radiation, an enhancement ratio (ER) of 1.7 (using tumor cure as an endpoint) was reported in a mouse tumor, whereas multiple doses of cisplatin used in combination with radiation further increased tumor response to treatment. These reports indicate the therapeutic advantage that can result from the concurrent administration of cisplatin and radiation [11]. In one study, cisplatin was delivered intratumorally by biodegradable polymer implants and was effective in potentiating the effect of both acute and fractionated radiation. For the fractionated treatments, the effect was maintained with increasing fraction numbers and treatment time [12] (Figure 6.4).

6.1.4.2 Mechanism of Radiosensitization by Cisplatinum

The basis for interaction of cisplatin and radiation has been attributed to several different mechanisms:

- Radiation induces free radicals and the subsequent formation of toxic platinum intermediates, which increase cell killing
- Ionizing radiation can increase cellular uptake of platinum

- Damage to DNA by ionizing radiation, which would normally be reparable, can become fixed and lethal through cisplatin's free electron–scavenging capacity
- Inhibition of DNA repair leads to an increased incidence of cell cycle arrest and apoptotic cell death after radiation (summarized by Wilson et al. [13])

It is debatable whether there is a direct interaction between cisplatin and radiation-induced lesions. The evidence that this does not occur, or if it does, it is not important for radiosensitization, comes from the results of *in vitro* studies that have shown that the best results are achieved by using low doses of the two agents:

- Radiosensitization of murine embryonic fibroblasts (MEF) cells was shown at 1 μg/mL of cisplatin, but an increase in concentration did not increase in radiosensitization but instead increased radioresistance [14]
- When OV-1063 and EMT-6 cell lines were preirradiated with 2 Gy, addition of the drug produced a clear additional effect but this was almost totally eliminated when cells were irradiated with a higher dose (6 Gy) [10]

These results argue against a simple model of physical interaction occurring because of the proximity of the lesions caused by the two modalities.

One explanation, which is supported by the timing of optimal exposure to the agents, is that synergy results from cisplatin's inhibition of radiation-induced DNA damage repair. This again is supported by experimental findings:

- In two cell lines, the murine mammary adenocarcinoma (EMT-6) and human ovarian carcinoma (OV-1063) cells, a 2-h postradiation drug exposure resulted in a supra-additive combined effect, whereas a 24-h preirradiation exposure or protracted postirradiation exposure yielded an additive or slightly subadditive response [10]
- In experimental tumors, the greatest dose-enhancement factors were observed when cisplatin was administered immediately before a daily fraction of radiation [15]

Results of early studies had suggested that the addition of cisplatin to radiation treatment inhibited the process of sublethal damage repair. The concept of sublethal damage repair or recovery (SLDR) was developed to explain the fact that the effect of a given dose of radiation is less if it is split into two fractions delivered a few hours apart.

The mechanisms of SLDR are complex but undoubtedly include different modes of DNA repair and modification of both homologous recombination and nonhomologous end-joining (NHEJ) have been implicated in the mechanism of synergism between radiation and cisplatin.

6.1.4.2.1 Nonhomologous End-Joining The involvement of NHEJ in the mechanism of cisplatin and radiation interaction is supported by a number of experimental findings:

- In one case, cross-resistance between ionizing radiation and cisplatin was associated with increased Ku80 activity [16].
- In a later study, it was shown that cisplatin-induced DNA damage reduced the ability of the DNA-dependent protein kinase (DNA-PK) to interact with duplex DNA molecules in vitro [17]. In this case, the Ku–DNA-binding subunits of DNA-PK had a reduced ability to translocate on duplex DNA containing cisplatin–DNA adducts, resulting in a decrease in the association of the p460 catalytic subunit of DNA-PK with the Ku–DNA complex. In addition, the catalytic subunits of DNA-PK that were bound with Ku at a DNA end containing cisplatin–DNA adducts had a reduced catalytic rate compared with heterotrimeric DNA-PK assembled on undamaged DNA. The position of the lesion from the terminus also influenced kinase activation, with maximal inhibition occurring when the lesion was closer to the terminus. These findings indicate that the presence of cisplatin adducts decrease the ability of Ku to translocate away from the terminus and results in the formation of inactive kinase complexes resulting in impaired DNA repair capacity.
- It was further shown that cisplatin radiosensitization was not evident in Ku80-deficient cells compared with their wild-type counterparts [14] and the results of split-dose experiments suggested that NHEJ played a significant role in SLDR.

6.1.4.2.2 Homologous Recombination The involvement of homologous recombination in the sensitization mechanism has been suggested from studies in repair-proficient wild-type and recombinational repair-deficient (*rad52*) strains of the yeast *Saccharomyces cerevisiae* [18]. Cisplatin exposure sensitized wild-type yeast cells with a competent recombinational repair mechanism but could not sensitize cells defective in recombinational repair, indicating that the radiosensitizing effect of cisplatin was because of inhibition of DNA repair processes involving error-free *RAD52*-dependent recombinational repair. In the yeast system, the presence or absence of oxygen during irradiation did not alter this radiosensitization. This study also suggested that excision repair was not involved in the sensitization mechanism because strains homozygous for *rad3-2* showed synergy between radiation and cisplatin.

6.1.5 Radiochemotherapy with Platinum Drugs

6.1.5.1 Cisplatin

Cisplatin (cis-diammine dichloroplatinum(II)) is one of the most commonly used drugs for concurrent chemoradiotherapy. Among earlier reviews summarizing clinical protocols that combine radiation and treatment with Pt drugs, this study in which cisplatin and hypofractionation were used to treat high-grade gliomas was included. The results showed that the regimen was able to increase palliation and improve the functional status of severely compromised patients [19]. Preclinical findings had suggested that cisplatin should be present at the tumor site during radiation to maximize the additive effect [20]. This conclusion was supported by the results of a clinical trial in which squamous cell carcinoma of the head and neck was treated with cisplatin and radiation, and the survival of responding patients was found to be increased when the drug and radiation were delivered concurrently [21]. An overview of chemoradiotherapy combinations involving cisplatin for the treatment of individual diseases is shown in Table 6.1.

6.1.5.2 Carboplatin

Limited data are available for carboplatin as a single agent and it is less well established as a radiosensitizer. Comparative studies have shown that there is no survival difference when compared with cisplatin, but chemoradiotherapy with carboplatin is superior to radiotherapy alone [22].

6.1.5.3 Oxaliplatin

Oxaliplatin currently has wide approval for the treatment of adjuvant and metastatic colorectal cancers when used in combination with 5-FU and folinic acid. Radiochemotherapy including oxaliplatin has also focused

TABLE 6.1 Summary of Diseases and Disease Entities for Which Cisplatin Is Used with Concurrent Radiation

Disease	Indication and Treatment	Platinum Drugs	Benefit
Head and neck cancer	Locally advanced HNC, primary or adjuvant treatment	Cisplatin	Improved organ preservation and survival compared with radiation alone
Non–small cell lung cancer	Stage IIIB nonoperable, nonmetastaic disease	Cisplatin, carboplatin/paclitaxel, cisplatin/etoposide	Curative approach in poor surgical candidates or IIIB disease
Small cell lung cancer	Limited stage disease	Cisplatin/etoposide	Curative in ~20% of patients
Esophageal cancer	Locally advanced disease	Cisplatin/5-FU	Survival benefits, increased cure rate
Gastric cancer	Adjuvant	Cisplatin	Some data indicate survival benefit
Cervical cancer	Primary modality	Cisplatin	Improved local and distal control, organ preservation
Bladder cancer	Primary modality	Cisplatin	Improved local control

Source: Based on information reviewed by Seiwert, T. et al., *Nat Clin Pract Oncol* 2007; 4(2):86–100.

on metastatic colorectal cancer. The addition of oxaliplatin to 5-FU-based radiochemotherapy treatment of colon cancer improved disease-free and overall survival versus 5-FU-based regimens alone. The combination of oxaliplatin, fluoropyrimidines, and radiation therapy in the treatment of rectal cancer was evaluated in phase II studies and the regimen showed good tolerability and encouraging response rates (results summarized by Czito and Willett [24]).

The results of two large European phase III trials reporting on the role of oxaliplatin in 5-FU-based chemoradiation regimens were less encouraging.

- In an Italian phase III trial, 747 patients with locally advanced rectal cancer were randomized to 5-FU-based chemoradiation with and without concurrent weekly oxaliplatin. Preliminary results showed that the addition of weekly oxaliplatin to 5-FU-based preoperative chemoradiation significantly increased grade 3 to 4 toxicity without significantly affecting pathologic downstaging rates [25].

- In a similarly large French study, 598 patients with locally advanced rectal cancer were randomized to receive concurrent radiation therapy and capecitabine versus concurrent radiation therapy, capecitabine, and oxaliplatin. Trial results showed higher grade 3 to 4 toxicity rates in patients receiving radiation therapy, capecitabine, and oxaliplatin versus radiation therapy and capecitabine, with no statistical differences in rates of pathologic complete response, although an improvement in circumferential radial margin involvement was seen [26].

A summary of details for ongoing or recently completed clinical trials involving oxaliplatin and radiation is shown in Table 6.2. Colorectal cancer was the most frequently studied while a number of other trials focused on pancreatic cancer. Radiochemotherapy protocols with oxaliplatin invariably included other chemotherapies. Oxaliplatin was most frequently combined with capecitabine or with 5-FU and third and fourth agents including cetuximab, bevacizumab, celecoxib, doxecetal, gemcitabine celecoxib, and raltitrexed.

6.1.5.4 Multiagent-Based Radiochemotherapy

In current radiochemotherapy protocols with concurrent treatment, it is extremely rare for cisplatin or any other drug to be used as a single agent. The combination of several drugs during chemoradiotherapy offers

TABLE 6.2 Summary of Ongoing or Recently Completed Clinical Trials Involving Concurrent Radiation and Oxaliplatin

Tumor	Treatment Phase	Number of Trials				Chemotherapy
		1	1/2	2	3	
Pancreatic cancer	RT	7		4		5-FU, gemcitabine, irinotecan, bevacuzimab, cetuximab FOLFOX6
Colorectal, rectal cancer	RT	8		15	1	Capecitabine, irinotecan, celecoxib, cetuximab, raltitrexed
Non–small cell lung cancer	RT			1		Capecitabine
Esophageal cancer	RT		1		1	Doxetaxel, capecitabine, lapatinib
Gastric cancer	RT	1		2		Capecitabine

Source: Data obtained from clinicaltrials.gov.

improved radiosensitization of tumor cells and, as already described, may help in drug targeting. The downside is that these effects might be at the expense of increased frequency of toxic effects.

6.1.5.4.1 Cisplatin and 5-FU Excellent results were obtained with a study of cisplatin and 5-FU with conventional once-daily or twice-daily radiotherapy for locoregionally advanced squamous cell head and neck cancer with 5-year overall survival at 65.7%. A comparison of the results from another study, which assessed similar patient populations, suggested that the cisplatin–5-FU chemoradiotherapy regimen may be superior to the cisplatin-based chemoradiotherapy regimen, on the basis of the survival data (65.7% at 5 years versus 37.0% at 3 years, respectively). The toxicity profile for 5-FU and cisplatin combination therapy was similar to that for cisplatin therapy alone. It was concluded that the combination of cisplatin and 5-FU with radiation therapy is highly efficacious [27].

6.1.5.4.2 Cisplatin and Paclitaxel Several multiagent chemoradiotherapy regimens were compared in a multicenter, randomized, phase II trial. Patients received cisplatin and paclitaxel chemoradiotherapy, cisplatin and 5-FU chemoradiotherapy, or 5-FU and hydroxyurea chemoradiotherapy (FHX) [28]. The results of this trial demonstrated equivalent 2-year overall survival for cisplatin and paclitaxel chemoradiotherapy and the FHX regimen (66.6% and 69.4%), with both of these regimens being superior to cisplatin and 5-FU chemoradiotherapy (57.4%).

Additional trials have investigated this combination regimen and consistently demonstrated good survival results [23]. Serious toxic effects were generally similar to those arising from cisplatin treatment alone or cisplatin and 5-FU chemoradiotherapy, with 67% of patients experiencing grade 3 or 4 toxicity. It was concluded that the combination of cisplatin and paclitaxel could be considered a safe and efficacious alternative to cisplatin-based chemoradiotherapy and should be considered for poor-risk patients and those who can tolerate a more intense treatment approach.

6.1.5.4.3 Tirapazamine and Cisplatin Tirapazamine synergizes with both radiation and cisplatin. A phase I trial in patients with T3/4 or N2/3 HNSCC (or both) showed a 3-year overall survival of 69%. The same combination therapy was also studied in patients with recurrent HNC treated with accelerated radiation; this regimen again showed signs of activity and tolerability [29]. Furthermore, a randomized, phase II study showed a trend toward the superiority of the tirapazamine and cisplatin chemoradiotherapy combination compared with the 5-FU and cisplatin-based regimen, although statistical significance was not reached. In subgroup analysis, tirapazamine was of particular benefit in tumors showing signs of hypoxia [30].

6.1.5.4.4 Cisplatin and Cetuximab The combination of cetuximab with cisplatin-based chemoradiotherapy is currently being investigated in more than 100 clinical trials (clinicaltrials.gov). Studies reported to date have had conflicting results. The results of one small study in which cetuximab and cisplatin-based chemoradiotherapy was assessed in locoregionally advanced head and neck cancer showed excellent 3-year overall survival of 76% [31]. This study was closed early, however, because of unexpected toxicity in five patients. Clinical trials of this regimen are ongoing.

6.1.5.4.5 Paclitaxel and Carboplatin Carboplatin is frequently used to circumvent cisplatin-related toxic effects. The combination of carboplatin and paclitaxel with radiation is one of the more widely used chemoradiotherapy approaches for head and neck cancer. The improved tolerability offered by this regimen, compared with that for cisplatin therapy, is a major advantage. Multiple clinical trials support the efficacy and tolerability both in newly diagnosed patients and those with recurrent disease. For example, in a trial of 52 patients with stage III or IV unresectable HNSCC who were treated with weekly carboplatin and paclitaxel in combination with hyperfractionated radiotherapy; overall 2-year survival was 63%. Although acute, grade 3 to grade 4 toxic effects were reported in 80% of patients, these side effects were generally more manageable than those reported with cisplatin therapy, with significantly less renal and neurological toxicities. On the basis of the excellent efficacy, this regimen is suggested as an acceptable treatment standard and an alternative to cisplatin-based regimens [23].

6.1.5.4.6 Carboplatin Plus 5-Fluorouracil The outcomes of two European trials examining the use of carboplatin and 5-FU-based chemoradiotherapy had encouraging results. In one trial, the group receiving the combined regimen had a significantly higher 3-year survival rate than those receiving radiation alone (51% versus 31%) and an improved locoregional control rate (66% versus 42%). In another trial, 240 patients with unresectable stage III or IV oropharyngeal and hypopharyngeal cancers were randomized to hyperfractionated accelerated radiation alone or combined with carboplatin and infusional 5-FU [32]. Overall survival at 1 year for the combined treatment group improved from 44% to 58%. Carboplatin-based chemoradiotherapy is better tolerated than cisplatin-containing regimens, with acute radiation-related toxic effects being common but described as manageable. Carboplatin and 5-FU chemoradiotherapy can be used as an alternative to cisplatin therapy, although the clinical evidence to support this approach is limited.

6.2 ALKYLATING AGENTS

The alkylating agents are a diverse group of anticancer agents with the common feature that they react through an electrophilic alkyl group or a substituted alkyl group to covalently bind to cellular nucleophilic sites. Electrophilicity is achieved through the formation of carbonium ion intermediates and can result in transition complexes with target molecules. Ultimately, reactions result in the formation of covalent linkages by alkylation with a broad range of nucleophilic groups, including bases in DNA, and these are believed to be responsible for ultimate cytotoxicity and therapeutic effect. Although the alkylating agents react with cells in all phases of the cell cycle, their efficacy and toxicity result from interference with rapidly proliferating tissues. A number of related alkylating drugs have been developed, with roles in the treatment of leukemias, lymphomas, and solid tumors. Most of the alkylating agents cause dose-limiting toxicities to the bone marrow and to a lesser degree the intestinal mucosa, with other organ systems also affected contingent on the individual drug, dosage, and duration of therapy. The main classes of alkylating agents that have been used in the clinic are shown in Table 6.3.

6.2.1 Temozolomide

A new class of alkylating drugs, the imidazotetrazinones, were synthesized in the late 1970s, and included the compound that would eventually be called temozolomide (TMZ). The imidazotetrazinones are related to dacarbazine (DTIC), a drug that was widely used for metastatic melanoma in the late 1970s and early 1980s, and also had some success in glioma. DTIC is a prodrug that requires enzymatic conversion to its active form, 5-(3-methyltriazen-1-yl)-imidazole-4-carboxamide (MTIC). The conversion occurs rapidly in mice, but minimally in humans [33], which probably contributed to its limited clinical efficacy. Because TMZ does not rely on an enzymatic reaction and is converted to MTIC at physiologic pH,

TABLE 6.3 Major Classes of Clinically Used Alkylating Agents

Drug	Structure	Main Therapeutic Use
Cyclophosphamide	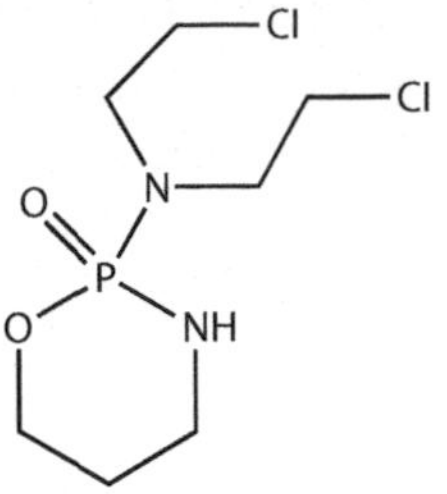	Lymphomas, leukemias, and solid tumors
Ifosfamide	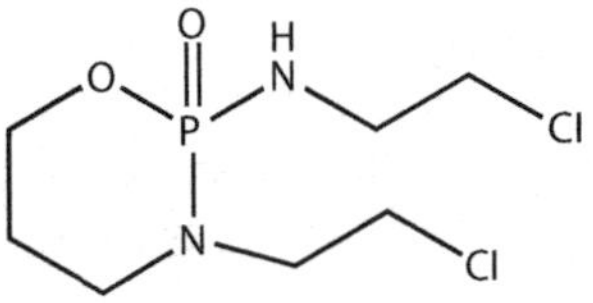	Testicular cancer, breast cancer, non-Hodgkins lymphoma, soft tissue sarcoma, osteogenic sarcoma, lung cancer, cervical cancer, ovarian cancer, bone cancer
Nitrosoureas carmustine	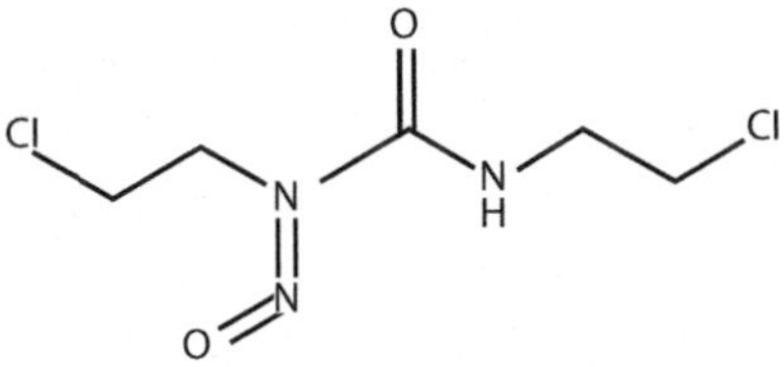	Glioma, glioblastoma multiforme, astrocytoma, multiple myeloma, lymphoma (Hodgkins, non-Hodgkins)
Streptotocin	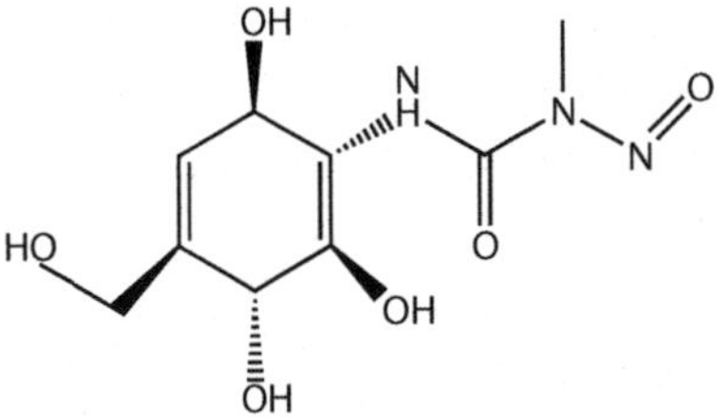	Cancers of the islets of Langerhans
Triazenes carbazine DITC	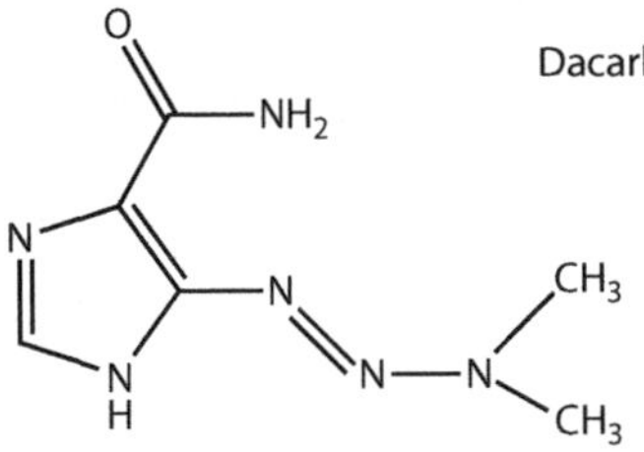 Dacarbazine	Malignant melanoma, Hodgkins lymphoma
Temozolomide	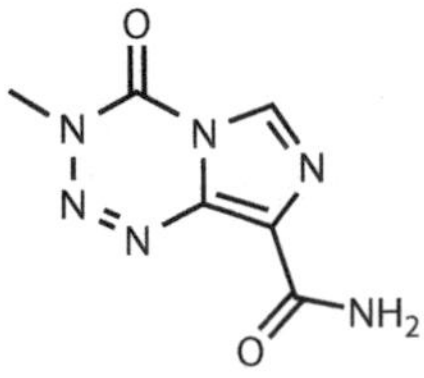	Glioblastoma, astrocytoma, metastatic melanoma

it was an attractive alternative to DTIC. TMZ also had the advantage of uniform tissue distribution (including brain) in mouse models and on this basis, the compound was advanced to human trials [34]. Although it was initially intended for use in patients with melanoma, TMZ has become the drug of choice for the treatment of glioblastoma.

6.2.1.1 Cytotoxicity of Temozolomide

TMZ is a small (194 Da), lipophilic prodrug that is rapidly metabolized at physiologic pH to its active metabolite, MTIC (methyltriazeno-imidazole-carboximide; Figure 6.5). TMZ is converted in the plasma to MTIC, its active metabolite, by a nonenzymatic chemical degradation process, entirely bypassing the liver, and avoiding interactions with other medications. MTIC is then irreversibly degraded to 4-amino-5-imidazole-carboxamide and a highly reactive methyldiazonium cation.

The methyldiazonium cation is a potent DNA methylating agent with three major products: N_7-methylguanine (70%), O^6-methylguanine (5%), and N_3-methyladenine (9%). O^6-methylguanine accounts for much of the cytotoxicity of TMZ: it incorrectly pairs with thymine during replication, triggering repeated attempts at MMR that generate DNA DSB (Figure 6.6).

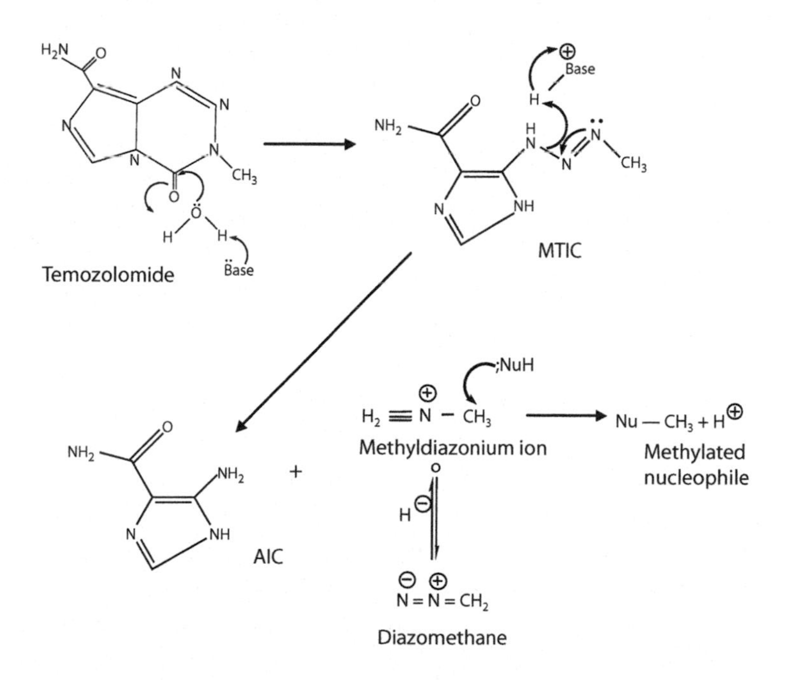

FIGURE 6.5 Decomposition of temozolomide in aqueous solution.

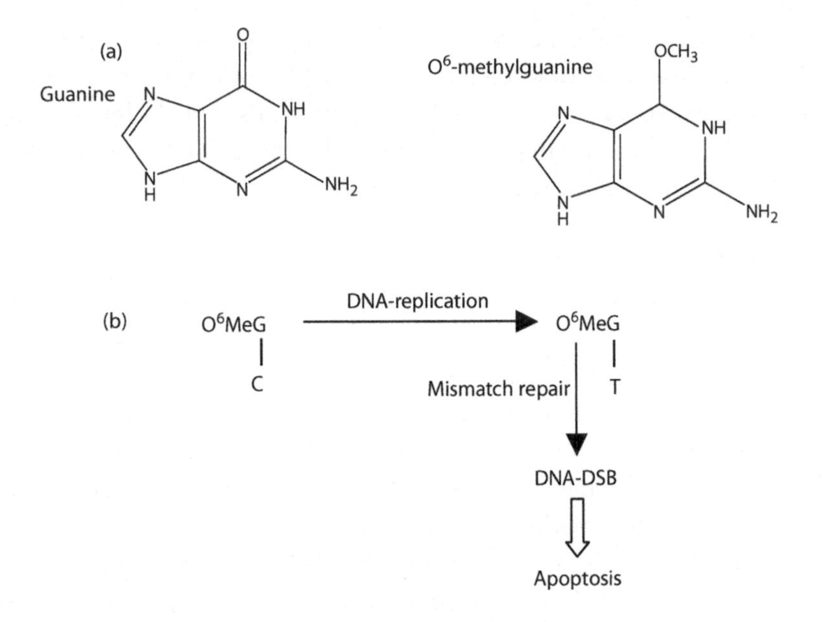

FIGURE 6.6 (a) Guanine and O^6-methylguanine, which accounts for much of the cytotoxicity of TMZ. (b) O^6–MeG incorrectly pairs with thymine during replication, triggering repeated attempts at MMR. Thymine is excised during MMR, but because O^6MeG is still present in the DNA, mispairing occurs again, and thymine is reinserted during the resynthesis of the double-stranded DNA, which remains a substrate for MMR. Futile removal and reinsertion of thymine can occur repeatedly eventually resulting in DSBs and apoptosis.

The enzyme O^6-methylguanine-DNA-methyltransferase (MGMT) repairs O^6-methylguanine by removing the methylating adduct and is an important mechanism of TMZ resistance. MGMT transfers methyl adducts at the O^6 position on guanine in DNA to an internal cysteine acceptor residue. This reaction results in irreversible inactivation of MGMT, requiring increased *de novo* protein synthesis to restore repair activity. Restoration of MGMT activity is a relatively rapid event, usually occurring within several hours, in peripheral blood cells [35] and in human malignant brain tumors [36]. Patients with tumors in which the MGMT gene is silenced by promoter methylation have a better prognosis and derive the most benefit from chemoradiation with temozolomide.

The therapeutic use of temozolomide is favored by its pharmacology. It crosses the blood–brain barrier and a significant proportion of the drug (~30%) rapidly enters the cerebrospinal fluid. Repetitive daily dosing is feasible because the drug does not have a long half-life and shows no significant accumulation with repeat administration.

6.2.1.2 Radiosensitization by TMZ: Preclinical Studies

Results of preclinical studies of the interaction of TMZ with radiation suggest that, at clinically relevant doses, TMZ is more effective when delivered at least 72 h before irradiation. The interaction with radiation is frequently additive rather than synergistic, and cellular sensitivity to TMZ largely predicts the effect of combination treatment. This is consistent with a model in which TMZ-induced base damage generates DSB through failed MMR during two cell cycles after exposure, thereby causing cytotoxicity [37]. Delayed induction of DSB by TMZ is difficult to measure by techniques such as γH2AX foci, probably because MMR-dependent evolution of DSB is asynchronous and accompanied by ongoing repair. In cell cycle experiments, however, it has been clearly shown that TMZ activates the G_2/M checkpoint after 48 h, consistent with indirect, delayed induction of DSB through MMR.

High doses of TMZ seem to have greater radiosensitizing potential and to interact with radiation at earlier time points. Early induction of signaling and checkpoint responses by high doses of TMZ [38] or the related methylating agent MNNG [39] have been reported, and increased apoptosis at 4 to 12 h when high-dose TMZ was given 2 h preradiation was also observed [40].

Because clinically, TMZ concentrations do not exceed 10 μM [41], it seems unlikely that TMZ directly induces DSB in patients and the therapeutic relevance of the experimentally observed early interactions with radiation are questionable. These results and others raise the question of whether TMZ, at clinically applicable doses, acts as a radosensitizer. In a detailed article challenging the assumption that TMZ is a radiosensitizer, Chalmers et al. [41] described the additive cytotoxic effects of radiation and TMZ in four glioma cell lines with varying p53 and MGMT status. Additive cytotoxic effects of radiation and TMZ were observed when TMZ was administered 72 h before radiation consistent with a model in which the induction of DSB by the two agents promotes additive cell killing. Scheduling was important, with increased effects on clonogenic survival, G_2/M checkpoint activation, and γH2AX foci induction being observed when TMZ preceded radiation treatment by 72 h. No evidence was found that TMZ is a radiosensitizer. Inhibition of ATM and ATR with caffeine abrogated G_2/M checkpoint activation and increased the cytotoxicity of combined treatment.

Determination as to whether or not TMZ acts as a radiosensitizer is not simply a case of nitpicking and semantics but has clinical implications. If

TMZ is a radiosensitizer, patients should be advised to take TMZ tablets at a specific time, such as1 h before radiotherapy. If it is not, as some of the experimental data suggests, it would be better if TMZ treatment were to commence at least 3 days before radiation to achieve maximum benefit, and there would be no need for patients to take chemotherapy tablets shortly before treatment. There is also a possibility that the interaction with TMZ is tumor-specific and that lower doses of TMZ are radiosensitizing for some tumors but not others.

6.2.1.3 Clinical Application of TMZ with Concomitant Radiation

In 2002, Stupp et al. [42] published the results of a pilot phase II protocol designed to test the safety and tolerability of concomitant radiation plus TMZ therapy at 75 mg/m^2 per day followed by adjuvant TMZ, dosed at 150 to 200 mg/m^2 5 days on/23 days off (Figure 6.7). During the concomitant phase, there was a modest increase in myelosuppression (6%), but the combination was otherwise well tolerated. The median survival was 16 months, with 58% and 31% survival at 1 year and 2 years, respectively. This study was quickly followed by a European Organization for

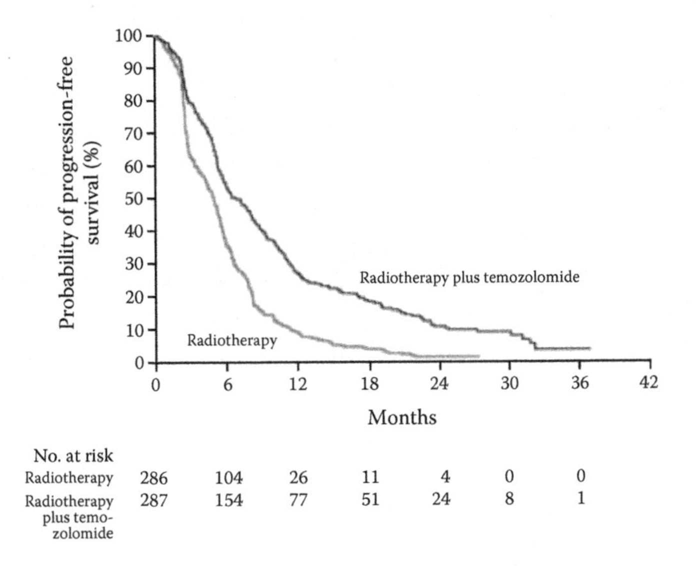

FIGURE 6.7 Kaplan–Meier estimates of overall survival according to treatment group. The hazard ratio for death among patients treated with radiotherapy plus temozolomide, as compared with those who received radiotherapy alone, was 0.63 (95% confidence interval, 0.52–0.75; $P < 0.001$). (From Stupp, R. et al., *Journal of Clinical Oncology* 20(5):1375–1582, 2002; Stupp, R. et al., *New England Journal of Medicine* 352(10):987–996, 2005. With permission.)

Research and Treatment of Cancer phase III trial (EORTC 26981), which compared surgery followed by radiotherapy plus concomitant TMZ to surgery followed by radiotherapy alone in patients with glioblastoma. There was a statistically significant increase in median survival of approximately 10 weeks (12.1 versus 14.6 months) [43] and apparent increase in 2-year survivors in patients receiving concomitant chemoradiation when compared with those receiving radiation alone (26.5% versus 10.4%). A trial comparing radiation alone to radiation with bischloroethylnitrosourea, also called carmustine (BCNU) or 1-(2-chloroethyl)-3-cycohexyl-1-nitrosourea also called lomustine (CCNU) demonstrated a 23% 2-year survival rate in 40- to 60-year-old patients with glioblastoma receiving combined therapy [44], but a significant number of those patients (~20% in both the CCNU and BCNU groups) had dose-limiting leukopenia, which restricted the number of patients that could safely be offered the regimen. Patients receiving BCNU or CCNU also had significant fatigue and nausea. In contrast, adding TMZ to radiotherapy caused no decrease in health-related quality of life.

Radiation with concomitant TMZ quickly became accepted as the standard of care for patients with newly diagnosed glioblastoma in Europe and the United States. No new cytotoxic agent has been more effective. Since its initial approval, there have been an large number of nonrandomized trials testing different dose regimens, including prolonged administration, 7 days on/7 days off, and a "dose-dense" 21-day on/7-day off regimen [45]. Some alternate regimens have shown promise, but only one has been tested against the standard regimen and this demonstrated no benefit to the alternate dose strategy [46]. One of the strongest predictive biomarkers for response to alkylating agents is MGMT methylation status. The MGMT gene encodes O^6-alkylguanine-DNA alkyltransferase (AGT), the DNA repair enzyme that reverses damage inflicted by TMZ. AGT is a "suicide enzyme"—one molecule of AGT removes only one alkyl molecule, inactivating itself in a stoichiometric fashion. In principle, hypermethylation of cytosine-phosphate-guanine islands in the MGMT promoter region leads to lower MGMT expression and should enhance the response to alkylating agents by inhibiting DNA repair and, in fact, MGMT promoter methylation is associated with a significantly higher median survival after therapy with TMZ (21.7 months versus 15.3 months in one trial [47]).

6.2.1.3.1 Improving Response to TMZ Radiation Therapy Since the introduction of temozolomide plus radiation, very little progress has been made in developing novel treatments for glioma and efforts have been made to improve the performance of standard treatment. One strategy to improve the efficacy of TMZ is to increase the effective dose of TMZ the tumor receives via alternate delivery mechanisms. Local delivery via wafer or convection-enhanced delivery could theoretically circumvent an intact blood–brain barrier. A polymer wafer (analogous to Gliadel) that combines TMZ with BCNU has been tested in rat models and may be more effective than either agent alone [48]. Slow-release microspheres containing TMZ have been used in animals, but not yet in humans [49]. Both of these options would be available to patients with surgical disease and near-total resections. PEGylation and targeted nanovesicles are two other potential systemic approaches that could increase TMZ delivery to tumor cells while minimizing effects on normal cells. Methods to improve the delivery of radiosensitizing molecules, including TMZ, will be discussed at greater length in a later chapter.

6.3 SUMMARY

Platinum compounds react within the cell to form DNA–protein cross-links as well as intrastrand and interstrand cross-links within DNA, and cellular attempts to repair these cross-links result in strand breaks. When used in combination with radiation, platinum agents increase the number of lethal DSBs in DNA. In addition, several alternative mechanisms have been proposed to underlie the synergistic effects of combining platinum compounds and radiation including enhanced formation of toxic intermediates through free radical formation, alteration in platinum pharmacokinetics, and inhibition of DNA repair.

The consequence of cisplatin adduct formation is a cascade of cellular events involving signaling pathways, checkpoint activation, DNA repair activity, and apoptosis. The interaction between cisplatin and radiation occurs on a number of levels, including enhanced formation of toxic platinum intermediates in the presence of radiation-induced free radicals; the capacity of cisplatin to scavenge free electrons formed by the interaction between radiation and DNA that would otherwise fixate reparable damage to DNA, radiation-induced increase in cellular cisplatin uptake; synergistic effects caused by cell cycle disruption; and the inhibition of radiation induced DNA lesion repair. *In vitro* studies have shown that the most effective combinations between cisplatin and radiation are achieved at lower doses of the two agents. This dose sensitivity argues against a simple model of physical interaction because of the proximity between the lesions caused by the two modalities. A plausible explanation, supported by the timing of optimal exposure to the agents, is that synergy is a consequence of cisplatin inhibition of radiation-induced DNA damage repair involved in the recovery from sublethal damage. Combined cisplatin and radiation treatment inhibits both critical pathways involved in DNA DSB repair, homologous recombination and nonhomologous end-joining.

Cisplatin is probably the most widely used anticancer agent in combination with radiation. Limited data are available for carboplatin as a single agent and it is less well established as a radiosensitizer. Radiochemotherapy including oxaliplatin has focused on metastatic colorectal cancer and the addition of oxaliplatin. Radiochemotherapy with platin drugs almost always involves additional chemotherapy. Combinations of cisplatin with 5-FU, paclitaxel, tirapazamine and cetuximab, and of carboplatin with 5-FU and paclitaxel are among the most common.

The alkylating agents are a diverse group of anticancer agents with the common feature that they participate in a reaction during which an electrophilic alkyl group or a substituted alkyl group covalently bind to cellular nucleophilic sites. This reaction results in the formation of covalent linkages by alkylation with nucleophilic groups, including bases in DNA, believed to be responsible for the cytotoxicity of these drugs. Temozolomide, from a class of alkylating drugs, the imidazotetrazinones, is a prodrug that requires conversion to its active form MTIC. TMZ is converted to MTIC in aqueous solution at physiologic pH, one of several aspects of its pharmacology, which favors therapeutic use. In addition, it crosses the blood–brain barrier and a significant proportion of the drug rapidly enters the cerebrospinal fluid. Repetitive daily dosing is feasible as the drug does not have a long half-life and shows no significant accumulation with repeat administration. Initially intended for use in patients with melanoma, TMZ has become the drug of choice for the treatment of glioblastoma.

High doses of TMZ have been demonstrated to have radiosensitizing potential and to cause early induction of signaling and checkpoint responses with increased apoptosis. TMZ at more clinically relevant (lower) doses seems to be more effective when delivered at least 72 h before radiation. The interaction with radiation is frequently additive rather than synergistic, and cellular sensitivity to TMZ is predictive of the effect of combination treatment. This is consistent with TMZ-induced base damage generating DSB through failed MMR during two cell cycles after exposure, thereby causing cytotoxicity.

Based on the outcome of several large clinical trials, the results of the first being published in 2002, radiation with concomitant TMZ has become accepted as the standard of care for patients with newly diagnosed glioblastoma in Europe and the United States. Compared with alkylators, which had been used previously (BCNU or CCNU), improvements were seen in survival with no decrease in health-related quality of life.

REFERENCES

1. Eastman, A., and Barry, M. Interaction of trans-diamminedichloroplatinum(II) with DNA: Formation of monofunctional adducts and their reaction with glutathione. *Biochemistry* 1987;26:3303–3307.
2. Plooy, A., van Dijk, M., Berends, F., and Lohman, P. Formation and repair of DNA interstrand cross-links in relation to cytotoxicity and unscheduled DNA synthesis induced in control and mutant human cells treated with cis-diamminedichloroplatinum(II). *Cancer Res* 1985;45:4178–4184.
3. Roos W., and Kaina, B. DNA damage-induced cell death by apoptosis. *Trends Mol Med* 2006;12(9):440–50.
4. Roos, W., and Kaina, B. DNA damage-induced cell death: From specific DNA lesions to the DNA damage response and apoptosis. *Cancer Lett* 2013;332:237–48.
5. Kelland, L. The resurgence of platinum-based cancer chemotherapy. *Nat Rev Cancer* 2007;7:573–584.
6. Galluzzi, L., Senovilla, L., Vitale, I. et al. Molecular mechanisms of cisplatin resistance. *Oncogene* 2012;31:1869–1883.
7. Siddik, Z. Cisplatin: Mode of cytotoxic action and molecular basis of resistance. *Oncogene* 2003;22:7265–7279.
8. Wheate, N., Walker, S., Craig, G., and Oun, R. The status of platinum anticancer drugs in the clinic and in clinical trials. *Dalton Trans* 2010;39(35):8113–8127.
9. Liu, D., He, C., Wang, A., and Lin, W. Application ofliposomal technologies for delivery of platinum analogs in oncology. *Int J Nanomed* 2013;8:3309–3319.
10. Gorodetsky, R., Levy-Agababa, F., Mou, X., and Vexler, A. Combination of cisplatin and radiation in cell culture: Effect of duration of exposure to drug and timing of irradiation. *Int J Cancer* 1998;75:635–642.
11. Yapp, D., Lloyd, D., Zhu, J., and Lehnert, S. The potentiation of the effect of radiation treatment by intratumoral delivery of cisplatin. *Int J Radiat Oncol Biol Phys* 1998;42(2):413–420.
12. Yapp, D., Lloyd, D., Zhu, J., and Lehnert, S. Tumor treatment by sustained intratumoral release of cisplatin: Effects of drug alone and combined with radiation. *Int J Radiat Oncol Biol Phys* 1997;39:497–504.
13. Wilson, G., Bentzen, S., and Harari, P. Biologic basis for combining drugs with radiation. *Semin Radiat Oncol* 2006;16:2–9.

14. Myint, W., Ng, C., and Raaphorst, G. Examining the non-homologous repair process following cisplatin and radiation treatments. *Int J Radiat Biol* 2002;78:417–424.
15. Kanazawa, H., Rapacchietta, D., and Kallman, R. Schedule-dependent therapeutic gain from the combination of fractionated irradiation and cisdiamminedichloroplatinum (II) in C3H/Km mouse model systems. *Cancer Res* 1988;48:3158–3164.
16. Frit, P., Canitrot, Y., Muller, C. et al. Cross-resistance to ionizing radiation in a murine leukemic cell line resistant to cis-dichlorodiammineplatinum(II): Role of Ku autoantigen. *Mol Pharmacol* 1999;56:141–146.
17. Turchi, J., Henkels, K., and Zhou, Y. Cisplatin-DNA adducts inhibit translocation of the Ku subunits of DNA-PK. *Nucleic Acids Res* 2000;28:4634–4641.
18. Dolling, J., Boreham, D., Brown, D., Raaphorst, G., and Mitchel, R. Cisplatin-modification of DNA repair and ionizing radiation lethality in yeast, *Saccharomyces cerevisiae. Mutat Res* 1999;433:127–136.
19. Hercbergs, A., Tadmor, R., Findler, G., Sahar, A., and Brenner, H. Hypofractionated radiation therapy and concurrent cisplatin in malignant cerebral gliomas. *Cancer* 1989;64:816–820.
20. Kallman, R. The importance of schedule and drug dose intensity in combinations of modalities. *Int J Radiat Oncol Biol Phys* 1994;28:761–771.
21. Glicksman, A., Slotman, G., Doolittle, C. et al. Concurrent cisplatinum and radiation with or without surgery for advanced head and neck cancer. *Int J Radiat Oncol Biol Phys* 1994;30(5):1043–1050.
22. Jeremic, B., Shibamoto, Y., Stanisavljevic, B., Milojevic, L., Milicic, B., and Nikolic, N. Radiation therapy alone or with concurrent low-dose daily either cisplatin or carboplatin in locally advanced unresectable squamous cell carcinoma of the head and neck: A prospective randomized trial. *Radiother Oncol* 1997;43:29–37.
23. Seiwert, T., Salama, J., and Vokes, E. The concurrent chemoradiation paradigm—General principles. *Nat Clin Pract Oncol* 2007;4(2):86–100.
24. Czito, B., and Willett, C. Beyond 5-fluorouracil: The emerging role of newer chemotherapeutics and targeted agents with radiation therapy. *Semin Radiat Oncol* 2011;21:203–211.
25. Aschele, C., Pinto, C., Cordio, S. et al. Preoperative fluorouracil (FU)-based chemoradiation with and without weekly oxaliplatin in locally advanced rectal cancer: Pathologic response analysis of the Studio Terapia Adiuvante Retto (STAR)-01 randomized phase III trial. *J Clin Oncol Rep* 2009;27:18S.
26. Gerard, J., Azria, D., Gourgou-Bourgade, S. et al. Comparison of two neo-adjuvant chemoradiotherapy regimens for locally advanced rectal cancer: Results of the phase III trial Accord 12/0405-Prodige 2. *J Clin Oncol* 2010; 28:1638–1644.
27. Adelstein, D. Multiagent concurrent chemoradiotherapy for locoregionally advanced squamous cell head and neck cancer: Mature results from a single institution. *J Clin Oncol* 2006;24:1064–1071.

28. Garden, A., Harris, J., Vokes, E. et al. Preliminary results of Radiation Therapy Oncology Group 97-03: A randomized phase II trial of concurrent radiation and chemotherapy for advanced squamous cell carcinomas of the head and neck. *J Clin Oncol* 2004;22:2856–2864.
29. Rischin, D. Phase I trial of concurrent tirapazamine, cisplatin, and radiotherapy in patients with advanced head and neck cancer. *J Clin Oncol* 2001;19:535–542.
30. Rischin, D., Hicks, R., Fisher, R. et al. Prognostic significance of [18F]-misonidazole positron emission tomographydetected tumor hypoxia in patients with advanced head and neck cancer randomly assigned to chemoradiation with or without tirapazamine: A substudy of Trans-Tasman Radiation Oncology Group Study 98.02. *J Clin Oncol* 2006;24:2098–2104.
31. Pfister, D., Su, Y., Kraus, D. et al. Concurrent cetuximab, cisplatin, and concomitant boost radiotherapy for locoregionally advanced, squamous cell head and neck cancer: A pilot phase II study of a new combined modality paradigm. *J Clin Oncol* 2006;24:1072–1078.
32. Staar, S., Rudat, V., Stuetzer, H. et al. Intensified hyperfractionated accelerated radiotherapy limits the additional benefit of simultaneous chemotherapy—Results of a multicentric randomized German trial in advanced head-and-neck cancer. *Int J Radiat Oncol Biol Phys* 2001;50:1161–1171.
33. Stevens, M., and Newlands, E. From triazines and triazenes to temozolomide. *Eur J Cancer* 1993;29A(7):1045–1047.
34. Newlands, E., Stevens, M., Wedge, S., Wheelhouse, R., and Brock, C. Temozolomide: A review of its discovery, chemical properties, pre-clinical development and clinical trials. *Cancer Treat Rev* 1997;23(1):35–61.
35. Kuhne, M., Riballo, E., Rief, N. et al. A double-strand break repair defect in ATM-deficient cells contributes to radiosensitivity. *Cancer Res* 2004;64:500–508.
36. Wedge, S., Porteous, J., Glaser, M. et al. *In vitro* evaluation of temozolomide combined with X-irradiation. *Anticancer Drugs* 1997;8:92–97.
37. Chakravarti, A., Zhai, G., Zhang, M. et al. Survivin enhances radiation resistance in primary human glioblastoma cells via caspase-independent mechanisms. *Oncogene* 2004;23:7494–7506.
38. Caporali, S., Falcinelli, S., Starace, G. et al. DNA damage induced by temozolomide signals to both ATM and ATR: Role of the mismatch repair system. *Mol Pharmacol* 2004;6:478–491.
39. Stojic, L., Cejka, P., and Jiricny, J. High doses of SN1 type methylating agents activate DNA damage signaling cascades that are largely independent of mismatch repair. *Cell Cycle* 2005;4:473–477.
40. Chakravarti, A., Erkkinen, M., Nestler, U. et al. Temozolomide-mediated radiation enhancement in glioblastoma: A report on underlying mechanisms. *Clin Cancer Res* 2006;12:4738–4746.
41. Chalmers, A., Ruff, E., Martindale, C., Lovegrove, N., and Short, S. Cytotoxic effects of temozolomide and radiation are additive- and schedule-independent. *Int J Radiat Oncol Biol Phys* 2009;75(5):1511–1519.

42. Stupp, R., Dietrich, P., Ostermann, K. et al. Promising survival for patients with newly diagnosed glioblastoma multiforme treated with concomitant radiation plus temozolomide followed by adjuvant temozolomide. *J Clin Oncol* 2002;20(5):1375–1382.
43. Stupp, R., Mason, W., van den Bent, M. et al. Radiotherapy plus concomitant and adjuvant temozolomide for glioblastoma. *N Engl J Med* 2005;352(10):987–996.
44. Nelson, D., Diener-West, M., Horton, J., Chang, C., Schoenfeld, D., and Nelson, J. Combined modality approach to treatment of malignant gliomas—re-evaluation of RTOG 7401/ECOG 1374 with long-term follow-up: A joint study of the Radiation Therapy Oncology Group and the Eastern Cooperative Oncology Group. *NCI Monogr* 1988;6:279–284.
45. Wick, W., Platten, M., and Weller, M. New (alternative) temozolomide regimens for the treatment of glioma. *Neuro Oncol* 2009;11(1):69–79.
46. Gilbert, M., Wang, M., Aldape, K. et al. RTOG 0525: A randomized phase III trial comparing standard adjuvant temozolomide (TMZ) with a dose-dense (dd) schedule in newly diagnosed glioblastoma (GBM) (abstract). *J Clin Oncol* 2011;29:2006.
47. Hegi, M., Liu, L., Herman, J. et al. Correlation of O^6-methylguanine methyltransferase (MGMT) promoter methylation with clinical outcomes in glioblastoma and clinical strategies to modulate MGMT activity. *J Clin Oncol* 2008;26(25):4189–4199.
48. Recinos, V., Tyler, B., Bekelis, K. et al. Combination of intracranial temozolomide with intracranial carmustine improves survival when compared with either treatment alone in a rodent glioma model. *Neurosurgery* 2010;66(3):530–537.
49. Dong, J., Zhou, G., Tang, D. et al. Local delivery of slow-releasing temozolomide microspheres inhibits intracranial xenograft glioma growth. *J Cancer Res Clin Oncol* 2012;138(12):2079–2084.

CHAPTER 7

Topoisomerase Inhibitors and Microtubule-Targeting Agents

7.1 RADIOSENSITIZATION BY DRUGS TARGETING DNA TOPOLOGY AND THE MITOTIC SPINDLE

The two groups of drugs described in this chapter are both cytotoxic and used extensively in chemotherapy and mixed modality treatments. The drugs targeting topoisomerases disrupt processes involved in the management of the topology of DNA as it goes through the processes of replication and transcription. The taxanes and Vinca alkaloids, on the other hand, are exceptional in chemotherapy in that the target is not any aspect of DNA structure or function but cytoskeletal elements, the microtubules that are mobilized prior to cell division to make up the mitotic spindle. Again, this affects the cell division and proliferation of tumor cells.

7.2 TOPOISOMERASES

DNA topoisomerases are ubiquitous and essential enzymes that have critical roles in the fundamental biological processes of replication, transcription, recombination, repair, and chromatin remodeling. The intertwining of complementary strands forming the DNA duplex is a structural basis for the storage and transmission of genetic information, but processes that require unwinding and rewinding of the DNA duplex can lead to topological entanglements and result in genome instability if left unresolved. DNA topoisomerases are the tools to resolve the problems of DNA entanglements by enabling topological transformations. Topoisomerases tackle these complex problems by a process of reversible transesterification. The tyrosine in the active site of these enzymes serves as a nucleophile by initiating the formation of a covalent adduct with the backbone phosphate of DNA, thus generating a transient break enabling topological transformations to occur. The second transesterification, a reversal of the first, reseals the DNA break and regenerates the free tyrosine.

Based on structure and mechanism, these topoisomerases are grouped into type I and type II, each of which is further divided into subfamily A and subfamily B. Type I enzymes carry out strand passage through a reversible single-strand break, whereas type II enzymes mediate strand transport through a double-strand DNA gate.

7.2.1 Type I Topoisomerases

Five human DNA topoisomerases, topoisomerase 1 (TOP1), TOP2α, TOP2β, TOP3α, and TOP3β, have been identified to date. Among these, TOP1 is the only type I DNA topoisomerase that has been demonstrated to be a molecular target of anticancer drugs.

Human TOP1, encoded by a single copy TOP1 gene on chromosome 20, is a monomeric 100 kDa protein. It relaxes both positively and negatively supercoiled DNA and requires no energy cofactor for its activity (reviewed in articles by Gellert [1] and Wang [2]). TOP1 is involved in the initiation and elongation of RNA transcription and DNA replication and, by regulating the supercoiling state of DNA, is essential for maintaining the stability of the genome. The presence of high levels of TOP1 in both slow-growing and rapidly proliferating tumor cells provides a basis for targeting TOP1 in cancer therapy.

7.2.2 Cytotoxicity of Drugs Targeting Topoisomerase 1

TOP1 drugs exert their biological effects by "poisoning" DNA through trapping a key reaction intermediate called the TOP1 cleavable complex. During its catalytic cycle, TOP1 transiently cleaves the DNA backbone to allow the passage of DNA strands and then reseals the DNA backbone in two successive transesterification reactions. A key covalent TOP1–DNA intermediate is formed between the tyrosine-723 residue of the enzyme and a 3′-phosphate at the break site during the transient DNA cleavage stage. TOP1-targeted drugs trap this key reaction intermediate, called the TOP1 cleavable complex [1,2]. The drug-trapped TOP1 cleavable complexes damage DNA through interactions with cellular processes including DNA replication.

The cytotoxic mechanism of camptothecin in mammalian cells is the fork collision model, which is shown in Figure 7.1 (reviewed in articles by Chen and Liu [3] and Li and Liu [4]). Based on studies using cell synchronization techniques and specific inhibitors of DNA polymerases, active DNA synthesis has been demonstrated to be directly involved and to contribute to the induction of the highly S phase-specific cytotoxicity of camptothecin [5,6]. It is hypothesized that the collision between the DNA replication machinery and the drug-trapped TOP1 cleavable complex

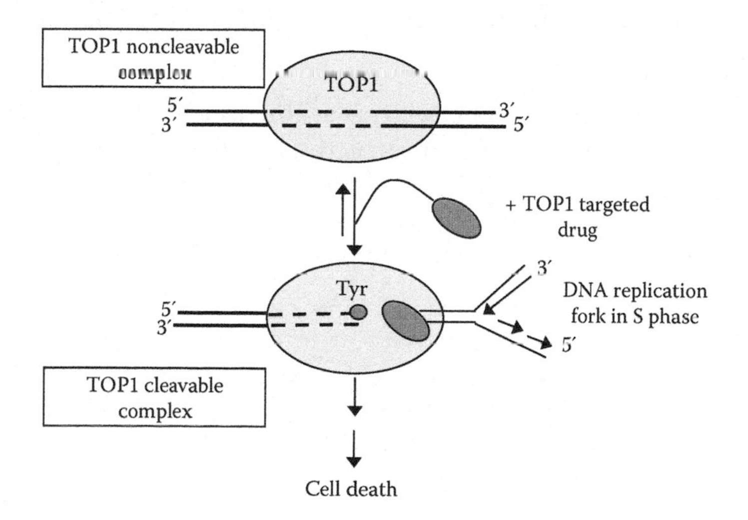

FIGURE 7.1 The S phase-specific cytotoxic mechanism of topoisomerase 1 (TOP1)-targeted drugs is explained by the fork collision model, according to which two different reaction intermediates, the "noncleavable complex" and the "cleavable complex," are formed upon the binding of TOP1 to DNA. In the absence of drugs, the equilibrium between these two reaction intermediates favors the noncleavable complex over the cleavable complex. By inhibiting the rejoining step, TOP1 drugs perturb this equilibrium by trapping a reversible TOP1-drug-DNA ternary reaction intermediate, called the "TOP1 cleavable complex." TOP1 drugs are highly S phase-cytotoxic and it is currently hypothesized that the collision between the replication machinery and the drug-trapped TOP1 cleavable complex leads to eventual downstream events including cell death.

leads to eventual downstream events including G_2 phase cell-cycle arrest and cell death [3,4,7].

7.2.2.1 Drugs Targeting Topoisomerase 1

An increasing number of anticancer drugs, including new camptothecin derivatives, DNA minor groove-binding drugs, and indolocarbazole derivatives have been demonstrated to exert their cytotoxic effect through interaction with TOP1.

- *Camptothecin,* originally isolated from the tree *Camptotheca acuminate* [8,9], is the best-characterized of the human TOP1-targeting drugs. The ring-closed lactone form of camptothecin derivatives is biologically active but not the ring-open carboxylate form (Figure 7.2).
- *Topotecan* (Hycamtin) is synthesized from camptothecin by the addition of a positively charged tertiary amine [8] and has improved solubility.
- *Irinotecan* (CPT-11; Camptosar) is synthesized as a prodrug, which is converted intracellularly to an active metabolite, SN-38, by carboxylesterase [9].

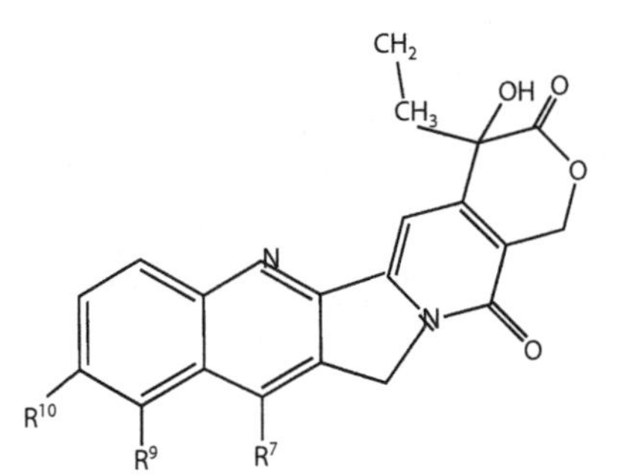

Name	R^7	R^9	R^{10}
Camptothecin	H	H	H
Topotecan	H	$(CH_2)^{-}N^{+}(CH_3)_2H$	$-C(=O)-$N–N
Irinotecan	H	H	OH

FIGURE 7.2 Chemical structures of some camptothecin derivatives.

Topotecan and irinotecan were approved by the Food and Drug Administration for the treatment of recurrent ovarian and colon cancers, respectively [10]. Subsequently, based on their demonstrated efficacies in clinical trials, the clinical usage of these two compounds has rapidly expanded to include a wide spectrum of human malignancies including cancers of the gastrointestinal tract, ovary, lung, cervix, and central nervous system [8,10].

It is believed that the broad-spectrum antitumor activity of camptothecins is due to their S phase-specific cytotoxicity [3,5,6], selective cytotoxicity against cancer cells [11,12], and the ability to overcome MDR1-mediated drug resistance [13].

- *DB-67* (7-*tert*-butyldimethylsilyl-10-hydroxy-camptothecin), one of the most active silatecans, is one of the latest generation of camptothecin derivatives undergoing preclinical investigation [9]. DB-67 has shown efficacy against human cancers including high-grade gliomas in animal models [14].

7.2.3 Mechanism of Radiosensitization by TOP1 Inhibitors

(The following is based on reviews by Wang et al. [2], Chen et al. [15,16], and the articles referenced therein.)

Understanding of the mechanism of radiosensitization by TOPI inhibitors has accumulated as a series of experimental steps were accomplished

- It was shown in early studies, using cultured cells and animal models, that camptothecin derivatives enhanced the cytotoxic effect of ionizing radiation. In one study, camptothecin derivatives radiosensitized human breast cancer MCF-7 cells in a schedule-dependent manner and radiosensitization was induced when drug treatment was given prior to, but not following, radiation.
- Other results suggested an important role for a stereospecific interaction between drug and TOP1 in initiating TOP1-mediated radiosensitization. This conclusion was supported by the fact that 20(*S*)-10,11-methylenedioxy-camptothecin, but not its 20(*R*)-isomer, radiosensitized MCF-7 cells and by results with Chinese hamster DC3F/C-10 cells, with drug-resistant mutant TOP1 for which the level of CPT-induced radiosensitization was significantly reduced. In agreement with this conclusion, DB-67 (a silate) was found to be

extremely potent in trapping TOP1 cleavable complexes *in vitro* and to be a strong radiation sensitizer of mammalian cells.

- Based on studies using DNA polymerase inhibitors and phase-specific cells selected by cell-cycle sorting techniques, the induction of TOP1-mediated radiation sensitization in mammalian cells was shown to be S phase-specific and to require active DNA synthesis. This conclusion was supported by results obtained in a cell-free SV40 DNA replication system, showing that interaction between DNA replication forks and TOP1–DNA cleavable complexes could lead to double-strand DNA breaks, arrest of the replication fork, and an aborted "cleaved" TOP1–DNA complexes. On the basis of these findings, a model of TOP1-mediated radiosensitization was proposed according to which drug-trapped TOP1 cleavable complex interacts with the replication fork during active DNA synthesis leading to the types of DNA damage that had been observed in the cell-free systems, and that one or some combination of these damages was responsible for TOP1-mediated radiosensitization.

- A significant observation was that TOP1-mediated radiosensitization but not cell-killing, induced by a 30-min camptothecin treatment, could be completely reversed over a period of 8 h in CHO cells. This finding is reminiscent of the repair of radiation-induced "sublethal damage" and suggests that the TOP1-mediated radiosensitizing activity may represent a form of "sublethal damage," which can be repaired in mammalian cells.

- Eukaryotic cells have evolved two major repair pathways for DNA double-strand breaks (DSB), homologous recombination and non-homologous end-joining (NHEJ). The involvement of the NHEJ pathway in TOP1-mediated radiation sensitization, but not in cytotoxicity, was demonstrated in a study of the role of Ku86 in DNA TOP1-mediated radiosensitization (Ku86 and Ku70 are subunits of Ku, a heterodimeric complex that acts as a DNA-binding subunit of the DNA-dependent protein kinase complex of the NHEJ pathway). It was found that a 30-min camptothecin treatment induced significantly greater radiosensitization in the Ku86-deficient Chinese hamster ovary xrs-6 cells than in the hamster Ku86-complemented xrs6 + hamKu86 cells, although the drug toxicity in the two cell lines was the same. This finding was confirmed in two pairs of transfectant

> sublines established from the Ku86-deficient Chinese hamster lung fibroblast XR-V15B cells, in which significantly higher levels of camptothecin-induced radiosensitization were observed in the vector-alone sublines than in the Ku86-complemented XR-V15B sublines whereas camptothecin treatments, ranging from 0.5 to 24 h, induced similar cytotoxicities in both vector-alone and Ku86-complemented sublines. Neither the DNA-damaging drugs etoposide and cisplatin nor the tubulin-binder vinblastine induced enhanced levels of radiosensitization in the Ku86-deficient cells, suggesting that Ku86 uniquely affects topoisomerase 1-mediated radiosensitization induced by camptothecin. Furthermore, cotreatment with DNA replication inhibitor aphidicolin abolished both camptothecin-induced cytotoxicity and radiosensitization in the vector-alone as well as in the Ku86-complemented subline cells, indicating that both events are initiated by replication-dependent topoisomerase 1-mediated DNA damages. Taken together, these results showed a role for Ku86 in modulating topoisomerase 1-mediated radiosensitization, but not cytotoxicity, in mammalian cells.

Figure 7.3 shows a model of TOP1-mediated cytotoxicity and radiosensitization based on the foregoing findings. In this model, the drug-trapped TOP1 cleavable complex initiates TOP1-mediated DNA damage by "interacting" with the replication fork during DNA synthesis. Double-strand DNA breaks, replication fork arrest, and an aborted "cleaved" TOP1–DNA complex may be generated. Dependence or nondependence on NHEJ double-strand DNA repair is the basis of the dissociation between the pathways leading to TOP1-mediated cytotoxicity and TOP1-mediated radiation sensitivity [16].

Studies of the radiation-sensitizing effect of TOP1 drugs in preclinical systems have contributed to the clinical application of combined modality therapy with radiation and TOP1-targeted drugs. For example, camptothecin derivatives were shown to induce radiation sensitization in cultured human breast cancer MCF-7 cells in a schedule-dependent manner that requires drug treatment prior to but not following radiation [17]. This observation indicates the importance of treating patients with TOP1 drugs prior to delivery of radiotherapy. Moreover, based on studies using DNA polymerase inhibitors and phase-specific cells sorted by cell-cycle sorting techniques, the induction of TOP1-mediated radiation sensitization in mammalian cells was shown to be an S phase-specific event that

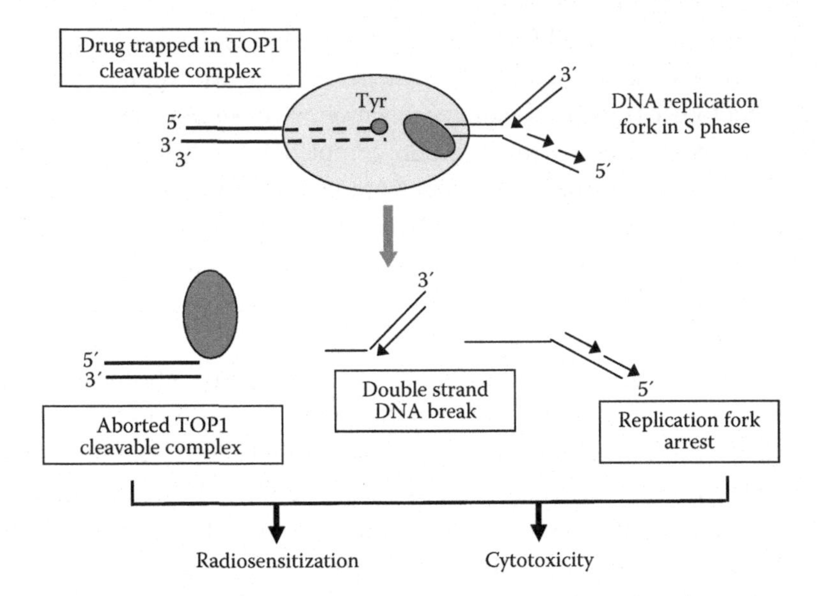

FIGURE 7.3 A proposed model for TOP1-mediated radiosensitization. The drug-trapped TOP1 cleavable complex may initiate TOP1-mediated RS by "interacting" with the replication fork during active DNA synthesis. As observed in a cell-free SV40 DNA replication system, at least three major biochemical events including double strand DNA breaks, arrest of replication fork, and an aborted "cleaved" TOP1–DNA complex can be generated. One or a combination of these three events may be responsible for the induction of TOP1-mediated RS. The current data indicate the involvement of a currently undefined repair process in the induction of TOP1-mediated RS and suggests that dissociation between the pathways leads to radiosensitization and cytotoxicity.

requires active DNA synthesis. This finding indicates a probable therapeutic advantage of TOP1 drugs in selectively radiosensitizing proliferating cancer cells that are actively synthesizing DNA.

7.2.4 TOP1 Inhibitors in Radiochemotherapy

Camptothecin derivatives including irinotecan and topotecan have demonstrated antitumor efficacy against cancers of the gastrointestinal tract, ovarian, lung, cervical, and central nervous system (CNS) tumors [8]. The clinical success attained with these agents as systemic therapy, encouraged an investigation of combined treatments. As described above, information about the timing and sequence of chemoradiation combination, as well as the cellular determinants for TOP1-mediated RS has been obtained through characterization of the radiosensitizing activity of camptothecin derivatives. The potential advantages of using TOP1-targeted drugs to enhance radiotherapy include tumor-selective targeting due to the elevated levels of TOP1 in cancer cells, an S phase-specific mechanism and the availability of molecular probes to monitor "drug-induced intermediates" *in vivo.*

7.2.4.1 Clinical Trials of Chemoradiation with Camptothecin Derivatives

7.2.4.1.1 Lung Cancer Clinical chemoradiation trials of camptothecin derivatives in lung cancer are the subject of a very recent review by Chen and coworkers [16]. Several clinical studies have demonstrated the feasibility and efficacy of TOP1 drugs in combination with thoracic radiotherapy for locally advanced non-small cell lung cancer (NSCLC).

- Weekly injection of irinotecan with concurrent thoracic radiotherapy for locally advanced NSCLC has been studied in a number of phase I and II trials. These studies established a maximum tolerated dose (MTD) of intravenous injection of irinotecan, administered weekly for 6 weeks concurrently with thoracic radiotherapy to 60 Gy. A generally good response rate of 58% to 79% was reported.

- Irinotecan, in combination with cisplatin-based chemotherapy and daily thoracic radiotherapy, has also been tested in patients with stage III NSCLC. An objective response rate of 60% was observed in the study.

- Single-agent topotecan, given by daily bolus injection on days 1 to 5, and days 22 to 26, was dose-escalated with concurrent daily thoracic radiotherapy in a phase I study for 12 patients with unresectable locally advanced NSCLC. A response rate of 17% with two complete responses was reported.

- Another phase I study was conducted by escalating both thoracic radiotherapy and infusion duration of topotecan at a constant dose of 0.4 mg/m^2 per day. The radiation dose and topotecan infusion duration were escalated in an alternating fashion at different dose levels. Studies reported well-tolerated side effects and recommended 60 Gy thoracic radiotherapy and 42-day duration of topotecan 0.4 mg/m^2 per day as the phase II regimen. A 43% response rate was reported for a total of 24 patients, including 22 patients with NSCLC.

- A number of clinical phase I/II studies with camptothecins, in combination with cisplatin-based chemotherapy and concomitant thoracic radiotherapy, have shown promising efficacy for small cell lung cancer (SCLC).
- A phase I study of irinotecan and cisplatin with concurrent split-course radiotherapy in limited-stage SCLC had an overall response rate of 94% with four complete responses was reported for 16 evaluable patients.

7.2.4.1.2 Brain The efficacy of irinotecan for CNS tumors has been demonstrated in clinical studies using different treatment schedules [15]. In a phase I/II trial of irinotecan and whole brain radiotherapy (WBRT) for patients with brain metastases from solid tumors, what were described as durable and impressive responses, including the complete disappearance of all evaluable lesions, were observed in patients receiving weekly irinotecan at 80 and 100 mg/m^2 dose levels in the phase I portion of the trial [15].

7.2.4.1.3 Head and Neck Cancer The results of two trials reviewed by Chen et al. [15] indicated that irinotecan and radiation in combination with docetaxel or 5-FU might potentially produce high response rates for locally advanced head and neck cancers. The dose-limiting toxicity was mucositis [18,19]. Ongoing and recently completed clinical trials with topotecan or irinotecan are summarized in Table 7.1 (www.clinicaltrials.gov).

7.2.5 Topoisomerase II

Topoisomerase II is a multisubunit enzyme that uses ATP to pass an intact helix through a transient double-stranded break in DNA to modulate DNA topology [4]. After strand passage, the DNA backbone is religated and DNA structure restored. The enzyme exists as two highly homologous isoforms, alpha and beta, which differ in their production during the cell cycle. The alpha isoform concentration increases twofold to threefold during G_2/M, and is higher in rapidly proliferating cells than in quiescent cell populations.

7.2.6 Inhibitors of Topoisomerase II

Topoisomerase II inhibitors are a diverse class of anticancer drugs that include the anthracyclines (doxorubicin and daunorubicin), etoposides,

TABLE 7.1 Ongoing Clinical Trials: Topoisomerase 1 Inhibitors and Concurrent Radiotherapy

Drug	Site of Tumor	No. of Trials	Phase 1	Phase 2	Phase 3	Additional Chemotherapy
Irinotecan	Esophagus	10	1	9		Cisplatin, bevacizumab, celecoxib, taxotere, carboplatin, sunitib, paclitaxel
	Lung	7	3	2	2	Cisplatin, celecoxib, vinorelbine, MMC, carbplatin, etoposide, erlotinib, bevacizumab
	Pancreas	4	1	3		Gemcitabine, celecoxib, oxaliplatin, leucovorin, 5FU
	Colorectal	6	1	4	1	Capecitabine, 5FU, leucovorin, cetuximab
	Sarcoma	1			1	Dactinomycin, cyclophosphamide, vincristine
	Brain	1		1		Temolozomomide
Topotecan	Lung	1	1			
	Rectum	2	2			Hycamtin
	Cervix	4	3	1		Cisplatin

and quinolones (Figure 7.4). Anthracyclines have effective and broad-spectrum antitumor activity but their clinical utility is frequently limited by systemic toxicity (i.e., cardiotoxicity with doxorubicin) or drug resistance (frequently mediated by *p*-glycoprotein; reviewed by Hande [20]).

7.2.6.1 Epipodophyllotoxins: Etoposide

Etoposide prevents topoisomerase II from religating cleaved DNA and thus introduces high levels of transient protein-associated breaks in the genome of treated cells. The alpha isoform of topoisomerase II seems to be the target of etoposide, although the beta enzyme, which does not change significantly during the cell cycle, could potentially be a target in slow-growing cancers. Two scissile bonds are formed for every topoisomerase II-mediated double-stranded DNA break and the results of DNA cleavage and ligation assay studies indicate a two-site model for the action of etoposide against human topoisomerase IIα, in which drug interactions at both scissile bonds are required to increase enzyme-mediated double-stranded DNA breaks.

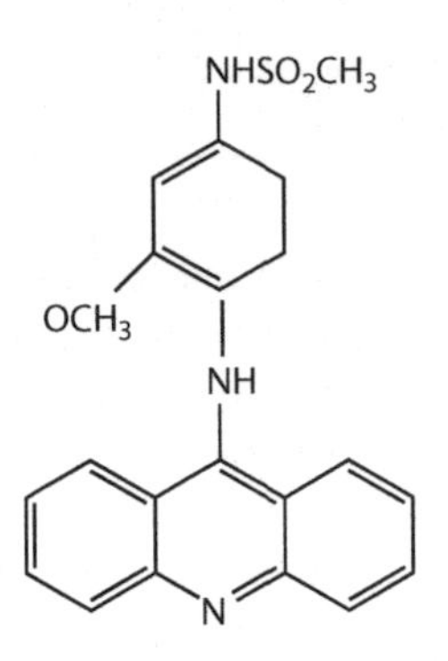

Amsacrine

Mitoxantrone

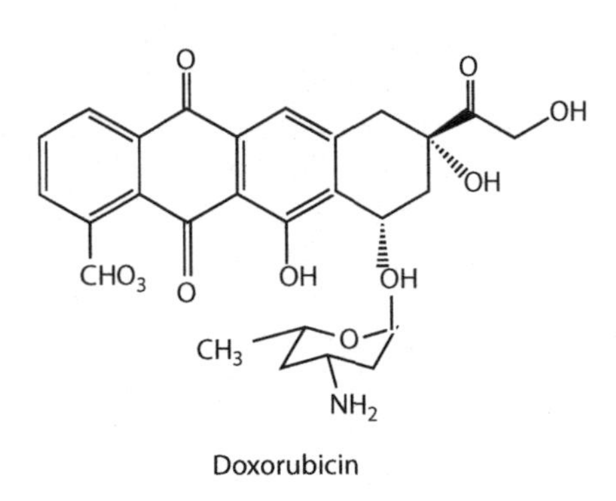

Doxorubicin

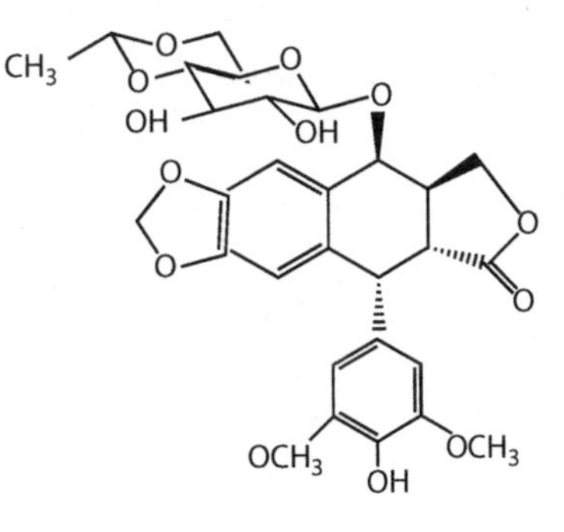

Etoposide

FIGURE 7.4 Chemical structure of topoisomerase II inhibitors.

Etoposide-induced DNA damage initiates cell signaling pathways that culminate in apoptosis. Caspase 2 has been identified as an important link between etoposide-induced DNA damage and the engagement of the mitochondrial apoptotic pathway [9]. Caspase 2 activates caspase 8, resulting in mitochondrial damage and subsequent downstream activation of caspases 9 and 3. Cells lacking caspase 3 are resistant to etoposide. Caspase 10 seems to trigger a feedback amplification loop that amplifies caspases 9 and 3. The tumor necrosis factor-related apoptosis inducing ligand (TRAIL) augments the expression of caspases induced by etoposide. Other cell-cycle control proteins are also important mediators in etoposide-induced apoptosis. p53, c-Myc, and BAFF have been identified as pathways utilized to arrest cell-cycle progression and induce apoptosis in certain cell lines exposed to etoposide. Etoposide activates two pathways, which lead to G_2/M arrest, one of which depends on the presence of p53 whereas the other is p53-independent. The presence of bcr-abl, which prolongs G_2/M arrest and allows for DNA repair mechanisms, decreases the cytotoxicity of etoposide whereas cells with dysfunctional early G_2/M checkpoint control (such as ataxia–telangiectasia mutated deficient fibroblasts) have increased chromosomal abnormalities following etoposide exposure.

7.2.6.1.1 Teniposide Teniposide is an analogue of etoposide that was approved for use in the United States in 1993 (10 years after etoposide was approved). Teniposide use has been limited primarily to the treatment of childhood lymphomas and leukemias, and for the treatment of CNS malignancies. However, it may have clinical efficacy equivalent to etoposide given its similar preclinical activity and toxicities. A few studies comparing the activity of these two agents have been performed.

7.2.6.2 Anthracyclines

Anthracycline antibiotics are commonly used antineoplastic agents with activity against breast cancer, leukemias, lymphomas, and sarcomas. The anthracyclines currently approved for use in the United States are doxorubicin, daunorubicin, epirubicin, and idarubicin. Anthracyclines inhibit topoisomerase II but also intercalate into DNA and form reactive metabolites that interact with many intracellular molecules. However, although the biological effects of anthracyclines may not be based solely on topoisomerase II activity, the ability of anthracycline analogues to poison topoisomerase II correlates well with the cytotoxic potential of the drug [21].

Anthracyclines trigger apoptotic cell death through complex signaling pathways. NFκB activation and IκBα degradation are early events triggered by anthracyclines. Cathepsin B is expressed through NFκB and TRAIL; p53 and the FAS/FAS–ligand system are additional pathways involved in anthracycline-induced apoptosis in various cell lines.

7.2.7 Radiosensitization by Inhibitors of Topoisomerase II

7.2.7.1 Etoposides

A number of experimental studies have been reported; the results of which cast light on etoposide radiosensitization.

- One of the earliest demonstrations of radiosensitization by topoisomerase II inhibitors described experiments using hamster lung fibroblast cells (V79 cells). Cells were exposed to etoposide or idarubicin for 24 h before or immediately after irradiation. Postradiation drug treatment resulted in radiosensitization, as demonstrated by a decrease in the mean inactivation dose, whereas exposure to either drug before irradiation resulted in no radiosensitization. Isobologram analysis revealed a synergistic interaction if etoposide exposure followed irradiation whereas the interaction resulting from the combination of irradiation and preradiation etoposide was within the envelope of additivity. Irradiation and postradiation idarubicin exposure also resulted in an additive interaction whereas preradiation idarubicin produced a subadditive response. In effect, synergistic interactions were more likely when the drug followed irradiation. Experiments at various ratios of radiation dose to drug concentration showed that the likelihood of synergism increased as the drug concentration increased relative to the radiation dose [22].

- In another study of the interaction between ionizing radiation and etoposide, again using asynchronous V79 fibroblasts, synergistic cell killing was observed when radiation was prior to or concomitantly with the drug. Kinetic analysis of radiation recovery suggested that etoposide targeted two interactive repair mechanisms. First, rapidly repairable radiation-induced DNA damage was fixed into lethal lesions by etoposide, giving a superadditive interaction with concomitant radiation–drug exposure. Second, cells arrested in G_2 phase following radiation were hypersensitive to the cytotoxic effect of etoposide. The shoulder of the radiation survival curve was abolished when γ-rays and drug were applied at 1-h intervals. This effect seemed to be correlated with a differential sensitivity of the various phases of the cell cycle to drug and radiation [23,24].
- A more recent study investigated GL331, a semisynthetic topoisomerase II inhibitor, which like etoposide is derived from the plant toxin, podophyllotoxin. GL331 was shown to exert cytotoxic effects on the T98G human glioma cells in a concentration-dependent and time-dependent manner. The combination treatments with radiation increased the dose-enhancement ratio to a much greater extent when GL331 treatment was preceded by radiation, with isobologram analysis revealing mainly supra-additive effects. GL331 and radiation separately caused the glioma cells to accumulate in the G_2/M phase of the cell cycle and combined treatment further increased the G_2/M fraction of the glioma cells. It was concluded that GL331

radiosensitization was sequence-dependent, with stronger cytotoxic effects when radiation was delivered concomitantly with the beginning of GL331 treatment. Such radiosensitization effects might be, at least partly, related to the increased cell accumulation in the G_2/M phase. In addition, post-irradiation GL331 caused a persistent inhibition of topoisomerase II function, such that the normally rapidly repairable radiation-induced DNA damage would be exacerbated by subsequent GL331 treatment and fixed into lethal lesions, resulting in supra-additive or synergistic cytotoxic effects [24].

Taken together, the results of these three studies suggested that etoposide is a nonintercalating topoisomerase II inhibitor that enhanced radiosensitivity through effects on radiation repair and the cell cycle.

7.2.7.2 Diverse Modes of Radiosensitization by Topo II Inhibitors

Topoisomerase II inhibitors are chemically diverse and it seems likely that radiosensitization does not always involve the same mechanism.

- Vosaroxin is a naphthyridine analogue structurally related to quinolone antibacterials, it intercalates DNA and inhibits topoisomerase II. In experiments where vosaroxin radiosensitized U251, DU145, and MiaPaca-2 cells; G_2 arrest was not observed at the doses of vosaroxin used, indicating that it was unlikely to be a mechanism of radiosensitization. From the other results, it seems that there was no significant initial increase in DNA damage based on γ-H2AX levels at early time points after radiation suggesting no increase in the actual number of DNA-DSBs but, at 24 h after radiation, there was an increase in γ-H2AX foci retention suggesting that vosaroxin inhibits the repair of radiation-induced DNA damage in U251 cells [25]. Vosaroxin also caused greater than additive growth delay in irradiated U251 tumors.

Amrubicin hydrochloride (AMR), an anthracycline derivative, exhibits antitumor effects by inhibiting topoisomerase II activity and suppressing DNA synthesis. AMR is metabolized to amrubicinol (AMROH) *in vivo* and AMROH has antitumor activity against human NSCLC, SCLC, and superficial bladder cancer, with anticancer effects that range from 10 to 100 times greater than AMR. There seems to be a unique correlation between the sensitivity to ionizing radiation and to topoisomerase II inhibitors, which stabilize the cleavable complex between the enzymes and DNA.

Experiments with amrubicin and amrubicinol in lung adenocarcinoma cells showed similar enhancement of radiosensitivity as that obtained with vosaroxin with enhancement ratios of 1.38 and 1.57, respectively. Radiosensitivity was enhanced by sequential combined treatment with AMR or AMROH by reduction of sublethal damage repair (SLDR). When the cells were treated with AMR or AMROH prior to fractionated irradiation, the cytocidal effect at 8 Gy on the survival curve was approximately 30 times greater than that with irradiation alone. As was the case with vosaroxin, there was no increase in apoptosis when cells were irradiated and treated with topoisomerase inhibitors but an increased proportion of necrotic cells after radiation and drug treatment with subadditive increases in the percentage of necrosis in the combination-treated cells [26].

7.2.8 Topoisomerase II Inhibitors in Concurrent Radiochemotherapy

Etoposide combined with other chemotherapy and concurrent radiotherapy has been used for a number of years and remains the standard approach for some forms of lung cancer, including locally advanced NSCLC and limited stage-SCLC (LS-SCLC). Cisplatin plus etoposide with concurrent twice-daily thoracic radiotherapy resulted in a 5-year survival rate of 25% for patients with LS-SCLC (reviewed by Ohe [27]). Rigas and Kelly [28] reported similar findings. The standard chemotherapy remains cisplatin and etoposide, but carboplatin is frequently used in patients who cannot tolerate or have a contraindication to cisplatin [29].

Thirty-eight clinical trials involving concurrent etoposide and radiotherapy are currently listed (clinicaltrials.gov). Thirty-four of these target lung cancer, and treatment is based on the standard radiochemotherapy protocol, which includes etoposide, with variations in terms of the radiation protocol and by the addition chemotherapy drugs or biological agents.

7.3 RADIOSENSITIZATION BY TARGETING MICROTUBULES

Microtubules are a fundamental component of the cytoskeleton in eukaryotic cells. They consist of filamentous and tube-like protein polymers composed of α- and β-tubulin heterodimers with a molecular weight of approximately 100 kDa. In living cells, microtubules are characterized by their dynamic alternating polymerization/depolymerization, which leads to net elongation/shortening of the filaments. During the cell cycle, microtubule dynamics are precisely controlled to regulate a variety of important processes, including cell shape maintenance, intracellular transportation, signal transduction, mitotic spindle formation, and cell division. Any disturbance of microtubule dynamics may lead to cell-cycle arrest or cell death.

The dynamics of microtubules comes from the continuous addition and loss of tubulin heterodimers at the polymer ends (dynamic instability). Microtubule filaments have the potential to grow and shrink; however, there is a behavioral difference between the end units: at the plus end, where β-tubulin is exposed, the addition and loss are more actively coupled to GTP hydrolysis than is the case at the minus end [30–32]. In an *in vitro* tubulin polymerization system, the growing and shrinking of the microtubule will continue until the system reaches equilibrium, which occurs when the level of free tubulin reaches a critical concentration. However, as the critical concentration for tubulin at the plus end is lower than that at the minus end, there exists a net addition of tubulin heterodimers at one end and balanced net loss at the other. This phenomenon is termed "treadmilling." Dynamic instability together with treadmilling contributes to microtubule dynamics [33].

7.3.1 Microtubule-Targeted Drugs

Microtubule-targeted drugs (MTAs) are arbitrarily classified into two large groups according to their apparent mechanism of action:

i. Microtubule-destabilizing agents (MDAs), such as the Vinca alkaloids and the colchicines, which prevent the polymerization of tubulin and promote the depolymerization of filamentous microtubules.

ii. Microtubule-stabilizing agents (MSAs), such as taxanes and epothilones, which promote the polymerization of tubulin to microtubules, further stabilizing them and thereby preventing depolymerization. Both of these microtubule-perturbing classes of agents have been successfully used in clinical cancer chemotherapies (Figure 7.5).

7.3.2 Microtubule-Stabilizing Drugs: Cytotoxic Effects

The structures of the two most important MSAs, paclitaxel and docetaxel, are shown in Figure 7.6. The perceived mechanism regarding the antimitotic

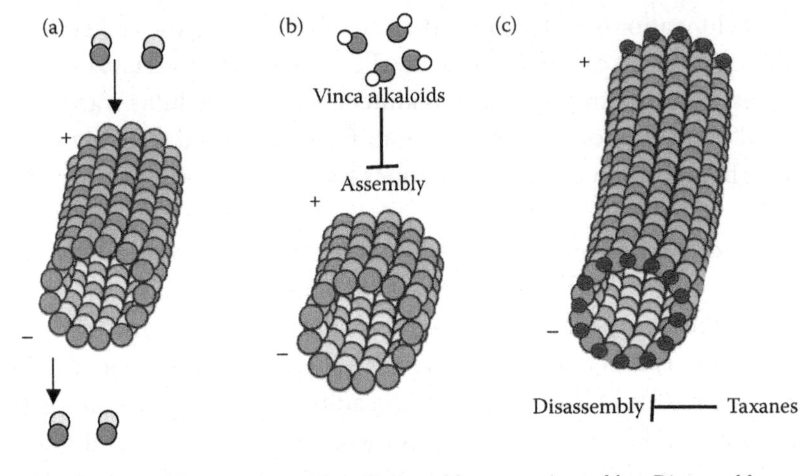

FIGURE 7.5 (a) Microtubules are highly dynamic, hollow fiber-like structures, which are composed of 13 filaments. Each filament is a tubulin polymer, with alternating α- and β-tubulin subunits. New tubulin subunits are added to the plus end of the microtubule while degradation occurs on the minus end. The regulation of both assembly and disassembly ensures proper microtubule dynamics. (b) Microtubule-targeting anticancer drugs affect the microtubule network in different ways. Vinca alkaloids bind to β-tubulin monomers, preventing them from being incorporated in growing microtubules. The balance between assembly and disassembly is distorted, and the cellular microtubule network collapses, causing mitotic failure. (c) Taxanes, such as paclitaxel and docetaxel, have the opposite effect. Instead of preventing microtubule assembly, these drugs interact with β-tubulin that has already been incorporated into microtubules. They stabilize the structure and prevent disassembly on the minus end. As a result, cells become overloaded with microtubules, again causing a mitotic crisis.

action of MSAs is that they interfere with the dynamics of spindle microtubules in the cell, so that the daughter chromosomes cannot align at the equatorial plate and move to the two poles of the mitotic spindle. Cells that fail to pass through mitotic checkpoints are arrested at the metaphase–anaphase transition and, as a consequence, are subjected to intrinsic mitochondrial apoptosis [32,34].

The addition of MSAs to the system significantly perturbs the balance between polymerized microtubules and free tubulin. Thus, MSAs at high micromolar concentrations reduce the critical concentration of tubulin, leading to an increase of microtubule polymer mass whereas MSAs

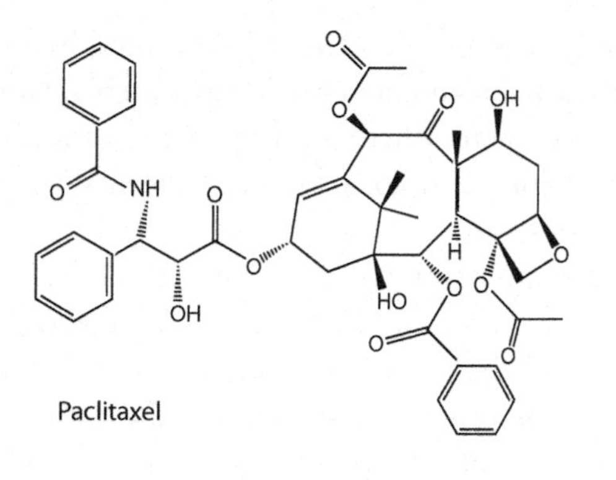

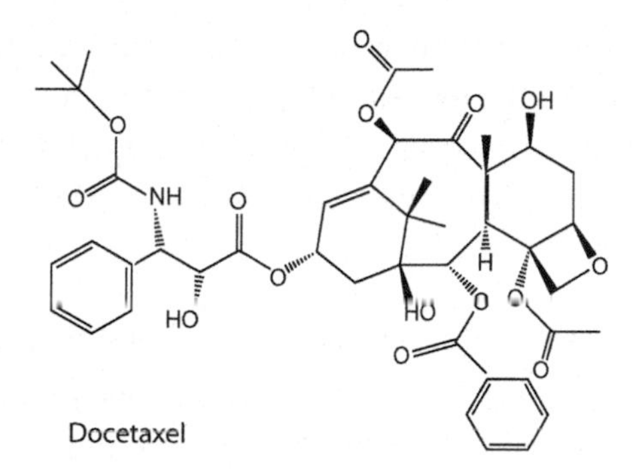

FIGURE 7.6 Chemical structure of taxanes used in radiochemotherapy.

at nanomolar or submicromolar concentrations suppress microtubule dynamics, including both dynamic instability and treadmilling, so that the formed filaments seem to be more stable. At the cellular level, the perturbation of microtubule formation by small molecules has been shown to affect downstream signaling pathways, leading to cell-cycle arrest and cell death, this was formerly believed to be the primary mechanism of action of the MSAs. However, cancer cell growth can be inhibited by several MSAs at much lower concentrations (nanomolar) than those needed for tubulin polymerization promotion or stabilization [35,36], and no significant increase in the mass of polymerized microtubules was observed in cells whose growth has been inhibited by MSAs. These findings could not be explained simply on the basis of direct correlation between microtubule formation/stabilization and cytotoxic activity observed and, in fact, it is the level of suppression of microtubule dynamics rather than the change of polymer mass that correlates with cell growth inhibition caused by MSAs at low concentrations [37,38]. In the cellular system, there are factors that intrinsically control microtubule dynamics. These include a family of proteins called microtubule-associated proteins, which bind to microtubules and stabilize them, whereas oncoprotein stathmin/oncoprotein [39] and some microtubule motor proteins (kinesins) are involved in destabilizing microtubules [40].

Cell-cycle arrest does not result in apoptosis unless the G_2/M checkpoint is disrupted. In the latter case, cells will either apoptose during mitosis upon activation of the spindle checkpoint, or following mitotic catastrophe, in which cells progress through an aberrant mitosis after which the cellular default apoptotic pathway is activated. Alternatively, mitotic slippage, in which cells exit from mitosis in the absence of chromosome segregation, can result in the generation of tetraploid cells, which apoptose in the presence of an intact tetraploid checkpoint. When this checkpoint is incapacitated, however, mitotic slippage can give rise to aneuploid cells. The cytotoxic effects of MSAs are summarized in Figure 7.7.

7.3.3 Radiosensitization by Microtubule-Stabilizing Drugs

Both paclitaxel and docetaxel have undergone extensive preclinical testing, both *in vitro* and *in vivo*, for their interaction with radiation (reviewed by Milas et al. [41] and Golden et al. [42]).

7.3.3.1 In Vitro *Studies*

It was shown that both agents sensitize tumor cells to radiation, although the effect did not always occur to the same extent. The degree of radioenhancement varied, ranging from as low as 1.1 to as high as 3.0 [43], depending on cell type, proliferation status, drug concentration, and duration of exposure to the drug. Enhancement was usually observed when the cells were incubated with the drug before irradiation, when moderate to high concentrations of drugs were present in culture medium for extended periods of time (24 h and longer), and when radiation was delivered at the time of the accumulation of cells in the G_2 and M phases of the cell cycle [43]. This implies that cells arrested in G_2/M, or a significant percentage of them, were not destined to die from the drug alone. On the other hand, an additive effect rather than radioenhancement was observed in cell lines whose cells entered mitotic arrest after taxane treatment and then underwent apoptotic death [44,45]. Unlike paclitaxel, docetaxel seems to radiosensitize cells through an additional cell-cycle mechanism (i.e., by being cytotoxic to radioresistant S phase cells) [45].

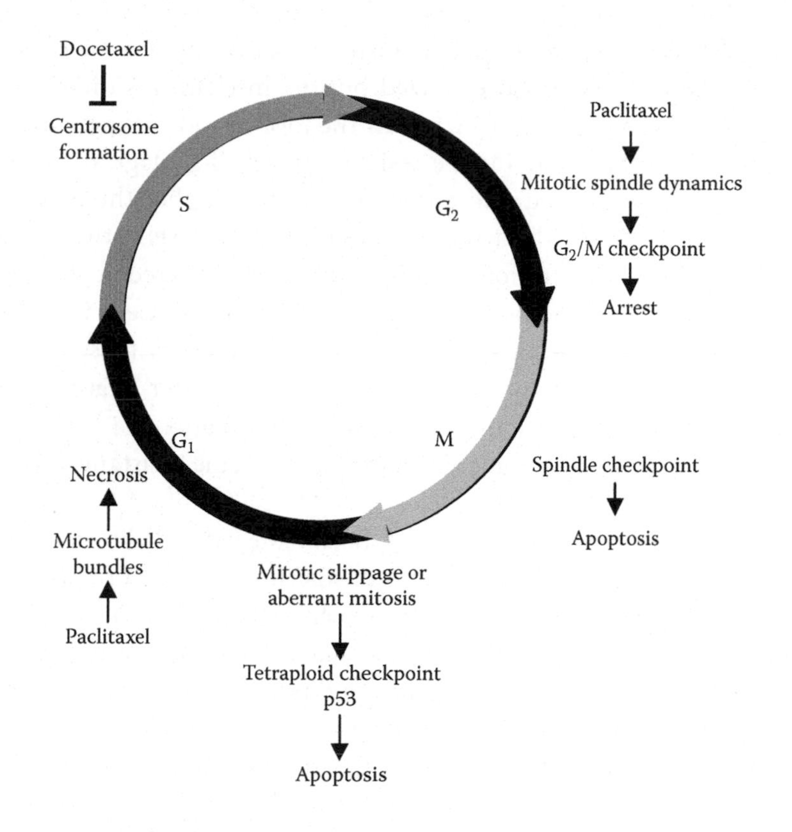

FIGURE 7.7 Mode of action of taxanes. Taxane-induced cytotoxicity. The cytotoxic effects of taxanes do not only depend on the specific drug, but also on the concentration that is used and on the presence of cell-cycle checkpoints in the cell that is being affected. After inhibition of mitotic spindle disassembly in G_2/M, cytotoxicity requires the activation of the spindle assembly checkpoint to cause mitotic arrest. Alternatively, aberrant mitosis (with improper chromosome segregation) or mitotic slippage (an improper exit from mitosis in the absence of chromosome segregation and cytokinesis producing tetraploid cells) may occur, both of which result in apoptosis in the presence of an intact p53 checkpoint. If all cell-cycle checkpoints are incapacitated, taxane treatment may result in aneuploidism.

7.3.3.2 In Vivo *Studies*

Both paclitaxel and docetaxel have been found to be effective in enhancing tumor radioresponse on the basis of the duration of radiation-induced tumor growth delay, tumor radiocurability rate, and delayed appearance of postradiation recurrences. Again, a range of enhancement factors was reported; from 1.2 to 3.3 [43,46]. At the level of the tumor, taxanes seem to affect radiosensitization by a number of mechanisms in addition to what is observed in tissue culture systems.

7.3.3.2.1 Mitotic Arrest Both taxane-resistant [46,47] and taxane-sensitive tumors [43,48] can be radiosensitized but the mechanisms differ in the two classes of tumors: mitotic arrest in the former and reoxygenation of hypoxic cells in the latter. Taxane-resistant tumors histologically display mitotic arrest several hours to a day after treatment, but these cells do not undergo apoptosis. The arrest reaches the highest level between 6 and 9 h after drug administration, and if radiation is delivered at that time, the highest degree of radiosensitization is achieved [49]. Cells in taxane-sensitive tumors are also arrested in mitosis but most of these cells die either by apoptosis or by necrosis within a few days after treatment. For these tumors, the taxane-induced response peaks between 1 and 3 days [43,48] and, in general, it is greater than that in taxane-resistant tumors.

7.3.3.2.2 O_2-Dependent Mechanisms of Taxane Mediated Radiosensitization Milas and colleagues [48] demonstrated experimentally that radiosensitization was dependent on the level of molecular oxygen within the tumor tissue. In their experiment, paclitaxel enhanced tumor responsiveness to radiation in mice under ambient conditions, but this enhancement was abolished when radiation was delivered after the induction of tissue hypoxia by leg clamping. Oxygen-dependent radiosensitization by paclitaxel seemed to be related to two processes: the selective killing of aerated cells resulting in their removal, and the greater accessibility of oxygen to previously hypoxic cells and the removal of irradiated cells reducing interstitial pressure, increasing capillary flow, and improving tumor oxygenation.

Experimentally, paclitaxel was shown to decrease the mean interstitial pressure by 36% and to improve tumor partial oxygen pressure by almost 100% in breast cancer patients treated with neoadjuvant chemotherapy in a phase II randomized clinical study [50]. Further studies demonstrated that the paclitaxel-associated increase in tumor blood flow and tumor oxygen delivery improved the efficacy of tumor cell killing with a subsequently delivered fraction of radiation [51]. This observation further emphasized the importance of the timing of taxane administration in relation to the delivery of fractionated radiation.

7.3.3.2.3 Angiogenesis Taxanes have been reported to inhibit angiogenesis as well as the prosurvival signaling pathways activated by VEGF and bFGF, further underscoring the complexity of the processes initiated when taxanes interact with tumor systems. When combined with radiation, paclitaxel inhibits endothelial cell proliferation, migration, and tubule formation at concentrations far less than those required for tumor cell killing [52]. In addition, docetaxel has been shown to inhibit endothelial cell proliferation, migration, tubule formation, and the number of newly formed blood vessels *in vivo* in a dose-dependent manner [53]. Taxanes were also able to reduce the expression of VEGF—a potent stimulator of angiogenesis and blood vessel permeability in the tumor microenvironment [54]. In this way, the antiangiogenic properties of taxanes may lead to an improvement in oxygenation by tumor blood vessel normalization and enhanced radiation-associated tumor cell killing.

7.3.3.2.4 Apoptosis Taxanes have been shown to radiosensitize tumor cells by apoptotically mediated mechanisms independent of p53 status [55,56]. In several tumor types (including colorectal and head and neck cancers), low-dose fractionated radiation enhanced the radiosensitization effect of both paclitaxel and docetaxel [56,57]. These studies also demonstrated that the taxanes radiosensitized irrespective of p53 status, by increasing the expression of p21 while simultaneously decreasing the expression of antiapoptotic proteins (bax, bcl-2, and NFκB; [56,57]). This occurs, however, only with fractionated radiation.

7.3.4 Clinical Application of Taxanes and Radiation

Taxanes are used extensively in the treatment of cancer. Paclitaxel is used to treat patients with lung, ovarian, breast, head and neck cancer, and advanced forms of Kaposi sarcoma. Docetaxel (generic or under the trade name Taxotere) is clinically well established and used mainly for the treatment of breast, ovarian, prostate, and NSCLC.

The effects of paclitaxel or docetaxel and concurrent radiotherapy have been investigated in a number of phase I/II clinical trials, often in conjunction with additional chemotherapy (reviewed by Golden et al. [42]). The published studies listed in this review were concerned with breast cancer (4) published between 1997 and 2012; cervical cancer (1) from 2010; head and neck cancer (8) from 2001 to 2011; and prostate cancer (5) from 2004 to 2012.

The referenced studies included the following:

- Anthracycline chemotherapy followed by paclitaxel is a standard adjuvant treatment for node-positive breast cancer. Concurrent paclitaxel chemotherapy and radiotherapy is being actively investigated in a number of trials. In one published study, concurrent paclitaxel chemotherapy and radiotherapy after breast-conserving surgery shortened total treatment time, provided excellent local control, and was well tolerated [58].

- Another published report describes a phase I trial of weekly docetaxel with concurrent three-dimensional conformal radiation therapy (3D-CRT) in unfavorable localized adenocarcinoma of the prostate. Assessment of the maximum tolerated dose of docetaxel indicated the feasibility of a dose of 20 mg/m^2. A prospective phase II trial with concurrent and adjuvant docetaxel combined with 3D-CRT and androgen deprivation in high-risk prostate cancer was designed. It was concluded that 3D-CRT with androgen deprivation and concurrent weekly docetaxel, followed by three cycles of adjuvant docetaxel, may be considered feasible in high-risk prostate cancer and deserved to be evaluated in a phase III randomized trial [59].
- A prospective phase II study to determine the response rate, toxicity, and 2-year survival rate of concurrent weekly paclitaxel and radiation therapy (RI) for locally advanced unresectable NSCLC was conducted. The weekly paclitaxel regimen was designed to optimize the radiosensitizing properties of paclitaxel. The survival outcome from

TABLE 7.2 Ongoing Clinical Trials with Paclitaxel and Concurrent Radiotherapy

	No. of	Phase			
Site of Tumor	**Trials**	**1**	**2**	**3**	**Additional Chemotherapy**
Lung	18	4	9	5	Cisplatin, MMC, irinotecan, tirapazamine, dasatinib gemcitabine, gefitinib, palefirmine, etoposide, thalidomide, lapatinib
Head and neck	13	1	11	1	Cetuximab, erlotonib, bevacizumab, FU, cisplatin, carboplatin, HU, Iressa
Esophagus	6	1	4	1	Gemcitabine, Abraxane, Tarceva, cisplatin, cetuximab, oxaliplatin, FU, celecoxib
Cervix	5	1	1	3	Cisplatin, carboplatin
Cervix	1			1	Dactinomycin, cyclophosphamide, vincristine
Pancreas	1		1		Cisplatin, irinotecan
Endometrium	2		1	1	Cisplatin, carboplatin, bevacizumuab
Thyroid			1		

TABLE 7.3 Ongoing Clinical Trials with Docetaxel and Concurrent Radiotherapy

Site of Tumor	No. of Trials	Phase			Additional Chemotherapy
		1	2	3	
Lung	18	2	14	2	Cisplatin, carboplatin, gefitinib, pemetred, filgastrim, pegfilgastrim
Head and neck	17	1	14	2	Cisplatin, cetuximab, bevacizumab, FU, carboplatin, gefitinib
Esophagus	5	1	2	2	Capecitabine, irinotecan, FU, cisplatin
Pancreas	2		2		Filgastrim, pegfilgastrim, gemcitabine, FU, leucovorin
Prostate	5	2	3		Casodex, Zoladex, Elgard, bicalutumide, leucoprolide
Breast	1	1			Cyclophosphamide

this regimen was considered to be encouraging and at least equivalent to that of other chemotherapy/radiation trials. Further clinical evaluation of weekly paclitaxel/RT in phase II trials was suggested [60].

There are currently approximately 110 clinical trials of paclitaxel and 90 trials of docetaxel with concurrent radiotherapy, which are ongoing or recently completed. To give an idea of the range of these trials, details from the first 50 in each group are listed in Tables 7.2 and 7.3.

7.3.5 Microtubule-Stabilizing Agents: The Vinca Alkaloids

7.3.5.1 Cytotoxicity

The Vinca alkaloids (Figure 7.8) target tubulin and microtubules depolymerizing microtubules and destroying mitotic spindles at high concentrations (10–100 nM), leaving dividing cancer cells blocked in mitosis with condensed chromosomes. At low but clinically relevant concentrations, vinblastine does not depolymerize spindle microtubules, yet it powerfully blocks mitosis (IC_{50} 0.8 nM) and cells die by apoptosis. Studies on the mitotic-blocking action of low Vinca alkaloid concentrations in cancer cells indicate that the block is due to the suppression of microtubule dynamics rather than microtubule depolymerization. Vinblastine binds to the β-subunit of tubulin dimers at the Vinca-binding domain, the binding site for some other chemotherapeutic drugs. The binding of vinblastine to soluble tubulin is rapid and reversible, and the binding of vinblastine induces a conformational change in tubulin for tubulin self-association (reviewed by Jordan and Wilson [32]).

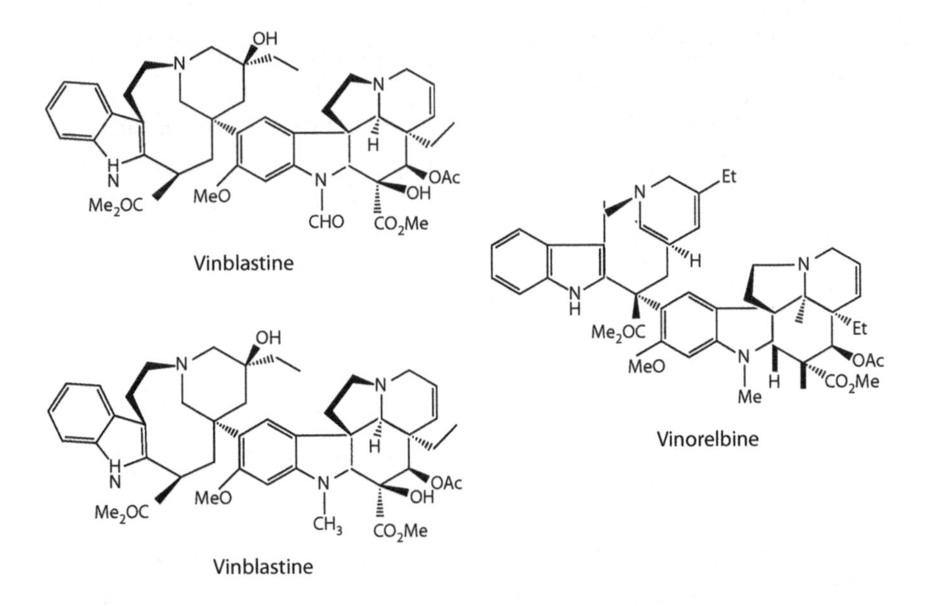

FIGURE 7.8 Chemical structures of Vinca alkaloids used in radiochemotherapy.

The ability of vinblastine to increase the affinity of tubulin for itself probably has a key role in the ability of the drug to stabilize microtubules kinetically. Vinblastine also binds directly to microtubules. *In vitro*, vinblastine binds to tubulin at the extreme microtubule ends with very high affinity, but it binds with markedly reduced affinity to tubulin that is buried in the tubulin lattice. The binding of one or two molecules of vinblastine per microtubule plus end is sufficient to reduce both treadmilling and dynamic instability by approximately 50%, without causing appreciable microtubule depolymerization. For example, the drug strongly reduces the rate and extent of microtubule growth and shortening, and increases the percentage of time the microtubules spend in an attenuated or paused state, neither growing nor shortening detectably. In mitotic spindles, slowing of the growth and shortening or treadmilling dynamics of the microtubules blocks mitotic progression. This suppression of dynamics has at least two downstream effects on the spindle: it prevents the mitotic spindle from assembling normally and it reduces the tension at the kinetochores of the chromosomes. Mitotic progress is delayed in a metaphase-like state with chromosomes often stuck at the spindle poles, unable to congress to the spindle equator. The cell-cycle signal to the anaphase-promoting complex to pass from metaphase into anaphase is blocked and the cells eventually die because of apoptosis.

7.3.5.2 Radiosensitization by the Vinca Alkaloids: Preclinical Studies

- Radiosensitization by vinorelbine, a semisysnthetic Vinca alkaloid, was demonstrated in human SCLC SBC-3 cells by clonogenic assay. SBC-3 cells were sensitized to radiation by vinorelbine using different schedules combining exposure to both. Flow cytometric analyses showed that the cells did not accumulate in the radiosensitive G_2/M phase of the cell cycle after concurrent treatment with vinorelbine and radiation. The results of an alkaline filter elution assay demonstrated that, in the presence of vinorelbine at 1 nM concentration, radiation-induced DNA strand breaks were not completely repaired at 24 h postradiation. It was concluded that human SCLC SBC-3 cells are sensitized to radiation by vinorelbine and that one possible mechanism could be the impairment of DNA repair following radiation-induced DNA damage [61].
- Clinical studies of gemcitabine and vinorelbine support the use of this combination concurrently with radiotherapy in the treatment of NSCLC. In an *in vitro* study, NCI-H460 cells were exposed to gemcitabine or vinorelbine (or both) simultaneously or sequentially, followed by irradiation with 0 to 10 Gy. Both drugs showed single-agent activity against NCI-H460 cells and targeted different phases of the cell cycle. When used in combination, they showed schedule-dependent interaction with an antagonistic effect observed with simultaneous exposure and the optimum scheduling being exposure to vinorelbine followed by gemcitabine 24 h later. Radiosensitization by gemcitabine was evident when radiation exposure was done immediately after 4 h of incubation whereas the radiosensitization effect of vinorelbine was observed for radiation given at 24 h post-incubation. Apoptosis induced by gemcitabine increased gradually, reaching 20% at 72 h posttreatment, whereas apoptotic cell death occurred early in vinorelbine-treated cells, reaching 40% at 24 h. It was concluded that the individual cytotoxic effects of gemcitabine and vinorelbine on NCI-H460 cells are phase-specific, and the combined effect of gemcitabine and vinorelbine is sequence-dependent. The radiosensitizing effects of both drugs seem to be related to enhanced apoptosis [62].
- Vinflunine (VFL) is a novel third-generation Vinca alkaloid with superior antitumor activity in preclinical models and an anticipated more favorable toxicity profile compared with other Vinca

alkaloids. The radiosensitizing properties of VFL and its cell-cycle effects in four human tumor cell lines (ECV304, MCF-7, H292, and CAL-27) were investigated. Twenty-four-hour treatment with VFL before radiation caused dose-dependent RS in all cell lines. The dose enhancement factor (DEF) ranged from 1.57 to 2.29 in the different cell lines and a pronounced effect was found in ECV304. A concentration-dependent G_2/M block was observed (starting at 4 h of incubation). After maximal G_2/M blockade, cells reentered the cell cycle, mainly by mitosis, with a small proportion of cells entering through the polyploid cell cycle. The polyploid cell population was most prominent after prolonged incubation with VFL, particularly in CAL-27 and ECV304, but was not observed in a tested normal fibroblast cell line (Fi 360). It was concluded that VFL has radiosensitizing potential but the mechanism is unclear [63].

7.3.5.3 Clinical Studies

The clinical use of Vinca alkaloids with other chemotherapy is extensive, and four of the most commonly prescribed—vinblastine, vincristine, vinorelbine (Navelbine), and vindesine—are used in the treatment of a wide range of solid tumors and hematological diseases (Table 7.4). Reports of these drugs used concurrently with radiation, however, are infrequent. This may be because although they are effective in a chemoradiation setting, they are not more effective than other more standard treatments such as cisplatin. The following reports have been published recently:

- Currently, concomitant chemoradiation using cisplatin is one of the standards of care for the management of head and neck cancer, but at the cost of increased acute toxicity. The aim of this study was to assess whether vinorelbine was less toxic and of at least comparable efficacy, if not better, compared with cisplatin. Vinorelbine was found to be as effective as cisplatin but was only marginally less toxic than cisplatin [64].
- In a small randomized trial, oral vinorelbine was tested against cisplatin as a radiosensitizing agent for the control of locoregional disease in locally advanced head and neck cancer. Patients were 65 years or older or diabetic and hypertensive; patients of any age were randomized to treatment by cisplatin or oral vinorelbine. Both drugs were administered weekly for six courses during pelvic external beam

TABLE 7.4 Ongoing Clinical Trials with Vinca Alkaloids and Concurrent Radiotherapy

Drug	Site of Tumor	Treatment	Phase 1	Phase 2	Phase 3	Additional Chemotherapy
Vinblatinee	Lung	CRT	1	1	1	Paclitaxel, carboplatin, cisplatin, efraproxiral, gemcitabine
	Melanoma	CRT		1		Aldesleukin, filgastrim dacabazone
Vincristine	Brain, CNS	CRT		3		Temozolomide, nedaplatin, carboplatin, filgastrim, cyclofosfamide
	Head and neck	CRT			2	FU, leucovorin, methotrexate
	Sarcoma	CRT	1		1	Irinotecan, bleomycin, busulfan, cyclofosfamide, dactinomycin, ifosfamide doxorubicin, melphalan
	Lymphoma	CRT		1		Cisplatin
Vinorelbine	Lung	CRT	2	1	3	Cisplatin, efaproxiral, gemcitabine, paclitaxel, pemetrexed, carboplatin, etoposide

radiotherapy and brachytherapy radiation. Efficacy and safety were assessed. Nineteen patients received oral vinorelbine, and 20 patients received cisplatin. Patients in both arms received a median of five applications of chemotherapy. Treatment was well tolerated in both arms, the most frequent toxicity being lymphopenia. At a median follow-up time of 16 months, there were no differences in either progression-free survival or overall survival between groups. It was concluded that these patient populations could safely be treated with either cisplatin or Navelbine as radiosensitizer. A larger randomized study would be needed to demonstrate the applicability of oral vinorelbine as an easier and practical alternative for cisplatin in cervical cancer [65].

7.4 SUMMARY

Radiosensitization by two disparate groups of chemotherapeutic drugs is described in this chapter. Inhibitors of topoisomerases disrupt processes involved in the management of the topology of DNA during replication and transcription whereas the targets for the taxanes and Vinca alkaloids are the microtubules, cytoskeletal elements that are mobilized prior to cell division to make up the mitotic spindle.

The cytotoxicity of drugs targeting topoisomerase 1 is a result of the poisoning of DNA through the trapping of a key reaction intermediate, called the TOP1 cleavable complex. A model of TOP1-mediated radiosensitization proposes that drug-trapped TOP1 cleavable complex interacts with the replication fork during active DNA synthesis and that some combination of radiation and drug-induced DNA damage is responsible for TOP1-mediated radiosensitization. Results of preclinical studies indicate TOP1-mediated radiation sensitization in mammalian cells to be an S phase-specific event that requires active DNA synthesis, suggesting that TOP1-targeting drugs might selectively radiosensitize proliferating cancer cells that are actively synthesizing DNA.

The most studied TOP1-targeting drugs, the camptothecin derivatives, irinotecan and topotecan, have shown clinical efficacy against a number of tumors. Clinical trials have demonstrated the feasibility and efficacy of TOP1 drugs in combination with thoracic radiotherapy for locally advanced NSCLC, and encouraging results have also been seen for brain cancers and for head and neck cancers. Investigation of the clinical application of TOP1 inhibitors and concurrent radiotherapy is being actively pursued.

Topoisomerase II is a multisubunit enzyme that uses ATP to pass an intact helix through a transient double-stranded break in DNA to modulate DNA topology. After strand passage, the DNA backbone is religated and DNA structure restored. Topoisomerase II inhibitors are a diverse class of anticancer drugs that include etoposides and anthracyclines, both of which induce DNA damage, initiating cell signaling pathways that culminate in apoptosis. Anthracyclines intercalate into DNA and induce signaling through several intracellular pathways, but the cytotoxic potential of the drug correlates well with the ability of anthracycline analogues to poison topoisomerase II. Based on experimental studies, radiosensitization by the nonintercalating topoisomerase II inhibitor results in the perturbation of radiation repair and the cell cycle. Etoposide combined with other chemotherapy and concurrent radiotherapy has been used for a number of years and remains the standard treatment for some forms of lung cancer including locally advanced NSCLC and LS-SCLC.

Microtubule-targeted drugs (MTAs) are arbitrarily classified into two large groups according to their apparent mechanism of action: the MDAs (Vinca alkaloids), which prevent the polymerization of tubulin and promote the depolymerization of filamentous microtubules; and the MSAs (taxanes), which promote the polymerization of tubulin to microtubules, further stabilizing them and thereby preventing depolymerization. Both categories of drug have been successfully used in clinical cancer chemotherapies.

MSAs interfere with the dynamics of spindle microtubules in the cell. Cells that fail to pass through mitotic checkpoints are arrested at the metaphase–anaphase transition and, as a consequence, are subjected to intrinsic mitochondrial apoptosis. In tissue culture, both paclitaxel and docetaxel sensitize tumor cells to radiation with the greatest sensitization being seen when radiation was delivered at the time of maximum accumulation of cells in the G_2 and M phases of the cell cycle. *In vivo*, both drugs have been found to enhance tumor radioresponse by a number of mechanisms in addition to what is observed in tissue culture systems including O_2-dependent mechanisms and inhibition of angiogenesis. Taxanes are used extensively in the treatment of cancer and are part of established protocols for several types of disease. The effects of paclitaxel or docetaxel and concurrent radiotherapy have been investigated in a number of phase I/II clinical trials, usually in conjunction with additional chemotherapy. There are a large number of ongoing phase I to III clinical trials involving docetaxel, paclitaxel, and other chemotherapy and concurrent radiotherapy.

The Vinca alkaloids target tubulin and microtubules, depolymerizing microtubules and destroying mitotic spindles at high concentrations, leaving dividing cancer cells blocked in mitosis with condensed chromosomes. At low but clinically relevant concentrations, depolymerization of spindle microtubules does not occur but the result may still be mitotic blockage and in either case the cells eventually die by apoptosis. Radiation sensitization has been demonstrated *in vitro* and *in vivo* with mechanisms including impairment of DNA repair and enhanced apoptosis. The clinical application of Vinca alkaloids in chemotherapy protocols is extensive but there is much less interest in the use of these drugs in radiochemotherapy, although the use of vinorelbine as a less toxic substitute for cisplatin has been investigated in a few trials.

REFERENCES

1. Gellert, M. DNA topoisomerases. *Annu Rev Biochem* 1981;50:879–910.
2. Wang, J. Cellular roles of DNA topoisomerases: A molecular perspective. *Nat Rev Mol Cell Biol* 2002;3(6):430–440.

3. Chen, A., and Liu, L. DNA topoisomerases: Essential enzymes and lethal targets. *Annu Rev Pharmacol Toxicol* 1994;34:191–218.
4. Li, T., and Liu, L. Tumor cell death induced by topoisomerase-targeting drugs. *Annu Rev Pharmacol Toxicol* 2001;41:53–77.
5. Hsiang, Y., Lihou, M., and Liu, L. Arrest of replication forks by drug-stabilized topoisomerase I-DNA complexes as a mechanism of cell killing by camptothecin. *Cancer Res* 1989;49:5077–5082.
6. Holm, C., Covey, J., Kerrigan, D., and Pommier, Y. Differential requirement of DNA replication for the cytotoxicity of DNA topoisomerase I and II inhibitors in Chinese Hamster DC3F cells. *Cancer Res* 1989;49:6365–6368.
7. Zhang, H., D'Arpa, P., and Liu, L. A model for tumor cell killing by topoisomerase poisons. *Cancer Cells* 1990;2:23–27.
8. Pantazis, P., Giovanella, B., and Rothenberg, M., eds. *The Camptothecins from Discovery to the Patient*. New York: New York Academy of Sciences, 1996.
9. Liehr, J., Giovanella, B., and Verschraegen, C., eds. *The Camptothecins—Unfolding Their Anticancer Potential.* New York: New York Academy of Sciences, 2000.
10. Takimoto, C., Wright, J., and Arbuck, S. Clinical applications of the camptothecins. *Biochem Biophys Acta* 1998;1400(1–3):107–119.
11. Pantazis, P., Early, J., Kozielski, A., Mendoza, J., Hinz, H., and Giovanella, B. Regression of human breast carcinoma tumors in immunodeficient mice treated with 9-nitrocamptothecin: Differential response of nontumorigenic and tumorigenic human breast cells *in vitro*. *Cancer Res* 1993;53:1577–1582.
12. Pantazis, P., Kozielski, A., Mendoza, J., Early, J., Hinz, H., and Giovanella, B. Camptothecin derivatives induce regression of human ovarian carcinomas grown in nude mice and distinguish between nontumorigenic and tumorigenic cells *in vitro*. *Int J Cancer* 1993;53:863–871.
13. Chen, A., Yu, C., Potmesil, M., Wall, M., Wani, M., and Liu, L. Camptothecin overcomes MDR1-mediated resistance in human KB carcinoma cells. *Cancer Res* 1991;51:6039–6044.
14. Pollack, I., Erff, M., Bom, D., Burke, T., Strode, J., and Curran, D. Potent topoisomerase I inhibition by novel silatecans eliminates glioma proliferation *in vitro* and *in vivo*. *Cancer Res* 1999;59(19):4898–4905.
15. Chen, A., Chou, R., Shih, S., Lau, D., and Gandara, D. Enhancement of radiotherapy with DNA topoisomerase I-targeted drugs. *Crit Rev Oncol/Hematol* 2004;50:111–119.
16. Chen, A., Chen, P., and Chen, Y. DNA topoisomerase I drugs and radiotherapy for lung cancer. *J Thorac Dis* 2012;4(4):390–397.
17. Chen, A., Okunieff, P., Pommier, Y., and Mitchell, J. Mammalian DNA topoisomerase I mediates the enhancement of radiation cytotoxicity by camptothecin derivatives. *Cancer Res* 1997;57:1529–1536.
18. Koukourakis, M., Bizakis, J., Skoulakis, C. et al. Combined irinotecan. *Anticancer Res* 1999;19:2309–2310.
19. Humerickhouse, R., Haraf, D., Stenson, K. et al. Phase I study of irinotecan (CPT-11), 5-FU, and hydroxyurea with radiation in recurrent or advanced head and neck cancer (abstract). *Proc Am Soc Clin Oncol* 2000;19:418a.
20. Hande, K. Topoisomerase II inhibitors. *Update Cancer Ther* 2008;3:13–26.

21. Tewey, K., Rowe, T., Yang, L., Halligan, B., and Liu, L. Adriamycin-induced DNA damage mediated by mammalian DNA topoisomerase II. *Science* 1984;226:466–468.
22. Haddock, M., Ames, M., and Bonner, J. Assessing the interaction of irradiation with etoposide or idarubicin. *Mayo Clin Proc* 1995;70:1053–1060.
23. Giocanti, N., Hennequin, C., Balosso, J., Mahler, M., and Favaudon, V. DNA repair and cell cycle interactions in radiation sensitization by the topoisomerase II poison etoposide. *Cancer Res* 1993;53(9):2105–2111.
24. Chen, Y., Lin, T.-Y., Chen, J.-C., Yang, H.-Z., and Tseng, S.-H. GL331, a topoisomerase II inhibitor, induces radiosensitization of human glioma cells. *Anticancer Res* 2006;26:2149–2156.
25. Gordon, I., Graves, C., Kil, W., Kramp, T., Tofilon, P., and Camphausen, K. Radiosensitization by the novel DNA intercalating agent vosaroxin. *Radiat Oncol* 2012;7:26–33.
26. Hayashi, S., Hatashita, M., Matsumoto, H., Shioura, H., Kitai, R., and Kano, E. Enhancement of radiosensitivity by topoisomerase II inhibitor, amrubicin and amrubicinol, in human lung adenocarcinoma A549 cells and kinetics of apoptosis and necrosis induction. *Int J Mol Med* 2006;18(5):909–915.
27. Ohe, Y. Chemoradiotherapy for lung cancer. *Expert Opin Pharmacother* 2005;6(16):2793–2804.
28. Rigas, J., and Kelly, K. Current treatment paradigms for locally advanced non-small cell lung cancer. *J Thorac Oncol* 2007;2(6 Suppl. 2):S77–S85.
29. Stinchcombe, T., and Gore, E. Limited-stage small cell lung cancer: Current chemoradiotherapy treatment paradigms. *Oncologist* 2010;15(2):187–195.
30. Altmann, K.-H., and Gertsch, J. Anticancer drugs from nature—Natural products as a unique source of new microtubule-stabilizing agents. *Nat Prod Rep* 2007;24:327–357.
31. Molodtsov, M., Ermakova, E., Shnol, E. et al. A molecular-mechanical model of the microtubule. *Biophys J* 2005;88:3167–3179.
32. Jordan, M., and Wilson, L. Microtubules as a target for anticancer drugs. *Nat Rev Cancer* 2004;4:253–265.
33. van Amerongen, R., and Berns, A. XR1-mediated thrombospondin repression: A novel mechanism of resistance to taxanes? *Genes Dev* 2006;20:1975–1981.
34. Jordan, M., Wendell, K., Gardiner, S., Derry, W., Copp, H., and Wilson, L. Mitotic block induced in HeLa cells by low concentrations of paclitaxel (Taxol) results in abnormal mitotic exit and apoptotic cell death. *Cancer Res* 1996;56:816–825.
35. He, L., Orr, G., and Horwitz, S. Novel molecules that interact with microtubules and have functional activity similar to Taxol. *Drug Discov Today* 2001;6:1153–1164.
36. Altmann, K. Microtubule-stabilizing agents: A growing class of important anticancer drugs. *Curr Opin Chem Biol* 2001;5:424–431.
37. Kamath, K., and Jordan, M. Suppression of microtubule dynamics by epothilone. *Cancer Res* 2003;63:6026–6031.
38. Honore, S., Kamath, K., Braguer, D. et al. Synergistic suppression of microtubule dynamics by discodermolide and paclitaxel in non-small cell lung carcinoma cells. *Cancer Res* 2004;64:4957–4964.

39. Aoki, S., Morohashi, K., Sunoki, T., Kuramochi, K., Kobayashi, S., and Sugawara, F. Screening of paclitaxel-binding molecules from a library of random peptides displayed on T7 phage particles using paclitaxel-photoimmobilized resin. *Bioconjug Chem* 2007;18:1981–1986.
40. Bhat, K., and Setaluri, V. Microtubule-associated proteins as targets in cancer chemotherapy. *Clin Cancer Res* 2007;13:2849–2854.
41. Milas, L., Mason, K., Liao, Z., and Ang, K. Chemoradiotherapy: Emerging treatment improvement strategies. *Head Neck* 2003;25:152–167.
42. Golden, E., Formenti, S., and Schiff, P. Taxanes as radiosensitizers. *Anti-Cancer Drugs* 2014;25(5):502–511.
43. Milas, L., Milas, M., and Mason, K. Combination of taxanes with radiation: Preclinical studies. *Semin Radiat Oncol* 1999;9:12–26.
44. Minarik, L., and Hall, E. Taxol in combination with acute and low dose rate irradiation. *Radiother Oncol* 1994;32:124–128.
45. Hennequin, N., Giocanti, N., and Favaudon, V. S-phase specificity of cell killing by docetaxel (Taxotere) in synchronized HeLa cells. *Br J Cancer* 1995;71:1194–1198.
46. Milas, L. Docetaxel-radiation combinations: Rationale and preclinical findings. *Clin Lung Cancer* 2002;3(Suppl. 2):S29–S36.
47. Milas, L., Mason, K., and Milas, M. Docetaxel and radiation in lung cancer: Preclinical investigations. *Adv Lung Cancer* 2001;3:6–8.
48. Milas, L., Hunter, N., Mason, K., Milross, C.G., Saito, Y., and Peters, L. Role of reoxygenation in induction of enhancement of tumor radioresponse by paclitaxel. *Cancer Res* 1995;55:3564–3568.
49. Mason, K., Kishi, K., Hunter, N. et al. Effect of docetaxel on the therapeutic ratio of fractionated radiotherapy *in vivo*. *Clin Cancer Res* 1999;5:4191–4198.
50. Taghian, A., Abi-Raad, R., Assaad, S. et al. Paclitaxel decreases the interstitial fluid pressure and improves oxygenation in breast cancers in patients treated with neoadjuvant chemotherapy: Clinical implications. *J Clin Oncol* 2005;23:1951–1961.
51. Milas, L., Hunter, N., Mason, K., Milross, C., and Peters, L. Tumor reoxygenation as a mechanism of taxol-induced enhancement of tumor radioresponse. *Acta Oncol* 1995;34:409–412.
52. Dicker, A., Williams, T., Iliakis, G., and Grant, D. Targeting angiogenic processes by combination low-dose paclitaxel and radiation therapy. *Am J Clin Oncol* 2003;26:e45–e53.
53. Sweeney, C., Miller, K., Sissons, S. et al. The antiangiogenic property of docetaxel is synergistic with a recombinant humanized monoclonal antibody against vascular endothelial growth factor or 2-methoxyestradiol but antagonized by endothelial growth factors. *Cancer Res* 2001;61:3369–3372.
54. Lissoni, P., Fugamalli, E., Malugani, F. et al. Chemotherapy and angiogenesis in advanced cancer: Vascular endothelial growth factor (VEGF) decline as predictor of disease control during taxol therapy in metastatic breast cancer. *Int J Biol Markers* 2000;15:308–311.

55. Dey, S., Spring, P., Arnold, S. et al. Low-dose fractionated radiation potentiates the effects of paclitaxel in wild type and mutant p53 head and neck tumor cell lines. *Clin Cancer Res* 2003;9:1557–1565.
56. Chendil, D., Oakes, R., Alcock, R. et al. Low dose fractionated radiation enhances the radiosensitization effect of paclitaxel in colorectal tumor cells with mutant p53. *Cancer* 2000;89:1893–1900.
57. Spring, P., Arnold, S., Shajahan, S. et al. Low dose fractionated radiation potentiates the effects of taxotere in nude mice xenografts of squamous cell carcinoma of head and neck. *Cell Cycle* 2004;3:479–485.
58. Chen, W., Kim, J., Kim, E. et al. A phase II study of radiotherapy and concurrent paclitaxel chemotherapy in breast-conserving treatment for node-positive breast cancer. *Int J Radiat Oncol Biol Phys* 2012;82:14–20.
59. Bolla, M., Hannoun-Levi, J., Ferrero, J. et al. Concurrent and adjuvant docetaxel with threedimensional conformal radiation therapy plus androgen deprivation for high-risk prostate cancer: Preliminary results of a multicentre phase II trial. *Radiother Oncol* 2010;97:312–317.
60. Choy, H., Safran, H., Akerley, W., Graziano, S., Bogart, J., and Cole, B. Phase II trial of weekly paclitaxel and concurrent radiation therapy for locally advanced nonsmall cell lung cancer. *Clin Cancer Res* 1998;4:1931–1936.
61. Fukuoka, J., Arioka, H., Iwamoto, Y. et al. Mechanism of vinorelbine-induced radiosensitization of human small cell lung cancer cells. *Cancer Chemother Pharmacol* 2002;49:385–390.
62. Zhang, M., Boyer, M., Rivory, L. et al. Radiosensitization of vinorelbine and gemcitabine in NCI-H460 non-small-cell lung cancer cells. *Int J Radiat Oncol Biol Phys* 2004;58(2):353–360.
63. Simoens, C., Vermorken, J., Korst, A. et al. Cell cycle effects of vinflunine, the most recent promising Vinca alkaloid, and its interaction with radiation, *in vitro*. *Cancer Chemother Pharmacol* 2006;58:210–218.
64. Sarkar, S., Patra, N., Goswami, J., and Basu, S. Comparative study of efficacy and toxicities of cisplatin vs vinorelbine as radiosensitisers in locally advanced head and neck cancer. *J Laryngol Otol* 2008;122:188–192.
65. Coronel, J., Cetina, L., Cantu, D. et al. A randomized comparison of cisplatin and oral vinorelbine as radiosensitizers in aged or comorbid locally advanced cervical cancer patients. *Int J Gynecol Cancer* 2013;23:884–889.

CHAPTER 8

Targeting the DNA Damage Response

ATM, p53, Checkpoints, and the Proteasome

8.1 THE DNA DAMAGE RESPONSE

Radiosensitization by targeting the DNA damage response—ATM, ATR, and associated pathways and checkpoints—is an intriguing option for therapy, particularly because this approach may achieve a degree of tumor specificity. Cell cycle checkpoint defects are known to decrease the ability to repair damaged DNA and to increase sensitivity to ionizing radiation (IR) and other DNA-damaging agents. The DNA damage response signal transduction network is shown schematically in Figure 8.1a.

The regulatory proteins involved in cell cycle progression, signal transduction, transcriptional regulation, DNA repair, and cell death are targeted by the ubiquitin-mediated proteasomal degradation system. The proteins targeted, which are of particular interest in the context of DNA damage response and DNA repair, are the subjects of this chapter and the succeeding chapter, and include $p21^{WAF}$, $p27^{KIP}$, p53, RAD51, cyclins (D,E,B), and PARP (Figure 8.1b). The BRCA1–BARD1 complex and the MDM protein, which target RAD51 and p53, respectively, are E3 protein ligases.

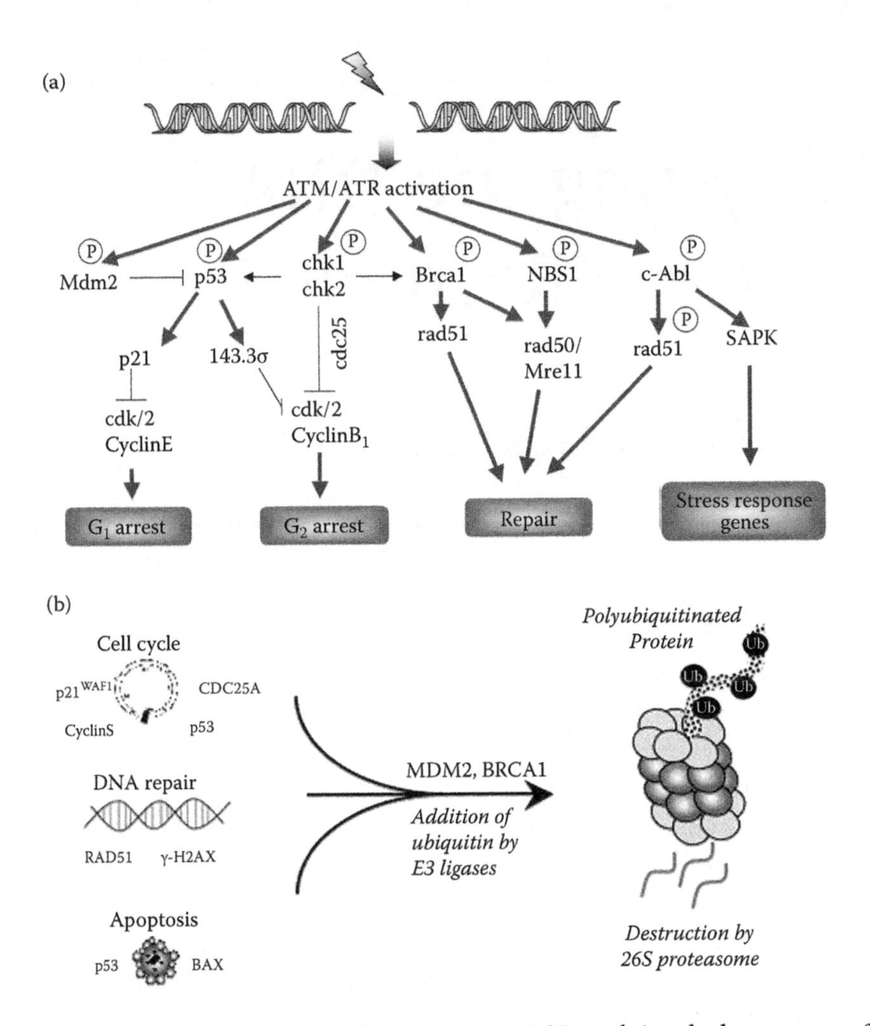

FIGURE 8.1 (a) ATM is activated in response to DSBs and signals the presence of DNA damage by phosphorylating downstream targets including histone H2AX and p53, 53BP1, CHK2, MDC1, NBS1, BRCA1, and SMC1 proteins. Downstream effectors of p53 are p21 and 14-3-3σ. p21 inhibits the activity of cdk2/cyclin E, and 14-3-3σ inhibits the activity of cdc2/cyclin B causing cell-cycle arrest, which is also mediated by activation of Chk1 and Chk2. c-Abl activates stress-activated protein kinase (SAPK) for transcriptional regulation of stress-response genes. Other proteins (BRCA1, NBS1) are involved in DNA repair. (b) Regulatory proteins involved in cell-cycle progression, DNA repair, signal transduction, and transcriptional regulation (e.g., cyclins, p53 and $p21^{WAF}$, RAD51, and BAX) are targeted for ubiquitin-mediated degradation of proteins by the 26S proteasome. Proteins that are targeted for destruction become polyubiquitinylated by E3 ligases. The BRCA1–BARD1 repair complex and the MDM2 protein (involved in p53 turnover) are E3 ligases.

8.2 ATM KINASE

In response to DNA breaks, normal human cells delay their progression through the G_1, S, and G_2 phases of the cell cycle, thus allowing time for repair of DNA damage. In the human disease, ataxia telangiectasia (A-T), a mutation of the ATM gene results in multiple cellular defects, including enhanced sensitivity to ionizing radiation, identifying ATM as a potential target for inhibitors sensitizing the tumor cell to radiotherapy. Cells obtained from patients with A-T display DNA damage checkpoint defects in G_1, S, and G_2 phases of the cell cycle, chromosomal instability, and radiosensitivity [1]. The ATM protein kinase has been characterized as a major regulator of the DNA damage response pathways, along with the closely related family members A-T and Rad3-related kinase (ATR) and DNA-dependent protein kinase (DNA-PK) [2]. ATM and the downstream targets of the ATM kinase are potential targets for the development of small-molecule inhibitors, which may radiosensitize tumor cells.

ATM is a member of the phosphatidylinositol 3-kinase-like family of serine/threonine protein kinases (PIKKs). Other members of this family include the ATM- and Rad 3-related (ATR) kinase that responds to single-stranded DNA, stalled or collapsed DNA replication forks, and the DNA-dependent protein kinase catalytic subunit (DNA-PKcs), which is a DNA double-strand break (DSB) repair protein. In an unperturbed cell, ATM exists as an inactive dimer (or higher-order oligomer). The introduction of DNA-DSB by ionizing radiation or other insults is sensed by the telomeric protein TRF2 and the MRE11–RAD50–NBS1 (MRN) complex causing the subsequent activation of the ATM kinase, which in turn activates signal transduction pathways essential for coordinating cell cycle progression with DNA repair. During this process, ATM, together with DNA-PKcs, phosphorylates the histone H2AX (called γ-H2AX when phosphorylated on serine residue 139) along megabase-length tracks surrounding a DNA break. Activated ATM phosphorylates several downstream substrates that contribute to the proper regulation of IR-induced arrests in G_1 phase (p53, Mdm2, and Chk2), S phase (Nbs1, Smc1, Brca1, and FancD2), and G_2 phase (BRCA1 and Rad17) of the cell cycle.

8.2.1 Inhibitors of ATM Kinase with Chemosensitizing and Radiosensitizing Capability

Compounds initially identified as ATM inhibitors (caffeine and wortmannin) were neither specific nor suitable for use *in vivo*. Radiosensitization of tumor cells by caffeine was first reported in the late 1960s [3], and later studies identified ATM, ATR, and, to a lesser extent, the catalytic subunit of DNA-dependent protein kinase (DNA-PK_{CS}) as the targets. Caffeine inhibits ATM and ATR activity *in vitro* at concentrations in the low millimolar range [4]. The sensitizing effect of caffeine was found to be greater in cells that did not express functional p53 [3], which was the case for many tumors, so that it seemed likely that ATM inhibition would selectively sensitize the large subset of tumors known to be in this category [4,5]. However, both caffeine and pentoxifylline induce systemic toxicity at the dose levels required for chemosensitization and radiosensitization, and the failure of these compounds to improve tumor response rates in clinical trials could be attributed to this and to the fact that only low serum levels could be achieved [4,6]. More recently, several other small-molecule inhibitors of ATM kinase with chemosensitizing capability have been developed (reviewed by Ljungman [7] and Alao [6]). Those that have been demonstrated to have potential as radiosensitizers, at least in *in vitro* systems, are listed below:

- *CP466722* was identified by screening a targeted compound library for potential inhibitors of the ATM kinase [8,9]. The compound is nontoxic and does not inhibit phosphatidylinositol 3-kinase (PI3K) or PI3K-like protein kinase family members.

  It does inhibit cellular ATM-dependent phosphorylation events, and disruption of ATM function results in cell cycle checkpoint defects and radiosensitization. Results of clonogenic survival assay indicated that radiosensitization required only transient inhibition of ATM. There would be a significant advantage if therapeutic radiosensitization similarly required ATM inhibition for only a short period because, for clinical application, drug pharmacokinetics is frequently a limiting factor (Figure 8.2).

- *KU55933* specifically inhibits ATM at nanomolar concentrations and sensitizes cancer cells to both IR and chemotherapeutic agents but does not increase the DNA damage sensitivity of cells lacking functional ATM [9]. Inhibition of ATM with KU55933 also prevented the DNA damage-induced degradation of cyclin D1 and induces apoptosis in cancer cell lines following drug-induced senescence [10]. An improved analog of KU55933 (KU60019) has Ki and IC_{50} values half those of KU55933 and is tenfold more effective than KU55933 at

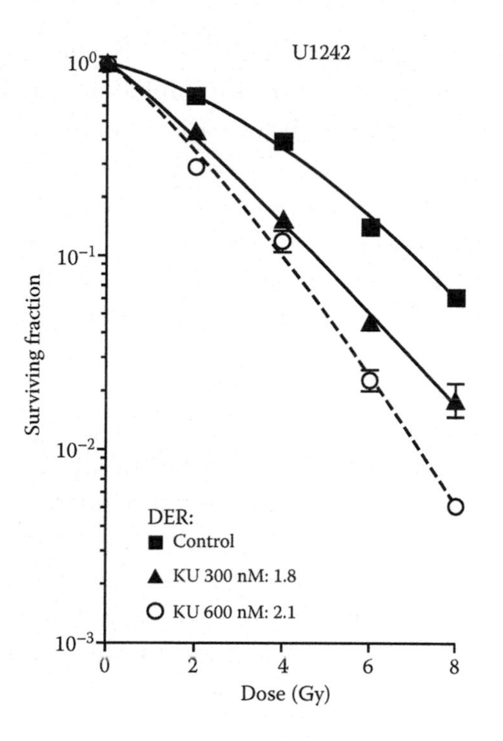

FIGURE 8.2 KU60019 potently radiosensitizes glioma cells at low concentrations. U1242 glioma cells were exposed to KU60019 at 0.6 or 0.3 μM. Surviving fractions were determined by crystal violet staining and colony counting. Data points, surviving cells plotted as a fraction of control (no IR); error bars; SEM; n = 3. Where error bars are not seen, they are obscured by symbols. DER: dose enhancement ratios of D37 values compared with irradiation-only control. (From Golding, S. et al., *Cell Cycle* 2012;11(6):1167–1173. With permission.)

blocking radiation-induced phosphorylation of key ATM targets in human glioma cells. KU60019 is a highly effective radiosensitizer of human glioma cells [11] (Figure 8.2).

- *CGK733* induces apoptosis in cancer cell lines following drug-induced senescence and inhibits the proliferation of otherwise untreated human cancer and nontransformed mouse fibroblast cell lines. Exposure of human cancer cell lines to CGK733 was associated with a rapid decline in cyclin D1 protein levels and a reduction in the levels of both phosphorylated and total retinoblastoma protein.
- *DMAG* and *17DMAG* are inhibitors of the molecular chaperone Hsp90, the targeting of which has been shown to cause radiosensitization by a number of mechanisms (Chapter 10). In one case,

exposure of MiaPaCa tumor cells to the Hsp90 inhibitor 17DMAG resulted in radiosensitization with inhibition of DNA repair. There was a reduction in ATM activation after irradiation, reduction in foci formation, and inhibition of DNA repair determined on the basis of duration of γH2AX foci and comet assay. Another HSP inhibitor DMAG was found to sensitize non-small cell lung cancer cells in culture associated with a reduction in ATM activation by radiation with maximum radiosensitization produced by pretreatment [12,13].

8.3 p53 IN THE DNA DAMAGE RESPONSE

It has been established that more than half of all human tumors have mutated p53 while the rest have lost normal p53 gene function with the result that p53-dependent pathways to cell-cycle arrest or cell death are deficient. Defects in p53-dependent signaling are known to be related to tumor resistance to radiation and to chemotherapy, making p53 and p53-dependent pathways prime targets for the improvement of cancer therapy.

8.3.1 p53 Function

The tumor suppressor protein p53 is a transcription factor that has an essential role in guarding the cell in response to various stress signals through the induction of cell cycle arrest, apoptosis, or senescence (Figure 8.3). Impairment of p53 function plays a crucial role in tumor evolution by allowing tumor cells to evade p53-dependent responses. p53 inactivation in tumors occurs through two general mechanisms. First, the inactivation of p53 function by point mutations in p53 itself, and second, through the partial abrogation of signaling pathways or effector molecules that regulate p53 activity. Experiments with a series of genetic models has established that the restoration of p53 activity in tumor cells could be an effective mode of cancer treatment.

The p53 protein acts as a transcription factor by forming a homotetramer on the target gene response elements. Its mRNA is ubiquitously expressed, but the expression levels vary depending on the stage of development and tissue type. In normal cells under nonstressed conditions, the biologically functional p53 protein has a short half-life and is hardly detectable by Western blotting. The protein level is, in fact, under the tight control of its negative regulator murine/human double minute 2 and 4 (MDM2/HDM2) via ubiquitination and, because MDM2 is also a target gene of p53, the system forms a negative feedback loop (Figure 8.4). Stress

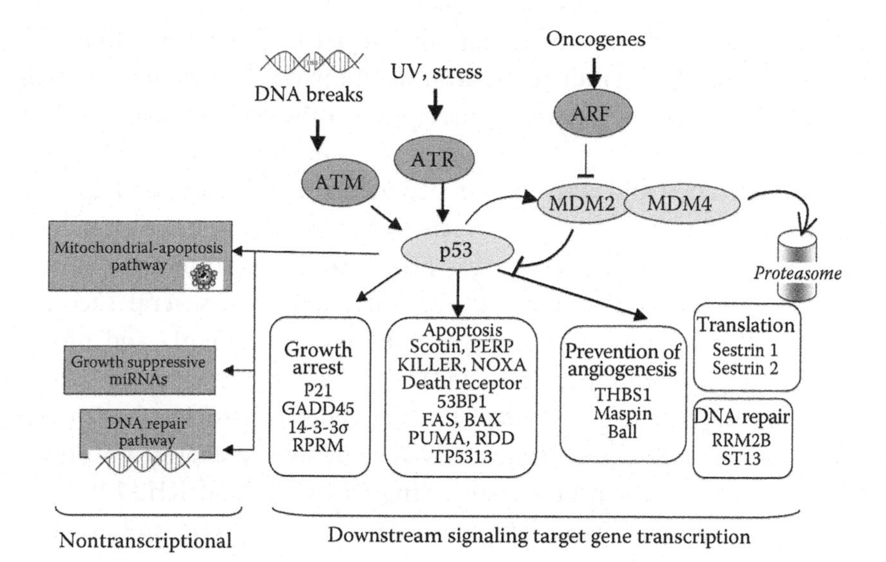

FIGURE 8.3 The p53 pathway. p53 is at the center of a complex web of biological interactions that translates stress signals into cell cycle arrest or apoptosis. Upstream signaling to p53 increases its level and activates its function as a transcription factor in response to a wide variety of stresses, whereas the downstream components execute the appropriate cellular response.

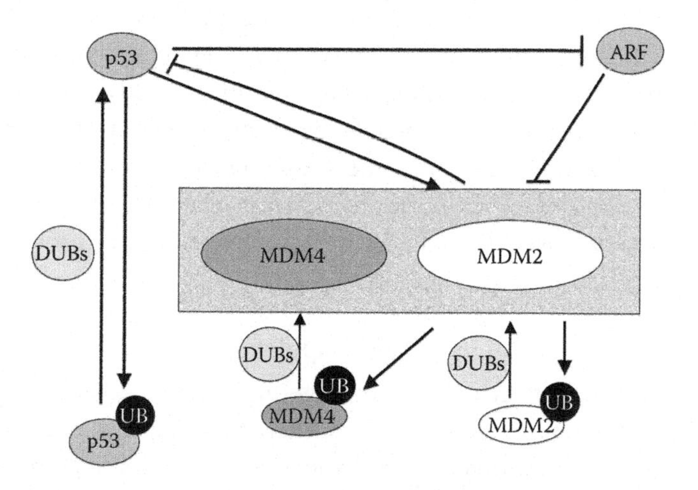

FIGURE 8.4 A negative feedback loop controls cellular levels of p53. In normal cells, p53 increases transcription of MdM2 over basal levels whereas MdM2 inhibits p53 function by modulating its transcriptional activity by preventing its interaction with the general transcription machinery. MdM2 also forms a heterodimeric complex with MdM4 that promotes the degradation of p53. UB, ubiquitin; DUBs, deubiquitinases.

signals, including chemo- or radiation-induced DNA damage, hypoxia, or nucleotide depletion induce the accumulation of p53 protein in normal cells, which in turn triggers the transcription of the different categories of p53 target genes.

The stabilization and activation of p53 following cellular stress leads to a number of cellular effects. p53 targets linked to cell-cycle arrest include p21, GADD45, and 14-3-3σ, whereas others are involved in DNA repair responses such as p53R2 and p48. p53 also regulates several secreted angiogenesis inhibitors including thrombospondin-1 (Tsp1) and brain-associated angiogenesis inhibitor (BAI).

Another important group of p53 target genes is associated with cell death. p53 has been shown to directly upregulate the expression of cell surface death receptor proteins including Fas/APO1 and KILLER/DR5, and to transcriptionally activate cytoplasmic proapoptotic proteins like PIDD and Bid. Mitochondrial proapoptotic proteins controlled by p53 include Bax, Bak, Puma, and Noxa. Downstream of mitochondria, p53 upregulates APAF1, which is involved in apoptosome formation and executioner caspase 6.

The downstream cellular effects of p53 activation in response to IR-induced DNA damage are dependent on several factors indicated by the fact that tumors originating from different tissues show different responses (apoptosis or growth arrest) to IR. Hematopoietic cells, for instance, undergo p53-dependent apoptosis more easily and efficiently than do fibroblasts. In lymphoid cells, blocking the apoptotic pathway by overexpression of the Bcl-2 oncogene causes a switch from p53-mediated apoptosis to senescence-like permanent arrest. The dose of IR also seems to be important, lower doses of IR initiate reversible cell growth arrest whereas higher doses trigger apoptosis.

8.3.2 Radiosensitization by p53 Manipulation

The overriding advantage of directly targeting mutant p53 is tumor specificity. The effectiveness of most chemotherapeutic agents and radiation therapy require a functional p53 pathway to be fully effective, thus restoring p53 function by whatever means will increase the sensitivity to chemotherapy and radiation therapy. In addition, mutant p53 is frequently overexpressed and posttranslationally modified in tumor cells, and the cellular environment of tumor cells also favors functional p53-induced apoptosis. Thus, stabilization of mutant p53 conformation by small molecules or peptide chaperones may selectively activate the apoptotic pathway

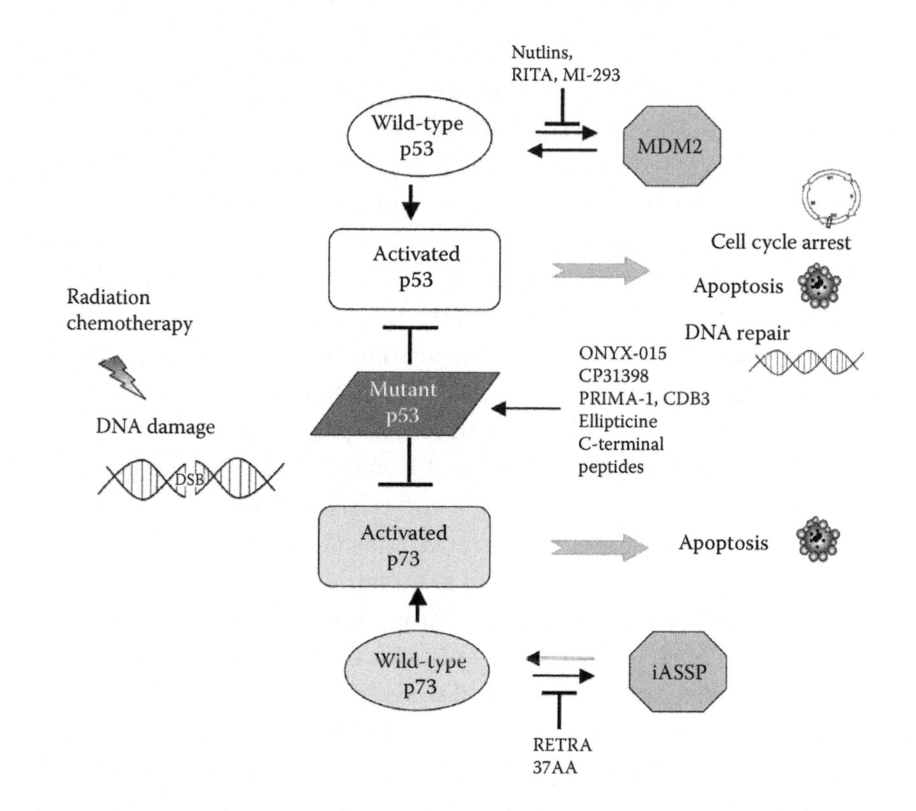

FIGURE 8.5 There are multiple approaches for targeting the p53-dependent tumor suppressor pathway with therapeutic agents. Currently, strategies to modulate p53 in cancer cells for improved cancer therapy include (1) delivery of wild-type p53 to cancer cells (rAd-p53); (2) elimination of mutant p53 with adenovirus (ONYX-015); (3) restoration of wild-type function of mutant p53 (CP-31398, PRIMA-1, CBD3, C-terminal peptides, ellipticine); (4) inhibition of degradation to stabilize p53 (nutlins, RITA, and MI-293); and (5) activation of other p53 family members to substitute p53 function (37AA and RETRA).

in the tumor cell. Current strategies to modulate p53 in cancer cells for the purpose of radiosensitization or chemosensitization are shown schematically in Figure 8.5.

8.3.2.1 Gene Therapy to Deliver Wild-Type p53

Various approaches have been explored with the intent to introduce an exogenous human p53 gene into p53-deficient cancer cells. Different protocols and vectors, including adenovirus, retrovirus, and vaccinia-derived vectors, have been developed. An early study used retrovirus-mediated gene transfer of wt p53 into human lung cancer cells to inhibit tumor growth *in vitro* and *in vivo* [14]. Another approach used human wt p53, delivered by replication-defective adenovirus (Ad-p53) and showed better transduction efficiency and lower toxicity. Preclinical studies *in vitro* and in animal models reported promising results, but clinical studies of the gene therapy approach in lung, bladder, ovarian, and breast cancers failed to demonstrate additional benefits compared with conventional treatments, and some trials were discontinued [15]. More positive results were shown for phase II and phase III clinical trials in China in patients with late-stage head and neck squamous cell carcinoma treated with recombinant Ad-p53 (rAd-p53) combined with radiotherapy [16].

The Ad-p53, under the brand name of Gendicine or Advexin, has been in clinical use in China since 2003 [17], and is in phase I to III clinical trial in the United States [18]. The results showed that Gendicine/Advexin is well tolerated in patients and efficacious in the treatment of numerous cancers, particularly head and neck cancer and lung cancer, as a single agent or in combination with chemotherapy or radiation therapy [16,18,19].

Currently, approximately 10 clinical trials (phases I–IV) involving some form of recombinant Ad-p53 are ongoing or recently terminated. These include single-agent trials and rAd-p53 in combination with surgery and chemotherapy. In one phase IV trial, rAd-p53 was combined with radioactive iodine, whereas in a phase II trial, Ad5CMV-p53 was combined with surgery, chemotherapy, and radiotherapy (www.clinicaltrials.gov).

8.3.2.2 Elimination of Mutant p53

This approach is exemplified by ONYX-015, an adenovirus lacking the E1B-55K gene product for p53 degradation, designed to selectively replicate within, and kill, p53-defective tumor cells. This virus is used not to deliver a therapeutic gene but to induce cell death by virus replication in cells with mutant p53. There is the potential to successfully combine radiation with cytolytic adenoviral therapy because it does not curtail viral replication. The E1B-deleted, replication-competent ONYX-015 (dl1520) adenovirus was originally described as being able to selectively kill p53-deficient cells due to a requirement of p53 inactivation for efficient viral replication, but later reports suggested that, in some cases, the specificity of ONYX-015 was independent of p53 gene status.

In a preclinical study [20], replication properties of ONYX-015 *in vitro* were examined using the RKO human colon carcinoma (*p53* wt) and RKO.p53.13 subclone expressing a dominant-negative form of p53, a pair of isogenic cell lines that differed only in their *p53* status. It was demonstrated that, although ONYX-015 can replicate in both *p53* wild-type and mutant cells *in vitro*, the virus shows significantly greater antitumor activity against mutant *p53* tumors *in vivo*. Moreover, ONYX-015 viral therapy can be combined with radiation to improve tumor control beyond that of either monotherapy.

A number of early clinical trials demonstrated that ONYX-015 as a single agent produced marginal benefit, but significant effects were seen when it was combined with chemotherapy (cis-platinum or 5-FU). Subsequently, the mechanism of the tumor selectivity of ONYX-015 was reexamined and the p53 protein no longer considered to be the main factor in tumor selectivity [21]. More recent clinical trials in China combining the administration of ONYX-015 with chemotherapy had encouraging results in patients with head and neck cancer [22] and, following a phase III clinical trial [22], an E1B-55k mutant adenovirus, mechanistically similar to ONYX-015, was approved by the Chinese Food and Drug Administration for its use in patients with head and neck cancer in combination with chemotherapy.

8.3.3 Small Molecules Targeting the p53 Pathway

Small molecules targeting p53 act by a number of mechanisms including the reactivation of mutant p53, the activation of wild-type p53, and the inhibition of wild-type p53 (Table 8.1).

8.3.3.1 Restoration of Wild-Type Function to Mutant p53

Small-molecule inhibitors or peptides have been developed that bind to mutant forms of p53 and revert them to wild-type conformation because, in cancer cells, mutant p53 is abundantly expressed, these drugs aim to selectively eliminate tumors by restoring p53 function. Some of these compounds have been shown to be effective in inducing cell-cycle arrest or apoptosis (or both) *in vitro*.

- *CDB3* was identified by a systematic search for molecules that stabilized the native form of p53. Based on the binding between the wild-type p53 core domain and the p53-binding protein 2 (53BP2), a nine-residue peptide (CDB3) was designed and shown to increase the thermostability of the p53 core domain *in vitro*. It was proposed that CDB3 had a "chaperone" function, binding and thermostabilizing newly synthesized p53 to allow proper folding. The interaction is transient and is later lost to enable p53-DNA binding. CDB3 induced

TABLE 8.1 Small Molecules Targeting p53: Activators, Reactivators, and an Inhibitor

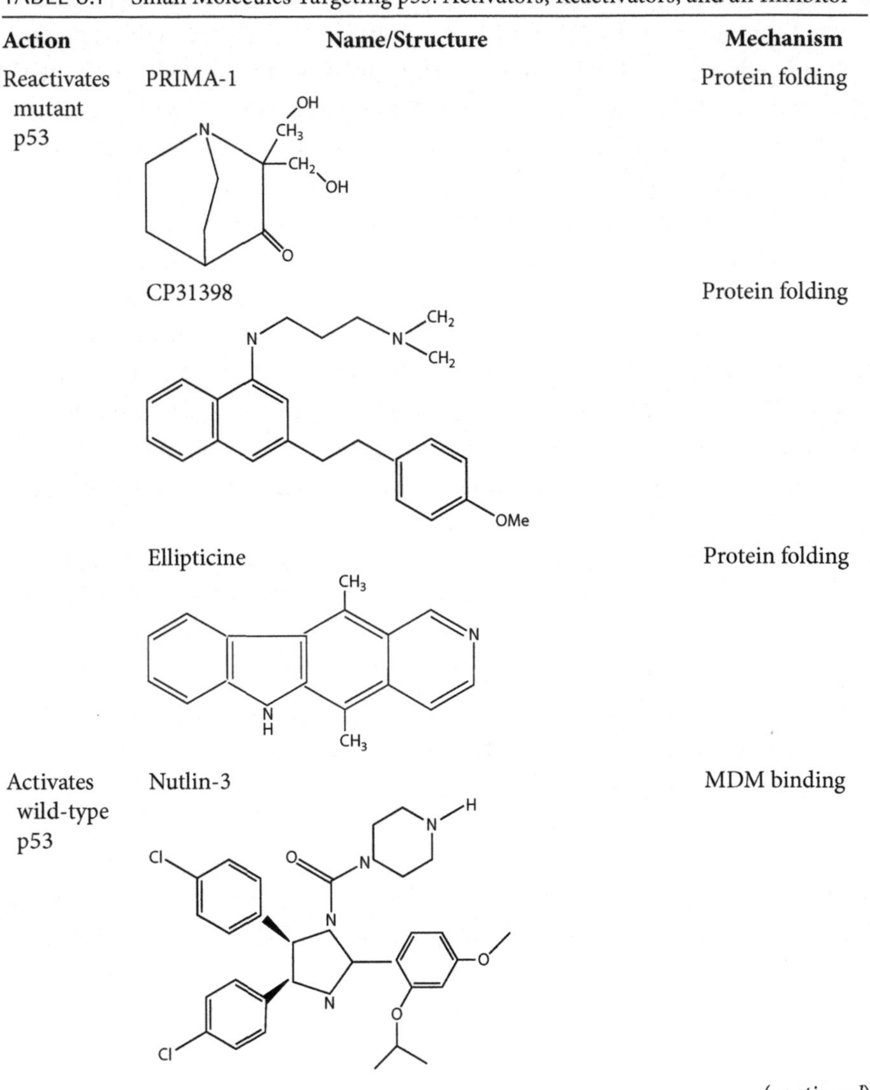

Action	Name/Structure	Mechanism
Reactivates mutant p53	PRIMA-1	Protein folding
	CP31398	Protein folding
	Ellipticine	Protein folding
Activates wild-type p53	Nutlin-3	MDM binding

(*continued*)

p53-dependent expression of MDM2, gadd45, and p21, accompanied by partial restoration of apoptosis. CDB3 also caused modest sensitization to radiation-induced apoptosis in HT116 human colon cancer cells expressing wild-type p53 [23].

- *C-terminal peptides* are a class of peptides derived from the negative regulatory domain of p53 within the carboxyl terminus that can restore wild-type activity to mutant p53. Short synthetic peptides

TABLE 8.1 Small Molecules Targeting p53: Activators, Reactivators, and an Inhibitor (Continued)

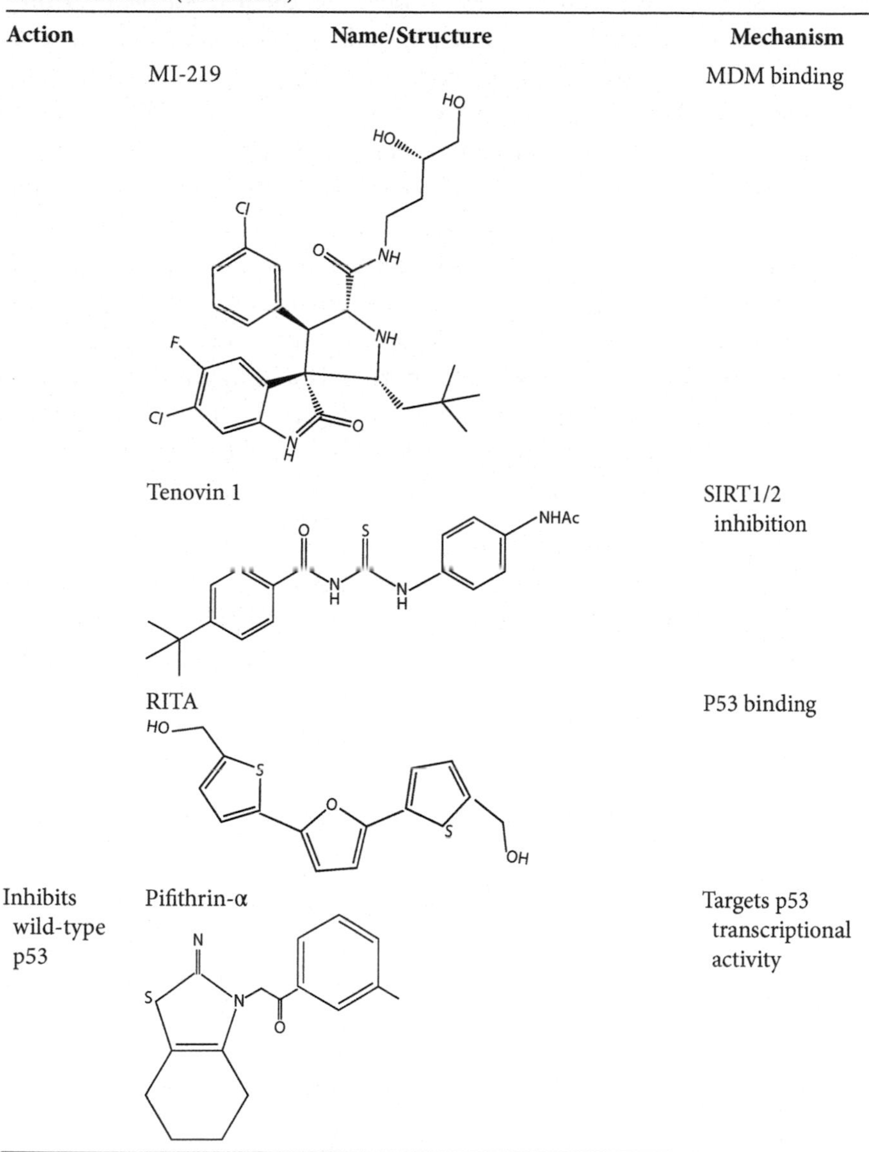

Action	Name/Structure	Mechanism
	MI-219	MDM binding
	Tenovin 1	SIRT1/2 inhibition
	RITA	P53 binding
Inhibits wild-type p53	Pifithrin-α	Targets p53 transcriptional activity

corresponding to certain fragments of the p53 sequence have proven to be effective in stabilizing p53 protein or restoring mutant p53. A positively charged peptide, corresponding to residues 361 to 382 of the p53 C-terminus, not only binds to the core domain of p53 thus enhancing the DNA-binding capacity of wild-type p53 but can also restore DNA binding of some p53 mutants [24].

- *PRIMA-1* and *MIRA-1.* PRIMA-1 (p53 reactivation and induction of massive apoptosis) and MIRA-1 were identified from NCI chemical library screening using a p53-null human osteosarcoma cell line engineered to artificially express the His-273 mutant of human p53 [25]. PRIMA-1 was found to activate transcription of p53 target genes only in cells with mutant p53 and not in controls, and was synergistic with Adriamycin to induce cell death in non-small cell lung cancer cells. PRIMA-1^{met}, an analogue of PRIMA-1 with improved efficacy, radiosensitizes prostate cancer cells [26] and, combined with cisplatin, was found to synergistically inhibit tumor growth in xenografts (Figure 8.6) [25]. PRIMA-1^{met} also radiosensitizes p53-null prostate cancer cells, under normoxia and hypoxia, suggesting that in this case, sensitization must be independent of mutant p53. The mechanism of radiosensitization of p53-null cells is unclear and may be cell line-specific.
- *Ellipticine* was isolated from an Australian evergreen tree of the Apocynaceae group as a potent antineoplastic agent. Its mechanism of action involves DNA intercalation, inhibition of DNA topoisomerase II activity, generation of free radicals, and induction of endoplasmic

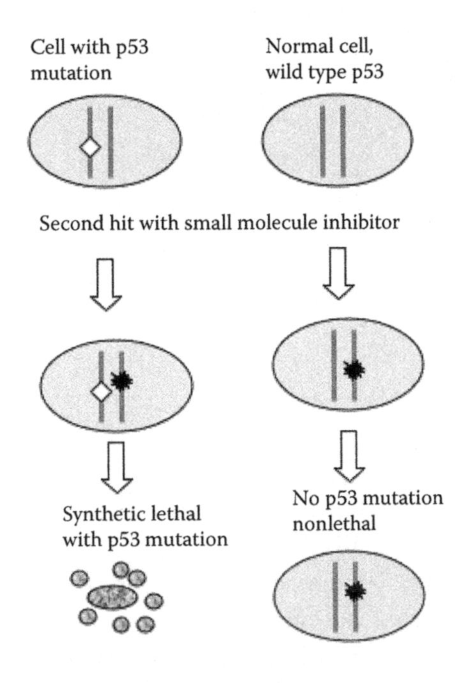

FIGURE 8.6 Synthetic lethality for p53 mutation.

reticulum stress. It has also been shown that Ellipticine can restore the wild-type conformation and increase the nuclear localization of p53 in cancer cells with a resultant increase in the transactivation of the p21 promoter. Increased nuclear p53 after ellipticine treatment was not associated with an increase in DNA DSBs, indicating that this effect relied on a mechanism independent of DNA damage. In addition, Ellipticine analogues have been demonstrated to be more potent against mutant p53 cells than against wild-type p53 cells.

- *CP-31398*, a styrylquinazoline, was the first compound reported to have the ability to alter a mutant p53 to a wild-type conformation and to rescue p53 function in some tumor cell lines and xenografts. Tumor cells with hot spot p53 mutations were found to be sensitive to CP-31398-induced cell-cycle arrest or apoptosis [27]. After several hours of CP-31398 treatment, significant p53 activity was detected by either immunoblotting p53 targets or a p53 reporter assay. Administration of CP-31398 suppressed the growth of p53 mutated tumor xenografts, including the DLD-1 colon carcinoma (mutation at position 241) and the A375 melanoma (mutation at position 249) in mice without obvious toxicity [27]. By promoting the active conformation of p53, not only can CP-31398 restore p53 function in mutant p53-expressing cells but also significantly increase the protein level and promotes the activity of wild-type p53 in multiple human tumor cell lines including ATM-null cells [28].

8.3.3.2 Targeting Mutant p53 to Kill: Synthetic Lethality for p53 Mutation

Synthetic lethality is a situation in which a cancer-associated mutation is present, which in itself is not lethal, but the mutated cancer cell will be inactivated if it is targeted by a second hit (Figure 8.6). In the case of p53, which is mutated in 50% of human cancer cells, compounds that selectively kill cancer cells harboring a mutant p53 constitute an important and novel class of anticancer drugs. The p53 synthetic lethal drugs, if identified and developed, can be used (1) for cancer treatment to selectively kill mutant p53-containing cancer cells and (2) for chemoprevention to eliminate mutant p53-containing cancer-prone cells at the early stage of carcinogenesis. Furthermore, p53 synthetic lethal drugs should have, in theory, minimal adverse effects, because normal cells do not contain a p53 mutation.

Small molecules that abrogate G_2/M checkpoint control can be considered to act through the synthetic lethal mechanism against cells with p53 mutations. Because p53-deficient cells already lack G_1 checkpoint control (because of loss of p53-mediated p21 induction in response to DNA damage), the abrogation of G_2 checkpoint control will selectively kill p53-deficient cancer cells through the induction of mitotic catastrophe.

This was demonstrated in an experiment in which G_2 arrest induced in CA46 cells by gamma irradiation was inhibited by treatment with UCN-01 in a dose-dependent manner. UCN-01 enhanced the cytotoxicity of gamma irradiation in p53-mutant CA46 and HT-29 cells but not in MCF-7 cells. The latter, with functional p53 protein, were more resistant to G_2 checkpoint abrogation by UCN-01 [29].

8.3.3.3 Modulation of p53 Level by Targeting p53 Regulators

Direct stabilization of p53 occurs by inhibition of degradation of the protein. Compounds that regulate p53 modificatory proteins indirectly affect p53 protein level. The most important group of compounds in this class are small molecules that disrupt the MDM–p53 interaction. The *MDM2* gene is overexpressed in a significant portion of human tumors including breast cancer. Based on this observation, it has been proposed that inhibition of MDM2 and p53 interaction might be an effective approach to promote p53 activity in such tumor cells, and several small-molecule and peptide inhibitors of the MDM2–p53 complex have been identified.

- *Nutlins* were the first class of small-molecule MDM2 inhibitors to be reported. Nutlins contain a cis-imidazoline core structure that can displace p53 from the complex as a result of their higher binding affinity to MDM2. One analogue, Nutlin-3, possesses potent antitumor activity in human cancers retaining wild-type p53, and the results of two recent *in vitro* studies suggest a role for Nutlin-3 as a radiosensitizer.

- Radiosensitization by Nutlin-3 was investigated by using three prostate cancer cell lines (p53 wild-type, p53-mutated, and p53-null) under oxic, hypoxic, and anoxic conditions. As a single agent, Nutlin-3 stabilized p53 and p21 levels and was toxic to wild-type p53 cells but had minimal toxicity toward p53-defective cells. Combined with radiation under oxic conditions, Nutlin-3 decreased clonogenic survival in all three cell lines. Anoxia induced p53 protein expression in wt-p53 cells and this was augmented by Nutlin-3 treatment, which was more effective as a radiosensitizer under hypoxic conditions, particularly in p53 wild-type [26].

- Radiosensitization by Nutlin-3 was found to occur in lung cancer cells with wild-type p53. There was increased apoptosis and cell cycle arrest after combined treatment with nutlin and radiation. In addition, combined treatment with Nutlin-3 and radiation decreased the ability of endothelial cells to form vasculature [30]. These and other examples of radiosensitization of human tumor cells by Nutlin-3 are summarized in Table 8.2.

- A small molecule named RITA (reactivation of p53 and induction of tumor cell apoptosis) was found to induce wild-type p53 accumulation as a result of its high-affinity binding to the NH_2-terminal domain of p53 [31], although a subsequent nuclear magnetic resonance study indicated that RITA might not block p53–MDM2 interaction *in vitro* [32].

- The *MI series.* The MI series of spiro-oxindole compounds, including MI-219, MI-63, and MI-43, were discovered through structure-based design by mimicking all four MDM2-contacting residues (Phe19, Trp 23, Leu 22, and Leu 26) on p53. The MI series of compounds, with MI-219 being the most potent, bind to Mdm2 with a high affinity. Drug-induced disruption of MDM2–p53 binding caused p53 accumulation, leading to the upregulation of many p53 target genes and to the induction of apoptosis in human cancer cell lines derived from breast, colon, and prostate cancers in a wt p53-dependent manner. MI-219 as a single agent also caused a complete inhibition of xenograft tumor growth *in vivo* at a dose that had no apparent toxicity to the tumor-bearing animal [33].

- MDM2 E3 ubiquitin ligase inhibitors. Blocking the ubiquitin ligase activity of Mdm2 prevents p53 degradation and thus indirectly elevates p53 levels.

  - *HLI98* belongs to a family of small molecules (HLIs) that inhibit the E3 activity of MDM2, thus increasing cellular p53 and selectively killing transformed cells expressing wild-type p53. A highly soluble derivative of the HLI 98 family (HL1373) has greater potency than the HLI98s in stabilizing Hdm2 and p53,

TABLE 8.2 Activity of Nutlin-3 and Ionizing Radiation against Human Cancer Cell Lines

Treatment			Cell Line		p53 Status	Effect	References
Nutlin	RT	CT					
1 μM nutlin-3, 48 h	0–6 Gy		Ca lung	H460	Wt	Radiosensitization, increased apoptosis and cell cycle arrest	Cao et al. [30]
				Val38	Mut	No effect	
				HUVEC	Wt	Increase in MDM2, p53, and p21 proteins. Reduced capacity to form vascular tubules	
10 μM nutlin-3, 3–6 h	0.02–8 Gy	10 μM RITA	Hu lung cancer	H1299	Wt	Accumulation of p53, HDM2, and iNOS after low-dose IR	Takahashi et al. [34]
				H1299	Mut	No response	
5 μM nutlin-3, 48 h	2 Gy		Prostate cancer	22RV1	Wt	Increased apoptosis and decreased clonogenic survival	Supiot et al. [26]
				DU145	Mut	No response	
				PC-3	Null	No response	
Nutlin-3	IR		Laryngeal cancer	17A, 17AS	Wt	Radiosensitization	Arya et al. [35]
			UM-SCC	11A, 12 81B	Mut/mut	No response	
				5, 10A	Nonsense mutation het and homozygous	No response	
Nutlin-3, 10 μM, 24 h	10 Gy	Cisplatin, 72 h	Human colon cancer	HCT116	Wt	Formation of tetraploid cells more resistant to IR and cisplatin-induced apoptosis	Shen et al. [36]
			Human osteosarcoma	U20S	Wt		
Nutlin-3, 5 μM	0–6 Gy		Human prostate cancer	LNCap	Wt	Radiosensitization attributable to increased induction of cellular senescence	Lehmann et al. [37]
				22Rv1	wt/mut	Modest radiosensitization at higher doses	
				DU145	mut/mut	No response	

activating p53-dependent transcription, and inducing cell death. HLI373 is effective in inducing apoptosis in several tumor cells lines that are sensitive to DNA-damaging agents [38].

- *SirT1* and *SirT2* are two members of the NAD^+-dependent class III histone deacetylases with a total of seven members in humans. SirT1 and SirT2 catalyze the reaction between an acetylated lysine with NAD^+, leading to the production of deacetylated lysine, 2′-*O*-acetyl-ADP-ribose and nicotinamide. It has been established that acetylation of p53 at lysine 382 enhances p53 DNA-binding activity, thus SirT1-catalyzed deacetylation of p53 at lysine 382 deactivates and destabilizes p53 [39].
- *Tenovin-1* and *Tenovin-6.* This series of compounds was identified in a cell-based screening for compounds that activate p53, using a p53-driven transactivation assay [40]. Tenovins stabilize wt p53, induce p53-dependent cell cycle arrest and apoptosis, and suppress xenograft tumor growth *in vivo* as a single agent [40]. They act by inhibiting the NAD-dependent deacetylase SirT1/T2, thus restabilizing and reactivating p53.

- *Nuclear export inhibitors (NEIs).* NEIs inhibit the nuclear export protein CRM1, which increases nuclear p53 level indirectly. NEIs have been synthesized based on leptomycin B (LMB), a specific inhibitor of the export protein CRM1. Blocking CRM1 shows therapeutic efficacy in inducing cancer cell apoptosis and inhibits tumor growth in wt p53 cancer models. New improved NEIs have shown an improved therapeutic ratio compared with the parent compound, LMB [41].

8.3.3.4 Activation of Other p53 Family Members: p63 and p73

p63 and p73 share significant sequence homology with p53 and can bind to the p53-specific DNA-binding sequence and transactivate transcription of p53 target genes. Their roles in tumorigenesis have been demonstrated by heterozygous deletion of either p63 or p73, which resulted in predispositions to various tumors in mice [42]. Unlike p53, p63 and p73 are rarely mutated in tumors but it has been shown that mutant p53 can bind and inactivate p73.

- *37AA* is a 37-amino acid peptide derived from the DNA binding domain of wild-type p53 that was, at first, found to selectively kill transformed cells independently of p53 [43]. Further analysis revealed that 37AA did not transactivate p53 target genes, instead it bound to iASPP, the inhibitory member of the ASPP family, and displaced iASPP from p73 in p53-null cells. p73 seemed to be critical to the cytotoxicity of 37AA. In a mouse xenograft model, an expression vector containing 37AA administered via dendrimer-based nanoparticles caused significant regression of human colorectal cancers with no systemic toxic effects [43]. Sensitization to radiotherapy or chemotherapy by 37AA has yet to be demonstrated.

- *RETRA* is a small molecule that was found to specifically suppress mutant p53-bearing tumor cells *in vitro* and in mouse xenografts. Further study showed that treatment of RETRA in cancer cells with mutant p53 resulted in the release of p73 from the mutant p53–p73 complex. The antitumor activity of RETRA was abrogated by inhibition with p73 siRNA, suggesting that p73 specifically mediates RETRA's apoptotic function [44].

8.3.4 Inhibitors of Wild-Type p53: Blocking Wild-Type p53 Activity to Prevent Damage to Normal Tissues during Cancer Treatment

An increase in the therapeutic ratio can also be achieved by the protection of normal tissues. One stratagem that has been investigated to achieve this is the use of pifithrin-α ("p-fifty-three inhibitor"), a small-molecule inhibitor of p53 that inhibits p53-dependent transactivation without affecting cell growth or survival rates and prevents apoptotic death in cells that would have undergone rapid p53-dependent apoptosis in response to treatments such as UV and ionizing radiation. PFT-α was found to protect mice against radiation-induced gastrointestinal syndrome [45]. The downside with this approach is that it will also reduce the anticarcinogenic effect of wt p53 and increase the probability of secondary malignancy, particularly after intensity-modulated radiotherapy.

8.4 TARGETING CELL CYCLE CHECKPOINT PROTEINS: CHK1 AND CHK2

Actively proliferating cells experience blocks in the cell cycle after exposure to ionizing radiation. These blocks that occur in the G_1, S, and G_2 phases following treatment with radiation and genotoxic drugs are referred to as checkpoints and are thought to allow for DNA damage repair to occur before further cell cycle progression. There has been considerable interest in targeting the molecular pathways involved with these checkpoints to inhibit repair in cancer cells.

8.4.1 Cell Cycle Control

(This section is based, in part, on reviews by Ashwell et al. [46], Alao [6], and Wilson [47].)

The cell cycle checkpoint pathways are operational during the entire cell cycle and, hence, may slow down the cell cycle at any point in the four phases, but the term *checkpoint* is usually interpreted to mean a response to DNA damage that causes the inhibition of the transition between successive phases at the G_1/S, intra-S, and G_2/M checkpoints. These checkpoints are distinct, but they all respond to DNA damage and have many proteins in common. The intra-S-phase checkpoint differs from the G_1/S and G_2/M checkpoints in that it also recognizes and deals with replication intermediates and stalled replication forks. The DNA damage response initiated at any phase of the cell cycle follows the pattern described earlier in this chapter. After the detection of DNA damage by sensor proteins, signal transducer proteins relay the signal to effector proteins. These effector proteins launch a cascade of events that causes cell cycle arrest, apoptosis, DNA repair, or activation of damage-induced transcription programs.

The ATM and ATR kinases that lie upstream in the DNA damage-response signal transduction network are central to the entire DNA damage response. Downstream of these proteins lie the CHK1 and CHK2 kinases, which carry out sections of the mammalian DNA damage response. An overview of these relationships is shown in Figure 8.7.

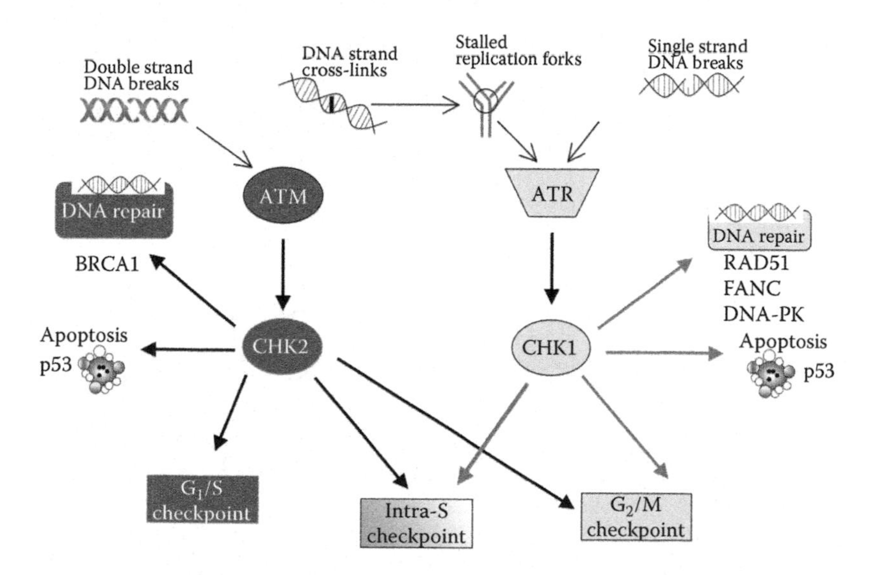

FIGURE 8.7 Roles of CHK1 and CHK2 in the DNA damage response.

8.4.1.1 CHK1: Role in G_2/M and S Phase Checkpoints

CHK1 regulates G_2/M and S phase cell-cycle checkpoints. It is expressed in the S and G_2 phases of proliferating cells and is absent or expressed at very low levels in quiescent and differentiated cells. CHK1 is activated by phosphorylation at serine (Ser) 345 and Ser317 in response to various types of DNA damage in mammals, including that induced by IR, UV, and topoisomerase inhibitors. For optimal activation of CHK1, the presence of other checkpoint proteins such as claspin is required. CHK1 phosphorylates Ser123 and several other serine residues of the phosphatase CDC25A, targeting it for ubiquitin-mediated degradation. As a result, CDC25A fails to dephosphorylate and activate CDK2 and CDK1 and the cells arrest in late G_1, S, and G_2 phases. Conversely, cells that fail to downregulate CDC25A after DNA damage do not arrest cell-cycle progression in an appropriate time frame. CHK1 can also phosphorylate Ser216 of CDC25C, which prevents its activation at G_2, at least partially, through 14-3-3σ-mediated translocation to the cytoplasm. Again, G_2 arrest ensues because of the inability of CDC25C to dephosphorylate and activate CDK1 (Figure 8.8).

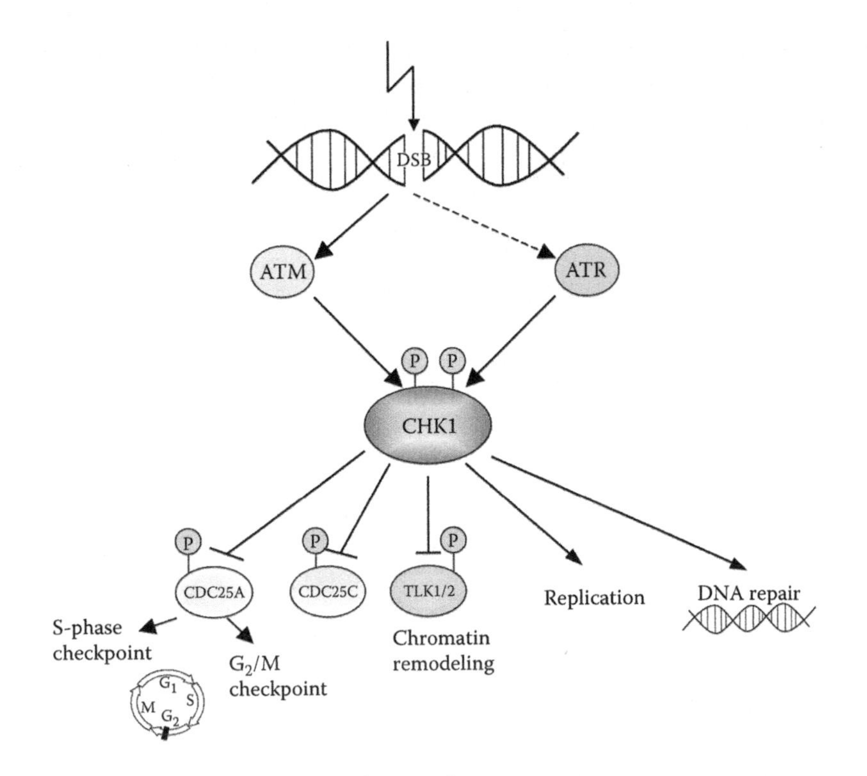

FIGURE 8.8 Functions and regulation of CHK1.

8.4.1.2 CHK2: The G_1 Checkpoint

CHK2 is mainly activated by DNA strand-breaking agents such as ionizing radiation and topoisomerase inhibitors through the ATM-dependent pathway. Other checkpoint proteins, such as 53BP1, MDC1, and the MRE11–RAD50–NBS1 complex are involved in modulating CHK2 activation. The transition from growth factor dependence in early G_1 to independence later in G_1 is coordinated by the retinoblastoma (Rb) and myc pathways. The molecular switch for this restriction point centers around the phosphorylation of Rb by cyclin D-CDK4 kinases triggering the release of E2F transcription factors and the myc-mediated activation of cyclin E and the consequent activation of CDK2 kinase, which is a prerequisite for the initiation of DNA replication. Cyclin E occurs late in G_1, somewhat after the restriction point. Thus, progression through G_1 can be blocked at both the restriction point (by preventing Rb phosphorylation) and nearer to the G_1/S border (by interference with cyclin E–CDK2 activity). This p53-independent checkpoint pathway results in a persistent G_1 block preventing S-phase entry.

In fact, there are two pathways leading to G_1 arrest. The first is the rapid response described above, whereas the second, in which p53 is the key protein, is a slower, response that may be irreversible. It is well known that loss of p53 leads to checkpoint abrogation, and p53 has long been considered to be a key factor in the G_1 response to ionizing radiation. The mechanism involves stabilization of p53 resulting in a rapid increase in protein level, which in turn mediates an increase in the cyclin-dependent kinase inhibitor, p21 targeting the CDKs required for entry into S phase. This pathway involves transcription and would be too slow to account for the rapid inhibition of CDK2, which is observed in response to genotoxic stress and in fact it seems that G_1 arrest results from dual processes. The upstream regulators, ATM and CHK2, are common to the immediate delay pathway and to the p53 response, but the latter is delayed because posttranslational modifications are required to activate sequence-specific DNA binding and transcription of the proteins involved. The dual pathways leading to G_1 arrest are summarized in Figure 8.9.

8.4.2 Small-Molecule Inhibitors of Checkpoint Kinases

The first small-molecule inhibitors used to investigate inhibition of CHK1 and CHK2, such as the staurosporine derivative UCN-01, had activity against multiple kinase targets. Second generation checkpoint kinase inhibitors that show increased selectivity for CHK1 and CHK2 over other

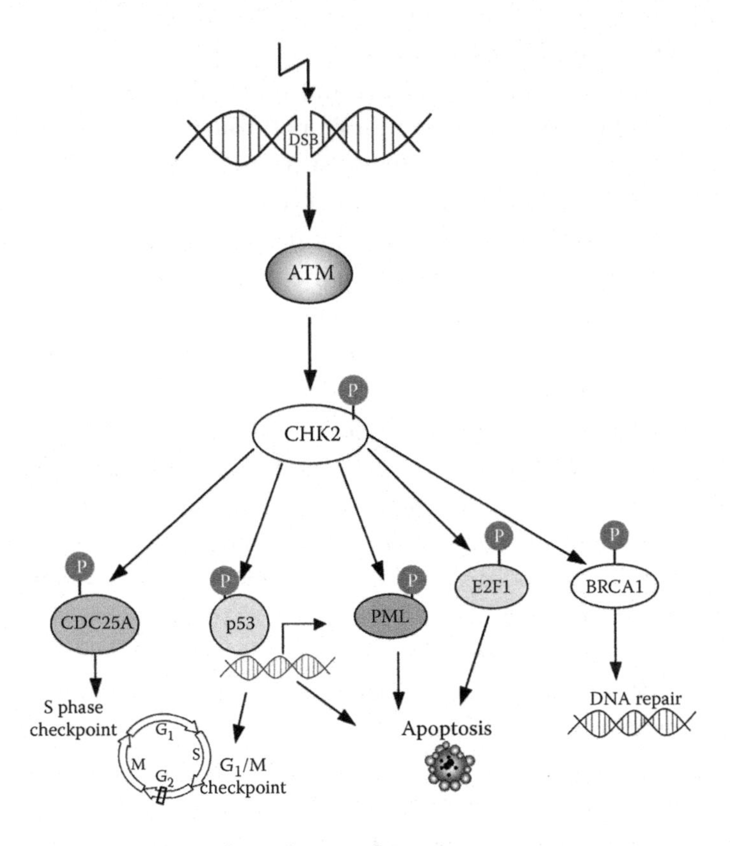

FIGURE 8.9 Functions and regulation of CHK2.

kinases have proven to be more effective. The structure and activity of selected checkpoint kinase inhibitors that have shown radiosensitizing activity are shown in Table 8.3.

8.4.2.1 Checkpoint Inhibitors and Ionizing Radiation

The results obtained combining radiation and checkpoint inhibitors have not always been easy to interpret. Interaction between the genetic background of tumor cells, inhibitor pharmacology, and scheduling of the combination of ionizing radiation and drugs all influence the outcome. Many human tumor cells have mutations or deletions in the p53 gene or other defects in the p53 signaling pathway impairing its role in the G_1 checkpoint with consequent greater dependence on the S and G_2 checkpoints. Although the functional status of the p53 pathway significantly affects the outcome of treatment with checkpoint-inhibitory drugs, the outcome may not always be predictable. (Checkpoint inhibitors are the subject of a recent review by Garrett and Collins [48].)

TABLE 8.3 Structure and Activity of Selected Checkpoint Kinase Inhibitors

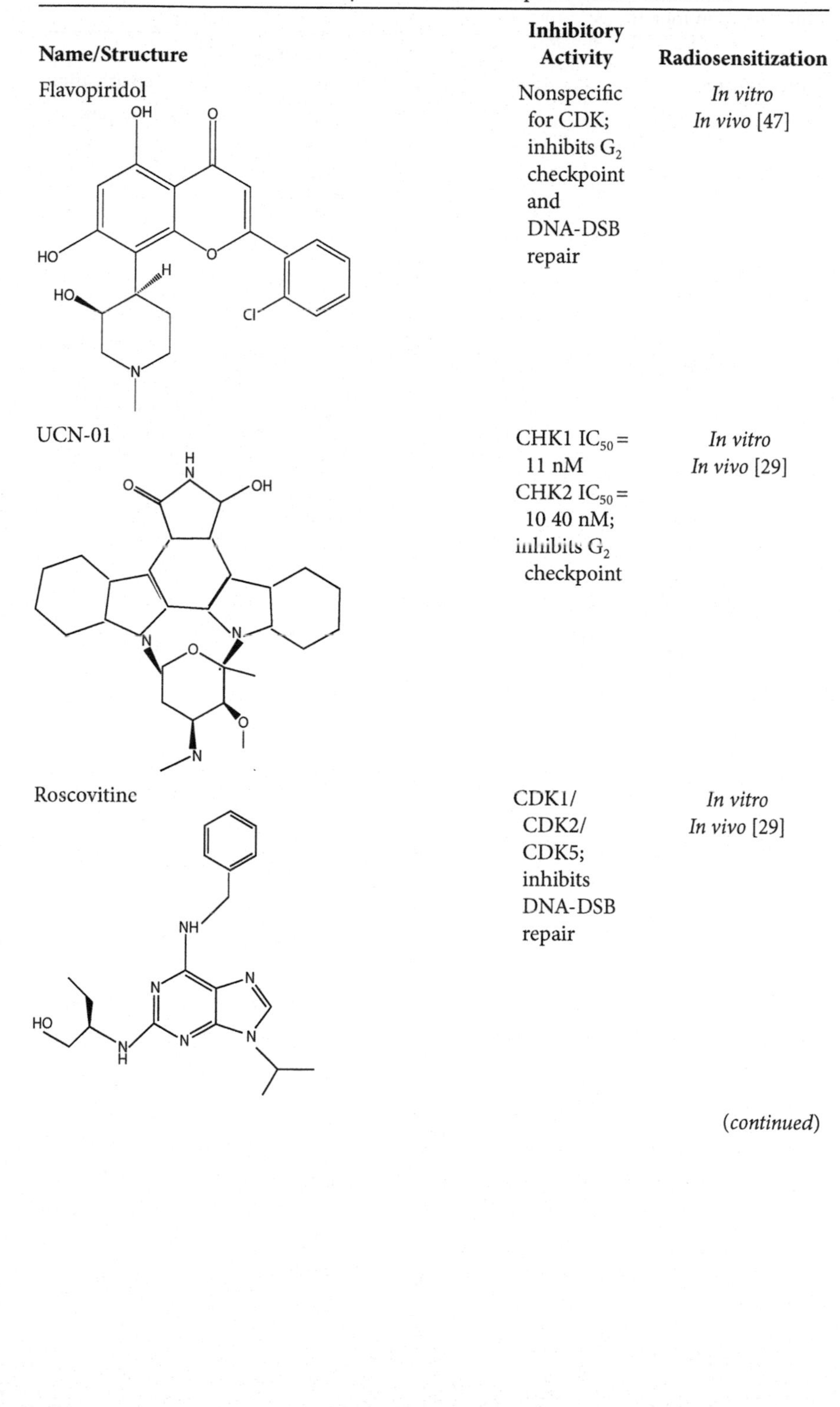

Name/Structure	Inhibitory Activity	Radiosensitization
Flavopiridol	Nonspecific for CDK; inhibits G_2 checkpoint and DNA-DSB repair	*In vitro* *In vivo* [47]
UCN-01	CHK1 IC_{50} = 11 nM CHK2 IC_{50} = 10 40 nM; inhibits G_2 checkpoint	*In vitro* *In vivo* [29]
Roscovitine	CDK1/ CDK2/ CDK5; inhibits DNA-DSB repair	*In vitro* *In vivo* [29]

(*continued*)

TABLE 8.3 Structure and Activity of Selected Checkpoint Kinase Inhibitors (Continued)

Name/Structure	Inhibitory Activity	Radiosensitization
CHIR-124 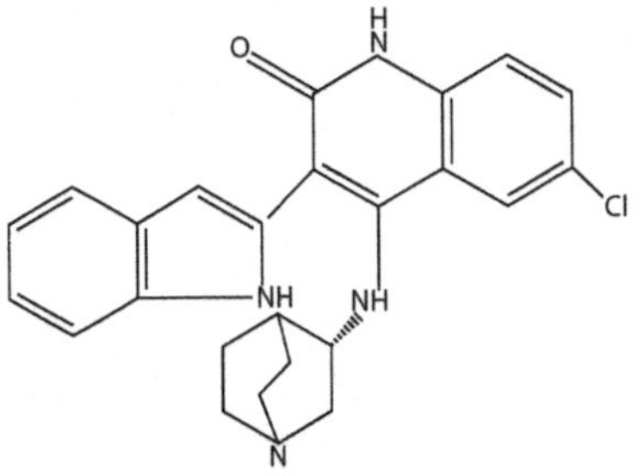	CHK1 IC_{50} = 0.3 nM CHK2 IC_{50} = 697 nM	*In vitro* [49]
CEP-3891	CHK1; inhibits S and G_2 checkpoints	*In vitro* [50]
AZD7762 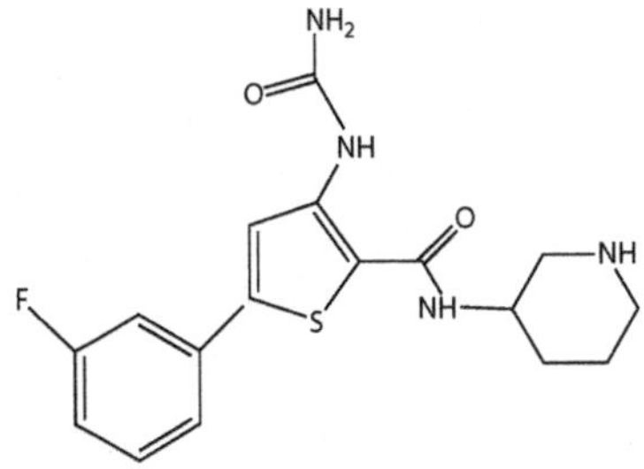	CHK1 k_i = 4 nM CHK2 IC_{50} < 10 nM	*In vitro* *In vivo* [51]
PV1019 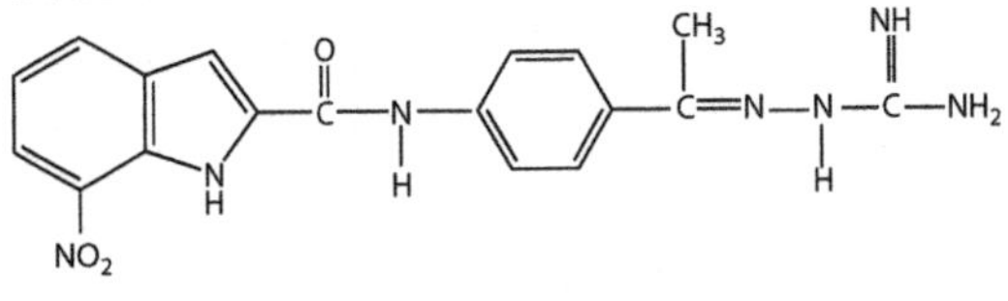	CHK1 IC_{50} = 15,730 nM CHK2 IC_{50} = 24 nM	*In vitro* [52]

8.4.2.1.1 Inhibition of CHK1 There are a number of reports of radiosensitization of human cancer cells *in vitro* and in xenograft tumors by inhibitors of CHK1.

- Flavopiridol is a nonspecific CDK inhibitor, one of the effects of which is to inhibit the G_2 checkpoint and DNA-DSB repair. Sensitization of esophageal and colon cancer cell lines to radiation has been demonstrated *in vitro*, whereas *in vivo* experiments showed an increase in growth delay in murine tumors and colon cancer xenografts [53].

- Another nonspecific inhibitor, UCN-01, was the subject of one of the first reports of CHK1 inhibition potentiating the effect of DNA-damaging agents in a p53-selective manner [28]. Comparison of clonogenic survival and cell cycle progression in matched p53-proficient and p53-deficient cells treated with IR and UCN-01 showed that when a p53-dependent G_1 arrest occurred in the first or second cell cycle after treatment, p53-proficient cells were afforded some protection against the treatment combination [50]. In contrast, when the cells were p53-deficient, UCN-01 sensitized colon, breast, lymphoma, cervix, and lung cancer cell lines *in vitro* and caused synergistic growth delay in murine fibrosarcoma following fractionated radiotherapy *in vivo* [52].
- CEP-3891 (undisclosed structure) abrogates the IR-induced S and G_2 cell cycle arrest and associated fragmentation of nuclei seen 24 h after IR in most cells. Nuclear fragmentation occurred as a result of defective chromosome segregation when irradiated cells entered their first mitosis, either prematurely without S and G_2 checkpoint arrest in the presence of CEP-3891 or after a prolonged S and G_2 checkpoint arrest in the absence of CEP-3891. CEP-3891 increased the overall cell killing after IR as measured by clonogenic assay [54].
- U251 human glioma cancer cells were sensitized to IR by the CHK2 inhibitor PV1019 [49].
- The dual CHK1/CHK2 inhibitor XL-844 radiosensitized p53-deficient HT29 colon cancer cells [55].
- The CHK1 inhibitor CHIR-124 effectively radiosensitized CHK2-deficient HCT116 cells [51].
- To date, the best performer among checkpoint inhibitors in terms of radiosensitization has been the dual CHK1/CHK2 inhibitor AZD7762. Significant results obtained with AZ7762 include:
    - AZD7762 sensitized MiaPaCa-2 pancreatic tumor cells to IR [56] and potentiated the effect of fractionated radiation in MiaPaCa-2 xenografts and in two patient-derived pancreatic tumor xenografts.
    - AZD7762 treatment enhanced the radiosensitivity of p53-mutated tumor cell lines (DMF = 1.6–1.7) to a greater extent than

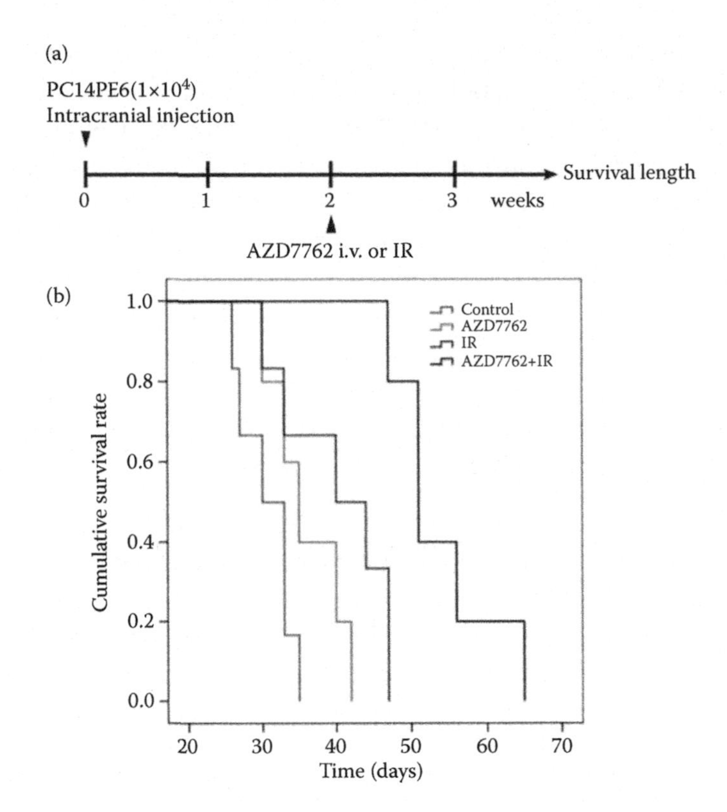

FIGURE 8.10 Effects of AZD7762 on PC14PE6 brain metastasis xenograft in response to IR. (a) Schedule of treatment. Whole brain irradiation (15 Gy) was performed at 1 h after intravenous injection of AZD7762 (25 mg/kg). (b) Kaplan–Meier plot comparing survival of PC14PE6 xenografts treated as control (vehicle), AZD7762 alone (25 mg/kg, i.v.), radiation alone (IR; 15 Gy), and combined AZD7762 (25 mg/kg, i.v. injection prior to radiation) and IR (15 Gy). (From Yang, H. et al., *Biochem Biophys Res Commun* 2011;406:53–58. With permission.)

it enhanced that of p53 wild-type tumor lines (DMF = 1.1–1.2). AZD7762 treatment alone exhibited little cytotoxicity to any of the cell lines and did not enhance the radiosensitivity of normal human 1522 fibroblasts. In an *in vivo* study, AZD7762 treatment abrogated radiation-induced G_2 delay, inhibited radiation damage repair (assessed by γ-H2AX), and suppressed radiation-induced cyclin B expression. HT29 xenografts exposed to five daily radiation fractions and twice-daily AZD7762 treatments exhibited significant enhancement of response compared with radiation alone [57].

- Radiosensitization by AZD7762 was also investigated in lung cancer cell lines and xenograft models of lung cancer brain metastasis. Clonogenic survival assays showed enhancement of radiosensitivity with AZD7762 after irradiation of various doses. AZD7762 increased ATR/ATM-mediated Chk1 phosphorylation, stabilized Cdc25A, and suppressed cyclin A expression in lung cancer cell lines. In xenograft models of lung cancer (PC14PE6) brain metastasis, AZD7762 significantly prolonged the median survival time in response to radiation (Figure 8.10 [58]). Depletion of Chk1 using shRNA also showed an enhancement of sensitivity to radiation in PC14PE6 cells. The results of this study support the fact that Chk1 can be a good target for enhancement of radiosensitivity.

8.4.2.1.2 Inhibition of CHK2 Experiments with different CHK2-selective inhibitors have given consistent results, showing that CHK2 inhibition protects normal murine and human T cells from radiation toxicity [53]. This is also the case for CHK2 knockout mice, which are radioresistant [59]. Thus, CHK2 inhibition might have radioprotective properties for normal tissue.

8.4.3 Clinical Trials of Checkpoint Kinase Inhibitors

To date, only CHK1-selective or dual CHK1/CHK2 inhibitors have been studied in the clinic and most phase I trials have been treatment combinations for which inhibition of CHK1 is predicted to be the mechanism of potentiation of coadministered DNA-damaging chemotherapy.

UCN-01 has been studied as a single agent and in combinations in many different phase I settings [60], but further progress has been hindered by poor pharmacokinetics and an unfavorable toxicity profile. The dual CHK1/CHK2 inhibitor XL-844 (EXEL-9844; structure undisclosed) entered phase I trials for leukemias and lymphomas as a single agent or in combination with gemcitabine, but development has been discontinued (www.clinicaltrials.gov).

The CDK inhibitor Flavopiridol was involved in a phase I trial in patients with locally advanced pancreatic cancer, which was completed in 2006. Similar to UCN-01, single-agent and combined modality phase I and II trials with flavopiridol showed dose-limiting toxicities [51].

The dual CHK1/CHK2 inhibitor AZD7762 has been administered in combination with gemcitabine and irinotecan for solid tumors in phase I trials that are now reported as complete (www.clinicaltrials.gov). PF-00477736 has shown selectivity for CHK1 over other kinases [60,61], with some partial responses and stabilization of disease; however, the development of this compound has been discontinued. There are no reports of any of the compounds listed above being involved in trials that included radiotherapy.

8.5 PROTEIN DEGRADATION BY THE UBIQUITIN–PROTEASOME SYSTEM

The proteasome is an intracellular complex, the function of which is to degrade unneeded proteins. The ubiquitin–proteasome system is the hub of the cellular proteolysis system using its degradative capacity to control and integrate numerous cellular processes. Protein degradation is as essential to the cell as is protein synthesis because it supplies amino acids for fresh protein synthesis and removes excess enzymes and transcription factors that are no longer needed. Proteasomes deal primarily with endogenous proteins including transcription factors, for instance, p53; cyclins, which must be destroyed to prepare for the next step in the cell cycle; proteins encoded by viruses and other intracellular parasites; and proteins that are folded incorrectly because of translation errors, are encoded by faulty genes, or have been damaged by interaction with other molecules in the cytosol (Table 8.4).

TABLE 8.4 Proteins Targeted by the Ubiquitin–Proteasome Pathway

Protein		Function
Cyclins	Cyclin A	Cell cycle regulation, S phase and mitosis
	Cyclin B	Cell cycle regulation, mitosis
	Cyclin D	Cell cycle regulation, G_1
	Cyclin E	Cell cycle regulation, G_1 and S
CDK inhibitors	p21	Cell cycle regulation
Transcription factors	E2A	Cell growth and differentiation
	E2F	Cell cycle regulation by gene expression control
Transcription factor inhibitors	IκB	Inhibits NF-κB, transcription factor for growth factors, cell adhesion molecules, angiogenesis factors, and antiapoptotic proteins
Oncogenes	c-Myc	Cell proliferation
	c-Fos	Cell proliferation, control of transcription factor AP-1
	cpJun	
Tumor suppressor proteins	p53	Cell cycle arrest, senescence, apoptosis
	Rb	Inactivation of transcription factor E2F, required for initiation of S phase
Apoptosis	Bax	Promotion of apoptosis
	MDM2	Promotion of apoptosis

8.5.1 Proteasome Structure and Function

The proteasome consists of a core particle (CP), a regulatory particle (RP), and the small protein ubiquitin. The CP is made of two copies of each of 14 different proteins, which are assembled in groups of seven; forming a ring with the four rings stacked on top of each other like a pile of doughnuts. There are two identical RPs, one at each end of the CP. Each RP consists of 14 different proteins (none of which are the same as those in the CP), six of these are ATPases. Some of the subunits have sites that recognize ubiquitin, a small protein (76 amino acids), conserved throughout evolution, which acts to target proteins for destruction. Proteins that are destined for destruction are conjugated to a molecule of ubiquitin, which binds to the terminal amino group of a lysine residue. Additional molecules of ubiquitin bind to the first, forming a chain and the whole complex attaches to ubiquitin-recognizing site(s) on the RP. The protein is unfolded by the ATPases of the RP using the energy generated by the breakdown of ATP. The unfolded protein is then translocated into the central cavity of the CP, where several active sites on the inner surface of the two middle molecules break specific bonds of the peptide chain. This produces a set of smaller peptides averaging about eight amino acids. These leave the CP and may be further broken down into individual amino acids by peptidases in the cytosol or in mammals, and incorporated in a class I histocompatibility molecule to be presented on the cell surface to the immune system as a potential antigen. The RP releases the ubiquitins for reuse (Figure 8.11).

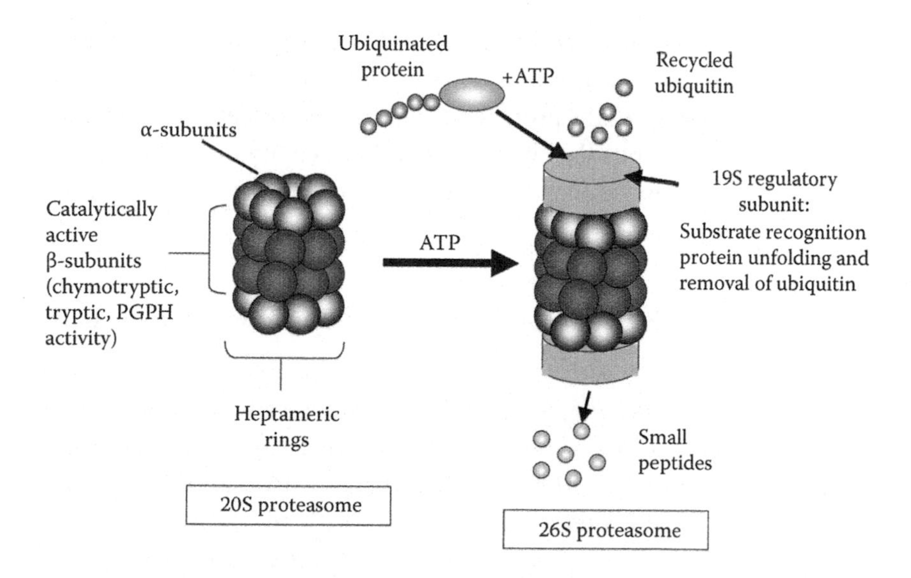

FIGURE 8.11 Schematic representation of a proteasome.

8.5.2 Mechanisms of Proteasome Inhibition

8.5.2.1 Targeting the 20S Proteasome

The majority of proteasome inhibitors target the 20S proteasome (the 20S CP). Peptide derivatives containing diverse reactive groups that work well as proteasome inhibitors include both synthetic peptides such as bortezomib and natural products like salinosporamide A. On the basis of their inhibitory mechanisms, the proteasome inhibitors are classified into five groups as listed below and in Table 8.5.

- *Peptide aldehydes.* These act against serine and cysteine proteases. MG-132 is a potent and selective inhibitor of the chymotrypsin-like activity of the proteasome, which is, however, unsuitable for therapeutic use.
- *Peptide boronates.* These have excellent inhibitory potency and selectivity toward the proteasome compared with other proteasome inhibitors. Unlike peptide aldehydes, boronates are not inactivated by oxidation and are not rapidly removed from the cell by the MDR system. Bortezomib, a proteasome inhibitor containing boronic acid, inhibits the proteasomal chymotrypsin-like activity most strongly and moderately inhibits its caspase-like activity. In 2003, bortezomib was approved by the FDA for the treatment of relapsed multiple myeloma and mantle cell lymphoma [62].
- *β-Lactones.* The first natural nonpeptidic proteasome inhibitor, lactacystin inhibits proteasome activity by binding with the N-terminal threonine residue in the proteasome via a stable covalent bond. Salinosporamide A inhibits the chymotrypsin-like activity of the proteasome and is 35 times more potent than the active principal of lactacystin.
- *Epoxyketones.* Epoxomicin, an Actinomyces metabolite, contains an α,β-epoxyketone moiety that forms an adduct with the N-terminal threonine residue in the proteasome, resulting in inactivation of the proteasome.
- *Macrocyclic vinyl ketones.* Syringolin A irreversibly inhibits all three types of proteasome activity, whereas glidobactin A, another microbial metabolite, inhibits the chymotrypsin-like and trypsin-like activities of the proteasome and reacts with the active sites of the threonine residues. Both syringolin A and glidobactin A inhibit proliferation and induce the apoptosis of malignant cells.

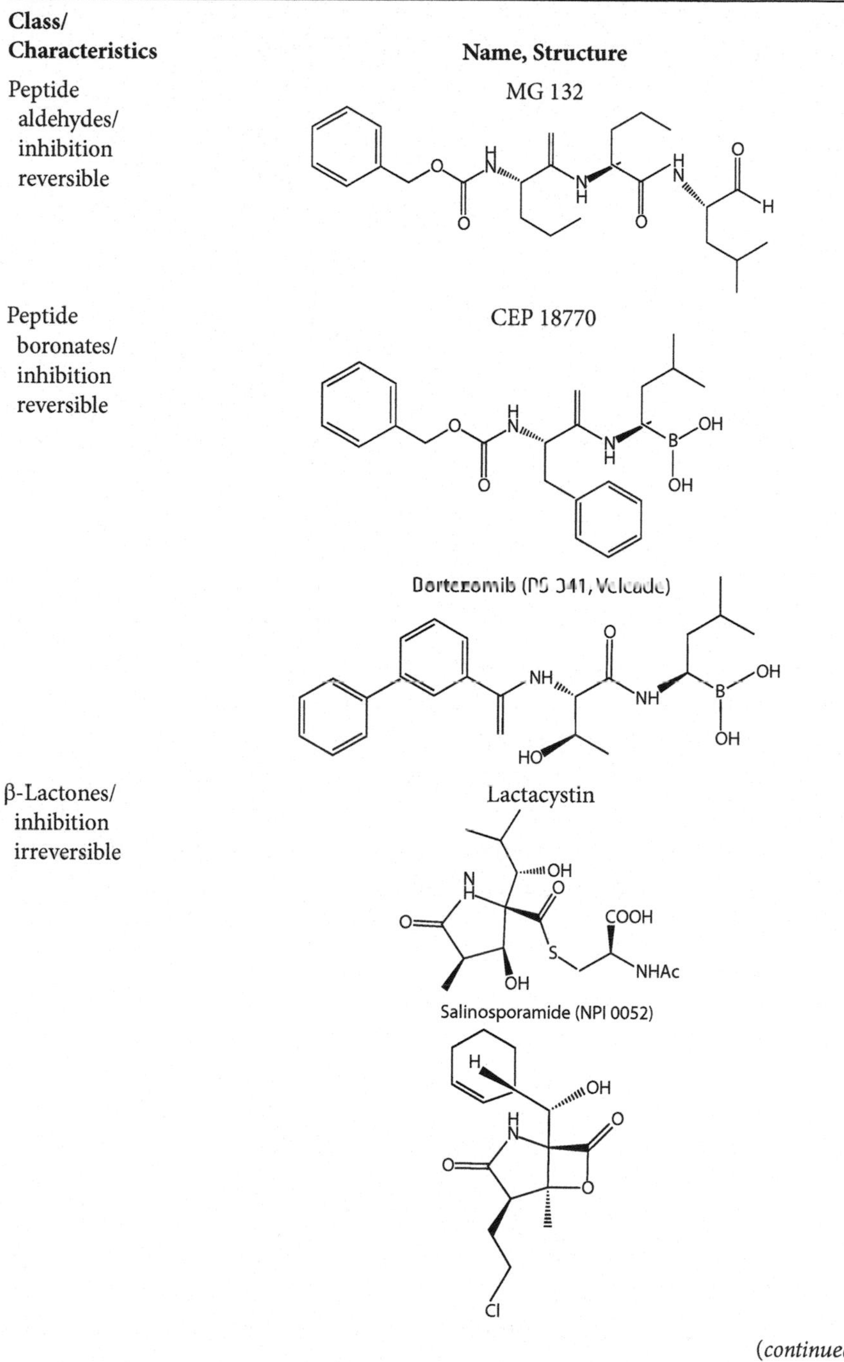

TABLE 8.5 Proteasome Inhibitors with Preclinical or Clinical Activity

Class/ Characteristics	Name, Structure
Peptide aldehydes/ inhibition reversible	MG 132
Peptide boronates/ inhibition reversible	CEP 18770
	Bortezomib (PS 341, Velcade)
β-Lactones/ inhibition irreversible	Lactacystin
	Salinosporamide (NPI 0052)

(*continued*)

TABLE 8.5 Proteasome Inhibitors with Preclinical or Clinical Activity (Continued)

Class/ Characteristics	Name, Structure
Epoxyketones/ inhibition irreversible	Epoxomicin Carfilzomib (PR-171) 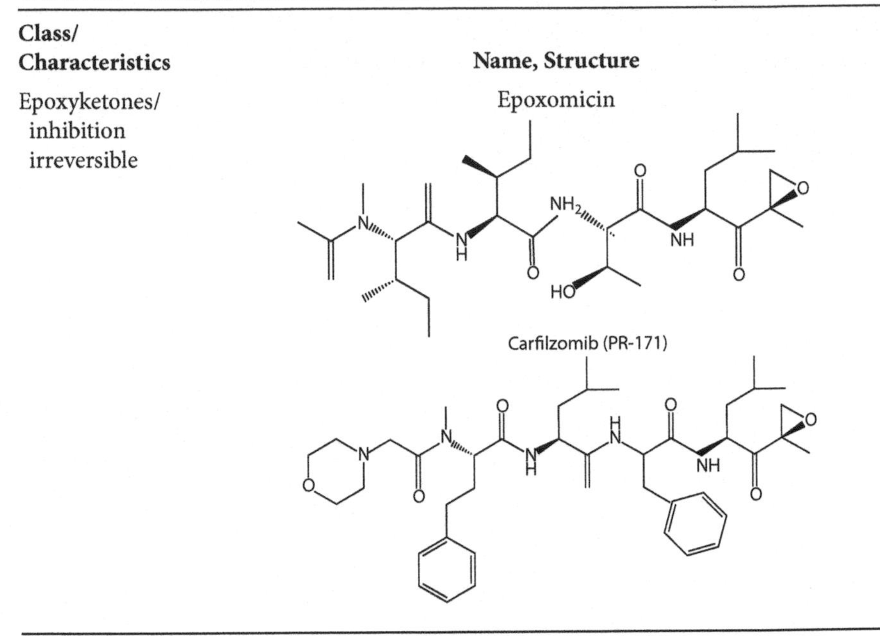

Among the five types of proteasome inhibitors described above, bortezomib was the first to enter clinical development and be approved for the treatment of multiple myeloma and mantle cell lymphoma, whereas three irreversible proteasome inhibitors, salinosporamide A (NPI-0052), carfilzomib (PR-171, an epoxomycin analogue), and CEP-18770 (a bortezomib analogue), are currently undergoing phase I and II clinical trials (clinicaltrials.gov).

8.5.2.2 Targeting the 19S Regulatory Particle

The 19S RP plays roles in the recognition of the ubiquitin chain of client proteins, removal of the ubiquitin chain, unfolding of the client proteins, and gate opening of the 20S proteasome. RP inhibitor peptoid-1 (RIP-1) was found to inhibit the unfolding activity of the 19S RP, resulting in inhibition of the degradation of ubiquitinated proteins.

8.5.2.3 Targeting the Delivery System Connecting the Ubiquitin System to the Proteasome

The ubiquitin–proteasome pathway consists of two systems, the ubiquitin system and the degradation system (the 26S proteasome). The former contains the ubiquitin-activating enzyme E1, ubiquitin-conjugating enzyme

E2, and ubiquitin ligase E3, and catalyzes the ubiquitination of client proteins. Recently, a third system, the so-called delivery system, has been proposed to function in the delivery of ubiquitinated proteins to the 26S proteasome. Compounds inhibiting the recognition/delivery system for ubiquitinated proteins could serve as a novel subclass of general antiproteasome chemotherapeutics.

8.5.2.4 Targeting the Ubiquitin System

In addition to inhibitors targeting the 20S CP, the 19S RP, and the delivery system, various inhibitors of the ubiquitin system, consisting of E1, E2, and E3 enzymes, have been developed. Among enzymes in the ubiquitin system E3s are a large family that recognizing a vast number of client proteins and targeting them for degradation. Among E3s, MDM2 or HDM2, a RING-type E3 for p53 protein, is frequently used as a target for inhibitor development. HDM2 is normally expressed at a low level, but is overexpressed in many human cancers. In Section 8.3.3.3, it was described how targeting MDM2/HDM2 is a promising stratagem to reactivate p53. Nutlin-3 is a MDM2 antagonist, which suppresses tumor progression in nude mice bearing subcutaneous human cancer xenografts [63], and recent studies also suggest a role for nutlins as radiosensitizers (Table 8.2). Also, in this category, the small molecule designated "reactivation of p53 and induction of tumor cell apoptosis" (RITA) inhibits p53–MDM2 interaction through its binding to p53 protein.

8.5.3 Effects of Radiation Combined with Proteasome Inhibitors

There are extensive reports in the literature of clinical and experimental studies of proteasome inhibitors (usually bortezomib) in cancer treatment. These deal predominantly with bortezomib used as a single agent, less frequently in combination with various chemotherapy drugs, and least frequently in combination with radiation, although the potential for proteasome inhibitors as radiosensitizers is frequently invoked.

8.5.3.1 Mechanisms of Radiosensitization by Proteasome Inhibitors

Radiosensitization by proteasome inhibitors has been attributed to several mechanisms that may act singly or in combination.

- Cancer cells frequently show high constitutive activity of the anti-apoptotic transcription factor nuclear factor κB (NF-κB), which results in their enhanced survival. Activation of NF-κB depends on the degradation of its inhibitor IκBα by the 26S proteasome and proteasome inhibitors induce apoptosis in cancer cells and, at nonlethal concentrations, sensitize cells to the cytotoxic effects of ionizing radiation and chemotherapeutic drugs.

- As has already been described, MDM2 or HDM2 is an E3 ligase. Targeting MDM2/HDM2 is one approach to radiosensitization by reactivating p53.
- The sensitizing role of proteasome inhibitors to DNA damaging agents can also be linked to the critical role of ubiquitin-mediated protein degradation in regulating the DNA damage response. In fact, proteasome-mediated protein degradation is needed for homologous recombination, the FA pathway, nucleotide excision repair, degradation of stalled RNA polymerases II, elimination of trapped DNA topoisomerase, and regulation of p53 (reviewed by Ljungman [7]). Because anticancer agents such as ionizing irradiation or chemotherapeutic agents generate DNA damage and induce DNA damage-dependent apoptosis, the proteasomal involvement in DNA repair pathways would argue for the synergistic interaction of proteasomal inhibitors with conventional therapeutic agents.

8.5.3.2 Experimental Findings

Published experimental results include the following:

- The extent of cell killing in irradiated colorectal cell lines showed an additive increase after pretreatment with PS-341 (borzetomib). In an *in vivo* experiment, the growth of LOVO tumors was reduced when radiation was combined with a single injection of PS-341 [64]. NF-κB activation was induced by radiation and inhibited by pretreatment with either PS-341 or IκBα super-repressor in all cell lines. Inhibition of radiation-induced NF-κB activation resulted in increased apoptosis and a decrease of cell growth and clonogenic survival. An 84% reduction in initial tumor volume was obtained in LOVO xenografts receiving radiation and PS-341.
    - Survival of cells from EMT-6 tumors after treatment of the host animals with various doses of radiation therapy was reduced when intraperitoneal injection PS-341 was combined with fractionated radiation. In another study by the same authors,

combined treatment with radiation and PS-341 was more effective than either modality alone against the Lewis lung carcinoma growing as a subcutaneous tumor or as lung metastases [65].

- The 20S core unit of the proteasome shares specific cleavage action sites with the HIV-I protease and the HIV-I protease inhibitor ritonavir has been shown to inhibit 20S proteasome function. The HIV-I PI saquinavir is less toxic than ritonavir, but like ritonavir, directly inhibits 20S and 26S proteasome function *in vitro.* Saquinavir treatment of PC-3 and DU-145 prostate cancer, U373 glioblastoma, and K562 and Jurkat leukemia cells resulted in apoptosis at concentrations that were similar to those needed to inhibit proteasome function. A short preincubation with saquinavir sensitized PC-3 and DU-145 prostate cancer cells to doses of ionizing radiation above 2 Gy with a change of the α/β ratio from 7.8 (PC-3) and 10.3 (DU-145) in control cells to 4.8 for both cell lines with saquinavir-treatment [66]. Interestingly, there have been reports of dramatically improved survival for AIDS patients experiencing HIV-related primary central nervous system lymphoma (PCNSL) treated with highly active antiretroviral therapy (HAART) and cranial irradiation when compared with cranial irradiation or HAART treatment alone [67].

- MG-132, an inhibitor of 26S proteasome activity, induced apoptosis in HD-My-Z Hodgkin cells and radiosensitized them, despite the fact that their constitutively active NF-κB levels are unaltered [68]. HD-My-Z are a moderately radiation-resistant cell line with a Surviving Fraction at 2 Gy of 0.56 ± 0.032. Transient inhibition of proteasome function by 3-h preincubation of HD-My-Z cells in 50 μM of MG-132 significantly sensitized them to ionizing irradiation, as measured by clonogenic survival (Figure 8.12).

8.5.4 Clinical Application of Proteasome Inhibitors

Bortezomib was approved by the FDA in 2006 for the treatment of relapsed/refractory multiple myeloma and mantle cell lymphoma. Results of numerous clinical trials attest to its efficacy and safety in both multiple myeloma and mantle cell lymphoma.

Various studies in which bortezomib was added to standard therapies as treatment of advanced solid tumors had generally disappointing results. Bortezomib failed to improve the efficacy of permetrexed in non-small

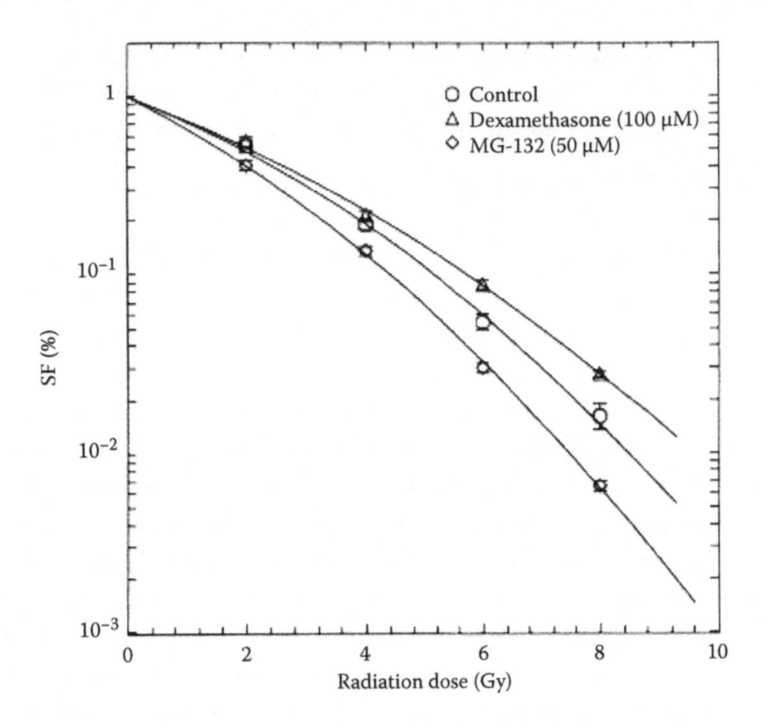

FIGURE 8.12 Clonogenic survival of HD-My-Z cells after treatment with MG-132 or dexamethasone. Intrinsic radiosensitivity of HD-My-Z Hodgkin lymphoma cells was decreased by inhibition of proteasome function by preincubation in MG-132 (50 mM) for 3 h before irradiation. Dexamethasone under the same conditions was a radioprotector. (From Pajonk, F. et al., *Int J Radiat Oncol Biol Phys* 47, 1025–1032, 2000. With permission.)

cell lung cancer, gemcitabine in pancreatic cancer, irinotecan in various advanced solid tumors, and carboplatin in ovarian cancer (these findings are summarized by Yang et al. [69]).

Although results to date have not been promising, bortezomib as a single agent and in combination with chemotherapy and radiotherapy in various solid tumors continues to be tested in trials. Bortezomib was given concurrently with reirradiation to patients with recurrent head and neck squamous cell carcinoma (HNSCC) and inhibition of the proteasome activity along with inhibition of NF-κB activation and induction of cell death after treatment were observed compared with pretreatment [70]. Two phase I trials studying the effectiveness, side effects, and best dose of bortezomib combined with radiation therapy or when given together with cetuximab and radiation therapy with or without cisplatin in treating patients with advanced head and neck cancer have been completed. Ongoing trials include a phase I study of bortezomib in combination with 5-fluorouracil and radiotherapy in treating patients with locally advanced and metastatic rectal cancer; a phase I/II trial studying the side effects and best dose of bortezomib, paclitaxel, and carboplatin when given with radiation therapy for treatment in patients with advanced non-small cell lung cancer; a phase II trial of combined treatment with BTZ, temolozide, and radiotherapy for brain and CNS malignancies; and a phase I study of BTZ and radiotherapy for metastatic squamous cell carcinoma of the head and neck (www.clinicaltrials.gov).

8.5.5 Natural Compounds with Proteasome-Inhibitory Effects in Clinical Trials

A number of natural products have been shown to have anticancer effects and, in some cases, to act as radiosensitizers or chemosensitizers (or both). Many of these, including green tea polyphenols, genistein, curcumin, shikonin, and disulfiram, have been shown to have activity as proteasomal inhibitors. These compounds will be discussed in more detail in Chapter 13, which deals with radiosensitization by phytochemicals.

8.5.6 Cancer-Initiating Cells Are Characterized by Low Proteasomal Activity

There is increasing evidence that solid cancers contain cancer-initiating cells (CICs), which are capable of regenerating a tumor that has been treated by surgery, chemotherapy, or radiotherapy. Reduced 26S proteasome activity is a general feature of CICs that might be exploited to identify, track, and potentially target these cells *in vitro* and *in vivo*.

Investigations with human cancer cell lines in tissue culture has demonstrated reduced proteasomal activity to be a characteristic of subpopulations from human cancer cell lines including glioma, breast, lung, and prostate cancer [71–74].

It has been further demonstrated that cancer cell lines with low proteasomal activity are more radioresistant than the bulk tumor cell population and that fractionated radiation can actually result in enrichment of the population with CICs [71,72,74]. The existence of a radioresistant phenotype for cancer cells with low proteasome subunit expression suggests that if this population of cells also exists in clinical samples, it could drive recurrences after chemotherapy or radiotherapy. The proteasome inhibitor, bortezomib, is in clinical use for patients experiencing multiple myeloma or mantle cell lymphoma. However, despite its excellent preclinical efficacy in animal models of other cancers, bortezomib failed to demonstrate antitumor activity as a single agent in patients with solid cancers in clinical trials, suggesting that CICs might be resistant to drugs targeting the proteasome [31–34].

8.6 SUMMARY

Following induction of DNA strand breaks by ionizing radiation, the ATM kinase is activated by sensing alterations in the chromatin structure induced by loss of DNA topology. Two very specific ATM inhibitors, KU55933 and CP466722, have been shown to be effective in rapidly sensitizing cancer cells to ionizing radiation. Importantly, these inhibitors are reversible, making it possible to selectively target ATM during radiotherapy.

The tumor suppressor p53 is an important regulator of both cell cycle checkpoints and induction of apoptosis after exposure to DNA-damaging agents. The p53 gene is mutated in more than half of all human cancers and is expressed at very high cellular levels when mutated. One strategy exploiting the high levels of p53 in cancer cells is to convert the mutated form of p53 into the wild-type conformation with the expectation that these compounds would be tumor-selective and cause cancer cells to induce apoptotic activity if p53 proteins were present. The reactivation of mutant p53 in cancer cells leading to apoptosis has been shown using both peptides and small-molecule inhibitors such as PRIMA-1, and this approach has shown promise in xenograft models. Another approach is to introduce wild-type p53 into tumor cells to enhance the therapeutic effects of radiation or chemotherapy. In cancer cells harboring wild-type p53, compounds that interrupt the MDM2–p53 circuit can increase the cellular levels of p53.

Two important phosphorylation substrates of ATM and ATR are the cell cycle checkpoint kinases CHK1 and CHK2. When activated after DNA damage, these proteins orchestrate arrests in the G_1/S, S, G_2/M, and M phases of the cell cycle. Inactivation of CHK1 may sensitize cancer cells to DNA-damaging agents by not allowing sufficient time for these cells to repair their DNA so that the cells enter S phase or mitosis with unrepaired DNA, and subsequent complications lead to cell death. CHK1 is also important for homologous recombination by activating RAD51 through phosphorylation. Inhibition of CHK1 has been shown to sensitize cells to ionizing radiation. Many drugs that inhibit CHK1 are especially effective in cells defective in G_1 arrest checkpoint such as p53 mutant cells.

In contrast to CHK1, CHK2 has been shown to stimulate DNA damage-induced apoptosis and may act as a tumor suppressor. Thus, it is possible that the targeting of CHK2 may make tumor cells less responsive to DNA-damaging therapies. There are ongoing clinical trials employing the drugs XL844 and AZD7762, which inhibit both CHK1 and CHK2, in combination with gemcitabine. The development of drugs that selectively inhibit CHK1 without affecting CHK2 would give stronger efficacy when combined with radiation or chemotherapy (or both). Available evidence thus strongly supports a role for CHK1 inhibition in the radiosensitization of cancer cells although it seems that the main effect for CHK2 inhibition might be as a radioprotectant at least of nonmalignant cells.

The ubiquitin–proteasome pathway regulates protein turnover in cells. Some proteasome inhibitors have shown utility as single agents and they preferentially kill tumor cells both *in vitro* and *in vivo* whereas proteasome inhibitors have also been shown to sensitize cancer cells to cisplatin and to radiation. The induction of the NF-κB survival pathway by DNA-damaging agents is dependent on proteosomal degradation of IκB, and proteasome inhibitors block NF-κB activation, leading to the sensitization of cells to radiation and chemotherapy. Ubiquitin-mediated protein degradation also plays a critical role in regulating the DNA damage response. Thus far, bortezomib is the most widely investigated proteasome inhibitor in clinical trials. It has demonstrated efficacy against hematological malignancies, in both single and combination regimens, and is FDA-approved for the treatment of multiple myeloma and mantle cell lymphoma. Bortezomib has failed to show significant efficacy for the treatment of solid tumors. Researchers continue to investigate bortezomib in combination with other modalities and new proteasome inhibitors with increased potency and decreased side effects.

REFERENCES

1. Kastan, M., and Lim, D. The many substrates and functions of ATM. *Nat Rev Mol Cell Biol* 2000;1:179–186.
2. Shiloh, Y. ATM and related protein kinases: Safeguarding genome integrity. *Nat Rev Cancer* 2003;3:155–168.
3. Gaudin, D., and Yielding, K. Response of a "resistant" plasmacytoma to alkylating agents and X-ray in combination with the "excision" repair inhibitors caffeine and chloroquine. *Proc Soc Exp Biol Med* 1969;131:1413–1416.
4. Sarkaria, J., and Eshleman, J. ATM as a target for novel radiosensitizers. *Semin Radiat Oncol* 2001;11:316–327.

5. Kawabe, T. G2 checkpoint abrogators as anticancer drugs. *Mol Cancer Ther* 2004;3:513–519.
6. Alao, J. The ATM regulated DNA damage response pathway as a chemo- and radiosensitisation target. *Expert Opin Drug Discov* 2009;4:495–505.
7. Ljungman, M. Targeting the DNA damage response in cancer. *Chem Rev* 2009;109:2929–2950.
8. Rainey, M., Charlton, M., Stanton, R., and Kastan, M. Transient inhibition of ATM kinase is sufficient to enhance cellular sensitivity to ionizing radiation. *Cancer Res* 2008;68:7466–7474.
9. Hickson, I., Zhao, Y., Richardson, C. et al. Identification and characterization of a novel and specific inhibitor of the ataxia-telangiectasia mutated kinase ATM. *Cancer Res* 2004;64:9152–9159.
10. Cresenzi, E., Palumbo, G., deBoer, J., and Brady, H. Ataxia telangiectasia mutated and p21CIP1 modulate cell survival of drug-induced senescent tumor cells: Implications for chemotherapy. *Clin Cancer Res* 2008;14:1877–1887.
11. Golding, S., Rosenberg, E., Adams, B. et al. Dynamic inhibition of ATM kinase provides a strategy for glioblastoma multiforme radiosensitization and growth control. *Cell Cycle* 2012;11(6):1167–1173.
12. Dote, H., Burgan, W., Camphausen, K., and Tofilon, P. Inhibition of Hsp90 compromises the DNA damage response to radiation. *Cancer Res* 2006;66:9211–9220.
13. Koll, T., Feis, S., Wright, M. et al. Hsp90 inhibitor, DMAG, synergizes with radiation of lung cancer cells by interfering with base excision and ATM-mediated DNA repair. *Mol Cancer Ther* 2008;7:1985–1992.
14. Fujiwara, T., Grimm, E., Mukhopadhyay, T., Cai, D., Owen-Schaub, L., and Roth, J. A retroviral wild-type p53 expression vector penetrates human lung cancer spheroids and inhibits growth by inducing apoptosis. *Cancer Res* 1993;53:4129–4133.
15. Zeimet, A., and Marth, C. Why did p53 gene therapy fail in ovarian cancer? *Lancet Oncol* 2003;4:415–422.
16. Peng, Z. Current status of gendicine in China: Recombinant human Ad-p53 agent for treatment of cancers. *Hum Gene Ther* 2005;16:1016–1027.
17. Pearson, S., Jia, H., and Kandachi, K. China approves first gene therapy. *Nat Biotechnol* 2004;22:3–4.
18. Vazquez, A., Bond, E., Levine, A., and Bond, G. The genetics of the p53 pathway, apoptosis and cancer therapy. *Nat Rev Drug Discov* 2008;7:979–987.
19. Bouchet, B., de Fromentel, C., Puisieux, A., and Galmarini, C. p53 as a target for anticancer drug development. *Crit Rev Oncol Hematol* 2006;58:190–207.
20. Rogulski, K., Freytag, S., Zhang, K. et al. *In vivo* antitumor activity of ONYX-015 is influenced by p53 status and is augmented by radiotherapy. *Cancer Res* 2000;60:1193–1196.
21. O'Shea, C., Johnson, L., and Bagus, B. Late viral RNA export, rather than p53 inactivation, determines ONYX-015 tumor selectivity. *Cancer Cell* 2004;6:611–623.

22. Khuri, F., Nemunaitis, J., Ganly, I. et al. A controlled trial of intratumoral ONYX-015, a selectively replicating adenovirus, in combination with cisplatin and 5-fluorouracil in patients with recurrent head and neck cancer. *Nat Med* 2000;6:879–885.
23. Supiot, S., Zhao, H., Wiman, K., Hill, R., and Bristow, R. PRIMA-1(met) radiosensitizes prostate cancer cells independent of their MTp53-status. *Radiother Oncol* 2008;86:407–411.
24. Selivanova, G., Ryabchenko, L., Jansson, E., Iotsova, V., and Wiman, K. Reactivation of mutant p53 through interaction of a C-terminal peptide with the core domain. *Mol Cell Biol* 1999;19:3395–3402.
25. Bykov, V., Issaeva, N., Shilov, A. et al. Restoration of the tumor suppressor function to mutant p53 by a low-molecular-weight compound. *Nat Med* 2002;8:282–288.
26. Supiot, S., Hill, R., and Bristow, R. Nutlin-3 radiosensitizes hypoxic prostate cancer cells independent of p53. *Mol Cancer Ther* 2008;7:993–999.
27. Foster, B., Coffey, H., Morin, M., and Rastinejad, F. Pharmacological rescue of mutant p53 conformation and function. *Science* 1999;286:2507–2510.
28. Wang, W., Takimoto, R., Rastinejad, F., and El-Deiry, W. Stabilization of p53 by CP-31398 inhibits ubiquitination without altering phosphorylation at serine 15 or 20 or MDM2 binding. *Mol Cell Biol* 2003;23:2171–2181.
29. Wang, Q., Fan, S., Eastman, A., Worland, P., Sausville, E., and O'Connor, P. UCN-01: A potent abrogator of G2 checkpoint function in cancer cells with disrupted p53. *J Natl Cancer Inst* 1996;88:956–965.
30. Cao, C., Shinohara, E., Subhawong, T. et al. Radiosensitization of lung cancer by nutlin, an inhibitor of murine double minute 2. *Mol Cancer Ther* 2006;5:411–417.
31. Issaeva, N., Bozko, P., Enge, M. et al. Small molecule RITA binds to p53, blocks p53-HDM-2 interaction and activates p53 function in tumors. *Nat Med* 2004;10:1321–1328.
32. Krajewski, M., Ozdowy, P., D'Silva, L., Rothweiler, U., and Holak, T. NMR indicates that the small molecule RITA does not block p53-MDM2 binding *in vitro*. *Nat Med* 2005;11:1135–1136.
33. Shangary, S., McEachern, D., Liu, M. et al. Temporal activation of p53 by a specific MDM2 inhibitor is selectively toxic to tumors and leads to complete tumor growth inhibition. *Proc Natl Acad Sci USA* 2008;105:3933–3938.
34. Takahashi, A., Matsumoto, H., and Ohnishi, T. Hdm2 and nitric oxide radicals contribute to the p53-dependent radioadaptive response. *Int J Radiat Oncol Biol Phys* 2008;71:550–558.
35. Arya, A., El-Fert, A., Devling, T. et al. Nutlin-3, the small-molecule inhibitor of MDM2, promotes senescence and radiosensitises laryngeal carcinoma cells harbouring wild-type p53. *Br J Cancer* 2010;103:186–195.
36. Shen, H., Moran, D., and Maki, C. Transient nutlin-3a treatment promotes endoreduplication and the generation of therapy-resistant tetraploid cells. *Cancer Res* 2008;68:8260–8268.

37. Lehmann, B., McCubrey, J., Jefferson, H., Paine, M., Chappell, W., and Terrian, D. A dominant role for p53-dependent cellular senescence in radiosensitization of human prostate cancer cells. *Cell Cycle* 2007;6: 595–605.
38. Kitagaki, J., Agama, K., Pommier, Y., Yang, Y., and Weissman, A. Targeting tumor cells expressing p53 with a water-soluble inhibitor of Hdm2. *Mol Cancer Ther* 2008;7:2445–2454.
39. Vaziri, H., Dessain, S., Eaton, E.N. et al. hSIR2(SIRT1) functions as an NAD-dependent p53 deacetylase. *Cell* 2001;107:149–159.
40. Lain, S., Hollick, J., Campbell, J. et al. Discovery, *in vivo* activity and mechanism of action of a small-molecule p53 activator. *Cancer Cell* 2008;13:454–463.
41. Mutka, S., Yang, W., Dong, S. et al. Identification of nuclear export inhibitors with potent anticancer activity *in vivo*. *Cancer Res* 2009;69:10–17.
42. Flores, E., Sengupta, S., Miller, J. et al. Tumor predisposition in mice mutant for p63 and p73: Evidence for broader tumor suppressor functions for the p53 family. *Cancer Cell* 2005;7:363–373.
43. Bell, H., and Ryan, K. Targeting the p53 family for cancer therapy: 'Big Brother' joins the fight. *Cell Cycle* 2007;6:1995–2000.
44. Kravchenko, J., Ilyinskaya, G., Komarov, P. et al. Small-molecule RETRA suppresses mutant p53-bearing cancer cells through a p73-dependent salvage pathway. *Proc Natl Acad Sci USA* 2008;105:6302–6307.
45. Komarov, P., Komarova, E., Kondratov, R. et al. A chemical inhibitor of p53 that protects mice from the side effects of cancer therapy. *Science* 1999;285:1733–1737.
46. Ashwell, S., Janetka, J., and Zabludoff, S. Keeping checkpoint kinases in line: New selective inhibitors in clinical trials. *Expert Opin Investig Drugs* 2008;17:1331–1340.
47. Wilson, G. Radiation and the cell cycle, revisited. *Cancer Metastasis Rev* 2004;23:209–225.
48. Garrett, M., and Collins, I. Anticancer therapy with checkpoint inhibitors: What, where and when? *Trends Pharmacol Sci* 2011;32:308–316.
49. Jobson, A., Lountos, G., Lorenzi, P. et al. Cellular inhibition of checkpoint kinase 2 (Chk2) and potentiation of camptothecins and radiation by the novel Chk2 inhibitor PV1019 [7-nitro-1H-indole-2-carboxylic acid {4-[1-(guanidinohydrazone)-ethyl]-phenyl}-amide]. *J Pharmacol Exp Ther* 2009;331:816–826.
50. Petersen, L., Hasvold, G., Lukas, J., Bartek, J., and Syljuåsen, R. p53-dependent G1 arrest in 1st or 2nd cell cycle may protect human cancer cells from cell death after treatment with ionising radiation and Chk1 inhibitors. *Cell Prolif* 2010;43:365–371.
51. Tao, Y., Leteur, C., Yang, C. et al. Radiosensitization by Chir-124, a selective CHK1 inhibitor. Effects of p53 and cell cycle checkpoints. *Cell Cycle* 2009;8:1196–1205.
52. Choudhury, A., Cuddihy, A., and Bristow, R. Radiation and new molecular agents part I: Targeting ATM-ATR checkpoints, DNA repair, and the proteasome. *Semin Radiat Oncol* 2006;16:51–58.

53. Camphausen, K., Brady, K., Burgan, W. et al. Flavopiridol enhances human tumor cell radiosensitivity and prolongs expression of gamma-H2AX foci. *Mol Cancer Ther* 2004;3:409–416.
54. Syljuåsen, R., Sørensen, C., Nylandsted, J., Lukas, C., Lukas, J., and Bartek, J. Inhibition of Chk1 by CEP-3891 accelerates mitotic nuclear fragmentation in response to ionizing radiation. *Cancer Res* 2004;64:9035–9040.
55. Riesterer, O., Matsumoto, F., Wang, L. et al. A novel Chk1 inhibitor, XL-844, increases human cancer cell radiosensitivity through promotion of mitotic catastrophe. *Invest New Drugs* 2011;29:514–522.
56. Morgan, M., Parsels, L., Zhao, L. et al. Mechanism of radiosensitization by the Chk1/2 inhibitor AZD7762 involves abrogation of the G2 checkpoint and inhibition of homologous recombinational DNA repair. *Cancer Res* 2010;70:4972–4981.
57. Mitchell, J., Choudhuri, R., Fabre, K. et al. *In vitro* and *in vivo* radiation sensitization of human tumour cells by a novel checkpoint kinase inhibitor, AZD7762. *Clin Cancer Res* 2010;16:2076–2084.
58. Yang, H., Yoon, S., Jin, J. et al. Inhibition of checkpoint kinase 1 sensitizes lung cancer brain metastases to radiotherapy. *Biochem Biophys Res Commun* 2011;406:53–58.
59. Takai, H., Naka, K., Okada, Y. et al. Chk2-deficient mice exhibit radioresistance and defective p53-mediated transcription. *EMBO J* 2002;21:5195–5205.
60. Ma, C., Janetka, J., and Piwnica-Worms, H. Death by releasing the breaks: CHK1 inhibitors as cancer therapeutics. *Trends Mol Med* 2011;17:88–96.
61. Blasina, A., Hallin, J., Chen, E. et al. Breaching the DNA damage checkpoint via PF-00477736, a novel small-molecule inhibitor of checkpoint kinase. *Mol Cancer Ther* 2008;7:2394–2404.
62. Bross, P., Kane, R., Farrell, A. et al. Approval summary for bortezomib for injection in the treatment of multiple myeloma. *Clin Cancer Res* 2004;10:3954–3964.
63. Vassilev, L., Vu, B., Graves, B. et al. *In vivo* activation of the p53 pathway by small-molecule antagonists of MDM2. *Science* 2004;303:844–848.
64. Russo, S., Tepper, J., Baldwin, A. et al. Enhancement of radiosensitivity by proteasome inhibition: Implications for a role of NF-κB. *Int J Radiat Oncol Biol Phys* 2001;50:183–193.
65. Teicher, B., Ara, G., Herbst, R., Palombella, V.J., and Adams, J. The proteasome inhibitor PS-341 in cancer therapy. *Clin Cancer Res* 1999;5:2638–2645.
66. Pajonk, F., Himmelsbach, J., Riess, K., Sommer, A., and McBride, W. The human immunodeficiency virus (HIV-1) protease inhibitor saquinavar inhibits proteasome function and causes apoptosis and radiosensitization in non-HIV-associated human cancer cells. *Cancer Res* 2002;62:5230–5235.
67. Hoffmann, C., Tabrizian, S., Wolf, E. et al. Survival of AIDS patients with primary central nervous system lymphoma is dramatically improved by HAART-induced immune recovery. *AIDS (Hagerstown)* 2001;15:2119–2127.
68. Pajonk, F., Pajonk, K., and McBride, W. Apoptosis and radiosensitization of Hodgkin cells by proteasome inhibition. *Int J Radiat Oncol Biol Phys* 2000;47:1025–1032.

69. Yang, H., Zonder, J., and Dou, Q. Clinical development of novel proteasome inhibitors for cancer treatment. *Expert Opin Investig Drugs* 2009;18(7): 957–971.
70. Waes, C.V., Chang, A., Lebowitz, P. et al. Inhibition of nuclear factor-kappaB and target genes during combined therapy with proteasome inhibitor bortezomib and reirradiation in patients with recurrent head-and-neck squamous cell carcinoma. *Int J Radiat Oncol Biol Phys* 2005;63:1400–1412.
71. Vlashi, E., Kim, K., Donna, L.D. et al. In-vivo imaging, tracking, and targeting of cancer stem cells. *J Natl Cancer Inst* 2009;101:350–359.
72. Lagadec, C., Vlashi, E., Donna, L.D. et al. Survival and self-renewing capacity of breast cancer initiating cells during fractionated radiation treatment. *Breast Cancer Res* 2010;12:R13.
73. Pan, J., Zhang, Q., Wang, Y., and You, M. 26S proteasome activity is downregulated in lung cancer stem-like cells propagated *in vitro*. *PLoS One* 2010;5:e13298.
74. Donna, L.D., Lagadec, C., and Pajonk, F. Radioresistance of prostate cancer cells with low proteasome activity. *Prostate* 2012;72(8):868–874.

CHAPTER 9

Radiosensitization by Inhibition of DNA Repair

9.1 OVERVIEW OF DNA REPAIR IN MAMMALIAN CELLS

Radiosensitization by targeting components of the DNA damage response, ATM, ATR, p53, and checkpoint proteins as well as the proteasomal system, were discussed in the previous chapter. In this chapter, radiosensitization by targeting DNA repair itself will be considered. There are multiple modes of DNA damage and repair in mammalian cells. After initial sensing, human DNA double-strand breaks (DSBs) are repaired mainly through two pathways that can both interact (and may compete) with each other across cell cycle transitions, homologous recombination (HR), and nonhomologous end-joining (NHEJ). HR is a template-guided, error-free pathway predominantly operating in the S and G_2 phases of the cell cycle and involves RAD51 (its paralogues RAD51B/C/D, XRCC2/3, and p53), RPA, BRCA2, BLM, and MUS81. In contrast, NHEJ does not depend on homologous sequences and can be precise or imprecise, depending on the structure of the DNA end. NHEJ is preferentially used during the G_1 phase of the cell cycle and involves the KU70/80, DNA-PKcs, Artemis, XLF, XRCC4, DNA ligase IV, ATM, p53, and MDM2 proteins. Cells defective in either HR or NHEJ proteins show increased rates of mutagenesis and chromosomal instability, which could relate to the propensity for acquired genetic instability during prostate carcinogenesis and tumor progression (Figure 9.1).

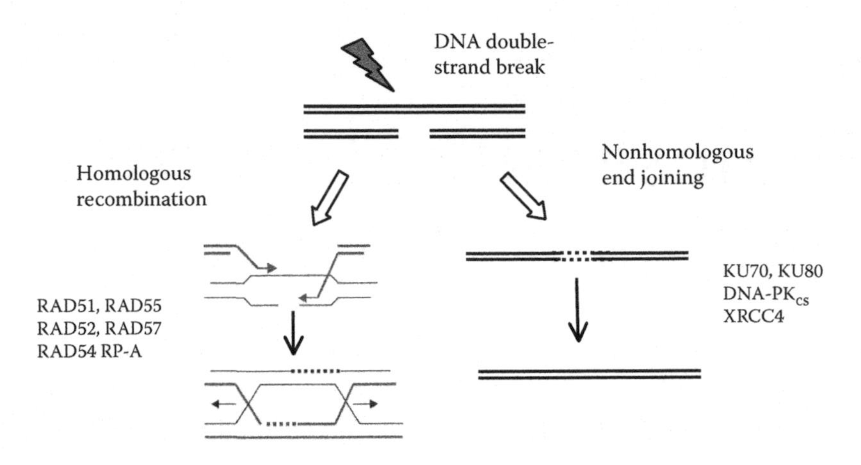

FIGURE 9.1 Schematic overview of homologous and nonhomologous DSB repair pathways. Left, homologous recombination (HR): the extended 3′ single-stranded tails of molecules generated by exonucleolytic processing of DSBs invade homologous intact donor sequences. Right, nonhomologous end joining (NHEJ).

9.2 DNA DOUBLE-STRAND BREAK REPAIR: NONHOMOLOGOUS END-JOINING

In mammalian cells, the majority of ionizing radiation (IR)-induced DSBs are repaired by NHEJ, which occurs mainly in the G_0 and G_1 phase of the cycle. NEHJ is responsible for the repair of DSBs induced by agents such as IR and by the programmed DSBs generated during the generation of T-cell receptors and immunoglobulin molecules through V(D)J recombination.

NHEJ involves the re-joining of two broken ends of the DNA in a sequence-independent manner and is often considered a more error-prone method of DSB repair because genetic information can be lost in the repair of staggered breaks. NHEJ requires the coordination of many proteins and signaling pathways (Figure 9.2).

The main proteins involved are DNA-dependent protein kinase (DNA-PK), consisting of a catalytic subunit (DNA-PKcs) and the heterodimer Ku70/Ku80, XRCC4, Artemis, and DNA ligase IV. DNA-PK belongs to a family of serine/threonine protein kinases, termed phosphatidylinositol-3-kinase-related kinases (PIKKs) comprising ataxia-telangiectasia mutated (ATM), ATM and Rad3-related (ATR), as well as DNA-PK. During NHEJ, the Ku70/80 heterodimer component of DNA-PK binds to the two DNA ends in a ring conformation, aligning the two DNA ends and subsequently activating the catalytic activity of DNA-PK, which promotes the ligation of DNA ends by the XRCC4–ligase IV complex. A recently identified

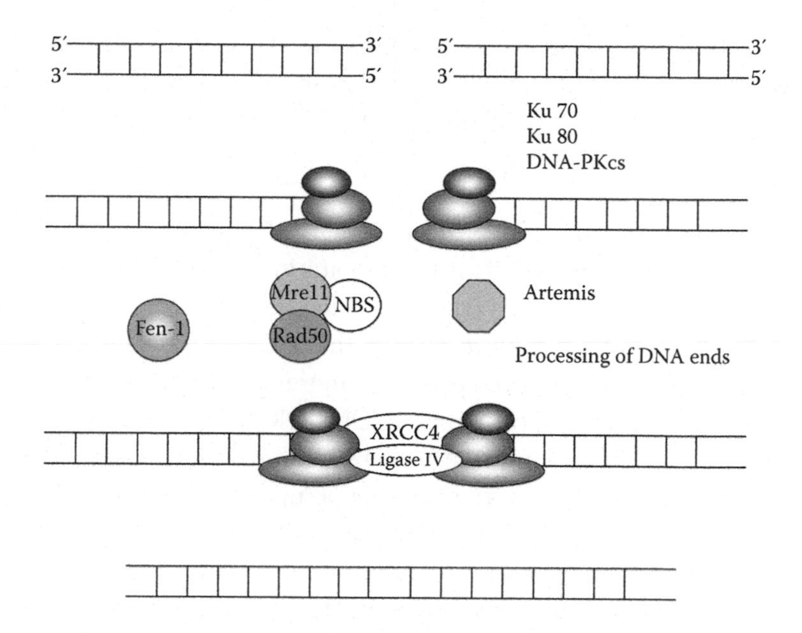

FIGURE 9.2 Nonhomologous end-joining. DNA-PKcs, KU70, KU80, DNA ligase IV, and XRCC4 are critical components of the NHEJ repair pathway. DNA-PK is composed of a heterodimeric DNA-binding component, named KU70/KU80, which binds to either blunt or staggered DNA ends at the DSB and recruits the large catalytic subunit kinase, DNA-PKcs, to the break. DNA-PKcs undergoes autophosphorylation after binding to the DNA break and may recruit additional proteins to the damaged site as potential phosphorylation substrates. The NBS1–MRE11–RAD50 protein complex and the Artemis protein are involved in processing of DNA ends during the initial binding and activation of DNA-PK kinase activity. The XRCC4–ligase IV heterodimer finally ligates the breaks to create intact DNA strands.

NHEJ component, XRCC4-like factor (XLF), stimulates the activity of the XRCC4–DNA ligase IV complex toward noncompatible DNA ends. Other proteins such as Artemis are required for the end-processing of a subset of IR-induced DSBs *in vivo* (Figure 9.2).

9.2.1 Targeting NHEJ for Radiosensitization

There has been considerable interest in the development of small molecules targeting the three major kinases involved in the DNA damage response and DSB repair, and based on evidence that targeting the NHEJ pathway could be a component of effective cancer treatment. Cells deficient in Ku70/80 or the catalytic subunit of DNA-PK (DNA-PKcs) are sensitive to DSBs induced by IR or chemotherapeutic drugs [1,2], suggesting that DNA-PK may constitute a good target for radiochemosensitization. Moreover, DNA-PK is upregulated in some cancers, implicating it as a factor required for tumor growth and survival [3].

9.2.1.1 Targeting DNA-PK

Preliminary investigations into the inhibition of DNA-PK used the broad-spectrum PIKK inhibitors caffeine, wortmannin, and LY294002 [4]; however, their lack of selectivity among the individual PIKKs has limited their clinical development. Nevertheless, the nonselective inhibitors were useful as a basis for the development of more specific DNA-PK inhibitors. Treatment with a flavone-based DNA-PK inhibitor, IC87361, led to tumor radiosensitization *in vitro* and *in vivo* without causing toxicity [5]. LY294002 was also used as a starting point for the development of a potent and selective DNA-PK inhibitor (NU7441) with an IC_{50} of 13 nM. NU7441 shows an excellent selectivity profile across the PIKKs and unrelated kinases and acts as a chemopotentiator both *in vitro* and *in vivo.* Pronounced radiosensitization was shown for MCF7 breast carcinoma cells following inhibition of DNA-PKcs by NU7441 (Figure 9.3). Cell

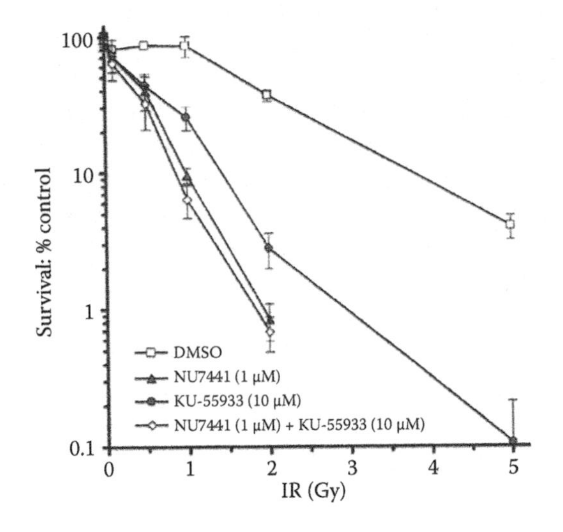

FIGURE 9.3 NU7441 and KU-55933 sensitize MCF7 breast carcinoma cells to ionizing radiation. MCF7 cells were incubated in a medium containing varying concentrations of KU-55933. MCF7 cells were incubated in a medium containing solvent alone (□), 10 mM KU-55933 (●), 1 mM NU7441 (▲), or both inhibitors (◇) for 1 h prior to γ-irradiation at doses between 0 and 5 Gy. Data are the mean of at least three independent experiments ±S.E. (From Cowell, I. et al., *Biochem Pharmacol* 2005;71:13–20. With permission.)

survival in a clonogenic assay following a dose of 2 Gy was significantly reduced, and the dose of radiation required to reduce cell survival to 10% of the control was reduced from 3.8 to 0.95 Gy [6]. NU7441 was also shown to be an effective radiosensitizer and chemosensitizer of other tumor cell lines and xenograft models [7].

Another drug, SU11752, was identified from a 3-substituted indolin-2-one library, as a 130 nM ATP-competitive inhibitor with radiosensitizing properties. This agent abrogates DNA-DSB repair without modulating the cell cycle and does not significantly affect ATM function or phosphoinositide-3 kinase activity [8]. Other small-molecule highly selective inhibitors of DNA-PKcs have been tested both *in vitro* and *in vivo*, including IC86621 and NU7026 [9,10]. Findings with these and other compounds with demonstrable radiosensitizing effect are summarized in Table 9.1. One compound listed in the table, BEZ 235, was developed as a

TABLE 9.1 Small-Molecule Inhibitors of NHEJ with Radiosensitizing Capability

Agent	Characteristics	Radiosensitization
NU7441 (KU-57788) (2-*N*-morpholino-8-dibenzothiophenyl-chromen-4-one)	IC_{50} = 14 nmol/L, at least 100-fold selectivity for DNA-PK compared with other PI3KK family kinases Limited aqueous solubility and oral bioavailability restrict further development	Potentiates IR and ET cytotoxicity by inhibition of DNA-PK; persistence of γH2AX foci after IR-induced DNA damage; prolongs G_2/M arrest; increases cytotoxicity of IR and ET in human colon cancer cell lines of differing p53 status (clonogenic assay) [7]
SU11752 [(3Z)-*N*-(3-chlorophenyl)-3-({3,5-dimethyl-4-[(4-methylpiperazin-1-yl) carbonyl]-1H-pyrrol-2-yl} methylene)-*N*-methyl-2-oxo-2,3-dihydro-1H-indole-5-sulfonamide]	Equally potent but more selective inhibitor of DNA-PK than wortmannin, inhibits DNA-PK by competing with ATP	Inhibitor of DNA DSB repair in cells and radiosensitizer (clonogenic assay). At concentrations inhibiting DNA repair, cell cycle progression and ATM kinase activity were unaffected [8]
Vanillin (3-methoxy-4-hydroxybenzaldehyde)	Selectively blocks NHEJ in human cell extracts by directly inhibiting DNA-PK	Subtoxic concentrations did not affect cell cycle progression. Weak radiosensitization of human tumor cells (clonogenic assay) [9]

(continued)

TABLE 9.1 Small-Molecule Inhibitors of NHEJ with Radiosensitizing Capability (Continued)

Agent	Characteristics	Radiosensitization
NU7026 LY293646 2-(4-morpholinyl)-4H-naphtho[1,2-b] pyran-4-one	Inhibits IR-induced DNA-DSB repair	Potentiates IR cytotoxicity (PF_{90} = 1.51 ± 0.04) in exponentially growing DNA-PK proficient but not deficient cells. Threefold reduction in PLDR in proficient cell lines [10]
IC87361 1-(2-hydroxy-4-morpholin-4-yl-phenyl)-ethanone	Enhanced radiation sensitivity in wt C57BL6 endothelial cells but not SCID cells lacking functional PK	Radiosensitizes wt but not SCID tumor microvasculature. LLC tumors treated with radiation and IC87361 showed greater TGD than tumors treated with radiation alone [5]
KU-55933 2-morpholin-4-yl-6-thianthren-1-yl-pyran-4-one	KU-55933 has IC_{50} value for ATM of 13 nM, *in vitro* and is specific with respect to other PIKKs	Sensitizes breast carcinoma cells to IR with apparent DNA repair deficit (persistence of IR-induced γH2AX foci) [6]
BEZ235	Orally available PI3K/mTOR inhibitor	Abrogates radiation-induced DSB repair. Radiosensitization and growth delay of xenografts. Radiation enhancement coincides with p53-dependent accelerated senescence phenotype accompanied by unrepaired DSBs [11]

Note: ET, etoposide; IR, ionizing radiation; PLDR, potentially lethal damage; SCID, severe combined immunodeficiency; TGD, tumor growth delay; wt, wild type.

PI3K/mTOR inhibitor and was subsequently shown to abrogate the repair of radiation-induced DSBs [11]. BEZ235 is currently involved in a number of phase I/II clinical trials but none of these involve a radiotherapy component (www.clinicaltrials.gov).

In addition to the small-molecule inhibitors, several biological agents have been developed to target DSB repair (Table 9.2). These include cetuximab (C225), an antibody to EGFR that blocks IR-induced activation of DNA-PK [12] and a peptide that disrupts the interaction between KU70 and DNA-PKcs [13]. Finally, a small hairpin RNA (shRNA) targeting PRKDC has been developed that causes changes in irradiated cells similar to those produced by BEZ235 [14].

TABLE 9.2 Inhibitors of NHEJ with Radiosensitizing Capability (Antibodies, Peptides, and siRNAs)

Cetuximab (EGFR-specific antibody C225)	Blocks IR-induced activation of DNA-PK essential for DNA repair after radiation. EGFR/DNA PK	Reduction in DNA repair (increased residual γ-H2AX-positive foci 24 h post-IR) in A549 and MDA MB 231 cells. Radiosensitization (colony formation) treated with C225 for 1 h before radiation [12]
Peptide representing the C terminus of Ku80	Selectively targeted and disrupted interaction between Ku complex and DNA-PKcs and the DNA binding activity of Ku	Inhibition of DNA-PK activity, reduction of DNA-DSB repair activity. Tumor cells (HN and MDA 231 modest sensitization when irradiated in the presence of target peptide [13]
SiRNA targeting PRKDC	PRKDC knockdown using siRNA	Causes accelerated senescence phenotype in irradiated cells similar to BEZ235 (Table 9.1). Inhibition of DNA-PK plus IR sufficient to induce accelerated senescence [11]
Aptamer-targeted DNA-PK shRNAs	Selective reduction of DNA-PK in prostate cancer cells	DNA-PK shRNAs, delivered by PSMA-targeting RNA aptamers, selectively reduced DNA-PK in PCa cells, xenografts, and human prostate tissues. Enhanced effects of IR in cellular and tumor models [14]

Note: ET, etoposide; IR, ionizing radiation; PLDR, potentially lethal damage; SCID, severe combined immunodeficiency; shRNA, small hairpin RNA; siRNA, small interfering RNA; TGD, tumor growth delay; wt, wild type.

9.3 DNA DOUBLE-STRAND BREAK REPAIR BY HOMOLOGOUS RECOMBINATION

HR repairs a DSB by using a homologous (i.e., identical sequence) section of undamaged DNA as the template to repair the broken DNA. Using DNA with the same sequence as a basis for repair ensures that the process can be accomplished without error. Briefly, single-strand regions are created around each side of the break, which are then coated with specialized proteins and the single-stranded nucleoprotein filaments thus formed invade undamaged double-stranded DNA on the neighboring sister chromatid, forming crossovers or bubble structures, which are then expanded by specialized enzymes called helicases. This process aligns an undamaged DNA template of the same base sequence around the break site, and DNA polymerases can then synthesize across the missing regions accurately repairing the break. To return the chromatin to its original configuration, the crossover structure that remains must be reversed. This is done with specialized nucleases (resolvases) that cut, or resolve, the junctions followed by reconnection or ligation with adjacent ends by a ligase (Figure 9.4).

9.3.1 Associated Gene Families

In addition to the genes and proteins listed above, two other gene families are involved, which are also known for their involvement in human repair deficiency syndromes, namely, *BRCA* genes 1 and 2 and the Fanconi anemia family [11,15]. Mutations or deletions in one or more of the genes in these families can compromise HR.

9.3.1.1 BRCA1

BRCA1 has multiple roles in HR and in many other processes. In association with its partner BARD1, BRCA acts as an E3 ubiquitin ligase 1 ubiquitinylating other proteins, thereby modifying their interactive properties and their function. BRCA1–BARD1 can form complexes with other proteins to perform different functions including involvement in replication inhibition (part of the intra-S checkpoint), with the MRN complex in NHEJ, with RAD52–BRCA2 in HR, and with RNA polymerase II in transcription. The ubiquitinylating function of BRCA1 is essential to each role.

9.3.1.2 BRCA2

The role of the BRCA2 protein is to regulate the binding of RAD51 to RPA-coated single-stranded DNA. Thus, BRCA2 is required for the localization of RAD51 to sites of DNA damage and this physical interaction between BRCA2 and RAD51 is essential for error-free DSB repair. Two different domains within BRCA2 interact with RAD51, one of these consists of eight BRC repeats, each of which can bind RAD51 whereas the other is a distinct domain, TR2, at the C terminus. It has been proposed that BRC repeats hold RAD51 in an essentially inactive monomeric form, and that when damage occurs, the BRCA2–RAD51 complex localizes to the site of DNA damage [16–18]. Once concentrated at a DSB, BRCA2 is able to load RAD51 onto the 3′-ssDNA overhangs, displacing RPA [19,20]. After RAD51 loading, BRCA2 functions to stabilize the resulting nucleoprotein filament, through the TR2 domain at its C-terminus [16]. This enables the filament to invade and pair with a homologous DNA duplex, initiating strand exchange between the paired DNA molecules (Figure 9.4). BRCA2 also interacts with PALB2, through which it localizes to DSBs together with

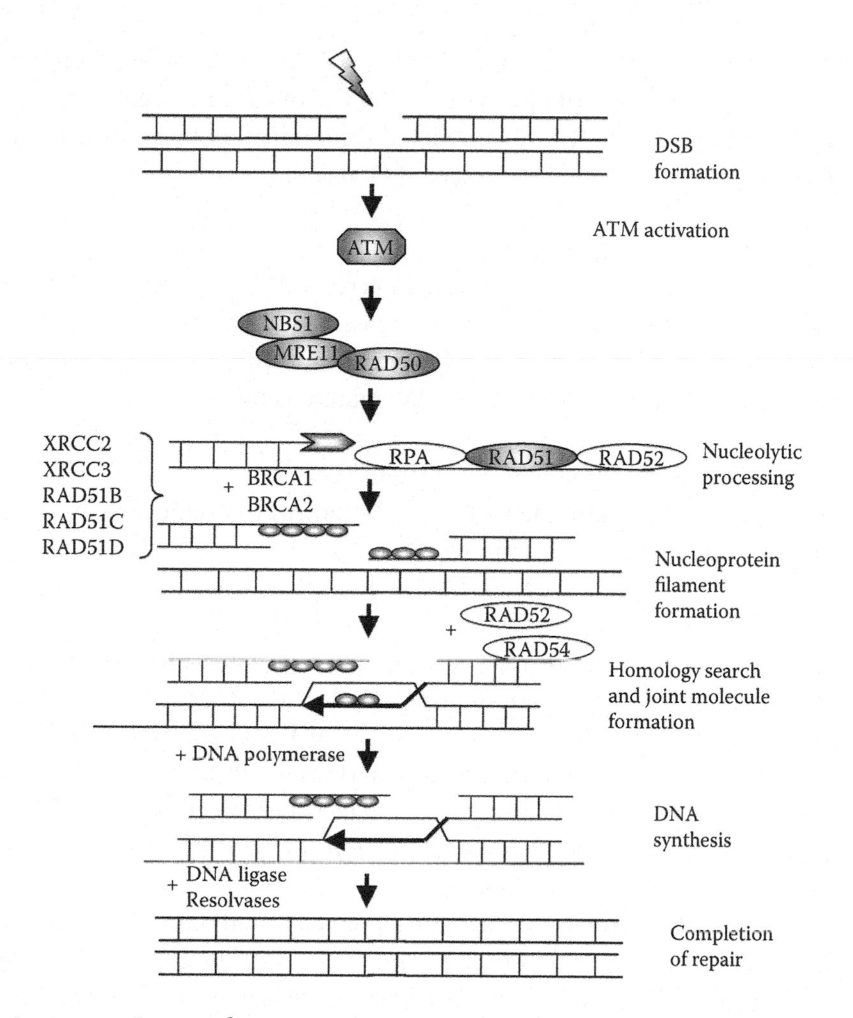

FIGURE 9.4 Repair of DNA-DSB by HR. The initiating step for HR is thought to be the processing of the 3′ end of the DNA break by the NBS1–MRE11–RAD50 protein complex. After binding of the RAD52 and RAD54 proteins, replication protein A (RPA) facilitates the assembly of the Rad51–BRCA2 complex on the single-strand 3′ DNA overhang to form a RAD51-nucleoprotein filament. The formation of the complex and invasion of the RAD51 filament is facilitated by the RAD51 accessory proteins, XRCC2, XRCC3, RAD51B, RAD51C, and RAD51D. The RAD51 nucleofilament DNA is then able to pair with a homologous region in duplex DNA forming a Holliday junction after alignment of sister chromatids. Complex chromatin alterations and configurations are required to unwind the DNA and allow for DNA strand exchange. After identification of the identical sister chromatic sequences, the intact double-stranded copy is then used as a template to repair the DNA break by subsequent DNA synthesis using DNA polymerases, ligases, and Holliday junction resolvases.

BRCA1. The BRCA2 DNA-binding domain (DBD) also stimulates homologous pairing and the strand-exchange activities of RAD51, suggesting that BRCA2 might facilitate RAD51-mediated recombination by binding to the dsDNA–ssDNA junction of the resected DSB [19].

9.3.1.3 Fanconi (FANC) Genes

Genes of this family have a role in HR and cells with *FANC* gene mutations all show increased sensitivity to DNA cross-linking agents, the repair of which depends on HR. However, mild or little increased sensitivity to ionizing radiation has been found in *FANC* mutant cells.

9.3.2 Targeting HR in Cancer Treatment

Downregulation of components of the HR pathway is achievable by both biological and chemotherapy approaches.

9.3.2.1 Non-Drug Approaches

9.3.2.1.1 Biological Agents That Downregulate Components of the HR Pathway

Collis et al. [21] showed the radiosensitizing effect of a RAD51-targeted ribozyme minigene in preclinical studies. In a similar study, it was found that siRNA targeting BRCA2 effectively radiosensitized cells *in vitro* and *in vivo*, suggesting that this type of approach could be beneficial clinically. Delivery of targeted oligonucleotides in a clinical setting might become an option with advances in delivery techniques.

siRNA targeting of RAD51 has been observed to radiosensitize lung, glioma, and prostate tumor cells *in vitro* and *in vivo* [22]. Targeting HR in this manner could exploit the therapeutic ratio because many normal tissues have a relatively slow cell turnover with an increased fraction of cells in the G_1 phase and are NHEJ-responsive, whereas malignant cells that have increased numbers of cells in the S and G_2 phases may be more amenable to HR targeting.

9.3.2.1.2 Hyperthermia Another approach to disrupting the process of HR could involve combining hyperthermia with radiotherapy. Heat has been shown to interfere with all aspects of the DNA-DSB repair process, including HR, although the specific mechanisms of heat-induced HR impairment are not yet known [23]. A number of clinical trials have shown the efficacy of combining hyperthermia and radiotherapy.

9.3.2.1.3 High LET Radiation From preclinical data, it seems unlikely that application of clinically relevant but nonconventional fractionation or dose-rate schemes will affect the repair of DNA-DSBs by HR. The use of high LET radiotherapy (i.e., neutrons, low-energy protons, alpha particles) with inhibitors of HR may be a different story. Several preclinical studies have shown that although high LET radiation does not have a significant effect on the sensitivity of cells deficient in NHEJ, cells deficient in HR are significantly more radiosensitive [24].

9.3.2.2 Drugs Targeting HR

A number of drugs commonly used in conjunction with radiotherapy have been shown to have an effect on HR. These include topoisomerase II inhibitors (e.g., etoposide), antimetabolites (i.e., gemcitabine, 5-fluorouracil), microtubule inhibitors (e.g., paclitaxel, vinorelbine), cross-linking agents (e.g., platinum compounds and mitomycin C), and alkylating agents (e.g., temozolomide). These were discussed in detail in Chapters 3 through 7 dealing with the combined use of radiotherapy and chemotherapy.

Most of the drugs found to be effective in reducing HR for chemosensitization and radiosensitization, either by accident or design, target the RAD51 protein. In many cases, however, other HR proteins are involved and the reduction in RAD51 may come about indirectly. Because RAD51 is commonly overexpressed in malignant cells and can lead to increased resistance to a variety of drugs and to ionizing radiation, this approach has had some success at the preclinical and clinical level.

9.3.2.2.1 Designer Inhibitors A 28–amino acid peptide derived from the BRC4 motif of BRCA2 tumor suppressor was found to selectively inhibit human RAD51 recombinase (HsRad51). Based on this approach, computer modeling was combined with *in vitro* biochemical testing to construct a highly efficient chimera peptide from eight existing human BRC motifs. The peptide obtained was approximately 10 times more efficient in inhibiting HsRad51–ssDNA complex formation than the original peptide [25]. To date, the *in vivo* performance of these model compounds as radiosensitizers or chemosensitizers has not been reported. Other successes in targeting RAD51 have been more serendipitous.

9.3.2.2.2 Berberine The main alkaloid component in Huang Lian and other medicinal herbs, berberine is commonly used for gastrointestinal discomfort in China. Laboratory studies have shown that berberine has antitumor activity for a wide variety of cancer cells, including glioblastoma, oral cancer, hepatoma, gastric cancer, prostate cancer, leukemia, and osteosarcoma. A number of cellular effects have been shown including inhibition of cell cycle progression, induction of apoptosis, suppression of the constitutive activation of nuclear factor-κB (NF-κB) in some cancer cells, induction of autophagy in lung cancer cells, and impairment of tumor growth by inhibiting angiogenesis [26].

Recently, berberine has been demonstrated to efficiently downregulate RAD51 expression in cancer cells while conferring radiosensitivity. In one study, berberine treatment was shown to lead to a reduction in RAD51 transcription, inhibition of RAD51 promoter activity and abrogation of the upregulation of RAD51 induced by ionizing radiation. The radiosensitizing effect of berberine was attenuated in cancer cells overexpressing exogenous RAD51. Downregulation of RAD51 resulted in failure to repair radiation-induced DSBs and reduction in survival of irradiated cells [26].

9.3.2.2.3 Gimeracil An inhibitor of dihydropyrimidine dehydrogenase (DPYD), gimeracil partially inhibits HR repair and has a radiosensitizing effect [27]. Investigation of the mechanism involved showed that DPYD depletion by siRNA significantly reduced the formation of radiation-induced foci of Rad51 and RPA, whereas it increased the number of foci of NBS1 and the number of foci showing colocalization of NBS1 and RPA in DPYD-depleted cells after radiation were significantly lower than in control cells. These results were interpreted to indicate that DPYD depletion affects the efficiency of DNA resection to generate a 3′ overhang in HR. In addition, the phosphorylation of RPA by irradiation was partially suppressed in the DPYD-depleted cells, contributing to partial inhibition of DNA repair by HR. DPYD depletion had a radiosensitizing effect and also enhanced sensitivity to cisplatin.

9.3.2.2.4 Drugs Targeting Heat-Shock Protein 90 A number of studies have shown that the radiation-modulating effect of the heat-shock protein 90 (HSP90) inhibitors and various targets have been proposed for their action [28]. In one case, tumor cells exposed to the HSP90 inhibitor 17-allylamino-17-demethoxygeldanamycin (17AAG) have reduced expression of BRCA1 and RAD51 but no alterations in Ku70 or ku80, suggesting that the radiosensitizing effect observed *in vitro* is caused by the impairment of HR and not NHEJ [29]. The feasibility of this approach has not been assessed at the clinical level.

9.3.2.2.5 Targeting c-Abl Kinase Imatinib was the first of a new group of drugs designed to act by inhibiting a specific class of enzyme, a receptor tyrosine kinase. Imatinib (Gleevec) and dasatinib (Sprycel) are inhibitors of c-Abl kinase. Gleevec is used in treating chronic myelogenous leukemia (CML), gastrointestinal stromal tumors (GISTs), and some other diseases, and by 2011, had been approved by the Food and Drug Administration (FDA) to treat 10 different cancers.

Imatinib mesylate is an inhibitor of c-ABL, c-KIT, and platelet-derived growth factor receptor tyrosine kinases. c-ABL kinase has been implicated in the radiation-induced assembly of RAD51 and RAD52. The treatment of cancer cells with imatinib decreases RAD51 expression as a consequence of c-ABL downregulation and sensitizes them to experimental chemotherapy and radiotherapy *in vitro* [30,31]. Decreased nuclear expression and chromatin binding of RAD51 protein was observed following imatinib treatment and reduced RAD51 expression correlated with decreased error-free HR [32]. Clonogenic survival experiments revealed increased cell killing for imatinib-treated cells in combination with ionizing radiation, gemcitabine, and mitomycin C, due in part to mitotic catastrophe. HR inhibition may be an additional mechanism of action for the chemosensitization and radiosensitization of solid tumors with imatinib with preservation of the therapeutic ratio [33].

Phase I/II clinical trials of imatinib and RT have suggested a radiation-modifying effect; however, no therapeutic gain has been realized yet. Clinical studies are now focusing on dasatinib in combination with RT in several early phase trials (reviewed by Barker and Powell [34]).

9.3.2.2.6 Targeting EGFR Tyrosine Kinase Erlotinib hydrochloride (trade name, Tarceva) is a drug used to treat non-small cell lung cancer, pancreatic cancer, and several other types of cancer. It is a reversible tyrosine kinase inhibitor acting on the intracellular tyrosine kinase domain of the epidermal growth factor receptor. Erlotinib has been found to have several effects that might modify radiation response including the attenuation of RAD51 expression [35]. In one study, it was confirmed that erlotinib attenuated the HR response *in vitro* through BRCA1 [36]. In fact, erlotinib has been the most extensively investigated novel agent postulated to modulate the HR pathway in combination with RT.

Results of three phase II trials of patients with newly diagnosed glioblastoma multiforme have been reported [37–39]. Compared with historical results using RT and temozolomide concurrently, only one of the trials seemed to show a survival advantage for patients treated with RT, temozolomide, and erlotinib [39]. A number of other phase I or II trials of erlotinib and RT have been completed, with mixed results in terms of toxicity and efficacy [34]. As of September 2012, approximately 60 clinical trials in various stages of completion were listed that combined erlotinib and radiotherapy with, in most cases, other chemotherapy agents involved. Twelve clinical trials were reported as phase III, the remainder as phase I/II (www.clinicaltrials.gov).

9.3.2.2.7 Targeting the Ubiquitin/Proteasome System Several proteasome inhibitors, including bortezomib, have been shown to modulate the biological response to IR and radiosensitization by inhibitors of the ubiquitin/proteasome system was discussed in detail in Chapter 8, Section 8.5.2. In the present context, some recent research [40] suggests that proteasome inhibitors block MDC1 degradation, which, in turn, leads to abrogation of BRCA1 recruitment and impairment in HR. Whether HR inhibition plays a significant role in the observed radiosensitizing effects of bortezomib and similar drugs.

9.3.2.2.8 Histone Deacetylase Inhibitors Histone deacetylase inhibitors (HDACIs) are discussed in more detail later in this chapter. They are included here because the results of some preclinical studies suggest that radiosensitization by these compounds may be partly attributable to impairment of HR. One example is PCI-24781 (formerly CRA-024781), a broad-spectrum phenyl hydroxamic acid HDAC inhibitor currently being evaluated in clinical trials in patients with neoplastic disease [41]. Treatment with PCI-24781 was shown to cause decreased expression of BRCA1, BRCA2, and RAD51 and, in a functional assay, HR was impaired in a dose-dependent fashion and a radiosensitizing effect was noted in tumor cells in culture. Studies with other HDACIs in conjunction with radiotherapy are ongoing and will be reviewed in detail in Section 9.5.3.

9.3.3 Clinical Relevance of Radiosensitization by DNA Repair Inhibition

HR is a pathway specific to S and G_2 phase cells, and is seen only in dividing cells. NHEJ, on the other hand, occurs in all phases of the cell cycle and is neither phase specific nor cycle specific. Important from the standpoint of radiotherapy is that NHEJ occurs in all cells and tissues, including slowly or nondividing systems. This, of course, includes the dose-limiting and late-reacting tissues that manifest the serious late effects of radiation therapy. From this point of view, targeting HR with inhibitory drugs holds less potential danger of causing late radiation damage and would be the preferred stratagem.

That being said, it is currently not clear how important HR-related factors are in the observed clinical effects of certain drug/radiation combinations. For several of the categories of drugs listed previously, radiosensitization has been shown in preclinical systems to be at least partially attributable to HR-related factors. Several drugs with preclinical evidence of affecting the HR pathway and modulating the IR response have been studied in clinical trials but none of these studies included procedures to establish direct correlation with HR. In other words, to date, at the clinical level, the direct relationship between targeting the HR mechanism as cause and enhancement of radiation response as effect has not been established. It was emphasized in a recent review [34] that the development of functional assays for measuring HR in clinical samples is needed before advances in this targeting strategy can be realized.

9.4 RADIOSENSITIZATION BY INHIBITORS OF POLY(ADP-RIBOSE) POLYMERASE

It is appropriate to follow the discussion on HR with a discussion of poly(ADP-ribose) polymerase (PARP) inhibitors. An important role in the regulation of DNA repair is played by members of the PARP family through involvement in single-strand break (SSB) repair and base excision repair (BER) but, under normal circumstances, PARP proteins do not contribute directly to DSB repair. In situations in which HR is defective, however, PARP inhibition can exacerbate the situation in such a way as to cause radiosensitization and chemosensitization and cell death.

9.4.1 Poly(ADP-Ribose) Polymerase

Members of the PARP family play an important role in the regulation of DNA repair. The PARP enzymes catalyze the synthesis of poly(ADP-ribose) polymers from nicotinamide adenine dinucleotide (NAD+). During this process PARP consumes NAD+ (nicotine adenine dinucleotide) to catalyze the formation of highly negatively charged poly(ADP-ribose) polymers of linear or branched structure with a length of 200 to 400 monomers, releasing nicotinamide as a by-product. Degradation of the polymers is performed by poly(ADP-ribose) glycohydrolase (PARG). PARG exhibits both endoglycosidic and exoglycosidic activity and the products of enzyme activity are mono(ADP-ribosyl)ated protein and mono(ADP-ribose) (Figure 9.5).

To date, 18 different PARPs have been described, sharing a conserved catalytic domain responsible for poly(ADP-ribose) synthesis. Although PARP-1 is known to play an important role in DNA repair, the role of other family members is less well understood. PARP-1 specifically binds

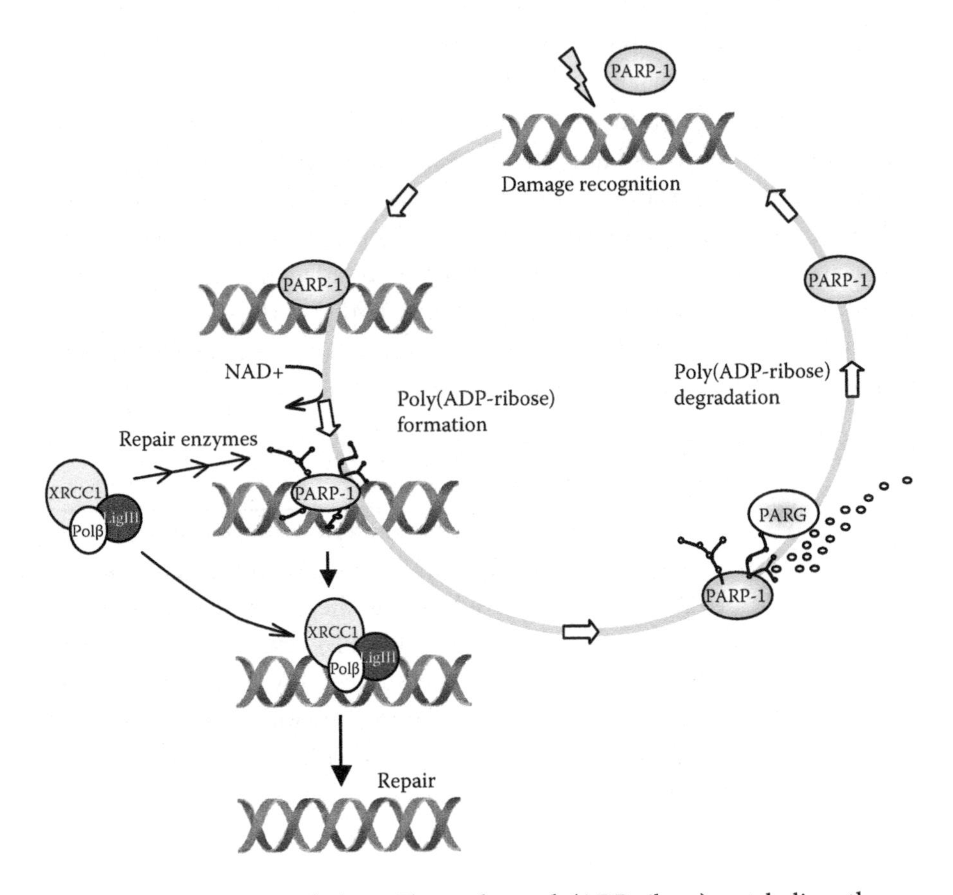

FIGURE 9.5 PARP metabolism. The nuclear poly(ADP-ribose) metabolic pathway. PARP-1 and PARP-2, possibly working as heterodimers as well as homodimers, rapidly recognize DNA strand interruptions. Binding to damaged DNA activates these PARPs to synthesize poly(ADP-ribose) on themselves (automodification) as well as on histones. This results in the local relaxation of chromatin and the recruitment of DNA repair enzymes, such as XRCC1 and DNA ligase III. Poly(ADP-ribosyl)ated PARP-1 and PARP-2 lose affinity for DNA, thus allowing access to repair proteins. PARG rapidly hydrolyzes poly(ADP ribose) using both endoglycosidic and exoglycosidic activities. This restores the affinity of PARP-1 and PARP-2 for DNA strand breaks that can initiate a new cycle of DNA damage recognition and PAR formation.

to SSBs formed in response to ionizing radiation or alkylating agents and upon binding is auto-poly(ADP-ribosyl)ated, allowing it to noncovalently interact with other proteins.

9.4.2 The Role of PARP in DNA Repair

PARP proteins participate directly or indirectly in several modes of DNA repair.

9.4.2.1 SSB Repair and BER

BER is an important pathway for the repair of SSBs involving the sensing of the lesion followed by the recruitment of a number of repair proteins. These include damage-specific DNA glycosylases, AP endonuclease 1, DNA polymerase β (Pol β), and the x-ray cross-complementing protein-1 (XRCC1-DNA ligase IIIα heterodimer). SSBs are generated during the initial stages of BER (recognition and removal of damaged bases by DNA glycosylase and AP endonuclease). These SSBs are recognized and processed by Pol β, and the DNA ends are finally sealed by DNA ligase IIIα (Figure 9.6).

Poly(ADP) ribose polymerase (PARP-1) is a critical component of the major short-patch BER pathway binding to DNA nicks and breaks and resulting in activation of catalytic activity, and causing poly(ADP)ribosylation of PARP-1 itself, as well as other acceptor proteins. In the short-patch BER pathway, the binding of PARP signals the recruitment of other components of DNA repair pathways and the modification of protein activity. The highly negatively charged poly-ADP-ribose (PAR) that is produced around the site of damage may also serve as an antirecombinogenic factor. PARP-1-mediated poly(ADP-ribosylation) recruits nuclear acceptor proteins, such as XRCC1, histones, and PARP-1 itself, to assemble repair complexes involved in the execution of DNA repair. Although PARP-1 itself plays no direct role involved in DNA damage processing by the BER pathway, it has been shown that PARP-1 protects excessive DNA SSBs from converting into DNA-DSBs and thus preserves them for subsequent repair by BER enzymes [42,43]. The role of PARP in BER is thus to extend the ability of BER enzymes to process DNA-SSBs, which arise directly from mutagen exposure or are produced during the processing of lesions following DNA damage.

Inhibition of PARP delays, but does not abolish, the repair of SSBs. In nonreplicating cells, this effect on SSB repair has a minimal effect on survival. In rapidly proliferating cells, however, PARP inhibition increases

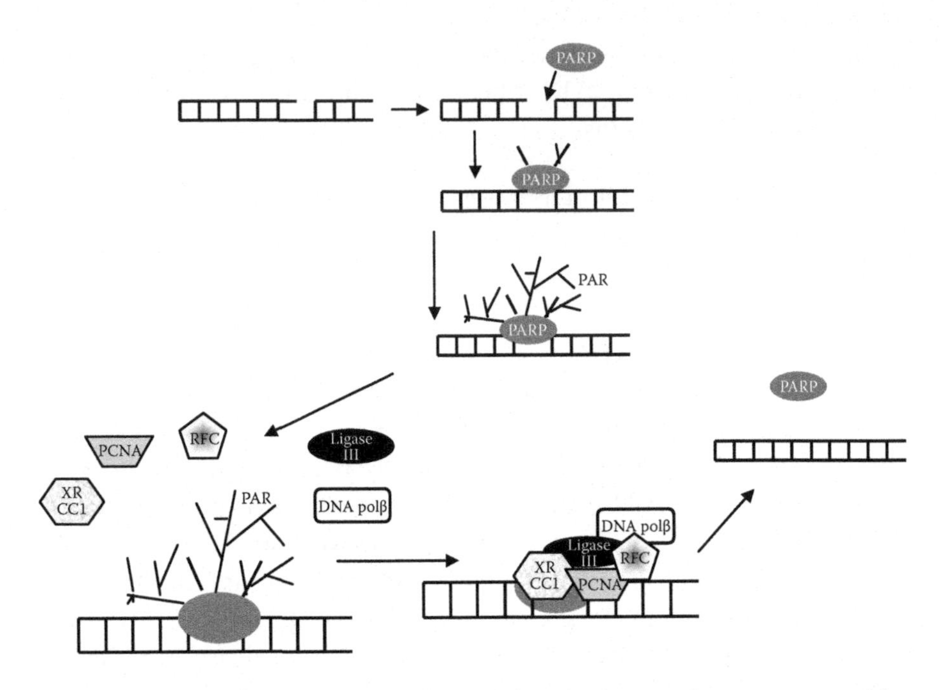

FIGURE 9.6 Interaction of BER and PARP. The initial step is the recognition of the SSB by PARP. PARP then synthesizes the PAR chain linking the PARP enzymes, histones, and other DNA-associated proteins. The next step is the recruitment of the DNA repair complex. After completion of SSB repair, PAR glycohydrolase is recruited and the PAR chain is degraded. This is followed by the release of the BER repair complex from the site of repair permitting the repair of another SSB.

radiosensitivity and the magnitude of the radiosensitizing effect in cell culture systems has been shown to correlate with the proportion of cells that are replicating [44,45].

Inhibition of PARP increases the level of unrepaired DSB in replicating cells by at least two mechanisms. First, the delayed repair of radiation-induced SSB increases the probability of unrepaired lesions colliding with the DNA replication machinery and generating excess DSBs. Second, the inhibition of the catalytic activity of PARP does not impede it from binding to DNA breaks but does prevent PARP from automodification by the addition of poly-ADP ribose. The unmodified PARP then remains bound to sites of DNA damage, interfering with downstream repair processes and amplifying the toxicity of the DSB that have been generated [46].

Defects in BER have been reported in a range of human tumors and seem to be relatively common [47,48]. It has been suggested that targeting this patient cohort would be an effective therapeutic stratagem.

9.4.2.2 Homologous Recombination

Cancer cells that are defective in some aspect of HR are sensitized to the lethal effects of PARP inhibitors. As described in Section 9.3.1, there is compelling evidence implicating BRCA1 and BRCA2 in the signaling of DNA damage to facilitate repair by HR. BRCA1- and BRCA2-deficient cells have been found to be extremely sensitive to PARP inhibitors and inhibition of poly(ADP-ribosyl)ation results in an increased level of DSBs.

Auto-poly(ADP-ribosyl)ation of PARP-1 and PARP-2 is essential for the release of the enzymes from DNA strand breaks and the recruitment of signal transduction and repair enzymes. In the presence of PARP inhibitors, failure to execute automodification causes PARP-1 and PARP-2 to stall on damaged DNA and hinders access of repair enzymes with the result that accumulating SSBs generate DSBs at DNA replication forks. This has serious consequences in BRCA-deficient cells because DSBs associated with replication forks are predominantly repaired by HR and abrogation of PARP-1 will increase the formation of DNA lesions that are potentially reparable by HR. In the absence of exogenous DNA damage, the continuous exposure of replicating cells to a PARP inhibitor has been shown to cause a significant increase in HR activity, presumably representing the repair of DNA replication forks that have stalled or collapsed after encountering an unrepaired SSB [49].

PARP inhibitors act to reduce the rate of repair of endogenously arising single-stranded lesions and obstruct the efficient resolution of stalled or collapsed replication forks by impeding the release of PARP molecules from damaged sites. HR is required for the repair of these lesions as shown by the extreme sensitivity to PARP inhibitors of HR-defective cells, such as breast or ovarian cancer cells that lack the HR proteins BRCA1 or BRCA2 [50,51].

9.4.2.3 Defects in Other DNA Repair Pathways

In cells that are deficient in core components of NHEJ (such as Ku70/80 and DNA ligase IV), an alternative, backup end-joining pathway exists that partly compensates for the NHEJ defect [52]. This pathway is dependent on PARP-1, which acts to bind and activate an alternative mode of end-joining that can be suppressed if PARP activity is inhibited [53]. Cells that are deficient in core NHEJ proteins are already radiosensitive and might thus be further sensitized by treatment with a PARP inhibitor. It is known that many cancers do exhibit some degree of DNA repair defect, it is uncommon for tumors to be profoundly deficient in one or more of the core NHEJ proteins. However, it is even less likely that nonreplicating cells are deficient in NHEJ, and thus it is possible that this difference could also be exploited for therapeutic advantage.

9.4.2.4 *Synthetic Lethality: Inhibition of PARP and Defects in DNA Repair*

As described in Section 8.3.3.2, synthetic lethality is the targeting of cells with one specific DNA repair defect by inhibiting a second DNA repair pathway, such as occurs when a mutation of either of two genes individually has no effect but the combination of two mutations can cause cell death [54].

A defect in HR when it occurs in *BRCA*-deficient cells, creates an opportunity for targeted treatments for patients with hereditary *BRCA1* and *BRCA2*-deficient cancers [55]. Deficiency of genes implicated in HR also confers sensitivity to PARP inhibitors [56], suggesting an approach that could be applicable in the treatment not only of hereditary *BRCA1*- and *BRCA2*-deficient cancers but of cancers with sporadic impairments of the HR pathway [57].

Demonstration of the sensitivity of *BRCA* mutant cells to PARP inhibition led to the initiation of clinical trials to test the efficiency of this approach. KU-0059436, a potent PARP inhibitor, is being tested as a single agent in phase I/II trials in patients with ovarian cancer carrying *BRCA* mutations and phase II trials of another PARP inhibitor, AG0146999, in combination with temolozide in patients with *BRCA* mutations, are also being undertaken [58,59]. Thus far, there have been no reports of exploitation of these relationships to selectively kill BRCA1/2 deficient tumors by using PARP inhibition, in combination with radiotherapy.

9.4.3 Inhibitors of PARP

PARP inhibitors are potent competitive inhibitors of NAD+. The earliest PARP inhibitors studied, nicotinamide (NA) and 3-aminobenzamide (3-AB), showed activity only at high (micromolar) concentrations, whereas the new generation drugs have higher potency often in the nanomolar range. These compounds are designed to mimic the moiety of NAD+ responsible for binding to the donor site of the catalytic domain of the PARP enzymes, thus ensuring specificity and potency. The majority of these drugs do not discriminate between PARP-1 and PARP-2 (Figure 9.7).

The cellular effects of PARP inhibitors vary according to the cellular environment. In particular, the presence and nature of DNA damage or metabolic

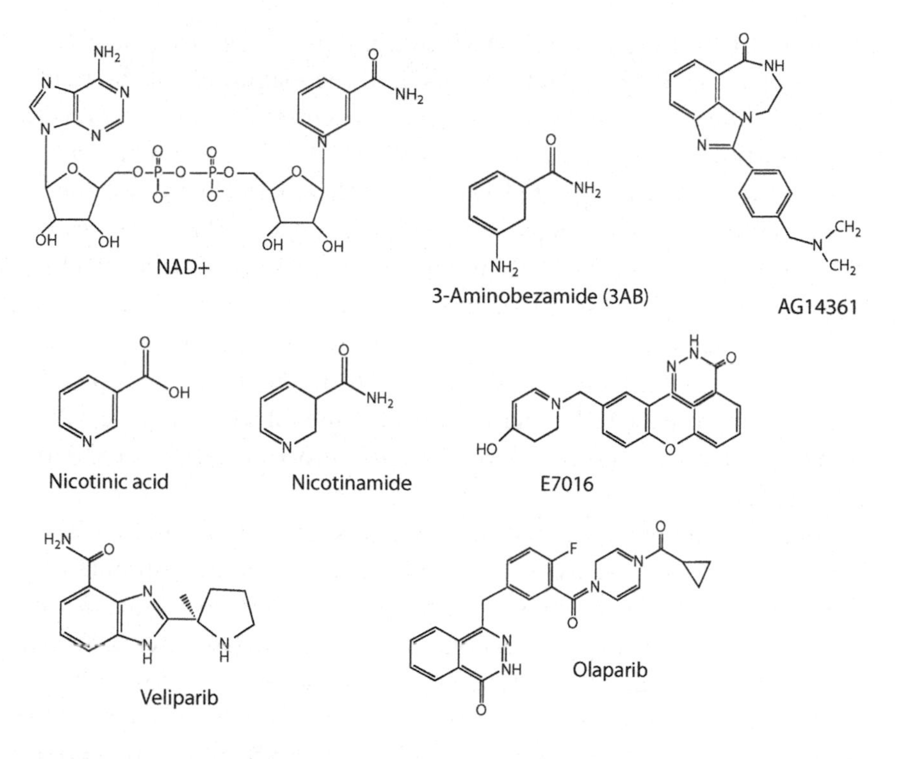

FIGURE 9.7 PARP inhibitors.

stress have an important effect on the consequences of PARP inhibition [60]. PARP is activated by DNA damage and acts to enhance repair and suppress potentially deleterious interactions between damaged sites. The reaction catalyzed by PARP consumes NAD+ and thus the repair and protection of DNA occurs at the expense of NAD+ depletion. NAD+ is essential for the synthesis of ATP, the energy source for an array of metabolic processes and high doses of DNA-damaging agents have been shown to profoundly reduce cellular NAD+ levels with reduction in the level of ATP. Under these circumstances, cell death may occur by necrosis rather than apoptosis, which is an energy-consuming pathway. One effect of pretreatment with a PARP inhibitor is to prevent NAD+ depletion and enable cells to execute apoptosis. In the case of rapidly replicating cancer cells, however, the effect of PARP inhibition is more apparent on DNA repair than on NAD+ metabolism.

9.4.3.1 Radiosensitization by PARP Inhibitors

The main effect of PARP inhibition is to delay the repair of SSBs, and a dysfunction in SSB repair has a minimal effect on the survival of nonreplicating cells. However, PARP inhibition does increase the radiosensitivity of rapidly proliferating cells, and the magnitude of the radiosensitizing effect in cell culture systems has been shown to correlate with the proportion of cells that are replicating [44,45]. PARP inhibition can increase the level of unrepaired DSBs in replicating cells by two mechanisms. First, the delayed repair of radiation-induced SSB increases the probability of unrepaired lesions disrupting the DNA replication machinery and generating excess DSBs, and second, although the inhibition of the catalytic activity of PARP does not impede binding to DNA breaks, it does prevent PARP from modifying itself by the addition of pADPr. Because unmodified PARP remains bound to sites of DNA damage, PARP inhibitors interfere with downstream repair processes and increase the toxicity of DSBs [46].

From the point of view of radiotherapy, it is very significant that PARP inhibitors selectively radiosensitize replicating cells because tumors usually contain a higher proportion of replicating cells than do normal tissues, and the cell-cycle checkpoint responses of these cells are often defective [61]. Moreover, some of the critical normal tissues that are dose-limiting in the clinic (e.g., brain and spinal cord) are composed almost entirely of nonreplicating G_0 or G_1 phase cells. Thus, PARP inhibitors can potentially increase the therapeutic index of radiation therapy for a variety of tumors by increasing damage in highly replicating tumor cells, but sparing noncycling normal tissues, which are responsible for the dose-limiting late damage after radiotherapy.

Not all critical normal tissues are nonproliferating, however, and it is possible that simultaneous inhibition of PARP and HR that results in radiosensitization of replicating cells could also carry a risk of inducing synthetic lethality in nonirradiated but proliferating tissues. There is no specific inhibitor of HR that could be used for experimental modeling of this situation; however, the HSP90 inhibitor 17-(allylamino)-17-demethoxygeldanamycin selectively accumulates in tumor cells and inhibits HR by downregulating Rad51 and BRCA2. Combined use of 17-(allylamino)-17-demethoxygeldanamycin and the PARP inhibitor olaparib has been reported to synergistically increase the radiosensitivity of proliferating glioma cell populations without affecting nonreplicating cells [62].

Recent *in vitro* and *in vivo* studies are listed in Tables 9.3 and 9.4. PARP inhibitors have shown dose-enhancement ratios ranging from 1.3 to more than 2 [45,56,60,63–68].

TABLE 9.3 Radiosensitization by PARP Inhibitors to Radiation (Experiments with *In Vitro* Systems)

Model	Drug	Conditions	Endpoint	Results	Reference
H460 (large cell lung cancer)	ABT-888	5 μM/L (0–6 Gy)	Clonogenesis Apoptosis (Annexin V/PI) endothelial damage	↓ Clonogenic survival vs. RT alone ↑ Apoptosis, inhibition of endothelial tubule formation	Albert et al. [56]
U87MG, T98G (human glioblastoma)	AZD2281 (Olaparib, KU-00594361)	μM/L 1 h pre + 3 or 24 h postirradiation	Clonogenic survival DNA repair (γH2AX)	↓ Clonogenic survival vs. RT alone ↓ DNA repair (replication dependent)	Dungey et al. [62]
Human lung cancer 1299 Prostate cancer DU145, 22RV1	ABT-888	5 μmol/L	Clonogenic survival Repair foci	↓ Clonogenic survival vs. RT alone ↓ Repair foci, effective in both oxic and hypoxic cells	Liu et al. [66]
U251, human glioblastoma MiaPiaCa pancreatic cancer, DU145 prostate cancer	E7016	3.5 μmol/L 6 h before RT (0–8 Gy)	Clonogenic survival Mitotic catastrophe Apoptosis (Annexin V)	↓ Clonogenic survival ↑ MI Apoptosis, no change	Russo et al. [64]
Human colon cancer LoVo, SW620	AG14361	0.4 μmol (8 Gy RT)	Clonogenic survival	Survival reduced by inhibition of potentially lethal damage	Calabrese et al. [57]
Human and mouse primary cells defective in Artemis, ATM, DNA ligase IV	AZD2281 (Olaparib, KU-0059456)	1 h preradiation and 22 h postradiation (0–8 Gy)	Clonogenic survival, alkaline comet assay, γH2AX	Survival ↓ by drug + RT in rapidly dividing and DNA repair–deficient cells	Löser et al. [63]

TABLE 9.4 Sensitization by PARP Inhibitors to Radiation *In Vivo*

Tumor Model	PARP Inhibitor	Conditions	Assays	Results	Reference
JHU006, JHU012 H + N xenografts	GPI-15427	30–300 mg/kg orally 1 h before 2 Gy	TGD apoptosis (TUNEL)	Inhibition of tumor growth, ↑ in apoptosis	Khan et al. [67]
Human lung cancer, H460, xenograft	ABT-888	25 mg/kg × 5 days; 1 h before 2 Gy	TGD k67 staining, apoptosis (TUNEL). CD34 staining for blood vessel density	TGD ↑ 6.5 days compared with radiation alone ↓ Tumor vasculature, proliferation, ↑ apoptosis	Chalmers et al. [68]
Human colon cancer, HCT 116, xenograft	ABT-888	25 mg/kg per day by osmotic pump 2 days preradiation; 2 Gy/day × 10 days	Survival of animal	↑ Mean survival time from 23 to 26 days for ABT-888 + RT vs. RT alone	Donawho et al. [65]
Human glioblastoma U251	E7016	30 mg/kg oral + TMZ 3 mg/kg oral; 4 Gy	TGD	TGD delayed by RT + drug compared with RT alone	Russo et al. [64]

9.4.3.1.1 PARP Inhibitors and Fractionated Radiation For clinical benefit, sustained enhancement of radiosensitivity during a fractionated schedule is essential. Results of several experimental studies with PARP inhibitors indicate that this may be attainable. In an *in vitro* study, KU-0059436 was shown to increase the radiosensitivity of four human glioma cell lines (T98G, U373-MG, UVW, and U87-MG) [45]. Radiosensitization was enhanced in populations synchronized in S phase and was further enhanced when the inhibitor was combined with fractionated radiation. Although fractionation (4×2 Gy) reduced cytotoxicity compared with single doses, the radiosensitizing effects of KU-0059436 were maintained with fractionation and the SER_{50} and SER_{37} values were greater for fractionated than for single-dose treatments. KU-0059436 delayed the repair of radiation-induced DNA breaks and was associated with a replication-dependent increase in γH2AX and Rad51 foci. These results demonstrated that KU-0059436 increases radiosensitivity in a replication-dependent manner that is enhanced by fractionation.

In a study with xenografted tumors, it was shown that AG14361 treatment before irradiation significantly increased the radiosensitivity of LoVo xenografts. Local tumor irradiation (2 Gy daily for 5 days) alone caused a 19-day tumor growth delay that was extended to 37 days in mice treated with AG14361 at 15 mg/kg before irradiation [57]. In another xenograft study, mice bearing H460 tumors irradiated with 2 Gy 1 h after ABT-888 treatment for 5 consecutive days to a total dose of 10 Gy, induced a longer duration tumor growth delay than did either agent used alone [56]. Similarly, fractionated treatment of xenografted glioblastoma U251, combined with the PARP inhibitor, increased the duration of tumor growth delay by a significant amount. Oral administration of the novel PARP inhibitor GPI-15427 before radiation therapy significantly reduced tumor volume (Figure 9.8). In the same study, it was found that the amount of double-strand DNA damage as measured by the comet assay was significantly elevated in tumor cells treated with a combination of GPI-15427 and radiation compared to control and radiation-alone groups. The apoptotic index detected with TUNEL staining in JHU006 and JHU012 xenograft tumors was also increased by combined treatment with radiation and GPI-15427 (Figure 9.8) [69].

The successful application, in an experimental setting, of PARP inhibitors with fractionated radiation to control tumor growth underlines the clinical potential of the combined treatment protocol.

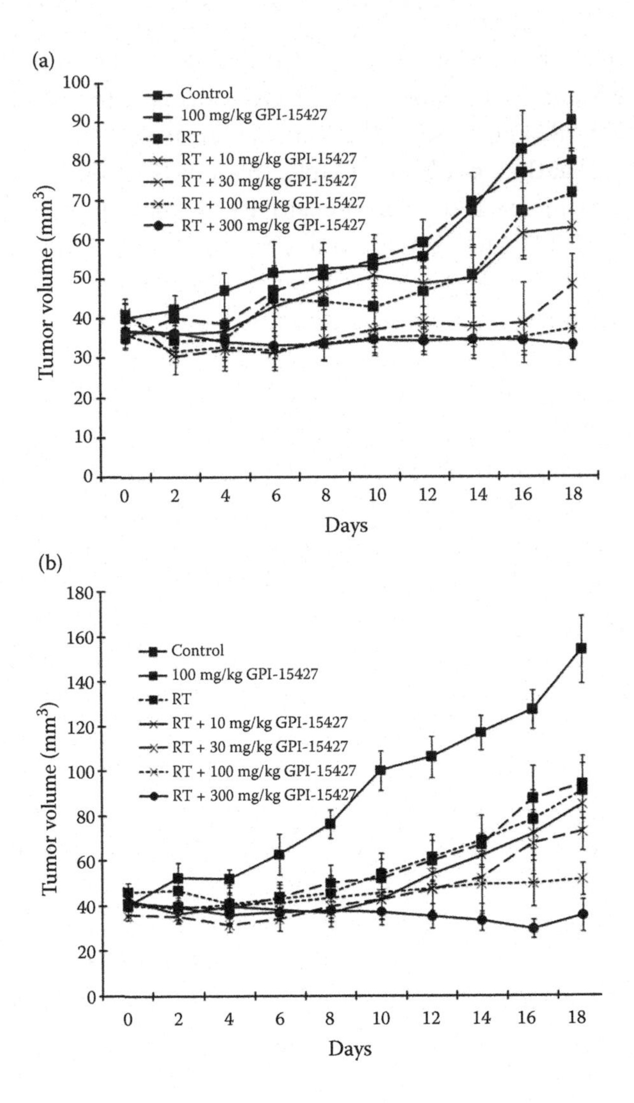

FIGURE 9.8 Oral GPI-15427 enhances the effect of radiation in xenograft head and neck tumors. (a) Tumor growth curves of JHU006 cell lines measured in two dimensions. Marked tumor growth diminution from day 4 was sustained through to day 18. (b) Tumor growth curves of JHU012 cell lines measured in two dimensions. Marked tumor growth diminution from day 2 was sustained through to day 18. (From Khan, K. et al., *Head Neck* 2010;32(3):381–391. With permission.)

9.4.3.1.2 PARP Inhibitors and Hypoxia It has been reported that chronic hypoxia can downregulate HR protein expression, leading to functional impairment of the HR pathway of DNA-DSB repair [70], and that chronically hypoxic cells that become HR-deficient can be hypersensitive to single-agent

PARP inhibition, in a similar manner to that reported for killing of cells that are genetically deficient in HR [50,51,71]. Liu et al. demonstrated that the PARP inhibitor ABT-888 acts as a radiosensitizing agent under hypoxic conditions. Using human prostate (DU-145, 22RV1) and non-small cell lung (H1299) cancer cell lines, it was observed that ABT-888 inhibited both recombinant PARP activity and intracellular PARP activity and ABT-888 was toxic to both oxic and hypoxic cells. When ABT-888 was combined with ionizing radiation, clonogenic radiation survival was decreased by 40% to 50% under oxic conditions, whereas under acute hypoxia, ABT-888 radiosensitized malignant cells to a level similar to oxic radiosensitivity [66]. The potential for ABT-888 to radiosensitize oxic and hypoxic malignant cells again suggests this agent as an adjunct for radiotherapy.

9.4.3.1.3 Protection of Normal Tissue against Chemotherapy-Induced and Radiotherapy-Induced Damage by PARP Inhibitors Several studies have demonstrated the ability of PARP inhibitors to protect against cardiotoxicity, nephrotoxicity, and intestinal damage induced by chemotherapeutics, presumably by interfering with the engagement of apoptosis induced by oxidative injury [72]. If PARP inhibitors could play a dual role as both potentiators of chemotherapy and radiotherapy and protectors against normal tissue toxicity, there would be great potential for improving the therapeutic ratio. The radiation tolerance of surrounding normal critical organs such as the spinal cord, lung, kidney, and small bowel is a significant limitation toward the delivery of a potentially curative dose of radiation to a tumor. Thus far, there are, however, no published reports of the protection of normal tissue against radiation damage by PARP inhibitors to match the reports of protection against chemotherapy agents.

9.4.3.2 Clinical Trials of PARP Inhibitors Combined with Radiotherapy

Despite encouraging results with preclinical studies and results becoming available from trials involving PARP inhibitors with other chemotherapy for the treatment of cancer, few clinical studies of PARP inhibitors in combination with radiotherapy have been undertaken. In a review by Chalmers et al. [73], two clinical trials in progress were described, both involved ABT-888 and targeted brain tumors:

- A phase I study combined escalating doses of oral ABT-888 with whole-brain radiation therapy (37.5 Gy in 15 fractions or 30 Gy in 10 fractions) in patients with cerebral metastases.

- A phase I/II study of the same drug in combination with radical radiation therapy with concomitant and adjuvant temozolomide in the first-line treatment of patients with glioblastoma multiforme. The phase I component seeks to establish the maximum tolerated dose of the PARP inhibitor in this setting, whereas the subsequent phase II study estimates the efficacy of this combination.

As of September 2012, only three studies were listed as ongoing on the www.clinicaltrials.gov database:

- A phase I study combining escalating doses of oral Iniparib with whole-brain radiation therapy in patients with cerebral metastases
- A phase I/II study of ABT-888 (Velaparib) in combination with radiation therapy in the treatment of locally advanced and inflammatory breast cancer
- A phase I study of the same drug with capecitabine and radiation in the treatment of locally advanced rectal cancer (www.clinicatrials.gov)

9.5 INHIBITORS OF HISTONE DEACETYLASE: TARGETING CHROMATIN MODIFICATION BY EPIGENETIC REGULATION OF GENE EXPRESSION

Epigenetic modulation of gene expression is an important regulatory process in cell biology. One form of epigenetic regulation involves the histone proteins that package DNA into regular repeating structures, the nucleosomes, each composed of a DNA strand wound around a core of eight histone proteins. The N-terminal tails of each histone extends outward through the DNA strand and amino acid residues on the histone tail can be modified by posttranslational acetylation, methylation, and phosphorylation. These modifications change the secondary structure of the histone protein tails in relation to the DNA strands, increasing the distance between DNA and histones, and increasing the accessibility of transcription factors to gene promoter regions [74]. Conversely, deacetylation, demethylation, and dephosphorylation of histones has the effect of decreasing access of transcription factors to promoter regions.

Alterations in the baseline activity of these proteins plays a role in multiple aspects of oncogenesis, including tumor differentiation, proliferation, and metastasis. Targeting these pathways is an important area of investigation, with HDACIs emerging as a novel class of molecularly targeted anticancer agents.

9.5.1 Structure and Biology of HDACs

An early observation was that many compounds with the ability to promote the differentiation of tumor cell lines, particularly those with a planar–polar configuration, induced accumulation of hyperacetylated histones, and histone hypoacetylation were also found to be associated with gene silencing, as in the inactivated female X chromosome [75]. Later experiments showed that treatment of cells with the short-chain fatty acid, sodium butyrate, caused hyperacetylation of histone octamers. This histone modification increased the spatial separation of DNA from histone and enhanced the binding of transcription factor complexes to DNA [76]. The first mammalian HDACs were cloned on the basis of their binding to known small-molecule inhibitors of histone deacetylation [75].

The acetylation status of histones is modulated by histone acetyltransferases (HATs), which add acetyl residues and by histone deacetylases (HDACs; Figure 9.9). There are at least 18 human HDACs, with varying functions, localization, and substrates (Table 9.5). The four classes of

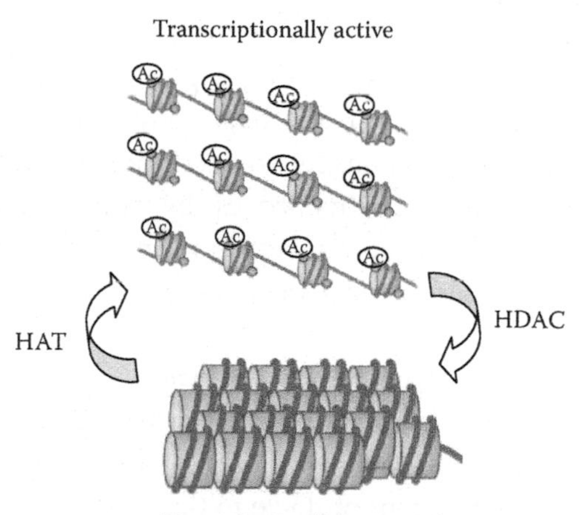

FIGURE 9.9 Histone acetylation. Acetylation of histone tails is mediated by HATs and results in an open modification of chromatin structure. It allows transcription factors to access the DNA and to initiate gene transcription. Conversely, gene repression is mediated via HDACs, which remove the acetyl groups from the histone tails, resulting in a closed chromatin structure.

TABLE 9.5 Characteristics of HDACs

Class	Enzyme	Zn_2 Dependence	Localization	Expression
I	HDAC1, HDAC2, HDAC3, HDAC8	Yes	Nucleus	Ubiquitous
IIa	HDAC4, HDAC5, HDAC7, HDAC9	Yes	Nucleus, cytoplasm	Tissue specific
IIb	HDAC6, HDAC10	Yes	Cytoplasm	Tissue specific
III	Sirtuins 1–7	No	Variable	Variable
IV	HDAC11	Yes	Nucleus, cytoplasm	Ubiquitous

HDACs are grouped by their homology to yeast proteins. Classes I, II, and IV all contain a zinc (Zn) molecule in their active site and are inhibited by the pan-HDAC inhibitors. The seven different class III HDACs (sirtuins), are homologous to the yeast Sir2, do not contain Zn in the active site, and are not inhibited by any current HDAC inhibitors [75].

9.5.2 Action of HDACs: Acetylation of Histone and Nonhistone Proteins

9.5.2.1 Histones

Posttranslational modifications of histones, including acetylation, methylation, and phosphorylation act as dynamic regulators of gene activity. In the case of acetylation, HATs catalyze the acetylation of lysines in core histones resulting in neutralization of the positive charges on histones, decreasing their interaction with the negatively charged DNA with the overall effect being to create relaxed, transcriptionally active, chromatin conformation. HDACs reverse this process by catalyzing the removal of acetyl groups from lysine residues, resulting in a more compacted and transcriptionally repressed chromatin structure.

9.5.2.2 Histone Modification and DNA Repair

Histone modifications play critical roles in numerous DNA repair pathways. The packaging of DNA into chromatin with highly compact and complex structures presents a significant obstacle to the access of repair machinery to damaged nucleic acid, and histone modifications regulate both the accessibility and turnover of chromatin. Thus, chromatin modification plays a critical role in DSB repair processing and both NER and DSB repair pathways require global and local chromatin relaxation to proceed. DSB detection induces an initial common response to HR and NHEJ characterized by the phosphorylation of histone H2AX. This modification leads to global and local relaxation events through the recruitment of histone-modifying and chromatin-modifying activities. All DNA repair pathways need to have an easy access to the lesions to proceed efficiently, but their requirements vary, involving either common or specific changes. Most histone modifications including histone acetylation, phosphorylation, and ubiquitylation are highly dynamic—allowing quick access to the lesion as well as a quick restoration of the chromatin after repair. It is conjectured that the deacetylation process is not required for targeting of the repair components but rather for facilitating the synapsis of broken ends by the repair machinery and for allowing the restoration of the chromatin after repair [77].

9.5.2.3 Nonhistone Proteins

In addition to regulating the acetylation state of histones, HDACs can bind to, deacetylate, and regulate the activity of a number of other proteins, including transcription factors [p53, E2F transcription factor 1 (E2F1), and NF-κB], and other proteins with diverse biological functions (α-tubulin, Ku70, and Hsp90). Nonhistone proteins known to be targets of HDACs are listed in Table 9.6. The effect of HDAC inhibition is amplified when the targeted protein affects multiple downstream cell regulatory processes. For example:

- ATM is activated when DSB-induced changes in chromatin structure are recognized by sensor proteins. After activation, ATM rapidly phosphorylates downstream substrates that are important in

TABLE 9.6 Nonhistone Proteins Targeted by HDACs

Function	Substrate
Cell mobility	α-Tubulin, Cortactin
Signaling mediators	β-Catenin, STAT3, Smad7, IRS-1
Oncogene	Bcl-6
DNA binding transcription factors	c-Myc, NF-κB, EF-1-3, GATA 1-3, HIF-1α POP-1, IRF2, IRF7, CREB, myo-D, p73, YY-1, MEF-2, SRY, EKLF, UBF
Steroid hormone receptors	Estrogen receptor-α, androgen receptor, glucocorticoid receptor
Transcription coregulators	HGM1, HGM2, CtBP-2 PGC-1α DEK, MSL-3
Chaperone	HSP-90
Nuclear import factors	Importin-α1, Importin-α7
Tumor suppressors	pRB, p53
DNA repair enzymes	Ku70, WRN, TDG, NEIL-2, FEN1
Viral proteins	EIA, L-HDAg, S-HDAg, T antigen, HIV TAT
Inflammation mediator	HMGB-1

DSB response pathways including p53, which has key roles in cell-cycle control and apoptosis; MDM2, CHK1, and CHK2, which are involved in cell-cycle regulation; and BRCA1 and NBS1, which are important in DSB repair. It has been shown that when using human fibroblasts, ATM interacts with HDAC1 both *in vitro* and *in vivo*, and that ionizing radiation enhances ATM-associated HDAC activity, suggesting that HDACs may have a critical role in ATM signaling in response to DNA damage [78].

- The p53 gene is negatively regulated by HDACs and the protein is a known substrate for HATs and HDACs [79]. In response to DNA damage, p53 is acetylated at lysine residues in the C-terminal region by the HATs, p300 and PCAF; this process, along with phosphorylation, is believed to stabilize the molecule, which accumulates in the nucleus where it regulates transcription [79]. It seems unlikely that HDAC inhibitors sensitize cells to the effects of radiation by modulating p53 because HDAC inhibitors have been shown to enhance the radiation sensitivity of cells regardless of their p53 expression status [80,81].
- DNA repair proteins: It has been shown that the radiation-induced increase in Rad51 and DNA-PKcs expression is markedly attenuated by pretreatment of a human prostate cancer cell line with HDAC inhibitor SAHA [82]. Another study has demonstrated that sodium butyrate decreases the expression of critical DNA repair proteins Ku70, Ku86, and DNA-PKcs in two melanoma cell lines [83].

Phosphorylation of H2AX to form γ-H2AX is one of the earliest events to occur after the induction of DSBs. Incubation of cells with HDAC inhibitors including MS-275 and sodium butyrate results in prolonged expression of induced γ-H2AX foci [81,83], indicating that HDAC inhibitor-mediated radiosensitization is associated with a decrease in the repair of DSBs. Furthermore, a study using the HDAC inhibitor Trichostatin A (TSA) demonstrated an increase in the number of radiation-induced γ-H2AX, suggesting that the change in chromatin structure caused by histone hyperacetylation results in an increase in the number of DSBs [84]. The effects of HDAC inhibition on nonhistone proteins is shown schematically in Figure 9.10.

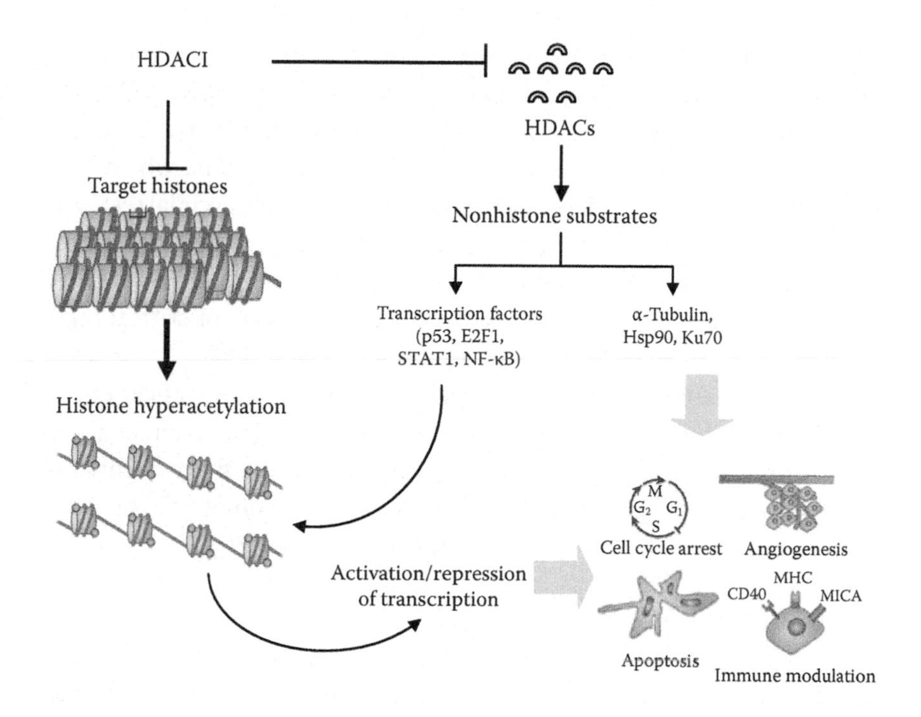

FIGURE 9.10 HDAC can bind to, deacetylate, and regulate the activity of a number of other proteins, including transcription factors (p53, E2F1, and NF-κB) and proteins with diverse biological functions (α-tubulin, Ku70, and Hsp90). Hyperacetylation of transcription factors with HDACIs can augment their gene-regulatory activities and contribute to the changes in gene expression observed after direct HDACI-mediated histone hyperacetylation. Hyperacetylation of proteins such as Ku70 and Hsp90 or disruption of protein phosphatase 1 (PP1)–HDAC interactions by HDACI might have no direct or indirect effect on gene expression but could be important for biological effects such as apoptosis and cell cycle arrest.

9.5.2.4 HDACs in Cancer

A common finding in cancer cells is a high level expression of HDAC isoenzymes and a corresponding hypoacetylation of histones [85]. In fact, loss of monoacetylation of lysine 16 and trimethylation of lysine 20 in the tail of histone H4 are considered, next to global DNA hypomethylation and CpG island hypermethylation, to be almost universal epigenetic markers of malignant transformation. Because levels of HDAC enzymes are generally elevated in cancer cells, it might be reasoned that HDAC inhibitors will exert greater effects on cancer cells than normal cells.

Acetylation of histones maintains chromatin in an open and transcriptionally active state. Deacetylation of histones by HDACs turns the chromatin into a closed formation that is not accessible for transcription factors. This situation can be reversed by HDAC inhibitors, which have been shown to induce differentiation, cell growth, cell-cycle arrest, and in certain cases apoptosis in numerous transformed cell lines in culture and in tumor cells in animal model systems [86,87]. In addition, HDAC inhibitors have also been shown to alter the transcription of critical regulators of invasion and to inhibit angiogenesis *in vivo* [87,88].

These observations support the interpretation of the anticancer effects of HDAC inhibitors as being caused in part by the accumulation of acetylated histones resulting in the altered transcription of a finite number of genes; the net result being the of activation or repression of specific genes resulting in antiproliferative or proapoptotic effects [87,89].

9.5.2.5 HDACs in Normal Cells

Histone hyperacetylation is induced in normal lymphocytes as well as liver and spleen HDACIs, suggesting that the radiosensitivity of normal tissues may also be enhanced. When HDAC inhibitors are used as a single modality, it is believed that the aberrant HDAC activity in tumor cells would make them much more susceptible to the cytotoxic/cytostatic effects of HDAC inhibitors than normal cells, and the results of most experimental studies support this [86]. In addition, *in vivo* administration to experimental animals of clinically relevant HDAC inhibitors using antitumor drug doses has been generally found to be without significant toxicities [90].

9.5.3 HDAC Inhibitors

HDACIs block the action of HDACs and return the chromatin to a more relaxed structure. The overall effect is to influence the transcription rate of large numbers of genes involved in all aspects of cell activity. The chemical formulas of clinically applicable HDAC inhibitors range from the benzamides (MS-275) to hydroxamates to aliphatic acids to cyclic tetrapeptides (Figure 9.11); therefore, it is likely that the process of HDAC inhibition will vary between different compounds and that each compound will have its own characteristic range of cellular activities.

More than 15 HDAC inhibitors have been tested in preclinical and early clinical studies. The common mechanism of action of these drugs is to bind

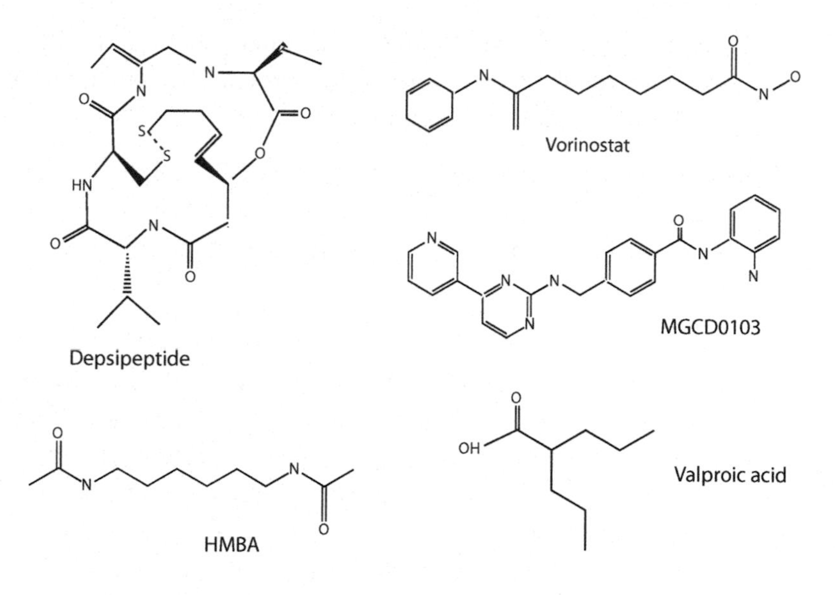

FIGURE 9.11 HDAC inhibitors.

a critical Zn_2 ion required for catalytic function of the HDAC enzyme [91]. These compounds were selected for their ability to inhibit histone deacetylation, but they vary widely in potency and HDAC isoenzyme specificity, and have variable effects on acetylation of nonhistone substrates [92]. The nonoverlapping effects of different inhibitors on transformed cells *in vitro* suggest that differing efficacies, toxicities, and therapeutic applications are to be expected. In tumor cells, HDACIs are believed to act primarily via the reactivation of previously silenced checkpoint and tumor suppressor genes, including p27kipl, P16INK4a, and p21WAF1. However, the mechanisms by which HDACIs enhance cell killing are difficult to establish because of the wide range of HDAC downstream targets and associated cellular processes involved (Figure 9.11).

There are four important classes of HDACI:

- *Short-chain fatty acids.* The least potent class of HDAC inhibitors is the aliphatic acid group of compounds, which are effective inhibitors of HDAC activity at millimolar concentrations The strong point of this class is the that they are well tolerated and, in some cases, have been used routinely for the treatment of other disorders. This class includes phenylbutyrate, which has advanced to clinical trials and valproic acid (VA), which is a well-established therapeutic for a variety of seizure disorders and bipolar (manic depressive) illness and is well tolerated. One member of this group, VA, is in a number of clinical trials as a cancer treatment.

- *Benzamides.* The benzamides (e.g., MS-275 and CI-994) [93], electrophilic ketones (e.g., trifluoromethyl ketones and α-ketoamides) [94] are classes of HDAC inhibitors exhibiting HDAC inhibition activity in the micromolar concentration range.
- *Hydroxamates.* The classic HDAC inhibitor, TSA, a Streptomyces metabolite that was originally developed as an antifungal agent, is a member of the hydroxamic acid class. TSA has been characterized extensively *in vitro* and numerous studies have demonstrated it to be a potent inhibitor of HDAC activity at nanomolar concentrations [95]. Vorinostat (suberoylanilide hydroxamic acid) is a synthetic hydroxamic acid that has been investigated at the preclinical and clinical level.
- *Cyclic tetrapeptides.* A separate class of HDAC inhibitors including depsipeptide and apicidin, that are also active at nanomolar concentrations [96,97].

9.5.3.1 Cytotoxicity of HDACIs

HDAC inhibitors alter the acetylation status of chromatin and other non-histone proteins, resulting in changes in gene expression, induction of cell death, apoptosis, cell cycle arrest, and inhibition of angiogenesis and metastasis. It has also been reported that HDAC inhibitors can induce polyploidy and aberrant mitosis such as mitotic slippage, and premature sister chromatid separation, which can lead to loss of cancer cell proliferation [98]. HDAC inhibitors mediate cell death through several pathways including cell growth arrest, effects on DNA repair and mitosis, induction of apoptosis, and antiangiogenic effects.

- *Cell cycle effects.* A number of studies have demonstrated that almost all HDAC inhibitors can inhibit cell growth by cell cycle arrest at G_0/G_1 or G_2/M checkpoints based on cell type or dose of HDAC inhibitor used. Protein p21 is most commonly reported to be upregulated by HDAC inhibitors in cancer cell lines. Dephosphorylation of pRb was also detected in human leukemia cells treated with LAQ 824, an HDAC inhibitor. HDAC inhibitors can also induce the downregulation of cyclin proteins, such as cyclin B1 (a regulator of G_2/M phase and the M phase transition), cyclin D_1 and D_2 (a regulator of G_1/S phase transition), and cyclin E to arrest the cell cycle [89,98,99].

- *DNA damage and repair.* Histone acetylation, induced by HDACI, causes structural alterations in chromatin, which can expose DNA to damaging agents, including ionizing radiation. At the same time, HDACIs can induce the accumulation of reactive oxygen species (ROS), which are a direct cause of DNA damage [81].

Lowered repair of DNA damage results from acetylation of proteins involved in DNA repair such as Ku70 [100], Ku80 [83], BRCA1 [101], Rad51 [82], and DNA-PK [82,102]. These proteins are involved in the two major pathways for the repair of DSBs, HR, and NHEJ. Additional suppression of DNA damage repair by HDACI is related to the direct inhibition of HDAC isoforms, preventing these from interaction with DNA damage sensor proteins such as ATM (interaction with HDAC1) [78] and 53BP1 (interaction with HDAC4) [45].

- *Apoptosis.* HDAC inhibitors typically induce cell death through the intrinsic apoptosis pathway. A number of studies demonstrate that HDAC inhibitors induce the intrinsic apoptosis pathway through inactivation of antiapoptotic and activation of the proapoptotic Bcl-2 family of proteins [103,104]. Antiapoptotic proteins of the Bcl-2 family, including Bcl-2, Bcl-xL, and Mcl-1, were downregulated by panobinostat (LBH589), an HDAC inhibitor in lung cancer cell lines [105]. The proapoptotic proteins of the Bcl-2 family, including Bak and BH3-only proteins (such as Bik, Bim, Bmf, and Noxa), were upregulated at messenger RNA (mRNA) or protein levels by HDAC-like tumor necrosis factor-related apoptosis inducing ligand (TRAIL) and Fas ligand (FasL) [58] This results in the activation of caspase-8 or caspase-10 and the initiation of the extrinsic apoptotic pathway [103].
- *Antiangiogenic effects.* Besides inducing cell death and cell-cycle arrest, HDACI also have antiangiogenic, anti-invasive, and immunomodulatory activities related to transcriptional changes. The antiangiogenic potential of these drugs has been attributed to the downregulation of proangiogenic genes such as vascular endothelial

growth factor (VEGF) [106,107] and chemokine (C-X-C motif) receptor 4 (CXCR4), and to the suppression of endothelial progenitor cell differentiation.

9.5.3.2 Radiosensitization by HDACIs

9.5.3.2.1 Effect of High (Cytotoxic) Doses of HDACIs High, cytotoxic concentrations of HDAC inhibitors can enhance the radiation sensitivity of cell lines in culture. In early experiments (reviewed by Karagiannis and El-Osta [84]) molecules such as sodium butyrate, phenyl butyrate, tributyrin, TSA, and SAHA as well novel HDAC inhibitors including the anilide-based MS-275, the cyclic tetrapeptide analogue bicyclic depsipeptide, and numerous hydroxamic acid analogues such as M344 were investigated for their radiation sensitizing properties *in vitro*. The most frequent observation was a synergistic effect after treatment of cells with the HDAC inhibitor and subsequent irradiation. The mechanisms that were invoked to explain these results were similar to those described previously for cell death, including modulation of cell cycle regulation, particularly G_1 phase arrest, inhibition of DNA synthesis, enhancement of radiation-induced apoptosis [22], and downregulation of survival signals [108].

9.5.3.2.2 Effect of Low Doses of HDACIs HDAC inhibitors can modulate the sensitivity of cells to ionizing radiation at low concentrations, which are nontoxic and do not induce cell-cycle arrest but are sufficient to alter histone acetylation. Because there is no reduction in survival associated with the use of the drug alone, this situation represents true radiosensitization. Explanations for this effect have centered on DNA damage and DNA damage signaling as well as activating intrinsic and extrinsic cell apoptotic death. As already described, HDACIs convert chromatin into an open structure that is more accessible to external damage; thus, in conjunction with ionizing radiation, HDAC inhibition can more efficiently induce potentially lethal DSBs.

Normal cellular responses to DSBs include detection and repair of these lesions by DNA damage sensor proteins and DNA repair proteins, and HDACI can perturb the interactions of proteins involved in these pathways. In the case of damage signaling, HDACI disrupts the association of HDAC enzymes with the sensor proteins ATM and 53BP1, and alters the acetylation status of proteins involved in both the HR (BRCA1, Rad51, and Rad50) [101,102] and NHEJ (Ku70, Ku80, and DNA-PK) [82,83,102,107] repair pathways.

9.5.3.2.3 Is Radiosensitization Selective for Cancer Cells? Munshi et al. [102] reported that sodium butyrate (NaB) treatment induced the radiosensitization of two melanoma cell lines, but caused no detectable effect on the radiosensitivity of a normal human fibroblast cell line. Consistent with these results, it was found that whereas HDAC inhibitors enhanced the *in vitro* sensitivity of tumor cells to DNA-damaging drugs, they had no effect on the drug sensitivity of normal breast or intestinal cells [78,90]. In a more direct comparison, the effects of HDAC inhibitors on radiation-induced skin injury were evaluated using a rat model [109]. It was found that phenylbutyrate (PB), TSA, and VA each suppressed the skin fibrosis associated with cutaneous radiation syndrome, an effect that was attributed to the suppression of radiation-induced aberrant expression of transforming growth factor β.

These results suggest that not only do HDAC inhibitors have no effect on the intrinsic radiosensitivity of normal fibroblasts *in vitro* but may also actually protect against radiation-induced injury *in vivo*. Here, again, there are conflicting reports; in one case, NaB treatment of normal human lymphocytes inhibited the repair of radiation induced DNA-DSBs as measured by premature chromosome condensation [110]. The relevance of this study is at least questionable because cell survival results were not reported and lymphocyte death after irradiation occurs primarily through apoptosis, whereas most solid tumor cells die as a result of mitotic catastrophe. There is, however, another experimental study (to be described in Section 9.5.3.3) in which cell survival in normal tissue was found to be reduced in some cases but not in others [111]. Taken overall, these data at least raise questions about the possible effects of HDAC inhibitors on the radio-response of normal tissue.

9.5.3.3 Experimental Studies of Radiosensitization by HDACIs Cell Lines

The scope of the experimental work at the *in vitro* level, which has been done on radiosensitization by HDACIs is shown in Tables 9.7a and b, which are largely based on information from articles by Banuelos et al. [112] and by De Schutter and Nuyts [22], and the references contained therein.

Some recent studies, which have been particularly revealing in terms of the mechanisms involved, are summarized in the following:

- Preclinical studies evaluating HDAC inhibitor-induced radiosensitization have largely focused on preirradiation drug exposure based on the assumption that enhanced radiosensitivity was mediated

TABLE 9.7a Preclinical Studies Using Human Tumor Lines with Hydroxymate and Benzamide HDAC Inhibitors in Combination with Radiotherapy

		Drug Class									
		Hydroxymates							Benzamides		
Tumor Type	**Tumor Name**	**TSA**	**SAHA**	**LBH 589**	**PCI 24781**	**LAQ 824**	**M344**	**CBHA**	**MS275**	**CI 994**	**SK7041**
Leukemia	K562	•									
Brain: medulloblastoma	DAOY	•	•								
	UW 228-2	•	•								
Brain: glioma	U373MG	••	•								
	U87MG	•									
	DU145	•	•								
	U251								•		
Melanoma	A375	••	•								
	MeWo	••									

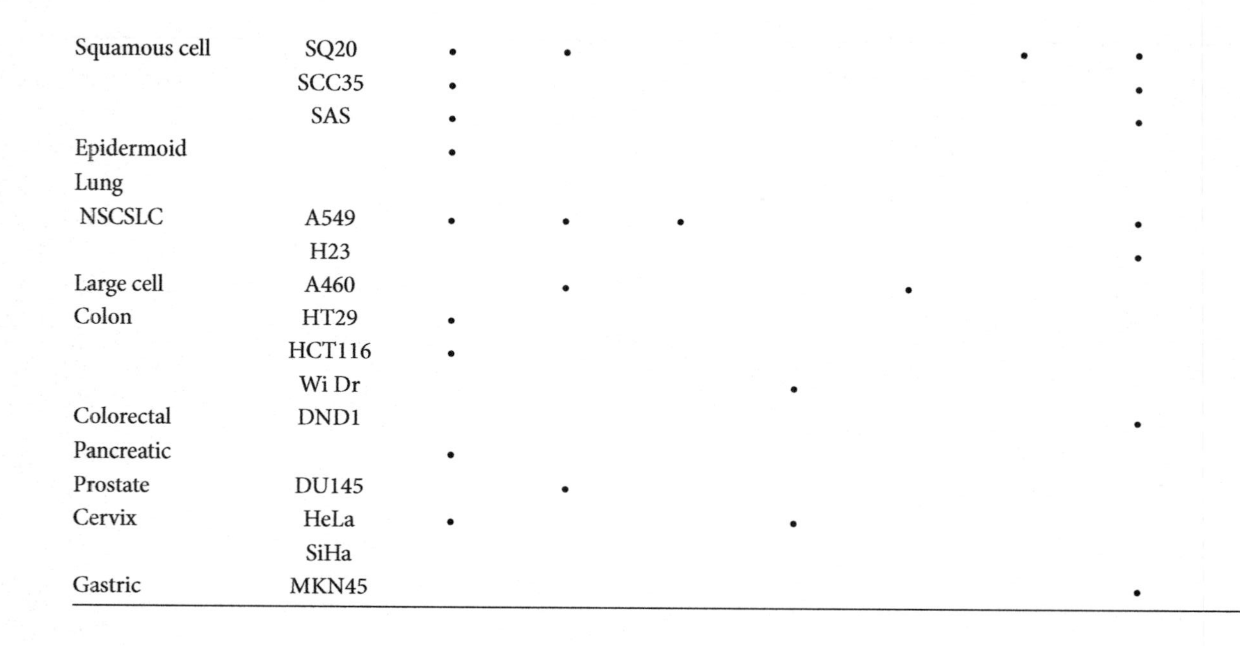

Squamous cell	SQ20	•	•				•	•			•
	SCC35	•						•			
	SAS	•						•			•
Epidermoid		•									
Lung											
NSCSLC	A549	•	•	•				•			•
	H23							•			
Large cell	A460		•			•					
Colon	HT29	•									
	HCT116	•									•
	Wi Dr				•						
Colorectal	DND1							•			
Pancreatic		•									
Prostate	DU145		•						•		
Cervix	HeLa	•			•						•
	SiHa										
Gastric	MKN45							•			

TABLE 9.7b Preclinical Studies Using Human Tumor Lines with Short-Chain Fatty Acid HDAC Inhibitors in Combination with Radiotherapy

		Drug Type							
		Short-Chain Fatty Acid							Cyclic Peptide
Tumor Type	**Tumor Name**	**VA**	**NaB**	**PB**	**PA**	**Tributyrin**	**ANI**	**ANI13**	**FK228**
Leukemia	K562	•							
	MOLT-4	•							
Brain:	DAOY		•						
medulloblastoma	UW 228-2		•						
Brain: glioma	U251	•		•			•	•	
	SF539	•							
	SKMG-3			•					
	U87MG			• •	•		•		
	DU145			•					
	SF188						•	•	
Melanoma	A375		• •			•			
	MeWo					•			
Breast	MCF7			•	•				
Nasopharynx	CNE2			•					
Colon	DLD1-A		• • •						•
	DLD1-D		• •						
	HCT116								
	Wi Dri			•	•				
Prostate	DU145		•	•					
	PC3M			•	•				
Cervix	SW756			•	•				
	ME180			•	•				
Gastric	MKN45								•
Squamous cell	SAS								
carcinoma	HSC-2								

by changes in gene expression. However, in one case, Chinnaiyan et al. [113] identified maximal radiosensitization when cells were exposed to HDAC inhibitors at both preradiation and postradiation, and these studies were expanded to determine whether postirradiation exposure alone affected radiosensitivity. The effect of the HDAC inhibitor VA on postirradiation sensitivity in human glioma cell lines U251 and SF539 was evaluated using clonogenic assay by exposing cells to VA for up to 24 h after irradiation. DNA damage repair was evaluated based on γH2AX and 53BP1 foci. Cell cycle phase distribution and acetylation of γH2AX were also investigated. The results indicated that VA did in fact enhance radiosensitivity for up to 24 h after irradiation and that exposure to VA after irradiation results in delayed repair kinetics.

- It was shown in prostate cancer cells that although valproic acid (VA) at low concentrations had minimal cytotoxic effects, it could significantly increase radiation-induced apoptosis. VA stabilized a specific acetyl modification (lysine 120) of the p53 tumor suppressor protein, resulting in an increase in its proapoptotic function at the mitochondrial membrane. This effect of VA was independent of any action of the p53 protein as a transcription factor in the nucleus because these effects were also observed in native and engineered prostate cancer cells containing mutant forms of p53 protein having no transcription factor activity. Transcription levels of p53-related or Bcl-2 family member proapoptotic proteins were not affected by VA exposure. The results suggest that, in addition to the nuclear-based pathways previously reported, lower concentrations of HDACIs may also result in radiosensitization through specific p53 acetylation and perturbation of mitochondrial-based apoptotic pathway(s) [114].
- As already noted, numerous studies have investigated the role of HDACI in tumor cells but less is known about their effects on normal tissue cells. In a study with human primary fibroblasts, the HDACIs SAHA, MS275, sodium butyrate, and VA were all found to increase radiosensitivity and reduce DSB repair capacity, However, the effect was not universal because HDACIs were found to be ineffective in radiosensitizing other classes of nonmalignant cells. Nevertheless, the authors of this study warn against the uncritical use of HDACIs, particularly in certain patient groups and in conjunction with radiotherapy [111].

TABLE 9.8 Radiosensitization of Human Tumor Models *In Vivo* by HDACIs

Drug	IR	Tumor	Endpoints and Results	Comments	Reference
Valproic acid, i.p. 2 × daily. 150 mg/kg, 3 days	4 Gy	Glioma, U251 xenograft	VA administration resulted in histone hyperacetylation in U251 xenografts, radiation-induced TGD. DEF = 2.6	VA-mediated increase in radiation-induced cell killing involved inhibition of DNA DSB repair	[115]
MS-175, 6 mg/kg, 2 × daily i.p., 3 days	6 Gy	Prostate DU145, s.c. in the leg	TGD: MS-275, 2.9 ± 2.7 days; IR, 1.9 ± 1.3 days; MS-275 + IR, 8.3 ± 3.9 days	Radiation delivered at time of maximum histone hyperacetylation. Growth delay for combined treatment greater than additive	[81]
Vorinostat, 75 mg/kg, oral	5 Gy	MDA-MB-231-BR, breast tumor brain metastatic cell line	TGD (survival), C, 13 days; IR, 16 days; Vorinostat, 16 days; combination, 27 days	Growth delay for combined treatment greater than additive	[116]
Vorinostat, 150 mg/kg, alternate days, three doses	1 Gy alt. days	NB1691^{luc} metastatic neuroblastoma	Tumor volumes quantified fluorescent imaging of whole mouse	Combination treatment decreased tumor volume compared to single modality	[117]
LAQ824[a]	5 × 3 Gy	H460, human lung cancer, large cell, s.c.	TGD: IR, 4 days; LAQ824, 7 days; combination, 19 days	Growth delay for combined treatment greater than additive	[118]
Valproic acid, i.p. 6 × 300 mg/kg, 3 days	10 Gy	DU 145, s.c.	TGD	Growth delay for combined treatment greater than either treatment alone	[114]
LBH589[a] 2 × 40 mg oral		H460, s.c. 5 × 2 Gy over 7 days	TGD: IR, 2 days; LBH589, 4 days; combination, 20 days	HDAC inhibition enhances the effects of IR on NSCLC tumor growth. Mice showed minimal toxicity from LBH589	[119]

Note: DEF, dose enhancement factor; i.p., intraperitoneal; s.c., subcutaneous; TGD, tumor growth delay.

[a] LBH589 and LAQ824 are cinnamic hydroxamic acid HDAC inhibitors.

TABLE 9.9 HDAC Inhibitors in Clinical Trials in Protocols Combining Radiotherapy

HDAC Inhibitor	Study Populations	Phase	Number
Vorinostat (suberoylanilide hydroxamic acid; SAHA)	Brain, CNS, and brain metastases	I/II	6
	Non-small cell lung cancer	I/II	4
	Pancreatic cancer	I/II	3
	Squamous cell carcinoma, head and neck	I/II	1
	Gastric cancer	I/II	1
	Pelvic cancer	I/II	1
Panobinostat, LBH589	Brain	I/II	1
	Esophagus, prostate, head and neck	I/II	1
Valproic acid	Brain tumors	I/II	1
	Cervical cancer	I/II	1
	Sarcoma	I/II	1

Source: www.clinicaltrials.gov, October 2012.

- Preclinical models. There have been numerous experiments with preclinical models of xenografted human tumors of different histologies that demonstrated radiosensitization by HDACIs. Some of these are summarized in Table 9.8.

9.5.3.4 Clinical Studies

The first HDAC inhibitor to receive FDA approval as a cancer therapy was suberoylanilide hydroxamic acid (SAHA; also known as vorinostat), which demonstrated efficacy in cutaneous T-cell lymphoma in clinical trials. Currently, there are a number of clinical trials involving HDAC inhibitors in the treatment of cancer. A small number are for HDACI as a single agent whereas a much larger number involve HDACI in combination with other drug(s) and biological agents. The most frequently used HDACI are vorinostat and VA. A much smaller group of trials include HDACIs in protocols involving radiotherapy. Table 9.9 lists ongoing or proposed clinical trials in which external beam radiotherapy is combined with HDACI alone or in combination with other treatments.

9.6 SUMMARY

Human DNA-DSBs are repaired through two pathways, HR and NHEJ. Cells deficient in the NHEJ proteins Ku70/80 or the catalytic subunit of DNA-PK are sensitive to DSBs induced by IR or chemotherapeutic drugs, suggesting DNA-PK as a good target for radiochemosensitization. Several potent small-molecule DNA-PK inhibitors have been developed and, in addition, some biological agents target DNA-PK indirectly including Cetuximab (C225), an antibody to EGFR that blocks IR-induced activation of DNA-PK.

A number of drugs commonly used in conjunction with radiotherapy have been shown to have an effect on HR, mostly by targeting the RAD51 protein, although in many cases, other HR proteins are involved and the reduction in RAD51 may be an indirect effect. RAD51 is commonly overexpressed in malignant cells and can lead to increased resistance to a variety of drugs and to ionizing radiation. Imatinib mesylate is an inhibitor of c-ABL kinase, which has been implicated in the radiation-induced assembly of RAD51 and RAD52; the treatment of cancer cells with imatinib decreases RAD51 and sensitizes them to experimental chemotherapy and radiotherapy. Erlotinib is a reversible tyrosine kinase inhibitor acting on the intracellular tyrosine kinase domain of EGFR, which attenuates the HR response through a BRCA1-mediated mechanism. Erlotinib combined with radiation and chemotherapy is being extensively investigated at the preclinical and clinical level.

The PARP family of proteins are characterized by the enzymatic property of poly(ADP-ribosylation), which modulates catalytic activity and protein–protein interactions. PARP-1 and PARP-2 are both DNA damage sensors and signalers of DNA damage. PARP inhibitors compete with NAD+ for PARP's catalytic site and prevents the synthesis of poly(ADP-ribose), disrupting the catalytic function of PARP but not interfering with its ability to bind to DNA. In the presence of a PARP inhibitor, PARP binds to sites of damage but fails to automodify, remaining bound to the lesion and preventing access of repair factors. Inhibition of PARP increases the prevalence of unrepaired lesions that can cause stalling and collapse of the DNA replication machinery and, if HR is also compromised as in BRCA1- or BRCA2-deficient cells, endogenous lesions go unrepaired and evolve into cytotoxic DSB.

Reports of preclinical studies support a role for PARP inhibitors as radiosensitizers. Tumor growth delay associated with radiation treatment was enhanced in a number of human tumor models and several studies showed that PARP inhibitors increase radiosensitivity in a replication-dependent manner that is enhanced by fractionation. It has also been demonstrated that the PARP inhibitor, ABT-888, acts as a radiosensitizing agent under hypoxic conditions. Despite encouraging preclinical results, very few clinical studies of PARP inhibitors with concurrent radiation are being undertaken; and there are only a small number of phase I and phase II trials ongoing.

DNA repair pathways require easy access to the lesions to proceed efficiently. Chromatin relaxation depends on the modification histones, which increase the spatial separation of DNA from histone and enable the binding of transcription factor complexes to DNA. The acetylation status of histones is modulated by HATs, which add acetyl residues, and by HDACs, which take them off. In addition to histones, HDACs can bind to, deacetylate, and regulate other proteins with diverse biological functions including transcription factors. HDAC inhibitors target histone deacetylation by a common mechanism but vary widely in potency, HDAC isoenzyme specificity, and acetylation of nonhistone substrates. Cytotoxicity of HDAC inhibitors involves several pathways including cell growth arrest, effects on DNA repair and mitosis, induction of apoptosis, and antiangiogenic effects. Radiosensitization by HDACIs can occur at low, nontoxic drug concentrations that do not induce cell-cycle arrest but are sufficient to alter histone acetylation. Conversion of chromatin to a more open structure makes it more accessible to external damage, including that by IR. In addition, HDACIs can perturb the interactions of proteins involved in normal cellular responses to DSBs mediated by DNA damage sensor proteins. Preclinical studies with xenograft tumors of different histologies clearly demonstrated radiosensitization by HDACIs. Many clinical trials of HDACIs (usually vorinostat or VA) in combination with other drug(s) and biological agents are ongoing or completed. A much smaller number are investigating protocols combining HDACIs and radiotherapy.

REFERENCES

1. Lees-Miller, S., Godbout, R., Chan, D. et al. Absence of p350 subunit of DNA-activated protein kinase from a radiosensitive human cell line. *Science* 1995;267:1183–1185.
2. Ouyang, H., Nussenzweig, A., Kurimasa, A. et al. Ku70 is required for DNA repair but not for T cell antigen receptor gene recombination *in vivo*. *J Exp Med* 1997;186:921–929.
3. Deriano, L., Guipaud, O., Merle-Beral, M. et al. Human chronic lymphocytic leukemia B cells can escape DNA damage-induced apoptosis through the non-homologous end-joining DNA repair pathway. *Blood* 2005;105:4776–4783.
4. Rosenzweig, K., Youmell, M., Palayoor, S., and Price, B. Radiosensitization of human tumor cells by the phosphatidylinositol3-kinase inhibitors wortmannin and LY294002 correlates with inhibition of DNA-dependent protein kinase and prolonged G2-M delay. *Clin Cancer Res* 1997;3:1149–1156.

5. Shinohara, E., Geng, L., Tan, J. et al. DNA dependent protein kinase is a molecular target for the development of noncytotoxic radiation-sensitizing drugs. *Cancer Res* 2005;65:4987–4992.
6. Cowell, I., Durkacz, B., and Tilby, M. Sensitization of breast carcinoma cells to ionizing radiation by small molecule inhibitors of DNA-dependent protein kinase and ataxia telangiectsia mutated. *Biochem Pharmacol* 2005;71:13–20.
7. Zhao, Y., Thomas, H., Batey, M. et al. Preclinical evaluation of a potent novel DNA-dependent protein kinase inhibitor NU7441. *Cancer Res* 2006;66:5354–5362.
8. Ismail, I., Martensson, S., Moshinsky, D. et al. SU11752 inhibits the DNA dependent protein kinase and DNA double-strand break repair resulting in ionizing radiation sensitization. *Oncogene* 2004;23:873–882.
9. Nutley, B., Smith, N., Hayes, A. et al. Preclinical pharmacokinetics and metabolism of a novel prototype DNA-PK inhibitor NU7026. *Br J Cancer* 2005;93:1011–1018.
10. Durant, S., and Karran, P. Vanillins—A novel family of DNA-PK inhibitors. *Nucleic Acids Res* 2003;31:5501–5512.
11. Azad, A., Jackson, S., Cullinane, C. et al. Inhibition of DNA-dependent protein kinase induces accelerated senescence in irradiated human cancer cells. *Mol Cancer Res* 2011;9(12):1696–1707.
12. Dittmann, K., Mayer, C., and Rodemann, H. Inhibition of radiation-induced EGFR nuclear import by C225 (Cetuximab) suppresses DNA-PK activity. *Radiother Oncol* 2005;76:157–161.
13. Kim, C., Park, S., and Lee, S. A targeted inhibition of DNA-dependent protein kinase sensitizes breast cancer cells following ionizing radiation. *J Pharmacol Exp Ther* 2002;303(2):753–759.
14. Ni, X., Zhang, Y., Ribas, J. et al. Prostate-targeted radiosensitization via aptamer-shRNA chimeras in human tumor xenografts. *J Clin Invest* 2011;121(6):2383–2390.
15. Zhang, J., and Powell, S. The role of the BRCA1 tumor suppressor in DNA double-strand break repair. *Mol Cancer Res* 2005;3:531–539.
16. Esashi, F., Galkin, V., Yu, X., Egelman, E., and West, S. Stabilization of RAD51 nucleoprotein filaments by the C-terminal region of BRCA2. *Nat Struct Mol Biol* 2007;14:468–474.
17. Davies, O., and Pellegrini, L. Interaction with the BRCA2 C terminus protects RAD51-DNA filaments from disassembly by BRC repeats. *Nat Struct Mol Biol* 2007;14:475–483.
18. Lord, C., and Ashworth, A. RAD51, BRCA2 and DNA repair: A partial resolution. *Nat Struct Mol Biol* 2007;14:461–462.
19. Yang, H., Jeffrey, P., Miller, J. et al. BRCA2 function in DNA binding and recombination from a BRCA2-DSS1-ssDNA structure. *Science* 2002;297:1837–1848.
20. Filippo, J.S., Chi, P., Sehorn, M., Etchin, J., Krejci, L., and Sung, P. Recombination mediator and Rad51 targeting activities of a human BRCA2 polypeptide. *J Biol Chem* 2006;281:11649–11657.

21. Collis, S., Tighe, A., Scott, S., Roberts, S., Hendry, J., and Margison, G. Ribozyme minigene-mediated Rad51 down-regulation increases radiosensitivity of human prostate cancer cells. *Nucleic Acids Res* 2001;29:1534–1538.
22. De Schutter, H., and Nuyts, S. Radiosensitizing potential of epigenetic anticancer drugs. *Anticancer Agents Med Chem* 2009;9:99–108.
23. Iliakis, G., Wu, W., and Wang, M. DNA double strand break repair inhibition as a cause of heat radiosensitization: Re-evaluation considering backup pathways of NHEJ. *Int J Hyperthermia* 2008;24:17–29.
24. Frankenberg-Schwager, M., Gebauer, A., and Koppe, C. Single-strand annealing, conservative homologous recombination, nonhomologous end joining, and the cell cycle-dependent repair of DNA double-strand breaks induced by sparsely or densely ionizing radiation. *Radiat Res* 2009;171:265–273.
25. Nomme, J., Renodon-Corniere, A., Asanomi, Y. et al. Design of potent inhibitors of human RAD51 recombinase based on BRC motifs of BRCA2 protein: Modeling and experimental validation of a chimera peptide. *J Med Chem* 2010;53:5782–5791.
26. Liu, Q., Jiang, H., Liu, Z. et al. Berberine radiosensitizes human esophageal cancer cells by downregulating homologous recombination repair protein RAD51. *PLoS One* 2011;6:e23427.
27. Takagi, M., Sakata, K., Someya, M. et al. Gimeracil sensitizes cells to radiation via inhibition of homologous recombination. *Radiother Oncol* 2010;96:259–266.
28. Camphausen, K., and Tofilon, P. Inhibition of Hsp90: A multitarget approach to radiosensitization. *Clin Cancer Res* 2007;13:4326–4330.
29. Noguchi, M., Yu, D., Hirayama, R. et al. Inhibition of homologous recombination repair in irradiated tumor cells pretreated with Hsp90 inhibitor seventeen-allylamino-seventeen-demethoxygeldanamycin. *Biochem Biophys Res Commun* 2006;351:658–663.
30. Chen, G., Yuan, S., Liu, W. et al. Radiation-induced assembly of Rad51 and Rad52 recombination complex requires ATM and c-Abl. *J Biol Chem* 1999;274:12748–12752.
31. Kubler, H., van Randenborgh, H., and Treiber, U. *In vitro* cytotoxic effects of imatinib in combination with anticancer drugs in human prostate cancer cell lines. *Prostate* 2005;63:385–394.
32. Choudhury, A., Zhao, H., Jalali, F. et al. Targeting homologous recombination using imatinib results in enhanced tumor cell chemosensitivity and radiosensitivity. *Mol Cancer Ther* 2009;8:203–213.
33. Choudhury, A., Cuddihy, A., and Bristow, R. Radiation and new molecular agents part I: Targeting ATM-ATR checkpoints, DNA repair, and the proteasome. *Semin Radiat Oncol* 2006;16:51–58.
34. Barker, C., and Powell, S. Enhancing radiotherapy through a greater understanding of homologous recombination. *Semin Radiat Oncol* 2010;20:267–273.
35. Chinnaiyan, P., Huang, S., Vallabhaneni, G. et al. Mechanisms of enhanced radiation response following epidermal growth factor receptor signaling inhibition by erlotinib (Tarceva). *Cancer Res* 2005;65:3328–3335.

36. Li, L., Wang, H., Yang, E., Arteaga, C., and Xia, F. Erlotinib attenuates homologous recombinational repair of chromosomal breaks in human breast cancer cells. *Cancer Res* 2008;68:9141–9146.
37. Peereboom, D., Shepard, D., Ahluwalia, M. et al. Phase II trial of erlotinib with temozolomide and radiation in patients with newly diagnosed glioblastoma multiforme. *J Neuro Oncol* 2010;98(1):93–99.
38. Prados, M., Chang, S., Butowski, N. et al. Phase II study of erlotinib plus temozolomide during and after radiation therapy in patients with newly diagnosed glioblastoma multiforme or gliosarcoma. *J Clin Oncol* 2009;27: 579–584.
39. Brown, P., Krishnan, S., Sarkaria, J. et al. Phase I/II trial of erlotinib and temozolomide with radiation therapy in the treatment of newly diagnosed glioblastoma multiforme: North Central Cancer Treatment Group Study N0177. *J Clin Oncol* 2008;26:5603–5609.
40. Shi, W., Ma, Z., Willers, H. et al. Disassembly of MDC1 foci is controlled by ubiquitin-proteasome-dependent degradation. *J Biol Chem* 2008;283:31608–31616.
41. Adimoolam, S., Sirisawad, M., Chen, J., Thiemann, P., Ford, J., and Buggy, J. HDAC inhibitor PCI-24781 decreases Rad51 expression and inhibits homologous recombination. *Proc Natl Acad Sci U S A* 2007;104:19482–19487.
42. Malanga, M., and Althaus, F. The role of poly(ADP-ribose) in the DNA damage signaling network. *Biochem Cell Biol* 2005;83:354–364.
43. Caldecott, K. Protein-protein interactions during mammalian DNA single-strand break repair. *Biochem Soc Trans* 2003;31:247–251.
44. Noel, G., Godon, C., Fernet, M., Giocanti, N., Mégnin-Chanet, F., and Favaudon, V. Radiosensitization by the poly(ADP-ribose) polymerase inhibitor 4-amino-1,8-naphthalimide is specific of the S phase of the cell cycle and involves arrest of DNA synthesis. *Mol Cancer Ther* 2006;5:564–574.
45. Dungey, F., Loser, D., and Chalmers, A. Replication-dependent radiosensitization of human glioma cells by inhibition of poly(ADP-ribose) polymerase: Mechanisms and therapeutic potential. *Int J Radiat Oncol Biol Phys* 2008;I72:1188–1197.
46. Godon, C., Cordelieres, F., Biard, D. et al. PARP inhibition versus PARP-1 silencing: Different outcomes in terms of single-strand break repair and radiation susceptibility. *Nucleic Acids Res* 2008;36:4454–4464.
47. Starcevic, D., Dalal, S., and Sweasy, J. Is there a link between DNA polymerase beta and cancer? *Cell Cycle* 2004;3:998–1001.
48. Hu, Z., Ma, H., Chen, F., Wei, Q., and Shen, H. XRCC1 polymorphisms and cancer risk: A meta-analysis of 38 case-control studies. *Cancer Epidemiol Biomarkers Prev* 2005;14:1810–1818.
49. Schultz, N., Lopez, E., Saleh-Gohari, N., and Helleday, T. Poly(ADP-ribose) polymerase (PARP-1) has a controlling role in homologous recombination. *Nucleic Acids Res* 2003;31:4959–4964.
50. Bryant, H., Schultz, N., Thomas, H. et al. Specific killing of BRCA2-deficient tumours with inhibitors of poly(ADP-ribose) polymerase. *Nature* 2005;434:913–917.

51. Farmer, H., McCabe, N., Lord, C. et al. Targeting the DNA repair defect in BRCA mutant cells as a therapeutic strategy. *Nature* 2005;434:917–921.
52. Bennardo, N., Cheng, A., Huang, N., and Stark, J. Alternative-NHEJ is a mechanistically distinct pathway of mammalian chromosome break repair. *PLoS Genet* 2008;4:e1000110.
53. Wang, M., Wu, W., Rosidi, B. et al. PARP-1 and ku compete for repair of DNA double strand breaks by distinct NHEJ pathways. *Nucleic Acids Res Res* 2006;34:6170–6182.
54. Hartwell, L., Szankasi, P., Roberts, C., Murray, A., and Friend, S. Integrating genetic approaches into the discovery of anticancer drugs. *Science* 1997;278: 1064–1068.
55. Fong, P., Boss, D., Yap, T. et al. Inhibition of poly(ADP-ribose) polymerase in tumors from BRCA mutation carriers. *N Engl J Med* 2009;361:123–134.
56. Albert, J., Cao, C., Kim, K. et al. Inhibition of poly(ADP-ribose) polymerase enhances cell death and improves tumor growth delay in irradiated lung cancer models. *Clin Cancer Res* 2007;13:3033–3042.
57. Calabrese, C., Almassy, R., Barton, S. et al. Anticancer chemosensitization and radiosensitization by the novel poly(ADP-ribose) polymerase-1 inhibitor AG14361. *J Natl Cancer Inst* 2004;96:56–67.
58. Yap, T., Boss, D., Fong, P. et al. First in human phase I pharmacokinetic (PK) and pharmacodynamic (PD) study of KU-0059436 (Ku), a small molecule inhibitor of poly ADP-ribose polymerase (PARP) in cancer patients (p), including BRCA1/2 mutation carriers. *J Clin Oncol* 2007;25:3529.
59. Tuma, R. Combining carefully selected drug, patient genetics may lead to total tumor death. *J Natl Cancer Inst* 2007;99:1505–1509.
60. Schlicker, A., Peschke, P., Burkle, A., Hahn, E., and Kim, J. Four-amino-1,8-naphthalimide: A novel inhibitor of poly(ADP-ribose) polymerase and radiation sensitizer. *Int J Radiat Biol* 1999;75:91–100.
61. Kastan, M., and Bartek, J. Cell-cycle checkpoints and cancer. *Nature* 2004;432:316–323.
62. Dungey, F., Caldecott, K., and Chalmers, A. Enhanced radiosensitization of human glioma cells by combining inhibition of poly(ADP-ribose) polymerase with inhibition of heat shock protein 90. *Mol Cancer Ther* 2009;8:2243–2254.
63. Löser, D., Shibata, A., Shibata, A., Woodbine, L., Jeggo, P., and Chalmers, A. Sensitization to radiation and alkylating agents by inhibitors of poly(ADP-ribose) polymerase is enhanced in cells deficient in DNA double-strand break repair. *Mol Cancer Ther* 2010;9(6):1775–1787.
64. Russo, A., Kwon, H., Burgan, W. et al. *In vitro* and *in vivo* radiosensitization of glioblastoma cells by the poly (ADP-ribose) polymerase inhibitor E7016. *Clin Cancer Res* 2009;15:607–612.
65. Donawho, C., Luo, Y., Penning, T. et al. ABT-888, an orally active poly(ADP-ribose) polymerase inhibitor that potentiates DNA-damaging agents in preclinical tumor models. *Clin Cancer Res* 2007;13:2728–2737.
66. Liu, S., Coackley, C., Kraused, M., Jalalic, F., Chan, N., and Bristow, R. A novel poly(ADP-ribose) polymerase inhibitor, ABT-888, radiosensitizes malignant human cell lines under hypoxia. *Radiother Oncol* 2008;88:258–268.

67. Khan, K., Araki, K., Wang, D. et al. Head and neck cancer radiosensitization by the novel poly(ADP-ribose) polymerase inhibitor GPI-15427. *Head Neck* 2010;32:381–391.
68. Chalmers, A., Lakshman, M., Chan, N., and Bristow, R. Poly(ADP-ribose) polymerase inhibition as a model for synthetic lethality in developing radiation oncology targets. *Semin Radiat Oncol* 2010;20:274–281.
69. Khan, K., Araki, K., Wang, D. et al. Head and neck cancer radiosensitization by the novel poly(ADP-ribose) polymerase inhibitor GPI-15427. *Head Neck* 2010;32(3):381–391.
70. Chan, N., Milosevic, M., and Bristow, R. Tumor hypoxia, DNA repair and prostate cancer progression: New targets and new therapies. *Future Oncol* 2007;3:329–341.
71. Chan, N., Koritzinsky, M., Zhao, H. et al. Chronic hypoxia decreases synthesis of homologous recombination proteins to offset chemoresistance and radioresistance. *Cancer Res* 2008;68:605–614.
72. Tentori, L., Leonetti, C., Scarsella, M. et al. Inhibition of poly(ADP-ribose) polymerase prevents irinotecan-induced intestinal damage and enhances irinotecan/temozolomide efficacy against colon carcinoma. *FASEB J* 2006;20:1709–1711.
73. Chalmers, A., Ruff, E., Martindale, C., Lovegrove, N., and Short, S. Cytotoxic effects of temozolomide and radiation are additive- and schedule-independent. *Int J Radiat Oncol Biol Phys* 2009;75(5):1511–1519.
74. Gregory, P., Wagner, K., and Horz, W. Histone acetylation and chromatin remodeling. *Exp Cell Res* 2001;265:195–202.
75. Lane, A., and Chabner, B. Histone deacetylase inhibitors in cancer therapy. *J Clin Oncol* 2009;27:5459–5468.
76. Lee, D., Hayes, J., and Pruss, D. A positive role for histone acetylation in transcription factor access to nucleosomal DNA. *Cell* 1993;72:73–84.
77. Escargueil, A., Soares, D., Salvador, M., Larsen, A., and Henriques, J. What histone code for DNA repair? *Mutat Res* 2008;658:259–270.
78. Kim, G., Choi, Y., Dimtchev, A., Jeong, S., Dritschilo, A., and Jung, M. Sensing of ionizing radiation-induced DNA damage by ATM through interaction with histone deacetylase. *J Biol Chem* 1999;274:31127–31130.
79. Luo, J., Su, F., Chen, D., Shiloh, A., and Gu, W. Deacetylation of p53 modulates its effect on cell growth and apoptosis. *Nature* 2000;408:377–381.
80. Zhang, Y., Adachi, M., Zhao, X., Kawamura, R., and Imai, K. Histone deacetylase inhibitors FK228, N-(2-aminophenyl)-4-[N-(pyridin-3-yl-methoxycarbonyl) amino-methyl]benzamide and m-carboxycinnamic acid bishydroxamide augment radiation-induced cell death in gastrointestinal adenocarcinoma cells. *Int J Cancer* 2004;110:301–308.
81. Camphausen, K., Burgan, W., Cerra, M. et al. Enhanced radiation-induced cell killing and prolongation of H2AX foci expression by the histone deacetylase inhibitor MS-275. *Cancer Res* 2004;64:316–321.
82. Chinnaiyan, P., Vallabhaneni, G., Armstrong, E., Huang, S., and Harari, P. Modulation of radiation response by histone deacetylase inhibition. *Int J Radiat Oncol Biol Phys* 2005;62:223–229.

83. Munshi, A., Kurland, J., Nishikawa, T. et al. Histone deacetylase inhibitors radiosensitize human melanoma cells by suppressing DNA repair activity. *Clin Cancer Res* 2005;11:4912–4922.
84. Karagiannis, T., and El-Osta, A. Modulation of cellular radiation responses by histone deacetylase inhibitors. *Oncogene* 2006;25:3885–3893.
85. Nakagawa, M., Oda, Y., Eguchi, T. et al. Expression profile of class I histone deacetylases in human cancer tissues. *Oncol Rep* 2007;18:769–774.
86. Kelly, W., O'Connor, O., and Marks, P. Histone deacetylase inhibitors: From targets to clinical trials. *Expert Opin Investig Drugs* 2002;11:1695–1713.
87. Marks, P., Richon, V., Miller, T., and Kelly, W. Histone deacetylase inhibitors. *Adv Cancer Res* 2004;91:137–168.
88. Kim, M., Kwon, H., Lee, Y. et al. Histone deacetylases induce angiogenesis by negative regulation of tumor suppressor genes. *Nat Med* 2001;7:437–443.
89. Butler, L., Zhou, X., Xu, W. et al. The histone deacetylase inhibitor SAHA arrests cancer cell growth, up-regulates thioredoxin-binding protein-2, and down-regulates thioredoxin. *Proc Natl Acad Sci U S A* 2002;99:11700–11705.
90. Atadja, P., Gao, L., Kwon, P. et al. Selective growth inhibition of tumor cells by a novel histone deacetylase inhibitor, NVP-LAQ824. *Cancer Res* 2004;64:689–695.
91. Finnin, M., Donigian, J., Cohen, A. et al. Structures of a histone deacetylase homologue bound to the TSA and SAHA inhibitors. *Nature* 1999;401:188–193.
92. Beckers, T., Burkhardt, C., Wieland, H. et al. Distinct pharmacological properties of second generation HDAC inhibitors with the benzamide or hydroxamate head group. *Int J Cancer* 2007;121:1138–1148.
93. Prakash, S., Foster, B., Meyer, M. et al. Chronic oral administration of CI-994: A phase I study. *Invest New Drugs* 2001;19:1–11.
94. Frey, R., Wada, C., Garland, R. et al. Trifluoromethyl ketones as inhibitors of histone deacetylase. *Bioorg Med Chem Lett* 2002;12:3443–3447.
95. Richon, V., Emiliani, S., Verdin, E. et al. A class of hybrid polar inducers of transformed cell differentiation inhibits histone deacetylases. *Proc Natl Acad Sci U S A* 1998;95:3003–3007.
96. Furumai, R., Matsuyama, A., Kobashi, N. et al. FK228 (depsipeptide) as a natural prodrug that inhibits class I histone deacetylases. *Cancer Res* 2002;62:4916–4921.
97. Singh, S., Zink, D., Liesch, J. et al. Structure and chemistry of apicidins, a class of novel cyclic tetrapeptides without a terminal alpha-keto epoxide as inhibitors of histone deacetylase with potent antiprotozoal activities. *J Org Chem* 2002;67:815–825.
98. Ma, Z., Ezzeldin, H., and Diasio, R. Histone deacetylase inhibitors: Current status and overview of recent clinical trials. *Drugs* 2009;69:1911–1934.
99. Dote, H., Cerna, D., Burgan, W. et al. Enhancement of *in vitro* and *in vivo* tumor cell radiosensitivity by the DNA methylation inhibitor zebularine. *Clin Cancer Res* 2005;11:4571–4579.
100. Subramanian, C., Opipari, A., Bian, X., Castle, V., and Kwok, R. Ku70 acetylation mediates neuroblastoma cell death induced by histone deacetylase inhibitors. *Proc Natl Acad Sci U S A* 2005;102:4842–4847.

101. Zhang, Y., Carr, T., Dimtchev, A., Zaer, N., Dritschilo, A., and Jung, M. Attenuated DNA damage repair by trichostatin A through BRCA1 suppression. *Radiat Res* 2007;168:115–124.
102. Munshi, A., Tanaka, T., Hobbs, M., Tucker, S., Richon, V., and Meyn, R. Vorinostat, a histone deacetylase inhibitor, enhances the response of human tumor cells to ionizing radiation through prolongation of gamma-H2AX foci. *Mol Cancer Ther* 2006;5:1967–1974.
103. Xu, W., Parmigiani, R., and Marks, P. Histone deacetylase inhibitors: Molecular mechanisms of action. *Oncogene* 2007;26:5541–5552.
104. Bolden, J., Peart, M., and Johnstone, R. Anticancer activities of histone deacetylase inhibitors. *Nat Rev Drug Discov* 2006;5:769–784.
105. Edwards, A., Li, J., and Atadja, P. Effect of the histone deacetylase inhibitor LBH589 against epidermal growth factor receptor-dependent human lung cancer cells. *Mol Cancer Ther* 2007;6:2515–2524.
106. Deroanne, C., Bonjean, K., Servotte, S. et al. Histone deacetylases inhibitors as anti-angiogenic agents altering vascular endothelial growth factor signaling. *Oncogene* 2002;21:427–436.
107. Goh, M., Chen, F., Paulsen, M., Yeager, A., Dyer, E., and Ljungman, M. Phenylbutyrate attenuates the expression of Bcl-X(L), DNA-PK, caveolin-1, and VEGF in prostate cancer cells. *Neoplasia* 2001;3:331–338.
108. Sah, N., Munshi, A., Hobbs, M., Carter, B., Andreeff, M., and Meyn, R. Effect of downregulation of survivin expression on radiosensitivity of human epidermoid carcinoma cells. *Int J Radiat Oncol Biol Phys* 2006;66:852–859.
109. Chung, Y., Wang, A., and Yao, L. Antitumor histone deacetylase inhibitors suppress cutaneous radiation syndrome: Implications for increasing therapeutic gain in cancer radiotherapy. *Mol Cancer Ther* 2004;3:317–325.
110. Stoilov, L., Darroudi, F., Meschini, R., van der Schans, G., Mullenders, L., and Natarajan, A. Inhibition of repair of X-ray-induced DNA double-strand breaks in human lymphocytes exposed to sodium butyrate. *Int J Radiat Biol* 2000;76:1485–1491.
111. Purrucker, J., Fricke, A., Ong, M., Rube, C., Rube, C., and Mahlknecht, U. HDAC inhibition radiosensitizes human normal tissue cells and reduces DNA double-strand break repair capacity. *Oncol Rep* 2010;23:263–269.
112. Banuelos, C., Banath, J., MacPhail, S., Zhao, J., Reitsema, T., and Olive, P. Radiosensitization by the histone deacetylase inhibitor PCI-24781. *Clin Cancer Res* 2007;13:6816–6826.
113. Chinnaiyan, P., Cerna, D., Burgan, W. et al. Postradiation sensitization of the histone deacetylase inhibitor valproic acid. *Clin Cancer Res* 2008;14:5410–5415.
114. Chen, X., Wong, J., Wong, P., and Radany, E. Low-dose valproic acid enhances radiosensitivity of prostate cancer through acetylated p53-dependent modulation of mitochondrial membrane potential and apoptosis. *Mol Cancer Res* 2011;9:448–461.
115. Camphausen, K., Cerna, D., Scott, T. et al. Enhancement of *in vitro* and *in vivo* tumor cell radiosensitivity by valproic acid. *Int J Cancer* 2005;114:380–386.

116. Baschnagel, A., Russo, A., Burgan, W. et al. Vorinostat enhances the radiosensitivity of a breast cancer brain metastatic cell line grown *in vitro* and as intracranial xenografts. *Mol Cancer Ther* 2009;8(6):1589–1595.
117. Mueller, S., Yang, X., Sottero, T. et al. Cooperation of the HDAC inhibitor vorinostat and radiation in metastatic neuroblastoma: Efficacy and underlying mechanisms. *Cancer Lett* 2011;306:223–229.
118. Cuneo, K., Fub, A., Osusky, K., Huamanib, J., Hallahan, D., and Geng, L. Histone deacetylase inhibitor NVP-LAQ824 sensitizes human non-small cell lung cancer to the cytotoxic effects of ionizing radiation. *Anti-Cancer Drugs* 2007;18:793–800.
119. Geng, L., Cuneo, K., Fu, A., Tu, T., Atadja, P., and Hallahan, D. Histone deacetylase (HDAC) inhibitor LBH589 increases duration of γ-H2AX foci and confines HDAC4 to the cytoplasm in irradiated non-small cell lung cancer. *Cancer Res* 2006;66(23):11298–11304.

CHAPTER 10

Targeting Growth Factor Receptors for Radiosensitization

10.1 EPIDERMAL GROWTH FACTOR FAMILY RECEPTORS

The ErbB family of receptor tyrosine kinases (RTKs), consists of four transmembrane tyrosine kinases: (EGFR)/(ErbB1)/HER1, HER2/(ErbB2), HER3 (ErbB3), and HER4 (ErbB4). The ErbB family receptors and ligands are shown in Figure 10.1. Activation of the ErbB family initiates a number of signaling pathways that ultimately affect fundamental processes including cell division, survival, and cell to cell interactions. Ligands for the ErbB receptors include epidermal growth factor (EGF), transforming growth factor (TGF), and amphiregulin binding to ErbB1; epiregulin and heparin-binding EGF-like growth factor binding to ErbB1 and ErbB4; and neuregulins 1 to 4 binding to ErbB3 and ErbB4 (Figure 10.1) [1]. Ligand binding results in ErbB receptor dimerization and the autophosphorylation of specific tyrosine residues of the intracellular tyrosine kinase domain, which in turn activates signaling pathways including those mediated by Ras, phosphatidylinositol-3-kinase (PI3K), signal transducer and activator of transcription 3 (STAT3), protein kinase C (PKC), and phospholipase D [1]. Signaling through the ErbB receptor family is dependent on the formation of homodimeric or heterodimeric combinations, with the biological activity of the ErbB receptors coming primarily from the

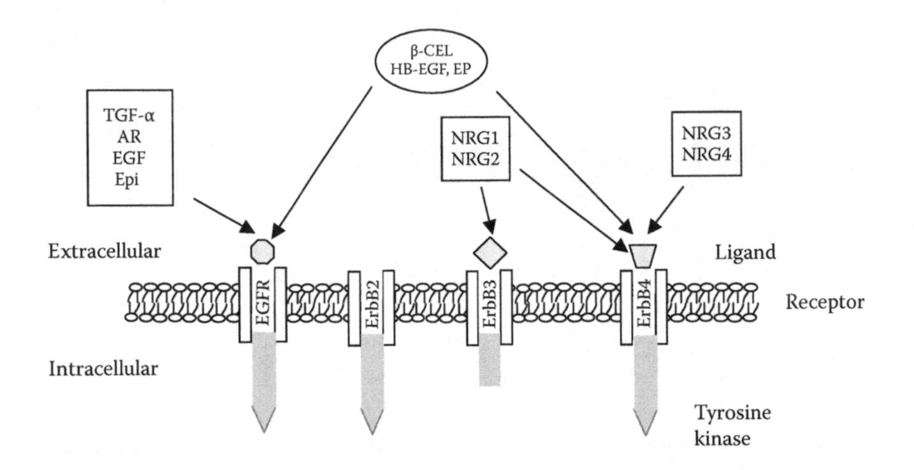

FIGURE 10.1 The ErbB/EGFR superfamily receptors and ligands. Four structurally related receptors are included. Of the four receptors in this family, three have intrinsic tyrosine kinase activity to allow downstream activation of pathways. ErbB3 lacks intrinsic tyrosine kinase activity; relying on phosphorylation through ErbB1 or ErbB2 heterodimerization. Currently, no ligands are identified for ErbB2, but it can form heterodimers with the remaining three receptors. AR, amphiregulin; β-CEL, β-cellulin; EP, epigen; Epi, epiregulin; HB-EGF, heparin-binding-EGF; NRG, neuregulin; TGF-α, transforming growth factor-α.

ErbB heterodimers with homodimers having significantly less activity [1,2]. The individual dimers are selective for activating the various signaling cascades [1]. In contrast to this ligand specificity of ErbB receptor activation, ionizing radiation is promiscuous, inducing the activation of all ErbB receptors within minutes of exposure to doses in the range of 2 Gy [3,4]. Because the ErbB receptors are important in tumor cell survival, their rapid response to radiation indicates a protective role in the cellular radioresponse and strongly suggests RTKs as a target for radiosensitization [3,4].

10.1.1 Receptor Activation by Ionizing Radiation

The mechanisms by which radiation activates transmembrane receptors and intracellular signaling pathways continue to be studied in a variety of experimental settings. There seem to be several cooperative processes, arising from responses to DNA damage as well as from distal consequences of ionizing events. It has been demonstrated for different tumor cell types *in vitro* that the EGF receptor (EGFR) is rapidly activated in response to radiation (reviewed by Valerie et al. [5]). Low-dose clinically relevant radiation exposure (1–2 Gy) activates ErbB1 and, by heterodimerization, other members of the ErbB receptor family (ErbB2, ErbB3, and ErbB4). Activation of ErbB1, ErbB2, and ErbB3 has been linked to downstream activation of intracellular signaling, including the RAF-1–MEK2–ERK1/2 and the PI3K phosphoinositide-dependent kinase-1–AKT pathways.

There are at least two mechanisms by which radiation exposure could activate plasma membrane receptors and intracellular signaling pathways. First, radiation causes DNA damage, which activates ATM/ATM and Rad3-related protein that, in turn, promotes the activation of receptors/intracellular signaling pathways and stimulates cell cycle checkpoints, p53 activity, and DNA repair complex function. Second, radiation generates ionizing events in the water of the cytosol that are amplified, probably through the mediation of mitochondria and which generate large amounts of reactive oxygen species (ROS) and reactive nitrogen species (RNS), one of the actions of which is to inhibit protein tyrosine phosphatase (PTPase) activities. In addition, radiation activates acidic sphingomyelinase and increases the production of ceramide. Inhibition of PTPases leads to a general derepression (activation) of receptor and nonreceptor tyrosine kinases and the activation of downstream signal transduction pathways. Radiation-induced ceramide has been shown to promote membrane-associated receptor activation by facilitating the clustering of receptors within lipid rafts [6].

How does EGFR become activated so quickly? The activity of tyrosine kinases and proteins regulated by tyrosine phosphorylation (e.g., ErbB1 and RAF-1) is held in check by the actions of PTPases. The relative activity of a PTPase is approximately one order of magnitude higher than that of the substrate (i.e., kinase) it dephosphorylates. PTPase activity is sensitive to oxidation or nitrosylation (or both) of a key cysteine residue in the active site, and thus, any agent that generates ROS or RNS has the potential to promote decreased PTPase activity and, thereby, increased tyrosine phosphorylation of multiple proteins [7]. Ionizing radiation induces small amounts of ROS by direct interaction with water, and these reactive species are magnified in a Ca^{2+}-dependent manner by mitochondria, generating more ROS and RNS, which act to inhibit multiple PTPase activities (Figure 10.2).

Initial radiation-induced activation of the ErbB receptors occurs approximately 0 to 10 min after exposure. However, the same receptors are reactivated approximately 60 to 180 min after irradiation. The primary

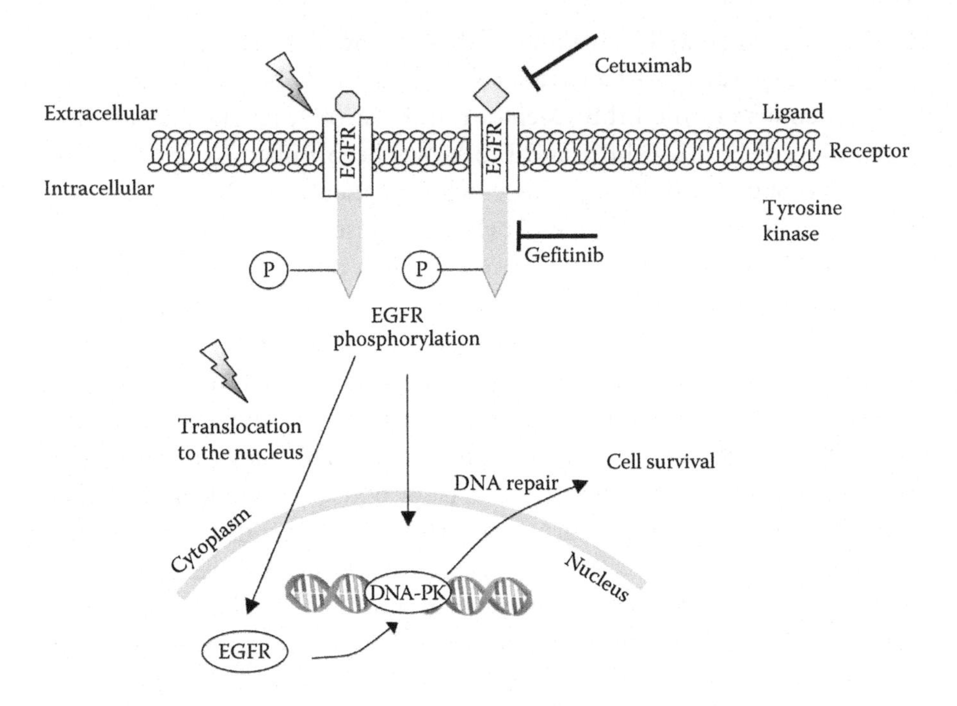

FIGURE 10.2 The effect of radiation and chemotherapy on EGFR signaling. After exposure to radiation or certain chemotherapeutic agents, epidermal growth factor receptor (EGFR) activates downstream signaling pathways promoting cell survival. Radiation stimulates the pathways activated by epidermal growth factor (EGF) and in addition can the translocation of phosphorylated EGFR (pEGFR) into the nucleus. This process coincides with the transport of Ku70/80 and protein phosphatase 1 into the nucleus resulting in increases in DNA-dependent protein kinase (DNAPK) levels, the repair of DNA-strand breaks, and cell survival. Cetuximab, but not gefitinib, blocks the nuclear transport of pEGFR.

mode of receptor activation at these later times occurs by paracrine/autocrine signaling [8]. The initial activation of ErbB1 and the ERK1/2 pathway was directly responsible for the cleavage, release, and functional activation of presynthesized paracrine ligands, such as transforming growth factor α (TGF-α), which feeds back to the irradiated tumor cell and potentially *in vivo* to unirradiated distal tumor cells, reenergizing the signaling system [9]. Increasing the radiation dose from 2 Gy up to 10 Gy enhances both the amplitude and duration of the secondary activation of ErbB1 and the secondary activation of the intracellular signaling pathways, suggesting that radiation can promote a dose-dependent increase in the cleavage of pro-TGF-α that reaches a plateau at approximately 10 Gy [8,9]. In contrast to the secondary receptor and pathway activations, primary receptor and signaling pathway activations seem to plateau at 3 to 5 Gy.

These observations suggest that ionizing radiation (IR) initiates a self-perpetuating loop of activity whereby radiation generates ROS/RNS signals within tumor cells that promote activation of growth factor receptors and signaling pathways, which in turn can promote the release of paracrine ligands from cells, leading to the reactivation of receptors and intracellular signaling pathways.

Autophosphorylation of EGFR-TK initiates a cascade of intracellular signaling pathways that play an important role in the regulation of cell proliferation, survival, adhesion, migration, and differentiation [10]. These pathways include the Ras-mitogen-activated protein kinase (MAPK) cascade, the PI3K cascade, the STAT cascade, as well as the PKC and phospholipase D cascades. In addition, activation of EGFR signaling can be mediated by ionizing radiation independent of ligand binding [10]. Exposure of tumor cells overexpressing EGFR to ionizing radiation activates survival and proliferation mechanisms predominantly through stimulated signaling via PI3K-AKT and Ras-MAPK [11,12]. Activation of these two pathways is thought to be the main causes of radioresistance of EGFR overexpressing tumors [10,12].

10.1.2 Overexpression Receptor RTKs on Cancer Cells

The aberrant expression of receptor RTKs, particularly ErbB1 and ErbB2, is associated with malignant transformation and tumor cell survival. There are several mechanisms that could account for aberrant EGFR activation in cancer. The most common mutant form of EGFR is EGFRvIII, which contains a large deletion in the extracellular domain leading to constitutive activation in the absence of ligand binding [13]. EGFR can also be dysregulated by other mechanisms including receptor overexpression, ligand-independent activation, and autocrine activation [13]. These changes lead to increased proliferation, angiogenesis, invasion and metastasis, and inhibition of apoptosis. EGFR-overexpressing cell lines have demonstrated radioresistance when compared with cells expressing normal levels of EGFR. Overexpression of EGFR occurs in more than 90% of cases of head and neck squamous cell carcinoma (HNSCC), and EGFR overexpression correlates negatively with locoregional control in HNSCC treated with radiotherapy (RT) [14,15]. Across several tumor types, EGFR overexpression has consistently been shown to be an indicator of poor prognosis, more advanced stage at presentation, and lower overall survival [15].

10.1.3 Mechanisms of Radiosensitization by EGFR Inhibition

An abundance of preclinical investigations have demonstrated augmentation of radiation response after the inhibition of EGFR signaling. Proposed mechanisms underlying this radiation-response enhancement include specific effects on cell cycle distribution, apoptosis, necrosis, DNA damage repair, angiogenesis, tumor cell motility, invasion, and metastatic capacity [16,17]. The persistence of reactive oxygen and nitrogen species can inhibit tyrosine phosphatase activities, thereby prolonging the phosphorylative activation of receptor and nonreceptor tyrosine kinases with concomitant effects on downstream signaling. Radiation can also activate acidic sphingomyelinase, resulting in increased production of ceramide.

10.1.3.1 Direct Inactivation of Tumor Cells

Recurrences after high dose irradiation often occur from a few surviving clonogenic cells. Thus, additional killing of a low number of clonogenic cells by a further treatment would be expected to have additive effects on local tumor control. Overall, evidence has accumulated supporting the view that direct inactivation of clonogenic tumor cells by anti-EGFR monoclonal antibodies (mAbs) very probably contributes to improved local tumor control after irradiation. For small-molecule TK inhibitors, however, the limited data available does not support this mechanism [18].

10.1.3.2 Cellular Radiosensitization through Modified Signal Transduction

As already described, autophosphorylation of EGFR-TK initiates a cascade of intracellular signaling pathways that play an important role in the regulation of cell proliferation, survival, adhesion, migration, and differentiation [10]. These pathways include the Ras-MAPK cascade, the PI3K cascade, the STAT cascade, as well as the PKC and phospholipase D cascades. Importantly, activation of EGFR signaling can also be mediated by ionizing radiation independent of ligand binding [3,10,17]. In fact, exposure of tumor cells overexpressing EGFR to ionizing radiation activates survival and proliferation mechanisms predominantly through stimulated signaling via PI3K-AKT and Ras-MAPK [3,16,19], and activation of these two pathways is thought to be an important cause of radioresistance of EGFR overexpressing tumors.

The signaling cascade comprising PI3K and the serine/threonine kinase, AKT, is a key [20] cytoprotective and survival mechanism for cancer cells [5,21,22]. It is one of the most frequently altered pathways in solid tumors, owing to aberrant activation of the receptor TK, mutation of Ras, elevation of PI3K/AKT protein level/activity, or loss of function of PTEN (phosphatase and tensin homologue deleted on chromosome 10), a PI3K antagonist [22]. Because EGFR activity is inducible by IR, even in the absence of ligand binding, PI3K/AKT signaling can play an important role not only in intrinsic but also in acquired tumor radioresistance.

10.1.3.3 Inhibition of DNA Repair

In mammalian cells, DSBs are mainly repaired through nonhomologous end-joining (NHEJ), which directly rejoins the two broken DNA ends using the key proteins Ku70/Ku80, the catalytic subunit DNA-PKcs, and the ligase IV–XRCC4 complex. EGFR has been shown to interact with DNA-PKcs as well as Ku70 and Ku80, and this interaction is enhanced by irradiation and by other stress signals [23]. The majority of these complexes are found in the nucleus and, in fact, EGFR can be bound to DNA and act either as a transcription factor or as a cofactor of DNA repair. There seems to be a cross-regulation between EGFR and DNA-PKcs, so that those cell lines showing an elevated level of EGFR are also characterized by a high level of DNA-PKcs [24]. Normally, the major part of DNA-Pkcs and of Ku70/80 are located in the nucleus. However, when EGFR is blocked by a specific antibody or tyrosine kinase inhibitor (TKI), a substantial amount of DNA-PKcs is retained in the cytoplasm due to a stalled nuclear import of the EGFR complex (Figure 10.3) [16,23,25]. In irradiated cells, this reduction in the nuclear activity of the DNA-PK complex leads to a depressed rate of repair as indicated by a reduced phosphorylation at Thr-2609 [25,26], which is essential for NHEJ. As a consequence, the kinetics of DSB repair is slowed down, finally resulting in an enhanced number of residual DSBs [25,26]. This effect is particularly observed for Ras-mutated tumor cell lines, most likely because of the specifically enhanced activation of the EGFR–PI3K–AKT pathway in these cells [26,27].

DNA repair might also be affected by EGFR inhibition via its interaction with XRCC1. This protein is known to be a cofactor of DNA-PKcs, but also to participate in base excision repair (BER), which is required for the

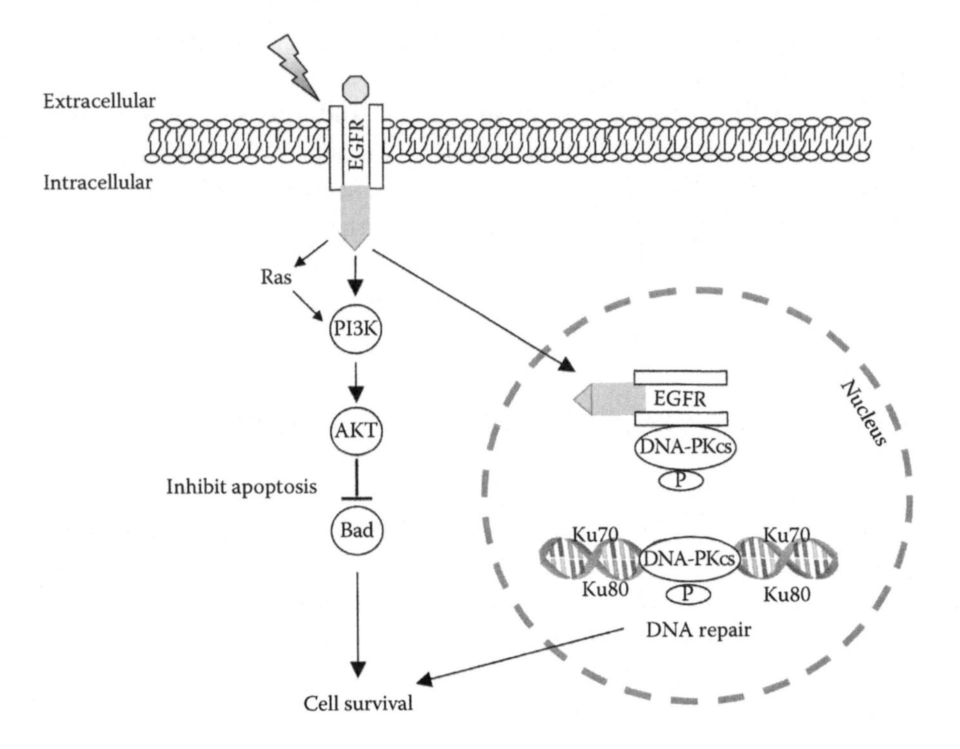

FIGURE 10.3 Signaling through EGFR and the PI3K/AKT pathways. Stimulation of the transmembrane EGFR and its associated TK by ligand binding or by IR leads to increased PI3K activity, either directly or through Ras. PI3K activates the serine/threonine protein kinase Akt. Activated Akt phosphorylates and regulates the function of many cellular proteins involved in prosurvival processes, including the suppression of IR-induced apoptosis before or after cell cycle arrest. In addition, IR causes a rapid nuclear translocation of EGFR independently of growth factor binding, resulting in its interaction with DNA-PK and the promotion of DSB repair.

repair of x-ray-induced single-strand lesions. The expression of XRCC1 is linked to EGFR, and a blockade of this receptor leads to downregulation of XRCC1 [27]. When BER is suppressed, there is an enhanced chance for clustered base damage to be converted into additional DSBs [20]. These effects might also contribute to the elevated number of residual DSBs after EGFR blockade in X-irradiated cells. There is extensive experimental data to support the contribution of DNA-DSB repair to the effects of EGFR inhibitors with predominant roles assigned to the repair proteins DNA-PKcs, Ku70, and Ku80.

10.1.3.4 Cell Cycle Effects

Both mAbs to EGFR and TKIs block tumor cells in the G_0/G_1 phase of the cell cycle [16]. After combined radiation–EGFR inhibitor treatment, the cell cycle may be prolonged by a radiation-induced G_2 block. Both these effects would contribute to a decrease of the radioresistant S phase fraction. which would sensitize tumor cells. However, the opposite effect would result from an increase in the proportion of radioresistant G_0/G_1 cells. Overall, the biological consequences of cell cycle perturbation by EGFR inhibitors during combined treatment schedules are unclear.

10.1.3.5 Inhibition of Proliferation and Repopulation

Antiproliferative effects of EGFR-TKIs and mAbs have been consistently demonstrated in preclinical experiments (reviewed in Baumann and Krause [17]). Inhibition of proliferation would be expected to lead to palliative effects of the treatment, that is, to tumor regression, prolongation of the time to recurrence, or slower tumor growth. In combination with radiotherapy, these effects might also translate into improved tumor cure if the repopulation of cancer stem cells, which can occur during fractionated radiotherapy, is affected. Preclinical data suggest that repopulation may be associated with an increase of EGFR expression, although this finding is not consistent for all tumor models [28]. Shortening of overall treatment time reduces the time available for repopulation, and has been demonstrated to improve local tumor control (reviewed by Baumann et al. [29]). Clinical data on patients with HNSCC show a correlation of a high EGFR expression with worse local control after long compared with short overall treatment times [30]. These data suggest that the EGFR may be involved in accelerated repopulation of clonogenic tumor cells and that EGFR inhibition might counteract this mechanism of radioresistance.

10.1.3.6 Effects on the Tumor Microenvironment

Hypoxic tumor and normal tissue cells have been shown to increase EGFR synthesis, and this increase was not abolished after the reoxygenation of the cells [31]. This effect may be tumor-dependent because upregulation of EGFR under hypoxic conditions was not confirmed in other experiments.

In addition to a possible direct link between hypoxia and EGFR, other interactions with angiogenesis exist as indicated by the antiangiogenic activity of EGFR inhibitors. *In vitro*, a reduced expression of the vascular endothelial growth factor (VEGF) protein has been reported after incubation with the anti-EGFR mAb C225 [32] or the TKI gefitinib [33]. *In vivo*, a significantly reduced blood vessel density was shown after mAb application [32] as well as after TK inhibition [33]. Both effects were also apparent after the combination of EGFR inhibitors with irradiation [16]. Theoretically, antiangiogenic effects could either improve or impair tumor oxygenation, and thereby modulate the effects of radiotherapy in opposite directions. In one experimental study using A431 tumors, the application of EGFR-TKI gefitinib without irradiation led to a significantly increased tumor oxygenation, measured by a reduction of the uptake of the PET hypoxia marker [18F]fluoroazomycin arabinoside (FAZA) [34,35]. Similarly, simultaneous application of the mAb C225 during fractionated irradiation improved reoxygenation of FaDu cancer stem cells *in vivo*, contributing to improved local control of FaDu tumors after combined treatment [35]. Overall, anti-EGFR mAbs as well as TKIs have been shown to express antiangiogenic activity in tumors, and a limited amount of experimental data suggests improved tumor oxygenation after EGFR-TK inhibition.

10.1.3.7 Prognostic Parameters for the Antitumor Effects of EGFR Inhibitors Alone or in Combination with Radiation

Response to anti-EGFR monotherapy, whether with RTK inhibitors (RTKIs) or with mAbs, varies between tumors and different types of disease and in some cases may be only marginal. Tumor characteristics that have been found to be indicators of the extent to which the tumor will respond to agents targeting EGFR include:

- *Overexpression of EGFR.* Elevated EGFR phosphorylation and signaling have been linked to tumor resistance to radiotherapy.
- *Mutation status of EGFR.* Several investigators have demonstrated that mutations or the number of gene copies of the EGFR correlated with response rate or survival after application of EGFR inhibitors alone or in combination with chemotherapy. In terms of mutated EGFR, preclinical studies using transgenic cells indicate that the type III EGFR variant (EGFRvIII) confers significant radioresistance to tumors, whereas dominant-negative tumor cells show a higher radiosensitivity [36].
- *Level of expression of other ErbB receptors.* Intracellular cross talk and the availability of alternate pathways are a complicating factor for the assessment of predictive parameters. For example, tumor cells after specific blockade of the EGFR-TK have been shown to

be able to form heterodimers between EGFR and cErbB2, which, despite the inhibition of the EGFR-TK, lead to inactivation of downstream PI3K/AKT signal transduction but not of the ras/raf/MAPK pathway [37]. This is in line with observations showing that cErbB2 overexpressing tumor cells are more resistant to EGFR-TKI [38], suggesting that simultaneous inhibition of several cErb receptors might be an approach to enhance the therapeutic effect.

- *The expression, activity, and mutations of molecules downstream of the EGFR may affect the efficacy of EGFR inhibitors.* Studies with a panel of tumor cell lines showed that Ras mutations could enhance the radiosensitizing but not the antiproliferative effect of EGFR-TKI *in vitro.* This supports the concept that EGFR inhibitors, which have not been proven useful in Ras-mutated tumors when given alone or combined with chemotherapy, may play an important role in modifying the effects of irradiation. There is a need for radiotherapy-specific experiments need to supplement standard preclinical evaluation procedures.

10.1.3.8 Effects on Normal Tissue

Studies in knockout mice and preclinical toxicology studies have shown that the major effects of inhibiting EGFR are shown by skin and gastrointestinal systems. Clinical studies using EGFR inhibitors such as cetuximab, gefitinib, and erlotinib have similarly shown skin and gastrointestinal toxicities [39]. In response to irradiation, activation of EGFR has been demonstrated in normal fibroblasts *in vitro* [40]. In oral mucosa, however, daily fractionated irradiation resulted in a decrease of the EGFR mRNA as assessed by reverse transcription polymerase chain reaction (RT-PCR), indicating the stabilization of the protein after irradiation. The role of these changes in EGFR transcription and protein in normal tissues is unclear. In parallel with what occurs in tumors, increased EGFR protein levels might be involved in the regulation of radiation-induced repopulation processes in turnover tissues; an important aspect of radiation tolerance during fractionated irradiation.

10.1.4 EGFR Inhibitors: mAbs and RTKIs

The preceding description of the role of EGFR in the radiation response referred frequently to results observed when EGFR was inhibited, and to the two approaches to EGFR inhibition that have been investigated in the laboratory and the clinic. In the first of these, mAbs are directed against the receptor to prevent ligands from binding to the extracellular domain of the receptor. This method prevents receptor dimerization and activation and ultimately induces receptor degradation. Several EGFR antibodies have been developed and are in various phases of clinical testing. The second approach to disrupting EGFR function involves the use of small-molecule TKIs, which bind to the ATP-binding pocket of the cytoplasmic domain of the receptor, preventing receptor phosphorylation and, ultimately, blocking downstream signaling cascades. Small molecules and antibody inhibitors of EGFR differ in the mechanism of action and pharmacology, and display different activity and toxicity profiles.

10.1.4.1 Monoclonal Antibodies

Monoclonal antibodies designed to target EGFR may be subclassified into those that are naked and those that are bispecific (binding two epitopes). The mechanisms of action of anti-EGFR mAbs include blockade of natural ligand binding, prevention of receptor activation and dimerization, and induction of receptor internalization and downregulation. Apart from their mechanisms of action, antibodies differ from small molecules in being highly specific to the target and having longer half-lives (days to 2–3 weeks). Cetuximab and panitumumab are two of the anti-EGFR mAbs that have been approved for cancer therapy. Cetuximab is a chimeric IgG_1 mAb, which is approved for the treatment of patients with squamous cell carcinoma of the head and neck (SCCHN) and metastatic colorectal cancer (mCRC). Panitumumab is a fully human IgG_2 mAb approved for the treatment of mCRC. Table 10.1 lists EGFR-targeting mAbs that have been shown to have radiosensitizing activity in preclinical models and have been evaluated clinically.

10.1.4.2 Small-Molecule TKIs

Small-molecule TKIs bind to the ATP-binding site of EGFR and inhibit EGFR autophosphorylation. TKIs can be subclassified into reversible and irreversible inhibitors and further stratified into those with a narrow versus broader spectra of kinase inhibition. Three orally administered anti-EGFR-TKIs have been approved for use in oncology: erlotinib, gefitinib, and lapatinib. Erlotinib and gefitinib are largely specific for EGFR, whereas lapatinib inhibits ErbB2 and, to a lesser extent, EGFR. They have relatively short half-lives and show variability in the clinic due to patient differences in bioavailability, metabolism, and, in some treatment schedules, the

TABLE 10.1 EGFR-Targeting mAbs in Clinical Use Which Have Shown Radiosensitizing Capability

Agent	Molecule	Specificity	Clinical Use	Radiosensitization
Cetuximab/ IMC-225, Erbitux (Imclone, Bristol Myers–Squibb)	Mouse–human chimeric IgG_2 mAb	EGFR-ECD	FDA-approved CRC SCCHN	Extensive preclinical data indicates EGFR inhibitors increase the intrinsic radiosensitivity of several types of cancer cells [41]
Panitumumab/ ABX-EGF (Abgen, Abgenix)	Fully human IgG_2 mAb	EGFR-ECD	FDA-approved (CRC)	Enhanced the antitumor efficacy of IR for HNSCC and non-small cell lung cancer cell lines and xenografts [42]
Nimotuzumab (YM Biosciences)	Humanized IgG_1 mAb	EGFR	I/II clinical trials	Enhanced the anti-tumor efficacy of IR for some human NSCLC cell lines *in vitro* and *in vivo*. Effect related to the level of EGFR expression Akashi 2008 [43]

Note: CD64, cluster of differentiation 64; CRC, colorectal cancer; ECD, extracellular domain; FDA, U.S. Food and Drug Administration; NSCLC, non-small cell lung cancer.

coadministration of drugs that induce or inhibit cytochrome P450. The oral route of administration and relatively short half-life facilitate dose modifications compared with antibodies. These drugs are approved for the treatment of patients with NSCLC (gefitinib and erlotinib), pancreatic carcinoma (erlotinib), and HER2-amplified breast carcinoma (lapatinib). In contrast with antibodies, limited activity has been seen in other tumor types, such as CRC or SCCHN. Nearly all the examples of small molecule TKIs cited in Table 10.2a [44–47] and b [48–50] have been shown to have radiosensitizing capability at the preclinical and clinical levels.

Both the antibody and small molecule EGFR inhibitors have shown a cytostatic effect on tumor cells *in vitro* by blocking cell

TABLE 10.2 Small-Molecule Inhibitors of EGFR Tyrosine Kinase in Clinical Use That Have Shown Radiosensitizing Capability

Agent	Molecule	Specificity	Status	Radiosensitization
(a) Reversible Inhibitors				
Gelfitinib (Astra Zeneca)	Anilinoquinazoline, reversible TKI (half-life 48 h)	HER1	FDA approved NSCLC	GBM line U251 expresses high levels of EGFR, and is hypersensitive to inhibition of the EGFR signaling pathway. Gelfitinib enhanced radiosensitivity, maximal effectiveness of combined treatments was dose-dependent and time-dependent [44]
Erlotinib (Genentech, OSIP, Roche)	Anilinoquinazoline, reversible TKI (half-life 36 h)	HER1	FDA approved NSCLC, pancreatic cancer	Radiosensitizing effect of erlotinib, was evaluated in three human cancer cell lines with different levels of HER1/EGFR expression. Extent of radiosensitization was proportional to HER1/EGFR expression, and to autophosphorylation of EGFR (HER1) [45]
Lapatinib (GlaxoSmithKline)	6-Thiazolyl-quinazoline, reversible TKI (half-life 24 h)	HER1/2	Approved (breast cancer)	Lapatinib combined with fractionated radiotherapy caused tumor growth inhibition in xenografted $EGFR^+$ and $HER2^+$ breast cancers. Inhibition of downstream signaling to ERK1/2 and AKT correlates with sensitization in $EGFR^+$ and $HER2^+$ cells, respectively [46]
BMS599626, AC480 Bristol Myers Squibb	4-Amino-pyrrolotriazine, reversible TKI	HER1/2/4	Phase I clinical trials	AC480 significantly enhanced the radiosensitivity of HN-5 cells, expressing both EGFR and Her2. Mechanisms included cell cycle redistribution and inhibition of DNA repair [51]

AEE788 (Novartis)	Pyrrolopyramidine	HER1/2 VEGFR2	Phase II clinical trials	Combined treatment effective *in vitro*/*in vivo* with DU145 prostate cancer model whereas PC-3 adequately treated with XRT alone. Correlated with differences in EGFR expression and showed effects on cell proliferation and vascular destruction [47]
		(b) Irreversible Inhibitors		
Pelitinib/EKB-569 (Wyeth)	3-Cyanoquinoline	HER1/2	1/II	EKB-569 radiosensitizes squamous cell carcinoma *in vitro*. Mechanism involves selective targeting of IR-induced NFκB-dependent survival signaling [48]
Canertinib/ci-1033 (Pfizer)	Aniloquinazoline	HER1/2/4	II	Caco-2 and LoVo cells, with high levels of EGFR and ErbB2 TK activity, were affected by CI-1033, SW620 cells, with low levels were not. Whereas CI-1033 produced only minimal radiosensitization in LoVo and Caco-2 cells *in vitro*, the combination caused prolonged suppression of tumor growth in both tumor types compared with either treatment alone [49]
BIBW 2992 (Boehringer Ingeheim)		Her1/2	II	BIBW 2669 and BIBW 2992 had clear antiproliferative effects *in vitro* and *in vivo*, but cellular radiosensitization was minimal. There was an effect of combined treatment on tumor growth delay *in vivo* cancer treatment [50]

cycle progression and proliferation. In some cases, the *in vivo* activity seems to be greater than might be anticipated from the *in vitro* results, and this may be attributable to additional direct or indirect effects on angiogenesis, invasion, and metastasis and to increased tumor cell death [49,52,53].

10.1.5 Preclinical Studies of Radiosensitization by EGFR Inhibitors

The isolation of EGF in 1965 and the subsequent purification of EGFR in 1980 led to the investigation of blockade of EGFR signaling as an antitumor stratagem. The capacity of EGFR inhibitors to augment chemotherapy response became established and, subsequently, basic experiments established that radiation exposure induced EGFR phosphorylation and that tumor cell proliferation could be blocked by the addition of an EGFR-signaling inhibitor. An inverse relationship was identified between the expression level of EGFR and response to radiation for tumor cells in culture and in xenograft model systems. Early reports of EGFR inhibitor modulation of radiation response mostly described experiments with the anti-EGFR mAb cetuximab, later radiation response enhancement using various EGFR inhibitors was established for a range of agents and across a spectrum of model systems.

10.1.5.1 mAb Inhibitors

Cetuximab has been shown to sensitize resistant tumor (HNSCC, lung, pancreatic, glioma) lines to IR. Pretreating A549 cells with cetuximab inhibits radiation-induced EGFR transport into the nucleus and abrogates the increase in DNA-PK activity, resulting in reduced DNA repair and increased DNA damage after radiation [23]. The G_0/G_1 cell cycle arrest induced by cetuximab enhances radiation-induced apoptosis in HNSCC cell lines [52].

In human tumor xenografts, the combination of cetuximab plus radiation has been shown to exert greater antitumor efficacy than either treatment alone. Combined cetuximab (0.2 mg once weekly for 4 weeks) and radiation (8 Gy administered 24 h after each cetuximab injection) produced complete regression of well-established SCC-1 and SCC-6 human squamous cell carcinoma xenografts for up to 100 days. In contrast, animals receiving either cetuximab or radiation alone experienced only short-lived tumor growth inhibition, followed by tumor growth at a rate comparable with controls (Figure 10.4) [16,52]. Molecular studies have demonstrated that combination treatment leads to the redistribution of cells to more radiosensitive points of the cell cycle, that is, G_1 and G_2-M

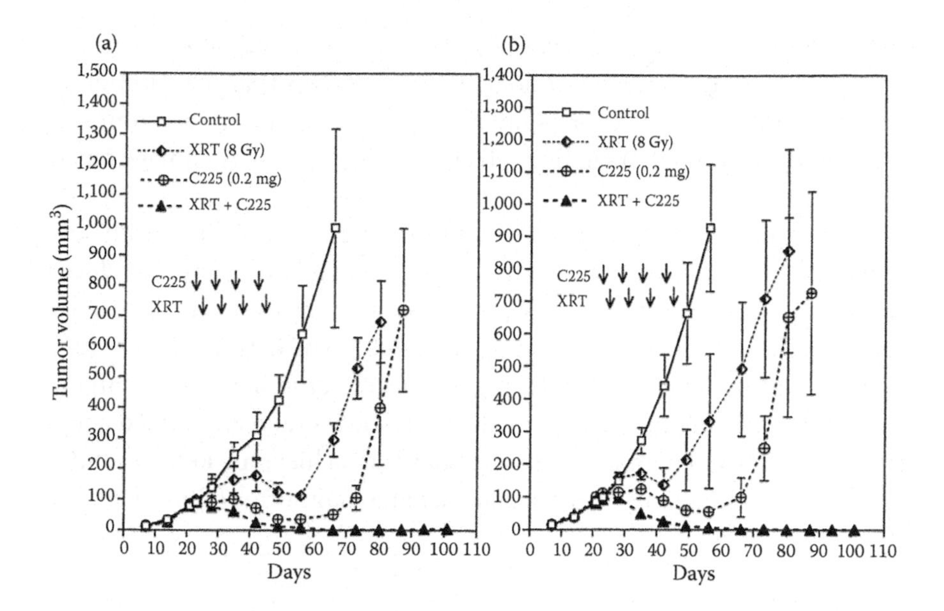

FIGURE 10.4 **Antitumor activity of C225 in combination with radiation in SCC** xenografts. SCC-1 cells (a) or SCC-6 cells (b) were injected s.c. into the flank of athymic mice. After 23 days (tumor mean size of 100 mm³), treatment was initiated by injecting 0.2 mg of C225 i.p. once a week for a total of four injections. The radiation (XRT)-treated group was exposed to a single 8-Gy fraction 24 h after each injection of C225. *Arrows*, specific days of C225 or XRT administration. Values represent mean tumor size (mm³) ± SE (n = 8/group). (From Huang, S., and Harari, P., *Clinical Cancer Research* 6:2166–2174, 2000. With permission.)

reduces DNA-PK activity and DNA repair capacity, and reduces tumor angiogenesis [16,52].

10.1.5.2 Small-Molecule RTKIs

Radiosensitization by several small molecule EGFR inhibitors has been investigated at the preclinical and clinical level.

10.1.5.2.1 Erlotinib This drug has been shown to enhance radiation response at several levels, including cell cycle arrest, induction of apoptosis, accelerated cellular repopulation, and DNA damage repair. Chinnaiyan et al. [54] reported that erlotinib modulated radiation response by influencing cell cycle kinetics and apoptosis and seemed to modify the effect of RT on EGFR autophosphorylation and *Rad51* expression [55]. In addition, erlotinib in combination with RT reduced the number of cells in S phase while increasing the level of apoptosis and promoting an increase insensitivity to RT. In tumor xenografts, erlotinib combined with RT inhibited tumor growth. Using microarray technology, it was shown that the addition of erlotinib influenced the expression of radiation response genes from several functional classes, including cell cycle arrest and DNA damage repair (Figure 10.5).

In another study involving three human cancer cell lines with low, moderate, and very high EGFR expression, the extent of erlotinib-induced radiosensitization was found to be proportional to the expression and autophosphorylation of EGFR [45]. The cell line A431, which expresses very high levels of EGFR, demonstrated the highest degree of radioresistance and in these cells, erlotinib increased the extent of G_1 arrest and augmented apoptosis. Erlotinib and high-dose RT were shown to achieve an additive antitumor effect in a xenograft model of glioblastoma multiforme (GBM) [56]. The

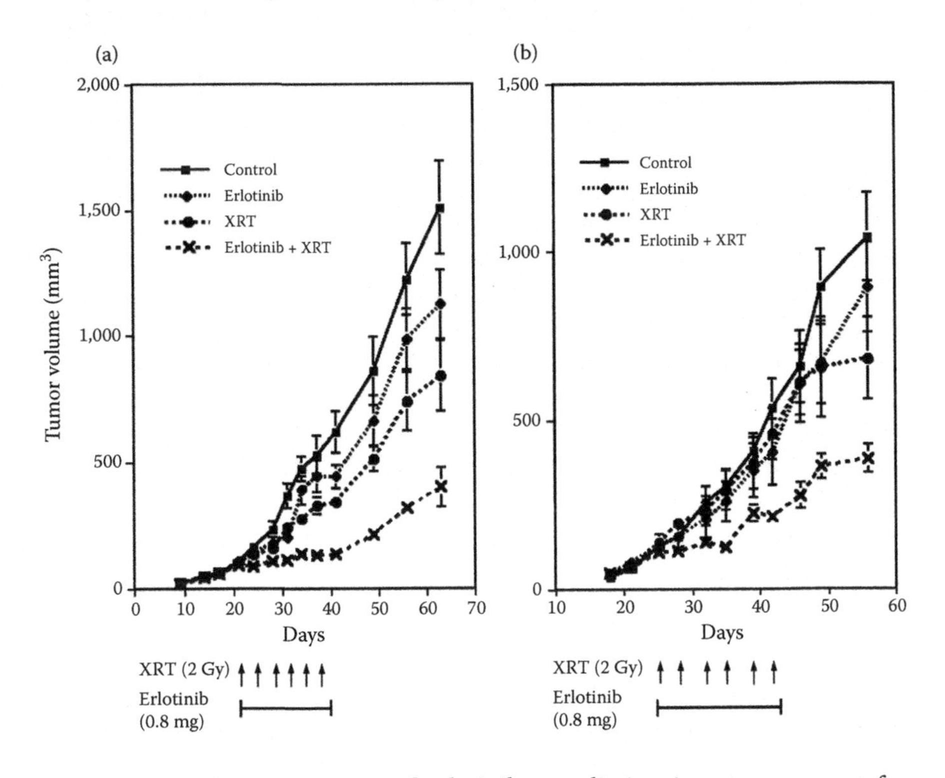

FIGURE 10.5 *In vivo* activity of erlotinib ± radiation in tumor xenografts. (a) H226 (10^6) or (b) UM-SCC6 (10^6) cells were injected s.c. into the flank of athymic mice. Mice were treated with erlotinib (0.8 mg daily via oral gavage), XRT (2 Gy fraction twice per week), or the combination for 3 weeks. *Points*, mean tumor size (mm³; $n = 6$ mice/group). (From Chinnaiyan, P. et al., *Cancer Research* 65:3328–3335, 2005. With permission.)

combination of erlotinib and high-dose RT (20 Gy/5 days) produced a reduction in survival greater than that seen for either treatment singly whereas the same result could not be achieved by low-dose radiation (12 Gy/12 days).

10.1.5.2.2 Gefitinib Studies demonstrating the synergism between gefitinib and radiotherapy have been undertaken in HNSCC, breast, NSCLC, ovarian, glioma, bladder, colon, esophageal, and biliary duct carcinoma cell lines and gefitinib has demonstrated a radiosensitizing effect in xenograft models of colon, NSCLC, and HNSCC (reviewed by Zaidi et al. [57]).

Although inhibition of EGFR radiosensitizes a number of tumor cell lines, not all cell lines are affected, and the lack of a radiosensitizing effect can be seen both for cells with vigorous expression of EGFR and for those in which it is not expressed at all. This result suggested that a more effective approach for nonresponding cell lines would be to target more than one of the ErbB receptors using a single or combination of agents [58]. Lapatinib, which inhibits ErbB1 and ErbB2, was used in the treatment of two breast carcinoma cell lines that overexpress both RTKs but only a modest degree of radiosensitization was achieved. Another dual inhibitor, BIBW-2992, was shown to induce tumor regression in xenograft and lung cancer models with better efficacy than erlotinib [59].

10.1.5.3 Preclinical Findings of the Effects of EGFR Inhibitors in Normal Cells

In normal cells, radiation induces EGFR activation in fibroblasts *in vitro* but EGFR inhibition is not radiosensitizing [60]. Consistent with these *in vitro* results, it was shown that the pharmacological inhibition of ErbB1 in a mouse model had no effect on the radioresponse of acutely responding oral mucosa. There seems to be a different situation in the lung where it was reported that the treatment of mice with the EGFR inhibitor ZD1839 enhanced the pulmonary fibrosis induced by the radiomimetic drug bleomycin. Experiments to determine whether ionizing radiation plus ZD1839 would produce the same effect have not been reported but it is significant that the treatment of patients with lung cancer with small-molecule inhibitors of EGFR has been associated with increased incidence of pulmonary fibrosis [61].

10.1.6 Clinical Applications of EGFR Inhibitors

Blockade of EGFR is the most widely used and successful of the targeted therapies for the treatment of cancer. Although the majority of approved treatments and successful trials have involved EGFR inhibitors and chemotherapy, there has also been extensive evaluation of treatments combining EGFR inhibitors with radiotherapy or radiochemotherapy.

10.1.6.1 Targeting the EGFR Receptor with mAbs

Monoclonal antibodies to EGFR were the first stratagem to be used to target EGFR, they have the advantage of being specific and long-lasting. Several treatments involving mAbs to EGFR are approved for use in different types of disease either as monotherapy or combined with chemotherapy. In one case, a mAb (cetuximab) was approved for use in combination with radiotherapy.

10.1.6.1.1 Cetuximab Two landmark studies evaluated the benefits of cetuximab (Erbitux) in patients with SCCHN in both the locally advanced (Bonner trial) and the recurrent or metastatic (EXTREME trial) settings. Erbitux was granted approval by the European Commission in November 2008 for the treatment of first-line recurrent or metastatic SCCHN based on the results of the EXTREME study. Cetuximab is also approved to be used alone or with other drugs to treat colorectalcancer that has metastasized.

Cetuximab was approved by the U.S. Food and Drug Administration (FDA) in March 2006 for use in combination with radiation therapy for treating SCCHN or as a single agent in patients who have had prior platinum-based therapy. Interim results of a phase III randomized trial, which was initiated in 2000, showed that adding cetuximab to primary radiotherapy increased overall survival in patients with locoregionally advanced SCCHN (LASCCHN) at 3 years. Five-year survival data was reported in 2010. For patients with LASCCHN, cetuximab plus radiotherapy significantly improved overall survival at 5 years compared with radiotherapy alone, confirming cetuximab plus radiotherapy as an important treatment option in this group of patients (Figure 10.6) [62]. Median overall survival in the radiotherapy-alone group was 29.3 months (95% CI, 20.6–41.4), compared with 49.0 months (32.8–69.5) in the cetuximab group. Five-year overall survival was 36.4% and 45.6%, respectively (hazard ratio [HR], 0.73; 95% CI, 0.56–0.95; $p = 0.018$).

The results of several relatively small studies that examined radiotherapy (RT) with cetuximab in stage III NSCLC were reviewed recently [63]. Three different strategies were pursued: (1) RT plus cetuximab (two studies), (2) induction chemotherapy followed by RT plus cetuximab (two studies), and (3) concomitant RT and chemotherapy plus cetuximab

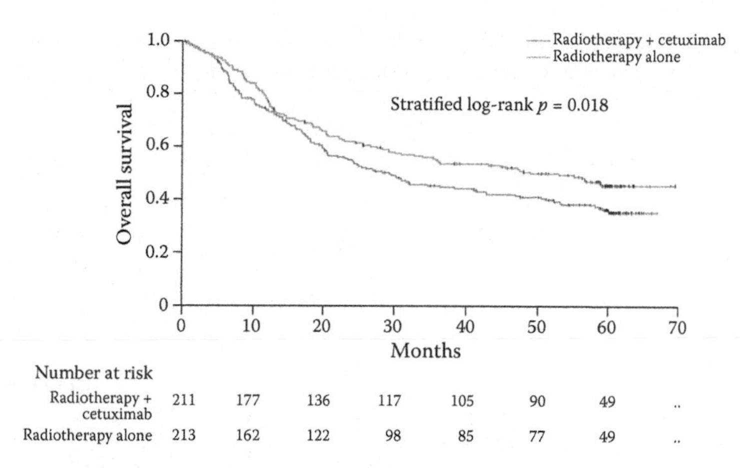

FIGURE 10.6 The 5-year survival data for patients with LASCCHN. Cetuximab plus radiotherapy significantly improved overall survival at 5 years compared with radiotherapy alone. (From Bonner, J. et al., *Lancet Oncology* 11:21–28, 2010. With permission.)

(two studies). Radiation doses were limited to 60 to 70 Gy. As a result of study design, in particular, the lack of randomized comparison between cetuximab and no-cetuximab, the efficacy results were difficult to interpret. However, strategies (1) and (3) seemed to be more promising than induction chemotherapy followed by RT and cetuximab. Toxicity and adverse events were more common when concomitant chemotherapy was given. Nevertheless, combined treatment seemed feasible.

Currently, there are numerous clinical studies being conducted to further evaluate cetuximab as an adjunct to RT with or without added chemotherapy for the treatment of a variety of cancers. Details of these are summarized in Table 10.3.

10.1.6.1.2 Panitumumab A fully human IgG_2 antibody that binds with high affinity to the EGFR and is approved for the treatment of recurrent colorectal cancer. It was approved by the U.S. FDA for the first time in September 2006, for "the treatment of EGFR-expressing mCRC with disease progression" despite prior treatment. Panitumumab was approved by the European Medicines Agency (EMEA) in 2007, and by Health Canada in 2008 for "the treatment of refractory EGFR-expressing mCRC in patients with nonmutated (wild-type) KRAS." Panitumumab was the first mAb to demonstrate the use of KRAS as a predictive biomarker. Panitumumab was evaluated in preclinical studies of HNSCC [42,64], which showed a

TABLE 10.3 Ongoing or Recently Completed Clinical Trials Involving mAb Inhibitors of EGFR with Concurrent Radiotherapy or Radiochemotherapy

			Phase				
Drug	**Tumor**	**Treatment**	**1**	**1/2**	**2**	**3**	**Chemotherapy**
Cetuximab (Erbitux)	SCCHN	RT		2	4	1	
		CRT	2	4	9	3	Doxetaxel, cisplatin, 5-FU
	Esophageal cancer	CRT			7		5-FU, cisplatin/oxaloplatin paclitaxel, irinotecan
	NSCLC	RT			1		
		CRT			1		
	Pancreatic cancer	CRT			1		UFT, leucovorin
	Anal cancer	CRT			1		5-FU, cisplatin
	Rectal cancer	CRT			1		Capecitabine
	Colon cancer	CRT				1	Irinotecan, oxaloplatin, leucovorin
	Cervical cancer	CRT			1		Cisplatin
	GBM	CRT			1		Cilengitide EMD
	Follicular lymphoma	CRT				1	Cyclophosphamide
Zalutumab	SCCHN	RT	1	1			
		CRT		1			Cisplatin
	NSCSC	CRT			1		
	Colorectal cancer	RT					

Source: www.clinicaltrials.gov.

favorable interaction between panitumumab and RT. A number of clinical trials are ongoing; the details of which are summarized in Table 10.4.

10.1.6.1.3 Zalutumumab A high-affinity, completely human IgG_1k mAb targeting EGFR derived from transgenic mice immunized with A431-derived EGFR. Zalutumumab has been studied in colorectal and lung cancer, but most clinical research has been focused on head and neck cancer. Details of ongoing clinical trials involving radiotherapy are summarized in Table 10.3.

TABLE 10.4 Clinical Trials Involving mAb Inhibitors of EGFR Receptor with Radiotherapy or Radiochemotherapy

Drug	Tumor	Treatment	Phase				Chemotherapy
			1	1/2	2	3	
Panitumumab	SCCHN	RT			2		
		CRT	1		2	2	Cisplatin, carboplatin, 5-FU, paclitaxel, docetaxel
	Esophageal cancer	RT			2		
		CRT			3		Cisplatin, docetaxel, oxaloplatin, leucovorin, 5-FU, carboplatin, paclitaxel
	Pancreatic cancer	RT		1			
		CRT			2		5-FU, capecitabine, gemcitabin
	Anal cancer	RT		1			
		CRT			1		Mitomycin C, 5-FU
	Rectal cancer	RT			1		
		CRT			2		Capecitabine
	Lung cancer	CRT			1		Carboplatin, paclitaxel
	Cervical cancer	CRT			1		Cisplatin
	Salivary gland cancer	RT			2		
Nimotuzumab	SCCHN	RT			2		
		CRT			3	1	Cisplatin, docetaxel
	Esophageal cancer	RT			1	1	
		CRT	1		1	1	Cisplatin, 5-FU
	Anal cancer	RT			1		
	Brain and CNS	RT			1		

(continued)

TABLE 10.4 Clinical Trials Involving mAb Inhibitors of EGFR Receptor with Radiotherapy or Radiochemotherapy (Continued)

Drug	Tumor	Treatment	Phase				Chemotherapy
			1	1/2	2	3	
	NSCLC, brain metastases	RT			1		
	NSCLC	RT		1			
	Cervical cancer	CRT			1		Cisplatin
	Gastric cancer	CRT			1		Capecitabine

Source: www.clinicaltrials.gov.

10.1.6.1.4 Nimotuzumab A humanized mAb that recognizes domain III of the extracellular region of the EGFR within an area that overlaps both the surface patch recognized by cetuximab and the binding site for EGF. The efficacy of nimotuzumab in combination with radiotherapy was assessed in a controlled, double-blind, randomized clinical trial conducted with 106 patients with advanced SCCHN, most of whom were unfit for chemoradiotherapy. The control group of patients received a placebo and radiotherapy. Treatment was safe and the most frequent adverse events consisted of grade 1 or 2 asthenia, fever, headache, and chills. No skin rash was detected. A significant improvement in complete response was found in the group of patients treated with nimotuzumab as compared with the placebo, and a trend toward survival benefit for nimotuzumab-treated subjects was also found. The survival benefit became significant when applying the Harrington–Fleming test, a weighted log-rank that underscores the detection of differences deferred on time. In addition, a preliminary biomarker investigation showed a significant survival improvement for nimotuzumab-treated patients as compared with controls for subjects with EGFR-positive tumors. All patients showed a quality of life improvement and a reduction of the general and specific symptoms of the disease [65]. Details of other ongoing and recently completed clinical trials of nimotuzumab with radiotherapy are summarized in Table 10.4.

10.1.6.2 Small-Molecule RTKIs

Receptor tyrosine kinase inhibitors have relatively short half-lives and show variability in the clinic due to patient differences in bioavailability and metabolism. The oral route of administration and relatively short half-life facilitate dose modifications compared with antibodies. Three orally administered anti-EGFR-TKIs have been approved for use in oncology.

10.1.6.2.1 Gefitinib (Iressa; ZD1839) A member of the anilinoquinazoline family specific to EGFR, gefitinib was the first TKI to be approved in NSCLC. It was initially used as monotherapy in patients with platinum-refractory NSCLC. After disappointing results in the Iressa Survival Evaluation in Lung Cancer study, it was withdrawn from the European Union and the United States in January 2005 [66].

Results of some phase I/II trials are summarized in Table 10.5 (reviewed by Zaidi et al. [57]). The majority of gefitinib phase I/II studies have been with concurrent chemoradiotherapy, making it difficult to judge the true clinical efficacy of gefitinib as a radiosensitizer.

Thirty-four ongoing or recently completed trials involving gefitinib and radiotherapy are listed (www.clinicaltrials.gov). The most frequently cited types of disease are NSCLC (11/34), squamous cell head and neck cancer (9/34), esophageal cancer (5/34), brain and CNS tumors (4/34), and metastatic brain disease (3/34). Gastric cancer and squamous cell skin cancer are also being investigated. Twenty-six trials involve chemoradiotherapy, with the drugs listed being those appropriate to treatment of the specific disease, whereas eight trials involve radiotherapy and gefitinib with no additional chemotherapy. There are 18 phase I/II trials and 14 phase II/III trials. Two of the trials listed included marker analysis.

10.1.6.2.2 Erlotonib hydrochloride (Tarceva) Following gefitinib in the evolution of small-molecule drugs targeting EGFR receptors, erlotonib is a drug used to treat non-small cell lung cancer, pancreatic cancer, and several other types of cancer. The U.S. FDA has approved erlotinib for the treatment of locally advanced or metastatic non-small cell lung cancer that has failed at least one prior chemotherapy regimen. In addition to being currently licensed as second-line therapy in NSCLC, erlotinib is FDA-approved as a combination therapy with gemcitabine in locally advanced pancreatic cancer.

There have been a number of clinical studies evaluating its properties as a sensitizer radiotherapy or concurrent chemoradiotherapy, some of which are summarized in Tables 10.5 and 10.6 [67]. There are also a number of ongoing or recently completed clinical trials listed for which the results are not yet available (www.clinicaltrials.gov). The most frequently cited

TABLE 10.5 Results of Clinical Trials Using Small Molecule RTKIs

Phase, Drug	Combined with	Tumor Site	Outcome
I/II Gefitinib	Concurrent, capecitabine, RT	Locally advanced rectal or pancreatic cancer	Diarrhea. 8/16 patients not allowing escalation above the lowest dose. 2/16 arterial thrombi [68]
I/II Gefitinib	Preoperative concurrent 5FU, RT	Locally advanced rectal cancer	Grade 3 diarrhea 5/41 Grade 3 skin toxicity 6/41 CR or PR 26/41 [69]
I Gefitinib	Gemcitabine, concurrent RT	Locally advanced pancreatic cancer	DLT not reached PR 1/18. SD 7/18 OS 7.5 m [70]
I/II Gefitinib	Concurrent RT	Locally advanced HNSCC	34/45 patients completed treatment; CR: 11, PR: 18 Little additional toxicity attributable to gefitinib [71]
I Erlotinib	Concurrent docetaxel, RT	Locally advanced HNSCC	DLT, mucositis. 1 patient died, 2 patients not evaluable. CR 15/23 patients. Despite the death it was felt to be tolerable regimen [72]
I Erlotinib	Concurrent cisplatin, RT	Locally advanced HNSCC	CR: 11/13 [73]
II Erlotinib	Concurrent cisplatin, RT	Locally advanced HNSCC	1 death. 14/31 Grade 3 in-field dermatitis. 25/31 completed treatment. CR 21/31 [74]
I Erlotinib	Gemcitabine, RT	Locally advanced pancreatic cancer	DLT: diarrhea and dehydration. PR: 6/13. Median survival of 14 months [75]
I Erlotinib	Temozolomide, RT	Glioblastoma multiforme	DLT not reached 20 patients Median survival 55 weeks [76]
I Erlotinib	Cisplatin concurrent RT	Locally advanced cervical cancer	DLT not seen CR: 5/6, PR: 1/6 Median survival 55 weeks
I Erlotinib	Cisplatin, 5FU, RT effect observed	Oesophageal carcinoma	11 patients. Dermatitis was the most common side effect [77]

types of disease are NSCLC (15/63), squamous cell head and neck cancer (16/63), pancreatic cancer (12/63), esophageal cancer (4/63), brain and CNS tumors (7/63), metastatic brain disease (6/63), and cervical cancer, skin cancer, rectal cancer (1/63 each) are also being investigated. Forty-two trials involve chemoradiotherapy with the drugs listed being those appropriate to treatment of the specific disease, whereas 21 trials involve

TABLE 10.6 Clinical Trials with Radiotherapy and Erlotinib

Phase	Tumor Site	Treatment	DLT	Efficacy
II ($n = 23$) randomized	Stage I-IIIA NSCLC	(1) RT vs. (2) RT + erlotonib	Grade 3 toxiticies (1) 4% pneumonitis (2) 8% dermatitis	RR (1) 55.5% (2) 83.3%
I single arm ($n = 11$)	NSCLC brain metastases	WBRT + concurrent erlotinib	Grade 3–5 toxicities. 18% Interstitial lung disease; 9% acneiform rash, 9% fatigue	PR 5/7 SD 2/7 (follow-up imaging in 7 pts)
Case reports ($n = 2$)	NSCLC recurrent brain metastases and parallel thoracic progression	WBRT + sequential erlotinib	No severe toxicities reported	Survival (1) >18 m (2) =15 m
I/II ($n = 31$) single arm	HNSCC	RT + cisplatin + erlotinib	Grade 3/4 toxicities: 52% in-field dermatitis 48% nausea; 39% vomiting; 35% dysphagia: 29% mucositis: 29% xerostoma	Pathologic CR 74.2%
I ($n = 23$) dose escalation	HNSCC	RT + docetaxel + erlotinib	DLT: Grade 3/4 mucositis 2; death 1	CR 15/18 pts
I ($n = 12$) dose escalation	HNSCC	RT + cisplatin + erlotinib	Grade 3/4 toxicities: mucositis (50%), anemia, syncope, constipation, dysphonia, dermatitis, respiratory infection (8%)	MTD: determined (RT = 63 Gy)

(*continued*)

TABLE 10.6 Clinical Trials with Radiotherapy and Erlotinib (Continued)

Phase	Tumor Site	Treatment	DLT	Efficacy
II (n = 48), single arm		RT + chemotherapy + erlotinib + bevacizumab	Grade 3/4 toxicities: during induction: 46% neutropenia; 14% mucositis; 14% diarrhea; 11% hand-foot syndrome; 6% neutropenic fever Local grade 3/4 toxicities during combined treatment: 76% mucositis/ esophagitis	RR = 77% 18 m PFS: 85%

Source: Based on a review by Mehta, V., *Frontiers in Ecology* 2:1–11, 2012.
Note: CR, complete response; DLT, dose-limiting toxicity; NSCLC, non-small cell lung cancer; OS, overall survival; PFS, progression-free survival; PR, partial response; RR, response rate; RT, radiotherapy; SD, stable disease; WBRT, whole-brain radiotherapy.

radiotherapy and erlotonib with no additional chemotherapy. There are 17 phase I trials, 10 phase I/II trials, 29 phase II trials, 5 phase III trials, and 2 phase III/IV trials. Eight of the trials listed involved marker analysis.

10.1.6.2.3 Lapatinib (Tykerb) A dual TKI of EGFR and HER2. In 2007, the U.S. FDA approved lapatinib in combination therapy for breast cancer in patients already using capecitabine. In January 2010, Tykerb received accelerated approval for the treatment of postmenopausal women with hormone receptor-positive metastatic breast cancer that overexpresses the HER2 receptor and for whom hormonal therapy is indicated. The only reported phase I trial combining it with concurrent cisplatin and radiotherapy was conducted in patients with locally advanced HNSCC. Of the 31 patients enrolled, 89% had a response (59% complete response, 30% partial response). Although dose-limiting toxicity was not observed with the escalated dose, mouth ulcers and radiation dermatitis were frequently encountered. A randomized phase II trial of chemoradiotherapy with either lapatinib (1,500 mg o.d.) or placebo is in progress in patients with EGFR overexpressing (immunohistochemistry 3+) stage III and IV HNSCC. Also listed as ongoing is a randomized phase III trial of adjuvant postoperative chemoradiotherapy with either lapatinib or placebo in patients with high-risk (margins < 5mm, nodal metastasis with extracapsular spread) surgically excised HNSCC. Currently active trials with lapatinib and concurrent radiation include five with head and neck cancer; one each with prostate cancer, GBM, and breast cancer; and two involving whole-brain irradiation for metastases from breast cancer.

10.1.6.2.4 Vandetanib (Caprelsa, AstraZeneca) A kinase inhibitor of a number of cell receptors, particularly EGFR, VEGF receptor (VEGFR), and the RET-tyrosine kinase, vandetanib is approved by the FDA for the treatment of late-stage (metastatic) medullary thyroid cancer in adult patients who are ineligible for surgery. The drug has undergone clinical trials as a potential targeted treatment for non-small cell lung cancer and there have been some promising results from a phase III trial with docetaxel. In late 2012, seven phase I or II clinical trials were listed involving vandetanib and radiotherapy. Of these, three were with brain or CNS tumors, two with head and neck cancers, one with NSCLC, and one with brain metastases from NSCLC. Four were with radiotherapy alone and three with chemoradiotherapy (www.clinicaltrials.gov).

10.2 ErbB2 (HER2)

The ErbB2 oncogene encodes a 185 kDa type I tyrosine kinase receptor and is a member of the EGFR family.

10.2.1 Basic and Preclinical Studies

There is no known ligand for ErbB2, but it is the preferred heterodimerization partner for other ErbB receptors because its extracellular domain remains in a fixed, open conformation that resembles a ligand-activated state. ErbB2-containing heterodimers are highly stable, demonstrating a low rate of ligand dissociation and prolonged and enhanced receptor signaling. Overamplification of the *ErbB2* gene results in the formation of a ligand-independent ErbB2 homodimer that is able to initiate downstream signaling cascades such as the PI3K and MAPK pathways (Figure 10.1).

ErbB2 is in fact the preferred partner for all the other ErbB proteins, an example being ErbB3, which contains six binding sites for the p85 catalytic subunit of PI3K, making it the most potent activator of the PI3K survival pathway among all EGFR family members. ErbB2/ErbB3 heterodimers are potent mediators of prosurvival signals and aberrant signaling from ErbB2/ErbB3 leads to deregulated PI3K signaling, a predictor of poor clinical outcome and resistance to various cancer therapies. There is redundancy and cross talk between the signaling networks that regulate growth and survival in epithelial tumors. In breast cancer, plasma membrane-associated estrogen receptor exerts its nongenomic biological effects through the PI3K–Akt and MAPK–Erk pathways linking estrogen receptor signaling to activation and signaling via EGFR family members. There is also cross talk between ErbB2 and the insulin-like growth factor 1 receptor (IGF1-R), a key regulator of cell growth and survival, which is also expressed in breast cancer.

10.2.2 Blockading ErbB2: Trastuzumab

The most widely used anti-ErbB2 therapy is the recombinant humanized antibody trastuzumab, which exerts its antitumor activity by binding to the extracellular domain, serving to both block heterodimerization and induce receptor internalization and subsequent degradation [78]. Clinically, trastuzumab (Herceptin) is currently routinely used in the treatment of $ErbB2^+$ breast cancer as well as for some other forms of primary and metastatic disease.

10.2.2.1 Radiosensitization by Trastuzumab

Trastuzumab has been shown to enhance radiosensitivity in breast, esophageal, and HNSCC cell lines [57]. In breast cancer cells with overexpression of ErbB2 receptor, trastuzumab decreases cell proliferation *in vitro* and reduces tumor formation in nude mice (Figure 10.7). Cell cycle analysis of irradiated cells showed that early escape from cell cycle arrest occurred in the presence of trastuzumab; suggesting that, in ErbB2-overexpressing cells, there may be insufficient time for completion of DNA repair before cell cycle progression occurs. Failure of cell cycle arrest has been linked to dysregulation of p21WAF1, a critical mediator of the cellular response to DNA damage. The results obtained with this breast cancer model indicated that, whereas the induction of p21WAF1 transcripts and protein product occurred at 6, 12, and 24 h after radiation treatment, increased levels of p21WAF1 transcript and protein were not sustained in ErbB2-overexpressing cells exposed to radiation in the presence of trastuzumab. Tyrosine phosphorylation of p21WAF1 is increased after treatment with radiation alone, but phosphorylation is blocked by combined treatment with antireceptor antibody and radiation. Thus, the failure of DNA repair in ErbB2-overexpressing cells treated with ErbB2 antibody may be related to dysregulation of p21WAF1 in the presence of the receptor antibody [79].

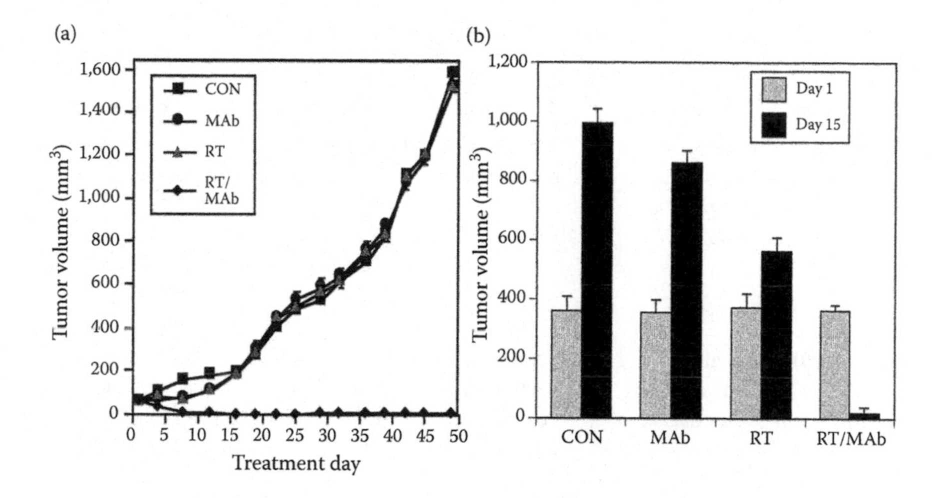

FIGURE 10.7 Combination treatment with antibody to HER2 growth factor receptor and radiation promotes remission of human breast cancer xenografts in nude mice. (a) Growth curves for MCF-7/HER2 tumors treated 14 days after subcutaneous implant of cells. Treatment groups: human IgG_1 control at 30 mg/kg (CON, ■); radiation 4 Gy + human IgG_1 (RT, ▲); rhuMAb HER2 at 30 mg/kg (MAb, ●); or combined radiation/rhuMAb (RT/MAb, ◆) therapy. (b) MCF-7/HER2 cells were injected s.c. After 35 days, mice were randomized on day 0 to groups of three to five animals on the basis of body weight and tumor nodule size, with tumors ranging in size from 350 to 400 mm^3. Treatment groups included human IgG_1 control (CON), radiation at 8 Gy with human IgG_1 (RT), and rhuMAb HER2 or combined radiation/rhuMAb (RT/MAb) therapy. Doses of antibody or IgG_1 were administered in divided doses on days 1, 4, and 7. Groups treated with radiation received an 8-Gy treatment at 4 h after administration of antibody or IgG_1 on day 1 only. Tumor volumes were recorded at day 1 and reassessed at 15 days. (From Pietras, R. et al., *Cancer Research* 59:1347–1355, 1999. With permission.)

10.2.2.2 Clinical Studies with Trastuzumab and Radiotherapy

Combined use of trastuzumab and radiotherapy has been investigated in a number of phase I/II trials in patients with breast cancer and esophageal cancer. Results of completed trials are summarized in Table 10.7. Results for the small-molecule TKI lapatinib (Tykerb), which has high specificity for ErbB1 and ErbB2, are also included in Table 10.7. Recent clinical data demonstrated that trastuzumab significantly improved survival for HER2-positive patients with brain metastases compared with patients who had not received trastuzumab (reviewed by Chargari et al. [80]). A retrospective

TABLE 10.7 Clinical Trials with Trastuzumab and Radiotherapy

Phase (No. of Patients)	Tumor Site	Treatment	Toxicities	Efficacy
I ($n = 30$)	Locally advanced esophageal cancer	T, concurrent paclitaxel, cisplatin, RT	Grade 3 esophagitis (9), nonmalignant pericardial effusion (1)	15/23 negative biopsies in patients who had repeat endoscopy (5 HER2$^+$, 10 HER2$^-$)
I/II ($n = 19$) 14/19 HER2$^+$	Locally advanced esophageal cancer	T, concurrent paclitaxel, cisplatin, RT	2/14 cases of grade 3/4 esophagitis	MS 24 months 2 year OS 50%. 8/14 HER2$^+$ patients with clinical CR
1/2 ($n = 15$)	Inoperable or chemoresistant BC locally, advanced disease	T, RT, amifostine, doxorubicin docetaxel		Complete responses in 5/7 with locally advanced disease. No high risk group patients relapsed within 2 years
I/II ($n = 12$)	HER2$^+$, CT-refractory, locally advanced or locoregionally recurrent BC	T, RT	Grade 3 toxicities skin ($n = 2$) and lymphopenia ($n = 1$). Delayed wound healing after surgery (1)	3/7 patients with mastectomy had complete or partial response. Significantly >than comparison cohort. MS = 39 months
II ($n = 33$)	Locally advanced HNSCC	T, lapatinib, RT	DLTs: perforated ulcer (1), transient elevation of liver enzymes (1), grades 3 to 4 mucositis, dermatitis, lymphopenia, neutropenia	Overall response rate 81% (65% at the recommended phase II dose)

Source: Based on information reviewed by Zaidi, S. et al., *Current Drug Discovery Technologies* 6:103–134, 2009.

Note: BC, breast cancer; CR, complete remission; HER2$^+$, HER2-positive tumors; OS, overall survival; RT, radiotherapy; T, trastuzumab.

TABLE 10.8 Summary of Ongoing or Recently Completed Clinical Trials Involving Trastuzumab and Radiotherapy or Radiochemotherapy

		Phase				
Tumor	**Treatment**	**1**	**1/2**	**2**	**3**	**Chemotherapy**
Breast cancer	RT	1		1	1	
	CRT	1		5	4	Carboplatin, cyclophosphamide, melphalan, thiotepa, doxorubicin, docetaxol, leucovorin methotrexate, paclitaxel
BC, brain metastases	RT	1	1	1		
	CRT				1	Chemotherapy
Pancreatic cancer	CRT			2		Gemcitabine
Esophagus					1	Carboplatin, paclitaxel
Bladder	CRT			1		Paclitaxel
Sarcoma	CRT					Chemotherapy

Source: www.clinicaltrials.gov.
Note: BC, breast cancer; CRT, chemoradiotherapy; RT, radiotherapy.

evaluation of whether trastuzumab should be continued after the diagnosis of patients with brain metastases in ErbB2-positive breast cancer showed that the median overall survival was significantly improved in patients continuing trastuzumab after WBRT (21 vs. 9 months; $p = 0.001$). These data suggested that trastuzumab should not be discontinued after brain progression but did not address the issue of potential toxicity [80]. In Table 10.8, information about ongoing or recently completed clinical trials is summarized. The majority of these are concerned with breast cancer.

10.3 INSULIN-LIKE GROWTH FACTOR 1 RECEPTOR

The IGF1-R is a transmembrane receptor tyrosine kinase from the insulin receptor family. Figure 10.8 shows the insulin receptor family and its ligands.

10.3.1 Basic and Preclinical Studies

IGF1 and IGF2 are recognized ligands of IGF1-R. activating downstream pathways including PI3K and Ras. IGF1 has been demonstrated to exert effects on cellular proliferation, anchorage-independent growth and metastasis, and to be antiapoptotic. IGF1-R overexpression is observed in prostate, breast, lung, and colon cancers and malignant glioma and has been associated with a poor prognosis [57]. It is also associated with both

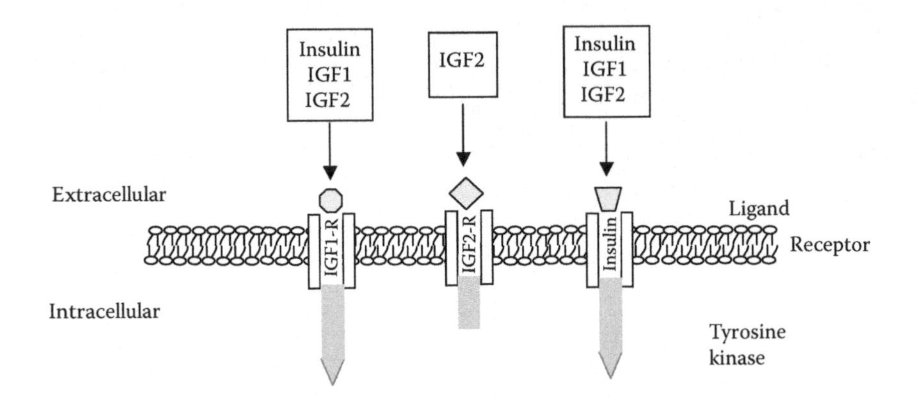

FIGURE 10.8 The insulin receptor family. Insulin and IGF bind to target receptors. Both the insulin and IGF1-R have an intracellular tyrosine kinase domain. IGF2-R is thought to mop-up IGF2 and reduces available ligand from binding to the other receptors. Insulin and IGF1 bind to both the insulin receptor and IGF1-R. IGF2 is able to bind to all three receptors. IGFBPs are a family of proteins that bind IGF1 or IGF2 hence altering their interaction with the insulin receptor family.

chemotherapy and radiotherapy-related treatment resistance [81,82], suggesting that disrupting the IGF1-R pathway might be an effective means of reducing radioresistance. Fibroblasts from IGF1-R-null mice embryos showed increased radiosensitivity, in comparison with wild-type cells [83] and an *in vitro/in vivo* study using antisense RNA to downregulate IGF1-R expression in mouse melanoma cells demonstrated increased radiosensitivity [84].

10.3.2 Cytotoxicity Effects and Radiosensitization by IGF Blockade

Monoclonal antibodies against the IGF1-R have been utilized preclinically in cancer cell lines and xenografts to block signaling and improve radiosensitivity. Mouse antihuman antibody to IGF1-R was shown to increase radiosensitivity in colon and lung cancer cells [82]. Tyrphostin, an IGF1-R TKI, has been shown to improve radiosensitivity in a human breast cancer cell lines [81]. Results of one *in vitro* study suggested a role for IGF1-R activation by radiation in modulating DNA repair. After the irradiation of an NSCLC-derived cell line, it was shown that IGF1-R was activated and that it promoted Ku-DNA binding (a key effector in NHEJ repair) within 4 h. Radiation-mediated Ku-DNA binding activation was reduced after the induction of IGF1-R blockade by Tyrphostin.

10.3.3 Clinical Effect of Blockading IGFR

There are currently a number of mAbs targeting IGFR in development for clinical use. Twenty-three phase I/II trials are listed using one or more of these agents, usually in conjunction with chemotherapy for the treatment of various cancers (www.clinicaltrials.gov). Monoclonal antibodies against the IGF1-R have been utilized preclinically in cancer cell lines and xenografts to block signaling and improve radiosensitivity.

10.4 SUMMARY

The rationale for targeting growth factor receptors for radiosensitization is based on the understanding that radiation activates transmembrane receptors and intracellular signaling pathways. In fact, IR initiates a self-perpetuating loop of activity whereby radiation generates ROS/RNS signals within tumor cells, which promote activation of growth factor receptors and signaling pathways that, in turn, can promote the release of paracrine ligands from cells, leading to the reactivation of receptors and intracellular signaling pathways. Exposure of tumor cells overexpressing EGFR to ionizing radiation activates survival and proliferation mechanisms, predominantly through stimulated signaling via PI3K-AKT and Ras-MAPK, and activation of these two pathways is thought to be the main cause of radioresistance for EGFR overexpressing tumors.

Two approaches to EGFR inhibition have been investigated in the laboratory and in the clinic. In the first of these, mAbs are directed against the receptor to prevent ligands from binding to the extracellular domain of the receptor. This method prevents receptor dimerization and activation, and ultimately induces receptor degradation. The second approach to disrupting EGFR function involves the use of small-molecule TKIs, which bind to the ATP-binding pocket of the cytoplasmic domain of the receptor, preventing receptor phosphorylation and ultimately blocking downstream signaling cascades. Both mAbs and RTKIs have been demonstrated to increase the effects of radiation in a number of preclinical systems.

Blockade of EGFR is the most widely used and successful of the targeted therapies for the treatment of cancer. Although the majority of approved treatments and successful trials have involved EGFR inhibitors and chemotherapy, there has also been extensive evaluation of treatments combining EGFR inhibitors with radiotherapy or radiochemotherapy. Cetuximab and panitumumab are the two anti-EGFR mAbs that have been approved for cancer therapy, whereas cetuximab is approved for use in combination with radiationtherapy for treating SCCHN. Three orally administered anti-EGFR-TKIs (gefitinib, erlotinib, and lapatinib) have been approved for use in oncology. Erlotinib and gefitinib are largely specific for EGFR, whereas lapatinib inhibits ErbB2 and, to a lesser extent, EGFR.

The ErbB2 oncogene is the preferred partner for all the other ErbB proteins, making it the most potent activator of the PI3K survival pathway among all EGFR family members. The most widely used anti-ErbB2 therapy is the recombinant humanized antibody trastuzumab, which binds to the extracellular domain, blocking heterodimerization and inducing receptor internalization and subsequent degradation. Combined use of trastuzumab and radiotherapy has been investigated in a number of phase I/II trials in patients with breast cancer and esophageal cancer. The small-molecule TKI lapatinib (Tykerb), which has high specificity for both ErbB1 (EGFR) and ErbB2 has also undergone early phase clinical evaluation with concurrent radiotherapy.

The IGF1-R is a transmembrane receptor tyrosine kinase from the insulin receptor family. Monoclonal antibodies against IGF1-R have been utilized preclinically in cancer cell lines and xenografts to block signaling and improve radiosensitivity.

REFERENCES

1. Yarden, Y., and Sliwkowski, M. Untangling the ErbB signalling network. *Nat Rev Mol Cell Biol* 2001;2:127–137.
2. Olayioye, M., Neve, R., Lane, H., and Hynes, N. The ErbB signaling network: Receptor heterodimerization in development and cancer. *EMBO J* 2000;19:3159–3167.
3. Schmidt-Ullrich, R., Valerie, K., Fogleman, P., and Walters, J. Radiation-induced autophosphorylation of epidermal growth factor receptor in human malignant mammary and squamous epithelial cells. *Radiat Res* 1996;145:81–85.
4. Bowers, G., Reardon, D., Hewitt, T. et al. The relative role of ErbB1-4 receptor tyrosine kinases in radiation signal transduction responses of human carcinoma cells. *Oncogene* 2001;20:1388–1397.
5. Valerie, K., Yacoub, A., Hagan, M. et al. Radiation-induced cell signaling: Inside-out and outside-in. *Mol Cancer Ther* 2007;6:789–801.
6. Galabova-Kovacs, G., Kolbus, A., Matzen, D. et al. ERK and beyond: Insights from B-Raf and Raf-1 conditional knockouts. *Cell Cycle* 2006;5:1514–1518.
7. Tonks, N. Protein tyrosine phosphatases and the control of cellular signaling responses. *Adv Pharmacol* 1996;36:91–119.
8. Khan, E., Heidinger, J., Levy, M., Lisanti, M., Ravid, T., and Goldkorn, T. Epidermal growth factor receptor exposed to oxidative stress undergoes Src- and caveolin-1-dependent perinuclear trafficking. *J Biol Chem* 2006;281(20):14486–14493.

9. Shvartsman, S., Hagan, M., Yacoub, A., Dent, P., Wiley, H., and Lauffenburger, D. Autocrine loops with positive feedback enable context-dependent cell signaling. *Am J Physiol Cell Physiol* 2002;282(3):C545–C559.
10. Nyati, M., Morgan, M., Feng, F., and Lawrence, T. Integration of EGFR inhibitors with radiochemotherapy. *Nat Rev Cancer* 2006;6:867–885.
11. El-Shewy, H., Kelly, F., Barki-Harrington, L., and Luttrell, L. Ectodomain shedding-dependent transactivation of epidermal growth factor receptors in response to insulin-like growth factor type I. *Mol Endocrinol* 2004;11:2727–2739.
12. Vrana, J., Grant, S., and Dent, P. Inhibition of the MAPK pathway abrogates BCL2-mediated survival of leukemia cells after exposure to low dose ionizing radiation. *Radiat Res* 1999;5:559–569.
13. Hirsch, F., Varella-Garcia, M., Bunn, P.A. et al. Epidermal growth factor receptor in non-small-cell lung carcinomas: Correlation between gene copy number and protein expression and impact on prognosis. *J Clin Oncol* 2003;21(20):3798–3807.
14. Milas, L., Fan, Z., Andratschke, N., and Ang, K. Epidermal growth factor receptor and tumor response to radiation: *In vivo* preclinical studies. *Int J Radiat Oncol Biol Phys* 2004;58(3):966–971.
15. Eriksen, J., Steiniche, T., Askaa, J., Alsner, J., and Overgaard, J. The prognostic value of epidermal growth factor receptor is related to tumor differentiation and the overall treatment time of radiotherapy in squamous cell carcinomas of the head and neck. *Int J Radiat Oncol Biol Phys* 2004;58(2):561–566.
16. Huang, S., and Harari, P. Modulation of radiation response after epidermal growth factor receptor blockade in squamous cell carcinomas: Inhibition of damage repair, cell cycle kinetics, and tumor angiogenesis. *Clin Cancer Res* 2000;6:2166–2174.
17. Baumann, M., and Krause, M. Targeting the epidermal growth factor receptor in radiotherapy: Radiobiological mechanisms, preclinical and clinical results. *Radiother Oncol* 2004;72:257–266.
18. Baumann, M., Krause, M., Dikomey, E. et al. EGFR-targeted anti-cancer drugs in radiotherapy: Preclinical evaluation of mechanisms. *Radiother Oncol* 2007;83:238–248.
19. Milas, L., Fan, Z., Mason, K., and Ang, K., eds. *Role of Epidermal Growth Factor Receptor and Its Inhibition in Radiotherapy*. Berlin, Heidelberg: Springer, 2003.
20. Purschke, M., Kasten-Pisula, U., Brammer, I., and Dikomey, E. Human and rodent cell lines showing no differences in the induction but differing in the repair kinetics of radiation-induced DNA base damage. *Int J Radiat Biol* 2004;80:29–38.
21. Rodemann, H., Dittmann, K., and Toulany, M. Radiation-induced EGFR-signaling and control of DNA-damage repair. *Int J Radiat Biol* 2007;83: 781–791.
22. LoPiccolo, J., Blumenthal, G., Bernstein, W., and Dennis P. Targeting the PI3K/Akt/mTOR pathway: Effective combinations and clinical considerations. *Drug Resist Update* 2008;1(11):32–50.

23. Dittmann, K., Mayer, C., Fehrenbacher, B. et al. Radiation-induced epidermal growth factor receptor nuclear import is linked to activation of DNA-dependent protein kinase. *J Biol Chem Biol Interact* 2005;280(31):182–189.
24. Um, J., Kwon, J., Kang, C. et al. Relationship between antiapoptotic molecules and metastatic potency and the involvement of DNA-dependent protein kinase in the chemosensitization of metastatic human cancer cells by epidermal growth factor receptor blockade. *J Pharmacol Exp Ther* 2004;311:1062–1070.
25. Dittmann, K., Mayer, C., and Rodemann, H. Inhibition of radiation-induced EGFR nuclear import by C225 (Cetuximab) suppresses DNA-PK activity. *Radiother Oncol* 2005;76:157–161.
26. Toulany, M., Kasten-Pisula, U., Brammer, I. et al. Blockage of epidermal growth factor receptor-phosphatidylinositol 3-kinase-AKT signaling increases radiosensitivity of K-RAS mutated human tumor cells *in vitro* by affecting DNA repair. *Clin Cancer Res* 2006;12:4119–4126.
27. Yacoub, A., McKinstry, R., Hinman, D., Chung, T., Dent, P., and Hagan, M. Epidermal growth factor and ionizing radiation upregulate the DNA repair genes XRCC1 and ERCC1 in DU145 and LNCaP prostate carcinoma through MAPK signaling. *Radiat Res* 2003;159:439–452.
28. Eicheler, W., Krause, M., Hessel, F., Zips, D., and Baumann, M. Kinetics of EGFR expression during fractionated irradiation varies between different human squamous cell carcinoma lines in nude mice. *Radiother Oncol* 2005;76:151–156.
29. Baumann, M., Saunders, M., and Joiner, M., eds. *Modified Fractionation.* London: Arnold, 2002.
30. Bentzen, S., Atasoy, B., Daley, F. et al. Epidermal growth factor receptor expression in pretreatment biopsies from head and neck squamous cell carcinoma as a predictive factor for a benefit from accelerated radiation therapy in a randomized controlled trial. *J Clin Oncol* 2005;23:5560–5567.
31. Laderoute, K., Grant, T., Murphy, B., and Sutherland, R. Enhanced epidermal growth factor receptor synthesis in human squamous carcinoma cells exposed to low levels of oxygen. *Int J Cancer* 1992;52:428–432.
32. Perrotte, P., Matsumoto, T., Inoue, K. et al. Anti-epidermal growth factor receptor antibody C225 inhibits angiogenesis in human transitional cell carcinoma growing orthotopically in nude mice. *Clin Cancer Res* 1999;5:257–265.
33. Huang, S., Li, J., Armstrong, E., and Harari, P. Modulation of radiation response and tumor-induced angiogenesis after epidermal growth factor receptor inhibition by ZD1839 (Iressa). *Cancer Res* 2002;62:4300–4306.
34. Solomon, B., Binns, D., Roselt, P. et al. Modulation of intratumoral hypoxia by the epidermal growth factor receptor inhibitor gefitinib detected using small animal PET imaging. *Mol Cancer Ther* 2005;4:1417–1422.
35. Krause, M., Ostermann, G., Petersen, C. et al. Decreased repopulation as well as increased reoxygenation contribute to the improvement in local control after targeting of the EGFR by C225 during fractionated irradiation. *Radiother Oncol Rep* 2005;76:162–167.

36. Lammering, G., Hewit, T., Holmes, M. et al. Inhibition of the type III epidermal growth factor receptor variant mutant receptor by dominant-negative EGFR-CD533 enhances malignant glioma cell radiosensitivity. *Clin Cancer Res* 2004;10:6732–6743.
37. Rajput, A., Koterba, A., Kreisberg, J., Foster, J., Willson, J., and Brattain, M. A novel mechanism of resistance to epidermal growth factor receptor antagonism *in vivo*. *Cancer Res* 2007;67:665–673.
38. Erjala, K., Sundvall, M., Junttila, T. et al. Signaling via ErbB2 and ErbB3 associates with resistance and epidermal growth factor receptor (EGFR) amplification with sensitivity to EGFR inhibitor gefitinib in head and neck squamous cell carcinoma cells. *Clin Cancer Res* 2006;12:4103–4111.
39. Yano, S., Kondo, K., Yamaguchi, M. et al. Distribution and function of EGFR in human tissue and the effect of EGFR tyrosine kinase inhibition. *Anticancer Res* 2003;23:3639–3650.
40. Gueven, N., Dittmann, K., Mayer, C., and Rodemann, H. Bowman-Birk protease inhibitor reduces the radiation-induced activation of the EGF receptor and induces tyrosine phosphatase activity. *Int J Radiat Biol* 1998;73:157–162.
41. Wang, M., Morsbach, F., Sander, D. et al. EGF receptor inhibition radiosensitizes NSCLC cells by inducing senescence in cells sustaining DNA double-strand breaks. *Cancer Res* 2011;71(19):6261–6269.
42. Kruser, T., Armstrong, E., Ghia, A. et al. Augmentation of radiation response by panitumumab in models of upper aerodigestive tract cancer. *Int J Radiat Oncol Biol Phys* 2008;72(2):534–542.
43. Akashi, Y., Okamoto, I., Iwasa, T. et al. Enhancement of the antitumor activity of ionising radiation by nimotuzumab, a humanised monoclonal antibody to the epidermal growth factor receptor, in non-small cell lung cancer cell lines of differing epidermal growth factor receptor status. *Br J Cancer* 2008;98(4):749–755.
44. Stea, B., Falsey, R., Kislin, K. et al. Time and dose-dependent radiosensitization of the glioblastoma multiforme U251 cells by the EGF receptor tyrosine kinase inhibitor ZD1839 ('Iressa'). *Cancer Lett* 2003;202(1):43–51.
45. Kim, J., Ali, M., Nandi, A. et al. Correlation of HER1/EGFR expression and degree of radiosensitizing effect of the HER1/EGFR-tyrosine kinase inhibitor erlotinib. *Indian J Biochem Biophys* 2005;42(6):358–365.
46. Sambade, M., Kimple, R.J., Camp, J. et al. Lapatinib in combination with radiation diminishes tumor regrowth in HER2+ and basal-like/EGFR+ breast tumor xenografts. *Int J Radiat Oncol Biol Phys* 2010;77(2):575–581.
47. Huamani, J., Willey, C., Thotala, D. et al. Differential efficacy of combined therapy with radiation and AEE788 in high and low EGFR-expressing androgen-independent prostate tumor models. *Int J Radiat Oncol Biol Phys* 2008;71(1):237–246.
48. Aravindan, N., Thomas, C., Aravindan, S., Mohan, A., Veeraraghavan, J., and Natarajan, M. Irreversible EGFR inhibitor EKB-569 targets low-LET γ-radiation-triggered rel orchestration and potentiates cell death in squamous cell carcinoma. *PLoS One* 2011;6(12):e29705.

49. Nyati, M., Maheshwari, D., Hanasoge, S. et al. Radiosensitization by pan ErbB inhibitor CI-1033 *in vitro* and *in vivo*. *Clin Cancer Res* 2004;10(2):691–700.
50. Schütze, C., Dörfler, A., Eicheler, W. et al. Combination of EGFR/HER2 tyrosine kinase inhibition by BIBW 2992 and BIBW 2669 with irradiation in FaDu human squamous cell carcinoma. *Strahlenther Onkol* 2007;183(5):256–264.
51. Torres M., Raju U., Molkentine D., Riesterer O., Milas L., Ang K. AC480, formerly BMS-599626, a pan Her inhibitor, enhances radiosensitivity and radioresponse of head and neck squamous cell carcinoma cells in vitro and in vivo. Invest New Drugs. 2011;29(4):554–61.
52. Harari, P., and Huang, S. Head and neck cancer as a clinical model for molecular targeting of therapy: Combining EGFR blockade with radiation. *Int J Rad Oncol Biol Phys* 2001;49:427–433.
53. Deutsch, E., Kaliski, A., Maggiorella, L., and Bourhis, J. New strategies to interfere with radiation response: "Biomodulation" of radiation therapy. *Cancer Radiother Oncol* 2005;9:69–76.
54. Chinnaiyan, P., Huang, S., Vallabhaneni, G. et al. Mechanisms of enhanced radiation response following epidermal growth factor receptor signaling inhibition by erlotinib (Tarceva). *Cancer Res* 2005;65:3328–3335.
55. Li, L., Wang, H., Yang, E. et al. Erlotinib attenuates homologous recombinational repair of chromosomal breaks in human breast cancer cells. *Cancer Res* 2008;68:9141–9146.
56. Sarkaria, J., Carlson, B., Schroeder, M. et al. Use of an orthotopic xenograft model for assessing the effect of epidermal growth factor receptor amplification on glioblastoma radiation response. *Clin Cancer Res* 2006;12:2264–2271.
57. Zaidi, S., Huddart, R., and Harrington, K. Novel targeted radiosensitisers in cancer treatment. *Curr Drug Discov Technol* 2009;6:103–134.
58. Britten, C. Targeting ErbB receptor signaling: A pan-ErbB approach to cancer. *Mol Cancer Ther* 2004;3(10):1335–1342.
59. Li, D., Ambrogio, L., Shimamura, T. et al. BIBW2992, an irreversible EGFR/HER2 inhibitor highly effective in preclinical lung cancer models. *Oncogene* 2008;27:4702–4711.
60. Gueven, N., Dittmann, K., Mayer, C., and Rodemann, H. Bowman-Birk protease inhibitor reduces the radiation-induced activation of the EGF receptor and induces tyrosine phosphatase activity. *Int J Radiat Biol* 1998;73:157–162.
61. Tofilon, P., and Camphausen, K. Molecular targets for tumor radiosensitization. *Chem Rev* 2009;109:2974–2988.
62. Bonner, J., Harari, P., Giralt, J. et al. Radiotherapy plus cetuximab for locoregionally advanced head and neck cancer: 5-year survival data from a phase 3 randomised trial, and relation between cetuximab-induced rash and survival. *Lancet Oncol* 2010;11:21–28.
63. Nieder, C., Pawinski, A., Dalhaug, A., and Andratschke, N. A review of clinical trials of cetuximab combined with radiotherapy for non-small cell lung cancer. *Radiat Oncol* 2012;11(7):3.
64. López-Albaitero, A., and Ferris, R. Immune activation by epidermal growth factor receptor specific monoclonal antibody therapy for head and neck cancer. *Arch Otolaryngol Head Neck Surg* 2007;133(12):1277–1281.

65. Rodríguez, M., Rivero, T., Bahi, R. et al. Nimotuzumab plus radiotherapy for unresectable squamous-cell carcinoma of the head and neck. *Cancer Biol Ther* 2010;9(5):343–349.
66. Chang, A., Parikh, P., Thongprasert, S. et al. Gefitinib (IRESSA) in patients of Asian origin with refractory advanced non-small cell lung cancer: Subset analysis from the ISEL study. *J Thorac Oncol* 2006;1(8):847–855.
67. Mehta, V. Radiotherapy and erlotinib combined: Review of the preclinical and clinical evidence. *Front Oncol* 2012;2:1–11.
68. Czito, B., Willett, C., Bendell, J. et al. Increased toxicity with gefitinib, capecitabine, and radiation therapy in pancreatic and rectal cancer: Phase I trial results. *J Clin Oncol* 2006;24(4):656–662.
69. Valentini, V., Paoli, A.D., Gambacorta, M. et al. Infusional 5-fluorouracil and ZD 1839 (gefitinib-Iressa) in combination with preoperative radiotherapy in patients with locally advanced rectal cancer: A phase I and II Trial (1839IL/0092). *Int J Radiat Oncol Biol Phys* 2008;72:644–649.
70. Maurel, J., Martin-Richard, M., Conill, C. et al. Phase I trial of gefitinib with concurrent radiotherapy and fixed 2-h gemcitabine infusion, in locally advanced pancreatic cancer. *Int J Radiat Oncol Biol Phys Med Biol* 2006;66(5):1391–1398.
71. Valentini, V. et al. ASCO Annual Meeting Proceedings Part I. *J Clin Oncol* 2006;24(18S):5543.
72. Savvides, P., Agarwala, S., Greskovich, J. et al. Phase I study of the EGFR tyrosine kinase inhibitor erlotinib incombination with docetaxel and radiation in locally advanced squamous cell cancer of the head and neck (SCCHN). *J Clin Oncol* 2006;24:5545.
73. Arias de la Vega, F., Herruzo, I., de las Heras, M. et al. Erlotinib and chemoradiation in patients with surgically resected locally advanced squamous head and neck cancer (HNSCC): A GICOR phase I study. *J Clin Oncol* 2008;26:6068.
74. Herchenhorn, D., Dias, F., Viegas, C. et al. PhaseI/II study of erlotinib combined with cisplatin and radiotherapy in patients with locally advanced squamous cell carcinoma of the head and neck. *Int J Radiat Oncol Biol Phys* 2010;78:696–701.
75. Iannitti, D., Dipetrillo, T., Akerman, P. et al. Erlotinib and chemoradiation followed by maintenance erlotinib for locally advanced pancreatic cancer: A phase I study. *Am J Clin Oncol* 2005;28(6):570–575.
76. Krishnan, S., Brown, P.D., Ballman, K. et al. Phase I trial of erlotinib with radiation therapy in patients with glioblastoma multiforme: Results of North Central Cancer Treatment Group protocol N0177. *Int J Radiat Oncol Biol Phys* 2006;65(4):1192–1199.
77. Dobelbower, M., Russo, S., Raisch, K. et al. Epidermal growth factor receptor tyrosine kinase inhibitor, erlotinib, and concurrent 5-fluorouracil, cisplatin and radiotherapy for patients with esophageal cancer: A phase I study. *Anticancer Drugs* 2006;17(1):95–102.
78. Pegram, M., Konecny, G., and Slamon, D. The molecular and cellular biology of HER2/neu gene amplification/overexpression and the clinical development of herceptin (trastuzumab) therapy for breast cancer. *Cancer Treat Res* 2000;103:57–75.

79. Pietras, R., Poen, J., Gallardo, D., Wongvipat, P., Lee, H., and Slamon, D. Monoclonal antibody to HER-2/neu receptor modulates repair of radiation-induced DNA damage and enhances radiosensitivity of human breast cancer cells overexpressing this oncogene. *Cancer Res* 1999;59:1347–1355.
80. Chargari, C., Idrissi, H., Pierga, J. et al. Preliminary results of whole brain radiotherapy with concurrent trastuzumab for treatment of brain metastases in breast cancer patients. *Int J Radiat Oncol Biol Phys* 2011;81(3):631–636.
81. Wen, B., Deutsch, E., Marangoni, E. et al. Tyrphostin AG 1024 modulates radiosensitivity in human breast cancer cells. *Br J Cancer* 2001;85(12): 2017–2021.
82. Cosaceanu, D., Carapancea, M., Castro, J. et al. Modulation of response to radiation of human lung cancer cells following insulin-like growth factor 1 receptor inactivation. *Cancer Lett* 2005;222(2):173–181.
83. Nakamura, S., Watanabe, H., Miura, M., and Sasaki, T. Effect of the insulin-like growth factor I receptor on ionizing radiation-induced cell death in mouse embryo fibroblasts. *Exp Cell Res* 1997;235(1):287–294.
84. Macaulay, V., Salisbury, A., Bohula, E., Playford, M., Smorodinsky, N., and Shiloh, Y. Downregulation of the type 1 insulin-like growth factor receptor in mouse melanoma cells is associated with enhanced radiosensitivity and impaired activation of Atm kinase. *Oncogene* 2001;20(30):4029–4040.

CHAPTER 11

Targeting Signaling Molecules for Radiosensitization

A VARIETY OF MOLECULES INVOLVED in transducing environmental signals through the cytoplasm to the nucleus have been shown to play a role in determining radiosensitivity. This chapter will deal with the downstream components signaling pathways initiated by growth factor receptors. Many of these molecules are mutated, abnormally expressed, or have alternative functions in neoplastic cells, they have received considerable attention as tumor-specific targets for radiosensitization.

11.1 Ras

These small GTP-binding proteins are an early component in a number of signal transduction pathways and play critical roles in the regulation of cell proliferation, differentiation, and oncogenic transformation. Mutated Ras has been detected in approximately 30% of human tumors with aberrant activity present in many more tumors due to abnormal upstream signaling activity.

11.1.1 Function of Ras

Ras is a proto-oncogene that regulates signal transduction pathways for many cellular functions including growth, motility, apoptosis, and differentiation by participating in signal transduction pathways downstream

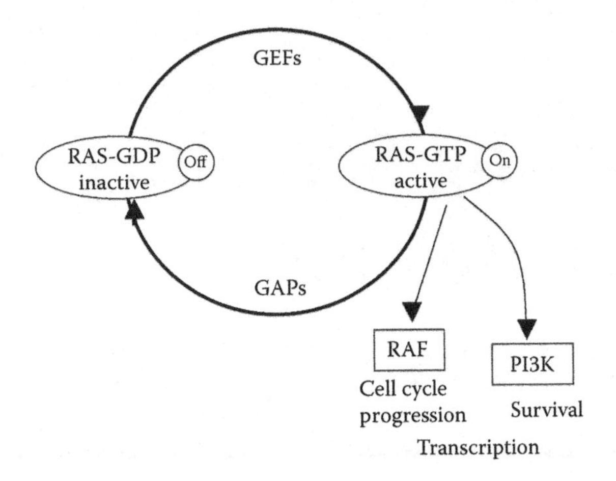

FIGURE 11.1 Ras protein activation and downstream signaling. Ras cycles between inactive GDP bound state and active GTP-bound state. The exchange of GDP for GTP is regulated by guanine nucleotide exchange factors (GEFs). GTP hydrolysis requires GTPase-activating proteins (GAPs), which enhance the weak intrinsic GTPase activity of Ras proteins. In its active form, Ras interacts with different families of effector proteins including RAF protein kinases and PI3K.

of the receptor tyrosine kinases. Following receptor activation, Ras is converted from its inactive GDP-loaded form to its active GTP-loaded form. Wild-type Ras exists in equilibrium between its active and inactive forms whereas mutated forms tend to remain activated and are found in up to 30% of most solid tumors. Ras-GTP acts on a group of small GTP-binding proteins, which are an early component in a number of signal transduction pathways (Figure 11.1).

11.1.2 Downstream Signaling from Ras

Upon activation, Ras can activate a number of different signal transduction pathways. The first RAS effector pathway identified was the RAF–MEK–ERK pathway. This pathway is an essential, shared element of mitogenic signaling involving tyrosine kinase receptors; leading to a wide range of cellular responses, including growth, differentiation, inflammation, and apoptosis. The RAF family of proteins (Raf-1, A-Raf, and B-Raf) are serine/threonine kinases that bind to the effector region of RAS-GTP, thus inducing translocation of the protein to the plasma membrane. The second best-characterized RAS effector family is phosphoinositide 3-kinases (PI3Ks). PIP3 stimulates the AKT/PKB kinase and several of the Rac-GEFs, such as Sos1 and Vav. AKT activation inhibits apoptosis by inhibiting the actions

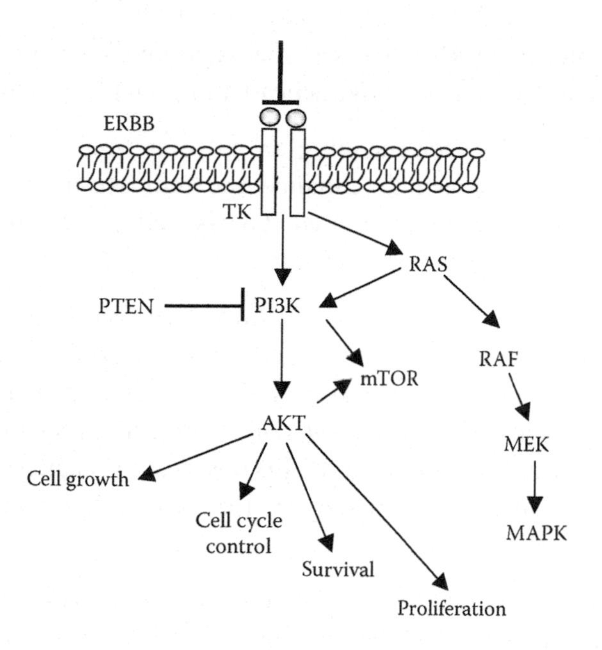

FIGURE 11.2 PI3K–AKT pathway and other pathways downstream of ERBB tyrosine kinase receptors. *Abbreviations*: AKT, protein kinase B; MAPK, mitogen-activated protein kinase. mTOR, mammalian target of rapamycin; PI3K, phosphatidylinositol-3-kinase; PTEN, phosphatase and TENsin homologue; TK, tyrosine kinase.

of Bad, caspase-9, and AFX. AKT further hinders apoptosis by phosphorylating the IκB repressor of NFκB. This pathway plays an important role as mediator of RAS-mediated cell survival and proliferation (Figure 11.2).

11.1.3 Ras and Radioresistance

There is ample evidence from *in vitro* systems that the expression of mutated Ras proteins is associated with a reduction in cellular radiosensitivity:

- In 1988, Sklar [1] showed that the resistance of NIH 3T3 cells to radiation was enhanced by the expression of a mutant H-ras or K-ras gene. In addition, the increased clonogenic survival was not simply a nonspecific consequence of transformation and was confirmed by showing that cell lines transformed by *v-myc*, *v-abl*, or *v-fms* were not more radioresistant than the parental cell lines.
- The genetic inactivation of oncogenic N-ras or K-ras in human colon tumor cell lines was found to increase radiosensitivity [2], whereas

small interfering RNA (siRNA)-mediated knockdown of K-ras had a similar effect radiosensitizing head and neck tumor cell lines [3,4].

- An adenoviral vector encoding an anti-Ras single-chain antibody fragment was shown to enhance the radiosensitivity of human tumor cells containing mutated Ras, as well as cells in which Ras protein was constitutively active [4].
- In studies with primary cell cultures, transfection of rat embryo fibroblast (REF) cells with oncogenic *ras* only produced a moderate increase in radioresistance. However, cotransfection of oncogenic *ras* with adenovirus EIA or the *v-myc* oncogene synergistically increased cell survival after radiation [5]. This observation suggested that radiation resistance, as mediated by *ras*, may be dependent on the genetic background of the cell.
- Another approach to defining the role of *ras* in radiation resistance is to assess the effect of inhibition of *ras* expression in mammalian tumor cell lines on radiosensitivity. Thus, in human colon cancer cell lines, deletion of the *K-ras* allele was shown to result in significantly decreased clonogenic survival after exposure to ionizing radiation (IR) [2].
- In experiments with human tumor cell lines, antisense oligonucleotides directed against H-ras increased the radiosensitivity of cells expressing mutant but not normal H-ras [6].

11.1.3.1 Which Downstream Pathway Mediates Radioresistance?

A causal link can be established between aberrant Ras activity and tumor cell radioresistance without defining the specific mechanisms involved. The mechanisms invoked have, for the most part, been limited to cytoplasmic signaling molecules focusing on the Ras pathway proceeding through Rafl/mitogen-activated protein kinase (MAPK) or PI3K/Akt. Activation of the MAPK pathway has been shown to mediate radioresistance in glioblastoma cells, whereas inhibition of this pathway was shown to enhance radiation-induced cell toxicity in both mammary carcinoma and prostate carcinoma cells [7]. This finding is not consistent for all cell lines, however, suggesting the existence of alternate pathways for Ras-mediated radioresistance [8]. A possible role for Raf-l, an intermediate in the MAPK pathway, in mediating radiation resistance was suggested by cell culture experiments demonstrating increased survival after irradiation when Raf-1 was inhibited using antisense oligonucleotides [9], and by the fact that transfection of a human squamous cell carcinoma cell line with constitutively active form of Raf-1 led to increased radioresistance [10]. However, experiments with PD98059, an inhibitor of MEK (downstream of RAF-1), in cells expressing mutant p53 and oncogenic *ras* have demonstrated no increase in radiosensitivity in clonogenic cell survival assays [11].

These data suggest that Raf-stimulated radioresistance is independent of MEK–ERK activation and that Ras-mediated radioresistance does not likely signal through MAPK. Furthermore, PI3K activity was found to be necessary for the radioresistance of Ras-transformed cells, implicating it as the critical effector pathway [10], which further suggested an essential role for Akt. The role of the PI3K–Akt–mTOR pathway in radiation response will be discussed in more detail in Section 11.2.2.

Contributing to the limited understanding of the mechanisms of Ras-mediated radioresistance is the complexity of the Ras signaling network [12]. In addition to the frequently studied Raf1 and PI3K are a number of Ras effector molecules, as well as various isoforms. These effectors activate a variety of distinct yet often interacting downstream signaling pathways ultimately influencing fundamental processes such as gene expression, second messenger activation, apoptosis, and cell cycle regulation.

11.1.4 Inhibitors of Ras

For Ras to function, it must first be activated via a posttranslational addition of a hydrophobic molecule to facilitate attachment to the cell membrane; a process that is undertaken by the enzyme farnesyltransferase. To date, strategies aimed at exploiting Ras as a target for tumor radiosensitization have focused on farnesyltransferase and geranyl-geranyl transferase, which mediate the prenylation of the carboxy terminus of Ras proteins, an event necessary for their attachment to the cell membrane and subsequent activation [13].

11.1.4.1 Radiosensitization of Tumor Cell Lines by Inhibition of Ras

Prenyl transferase inhibitors (PTIs) were shown to enhance the *in vitro* radiosensitivity of a variety of human tumor cell lines including pancreatic, lung, colon, and breast carcinomas, as well as gliomas [14], which expressed oncogenic forms of Ras. Importantly, PTIs had no effect on tumor cells that contained wild-type Ras or on normal fibroblasts [11]. In a leg tumor xenograft model, PTI delivery significantly enhanced the *in vivo* radioresponse of H-ras mutant T24 bladder carcinoma cells and the K-ras mutant MiaPaca-2 and PSN-1 pancreatic carcinoma cells [15].

On the basis of such results, PTIs have received considerable attention as "anti-Ras" drugs; a number have undergone clinical evaluation in combination with radiotherapy in Section 11.1.4.3. However, although PTIs inhibit Ras prenylation and activation, they also inhibit the prenylation of a variety of other proteins. More than 100 proteins have been identified as requiring the posttranslational modification of prenylation [16]. The consequence of inhibiting the prenylation of proteins other than Ras on radiosensitivity has not been completely defined. However, among the proteins whose prenylation and activity are inhibited by PTIs are RhoB and Rheb. Inhibition of RhoB using a genetic approach and by PTIs was found to enhance the radiosensitivity of glioma cells [14], while inhibition of Rheb prenylation by a PTI reduced its activity along with that of mTOR and enhanced the antitumor activity of taxane and tamoxifen [17]. The effects of inhibiting Rheb on radioresponse has not been specifically tested but mTOR inhibition has been associated with an increase in radiosensitivity. The lack of specificity of the PTIs for Ras clearly complicates the interpretation of experimental and clinical studies.

11.1.4.2 Evaluation of Ras Inhibitors in Preclinical Models

Results of studies of farnesyl transferase inhibitors (FTIs) using tissue culture or xenograft models are summarized in Table 11.1. As described in Section 11.1.4.2, PTIs have been shown to enhance the *in vitro* radiosensitivity of a variety of human tumor cell lines including pancreatic (T23), lung, colon, and breast carcinomas [18], which expressed oncogenic forms of Ras. Significantly, PTIs had no effect on tumor cells that contained wild-type Ras or on normal fibroblasts [11]. In xenograft models, PTI delivery significantly enhanced the *in vivo* radiosensitivity of H-ras mutant T24 bladder carcinoma cells [19], the K-ras mutant MiaPaca-2, and PSN-1 pancreatic carcinoma cells [15].

11.1.4.3 Clinical Trials with Ras Inhibitors

Despite impressive preclinical data, monotherapy with FTIs in human clinical trials has shown limited antitumor activity in hematopoietic cancers and, generally, no or very little activity in solid tumors. This conclusion is based on the results of clinical studies, starting in 2000, which evaluated four FTIs in at least 75 clinical trials (tipifarnib, lonafarnib, BMS-214662, and L-778123). In 64 of these studies (with 35 being phase I trials), the clinical response was determined (reviewed by Berndt et al. [20]). When used in combination with other agents, FTIs have fared slightly better. For example, phase I studies

TABLE 11.1 Ras Inhibitors Preclinical Models

Tumor Model	Treatment	Results	Conclusions	Reference
Bladder, breast, colon, cervix, lung cell lines.T24, H5578T (act K-ras) SW480, A549 (act H-ras) SKBr-3, HT29, HeLa (wt)	RT plus FTI-277 and GGTI-298	Clonogenic assay. Sensitization by PTIs in cell lines with oncogenic H-ras, K-ras mutations. No sensitization in cell lines expressing wild-type ras	Inhibition of oncogenic activity human tumor cells radiosensitizes these cells, demonstrate the potential of using PTIs with RT for cancer treatment	[18]
Human xenografts, T24 (activated *ras*) HT29 (wt *ras*)	PTI, FTI-276, L 744,832 continuous administration for 7 days	For T24 FTI + RT, reduced clonogenic survival and slowed tumor regrowth to greater extent than either treatment alone	Synergistic action of FTI and radiation against tumors with activated ras	[19]
Seven pancreatic cancer lines (activated ras), two cell lines with wt *ras*	siRNA knockdown of *K-ras*. Panel of PTIs of differing specifications for FT and GGT	Sensitization of pancreatic Ca lines *in vitro* and after radiation *in situ* (clonogenic assay)	Inhibition of activated K-ras promotes radiation superadditive killing pancreatic carcinoma	[15]
Radioresistant hepatocarcinoma in C3H/HeJ mice (wt *ras*)	PTI LB42907 60 mg/kg, twice daily for 30 days + 25 Gy radiation	RT + FTI increased radiation-induced apoptosis; ↓ ras by FTI greatest at 4 h posttreatment. ↑ tumor response to irradiation (TGD) (ER = 1.32)	FTI + RT potentially effective in absence of ras mutation if FTI inhibits ras activity and other proteins requiring farnesylation for activity	[3]
Radioresistant glioma (U87, SF763, RS SF767, U251-MG. U87 transfected with dominant negative forms of Ras or Rho-B	FTI R115777 48 h before 2 Gy	RR glioma survival reduced 45% by drug, RS glioma survival unchanged. Blocking RhoB reduced U87 survival after radiation	Rho-B, not Ras implicated in U87 resistance	[14]

TABLE 11.2 Clinical Trials with Ras Inhibitors

Phase	Treatment	Tumor Site	Outcome
I	FTI L778,123 (continuous infusion) and RT	3 patients with HNSC, 6 patients with NSCLC	DLT of neutropenia (1 patient) CR: 3 patients with NSCLC, 2 patients with HNSCC PR: 1 patient with NSCLC
I	L778,123 and radiation	12 patients with locally advanced pancreatic cancer	C in 2 (1 patient with diarrhea, one patient with gastrointestinal hemorrhage)/4 patients with escalated dose; PR: 1 patient at the lower dose [21]
I	FTI tipifarnib (R115777) and concurrent RT followed by adjuvant use until disease progression	13 patients with glioblastoma multiforme	DLT at the escalated dose (1 death, 2 with pneumonitis). PR in 1/9 patients, SD in 4/9 patients, and PD in 3/9 patients had progressive disease [22]
I/II	Lovostatin + RT	Anaplastic glioma, glioblastoma multiforme	Minor response in 2/9 patients, PR in 2/9 patients, and progressive disease in 5/9 patients [23]
I (Multi-institutional)	FTI L778,123 + RT	NSCLC, HNSC	Antitumor activity similar to standard radiochemotherapy. Q–T prolongation in EKG measurements, halted further clinical evaluation [24]
I	FTI R115777, with radiation	Glioblastoma multiforme	No unusual or excessive normal tissue reaction

Note: CR, complete response; DLT, dose-limiting toxicity; FTI, farnesyl transferase inhibitor; HNSCC, head and neck squamous cell carcinoma; NSCLC, non-small cell lung cancer; PD, progressive disease; PR, partial response; RT, radiotherapy; SD, stable disease.

TABLE 11.3 Ongoing Clinical Trials with FTIs (www.clinicaltrials.gov)

Phase	FTI	Tumor Site	FTI Combined with
II	Tipifarnib (FTI R115777)	Locally advanced pancreatic cancer	RT, gemcitabine, paclitaxel
I	Tipifarnib (FTI R115777	Juvenile myelomonocytic leukemia in children	13-Cis retinoic acid, cytosine arabinoside, fludarabine, hematopoietic stem cell transplant
I	Tipifarnib (FTI R115777	Locally advanced pancreatic cancer	RT
	Tipifarnib (FTI R115777	Lung (stage III NSCSC)	RT, carboplatin, paclitaxel

based on a combination of tipifarnib with gemcitabine and cisplatin have shown some promise in advanced solid tumors (33.3% complete response rate or 26% partial response rate). Similarly, in phase II neoadjuvant settings, tipifarnib increases the rate of pathological complete responses from the historical 10% to 25% when combined with chemotherapy (doxorubicin and cyclophosphamide) in patients with locally advanced breast cancer.

The results of some published clinical trials in which Ras-targeted drugs were combined with radiotherapy are summarized in Table 11.2, and details of ongoing or recently completed trials are shown in Table 11.3. Clinical evaluation of PTI/radiotherapy combination in cancer treatment has thus far shown little evidence that an enhancement of radiosensitivity occurred in any of the tumor subtypes studied [25]. As indicated by the results of *in vitro* studies, Ras-mediated radioresistance is likely to be highly dependent on the genetic background of the tumor [25], and recent and ongoing trials have mostly incorporated biomarker analysis. However, an additional weakness in the clinical trials putatively targeting Ras has been the trials' sole reliance on PTIs. The large number of proteins other than Ras whose function may be affected is likely a complicating factor, as are the redundancies/interactions between the two forms of Ras prenylation, that is, those mediated by the farnesyltransferase and those mediated by geranyl-geranyl transferase. Thus, in addition to a thorough mechanistic understanding, a critical aspect in translating molecular target results from the laboratory to the clinic is the availability of a specific and effective targeting agent.

11.2 CYTOPLASMIC SIGNALING DOWNSTREAM FROM Ras

As described previously, a causal link between aberrant Ras activity and tumor cell radioresistance is well established. Investigations have, for the most part, focused on cytoplasmic signaling routed through the Ras pathways and proceeding through the Raf1/MAPK and PI3K/Akt routes. Data generated from both human and rodent models suggested that although Raf may be involved, Ras-mediated radioresistance does not involve MAPK [8,11]. PI3K activity was found to be necessary for the radioresistance of Ras-transformed cells, implicating it as the critical effector pathway [8,11], which further suggested an essential role for Akt.

11.2.1 The PI3K–Akt–mTOR Pathway

The PI3K pathway is an important signal transduction pathway commonly activated in cancer (Figure 11.3). There are three PI3K classes, with different structures and characteristics; class I can be further subdivided into class Ia and class Ib. Most of the oncogenic mutations occur in the p110α subunit of class Ia PI3Ks, and this group has received the most attention. The PI3K pathway can be activated by upstream receptor tyrosine kinases leading to phosphorylation of phosphatidylinositol-4,5-bisphosphate (PIP2) to generate phosphatidylinositol-3,4,5-trisphosphate (PIP3). PIP3 can be dephosphorylated by the phosphatase and tensin homologue (PTEN), which terminates PI3K signaling. The accumulation of PIP3 activates a signaling cascade starting with the phosphorylation (activation) of the protein serine-threonine kinase Akt at threonine 308 by phosphoinositide-dependent kinase 1 (PDK1). Through several downstream proteins, the activation of Akt results in the stimulation of protein translation and cell proliferation, and the inhibition of apoptosis. Through tuberous sclerosis complex (TSC) and RAS homologue enriched in brain (RHEB), Akt activates the mammalian target of rapamycin (mTOR), which leads to phosphorylation of its downstream targets, the translation regulators, 4EBP1 and S6K1. mTOR is segregated in at least two complexes: the raptor-containing complex mTORC1 and the rictor-containing mTORC2. Only mTORC1 is sensitive to inhibition by rapamycin, although prolonged rapamycin treatment can also partially inhibit mTORC2 activity. mTORC1 integrates proliferative signals from the PI3K pathway with an energy-sensing pathway, which is triggered by the availability of ATP and amino acids, and is linked to mTOR by the Peutz–Jeghers tumor-suppressor serine-threonine kinase 11 (STK11). mTOR residing in the mTORC2 complex is an important effector of pathways regulating cell growth and survival; it phosphorylates and activates Akt at the Ser473 position, placing the mTOR protein upstream as well as downstream of Akt (Figure 11.3).

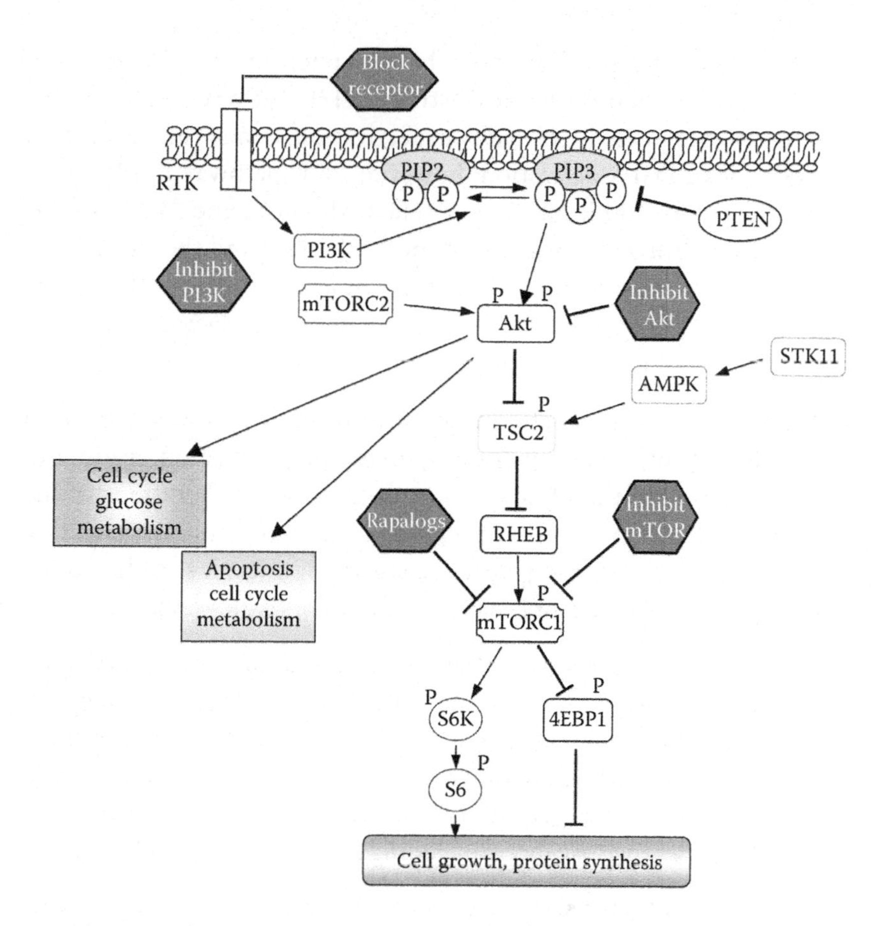

FIGURE 11.3 The PI3K pathway and main targets for therapeutic intervention. The PI3K pathway integrates growth factor stimulation (RTK) with the availability of nutrients (AMPK) and can be activated in cancer by amplification or mutation of upstream RTKs, such as HER2 and EGFR, by downstream activating mutations or by inactivating deletions or mutations (e.g., PTEN). The PI3K pathway can be inhibited at several levels for therapeutic gain.

11.2.2 Activation of the PI3K Pathway in Cancer

Activation of the oncogenic PI3K pathway can occur through several mechanisms:

- Hyperactivation of upstream RTKs, such as the human epidermal growth factor receptor 2 (ErbB2) and the epidermal growth factor receptor (EGFR). Both ErbB2 and EGFR have been targeted successfully by antibodies and small-molecule inhibitors (see Chapter 10).

ErbB2-positive cancers seem to be dependent on PI3K activation, because inhibition of the downstream PI3K pathway leads to substantial apoptosis. PI3K pathway activation is also an important mediator of EGFR activation; however, PI3K pathway inhibition is insufficient for full inhibition of cells with oncogenic EGFR activation. Simultaneous inhibition of the MEK/ERK and the PI3K pathways is required for apoptosis to occur at levels similar to those seen with EGFR inhibitors in EGFR-dependent cells, indicating that both pathways are important [26].

- The first identified genetic mechanism of PI3K pathway activation was the loss of PTEN function by mutation or deletion, leading to accumulation of the PI3K product PIP3. Somatic alterations of the PTEN gene occur in a diverse range of human cancers, including endometrial, brain, and prostate cancers; in mice, loss of heterozygous PTEN leads to neoplasia in a number of tissues [27].
- The PIK3CA gene, encoding the p110α subunit of class IA PI3K, has been found to contain activating mutations in a significant proportion of colorectal cancers [28] and to be commonly mutated in other cancer types including breast, endometrial, lung, and cervical cancer. Mutations in the p85α regulatory subunit of PI3K (PIK3R1) have been reported for a range of cancers including glioblastoma and colon cancer.
- Although oncogenic activation of KRAS is believed to exert its functions mainly through the MAPK pathway, it has become clear that the PI3K pathway is an important effector of mutant KRAS as well. RAS can bind and activate PI3K and mice carrying a PIK3CA mutation that blocks the interaction with RAS become resistant to RAS oncogene-induced tumorigenesis [29].

11.2.3 Inhibitors of the PI3K–Akt–mTOR Pathway

As previously described, the PI3K pathway can be activated by amplification or activating mutation of upstream receptor tyrosine kinases, and by mutations or deletions downstream in the pathway (Figure 11.3). Similarly, inhibition can occur at several levels in the pathway. Inhibition of upstream components of the pathway effectively block the activity of the PI3K pathway; an example being trastuzumab, a monoclonal antibody targeting ErbB2, which has been one of the most successful and widely used targeted therapies. However, many ErbB2-positive cancers are not sensitive to ErbB2-based therapies or become resistant during treatment; activation of the pathway downstream being one of the causes of resistance.

11.2.3.1 Targeting PI3K

Clinically relevant methods of inhibiting PI3K pathway by use of pharmacological compounds are difficult to achieve because of the ubiquitous presence of PI3K in many mammalian cells, which may render such compounds significantly toxic. Compounds that target specific isoforms of PI3K have been introduced in an attempt to confer improved selectivity of the compound for appropriate cellular targets. For example, IC486068 from ICOS targets the p110δ isoform of the PI3K proteins. Combination of IC486068 with conventional fractionated doses of IR led to increased apoptosis, decreased clonogenicity, and decreased migration of human endothelial cells *in vitro* and to tumor vasculature destruction *in vivo* [24]. This finding translated to increased tumor growth delay in xenograft models of GL261 and LLC *in vivo* [24]. Although some drugs only target specific PI3K isoforms, others have a broader spectrum including both p110α and mTOR. Several PI3K inhibitors are in clinical development. Phase I/II trials assessing the dual PI3K/mTOR inhibitors BEZ235, BGT226, and XL765, and the PI3K inhibitor XL147 are being undertaken.

11.2.3.2 Targeting Akt

Akt is the most important downstream target of PI3K, and several inhibitors of Akt have been developed. Perifosine is a phosphatidylinositol analogue and has been tested in several phase I and II trials. MK2206, an allosteric inhibitor of Akt1 and Akt2, is being clinically tested in combination with various chemotherapy regimens, ErbB2 inhibitors, and with the MEK inhibitor AZD6244. The combination of a downstream PI3K pathway inhibitor with a MEK inhibitor is significant because it has become apparent that the PI3K and the MAPK pathways are interconnected, and inhibition of one pathway can be bypassed by the activation of the other.

11.2.3.3 Targeting mTOR

mTOR inhibitors can be divided into two classes: the rapalogs, of which rapamycin is the prototype, and the small-molecule mTOR inhibitors. Rapalogs form a complex with FK506 binding protein 12 (FKBP12), binding and inhibiting mTORC1. Three rapalogs are under investigation in clinical cancer trials: the rapamycin prodrug temsirolimus and the rapamycin analogues everolimus and deforolimus.

As a class of agents, rapalogs have been reported to have activity against a wide range of malignancies, including mantle cell lymphoma, sarcoma, and renal cancer [30]. Some encouraging results have been achieved in the treatment of cancer patients with rapalogs but, in most cases, a clear improvement in survival has not been shown [31]. A major reason for the lack of effect might be the existence of a negative feedback loop from mTORC1, through S6K1, to upstream PI3K pathway activity [32]. The existence of feedback between mTORC1 and AKT has led to the development of compounds that can directly inhibit mTOR kinase activity in both complexes, including AZD8055 and OSI-027, which are being tested in phase I/II trials.

11.2.4 The PI3K Pathway and Radiation Response

As already described, a causal link between aberrant Ras activity and tumor cell radioresistance is well established. Investigations of human and rodent models focused on cytoplasmic signaling through the Raf1/MAPK and PI3K/Akt pathways downstream of Ras have indicated that although Raf may be involved in radioresistance, Ras-mediated radioresistance does not involve MAPK. Thus, PI3K activity was found to be necessary for the radioresistance of Ras-transformed cells, implicating it as the critical effector pathway and suggesting, in some circumstances, an important role for Akt. Inhibition of Akt has been reported to modestly enhance the radiosensitivity of some tumor cell lines [3], but to be ineffective in others. Radiosensitization by inhibition of PI3K/Akt signaling has been shown to be brought about by a number of mechanisms.

11.2.4.1 Tumor Cell Apoptosis

One of the characteristics of cancer cells is their ability to evade programmed cell death. It has been proposed that activated Akt promotes survival of cells exposed to IR through inhibition of apoptosis. Apoptosis induced by DNA damage is typically associated with the activation of a family of proteases, the caspases, as a result of a sequence of mitochondria-mediated events. Antiapoptotic activity of Akt has been shown to result from Akt-dependent phosphorylation of proapoptotic Bad and caspase-9 proteins causing their inactivation.

11.2.4.2 Destruction of the Tumor Vasculature by Radiation-Induced Apoptosis

The tumor microenvironment, in particular, the tumor vasculature, has been established as an important target for the cytotoxic effects of radiation therapy. Endothelial cells, although sensitive to IR at higher doses, are less sensitive to RT doses in the clinical dose range of 2 to 3 Gy, in part, because of the activation of the PI3K/Akt cell viability signaling. PI3K/Akt pathway activated in endothelial cells exposed to IR thus presents an important molecular target for the development of radiation. Several preclinical studies have used the inhibition of the PI3K/Akt activation as a stratagem for tumor radiosensitization by targeting the tumor vasculature (reviewed by van der Heijden and Bernards [31]).

11.2.4.3 DNA Double-Strand Break Repair

Nonhomologous end-joining (NHEJ) is responsible for the repair of a significant proportion of DNA double-strand breaks (DSB) caused by irradiation. An important protein involved in this repair machinery is the DNA-dependent protein kinase catalytic subunit (DNA-PKcs) and EGFR-initiated signaling by the PI3K–Akt pathway has been shown to be involved in the regulation of DNA-PKcs and of DNA repair (see Chapter 10). Stimulation of EGFR, after homodimerization or heterodimerization with other members of the ERbB family, results in an increase in the nuclear content of DNA-PK subunits, thereby enhancing the DNA-PK-dependent NHEJ system. Blockage of the EGFR tyrosine kinase domain by the pyrimido-pyrimidine BIBX1382 had a cytotoxic effect in K-RAS-mutated cells (A549) but not in wild-type K-RAS (FaDu) cells, which were resistant, emphasizing the importance of K-RAS in these repair processes [33]. Selective inhibition of this pathway by tyrosine kinase inhibition in combination with radiotherapy leads to an increase in residual DNA DSBs, indicative of impaired DNA repair in K-RAS-mutated cells. The targeting of Akt activity by siRNA sensitizes human tumor cells to IR [34]. RAS activation by mutation or by receptor tyrosine kinase activity is a frequent event in human tumors, suggesting that the PI3K/Akt-mediated repair of DNA damage might be an important mechanism of radioresistance.

11.2.4.4 Hypoxia: Inhibition of Hypoxia-Inducible Factor 1α Induction and Normalization of Tumor Vasculature

Cancer cells have the ability to undergo genetic and adaptive changes in response to hypoxia that allows them to survive and even proliferate under hypoxic conditions. The hypoxia-inducible factor 1 (HIF-1) protein is a key transcription factor induced by hypoxia, which modulates the expression of genes and their protein products involved in tumor growth and apoptosis. Upon activation, HIF-1 leads to upregulation of many gene products, including vascular endothelial growth factor (VEGF), the glucose transporters (GLUT-1 and GLUT-3), and carbonic anhydrase IX. Under normoxic conditions, HIF-1α is rapidly deactivated by binding to the von Hippel–Lindau protein or by a factor inhibiting HIF-1, preventing further transcription. However, under hypoxic conditions, transcriptional activity is regulated through the PI3K/Akt pathway and activation of this pathway results in increased transcription and expression of HIF-1. PI3K–Akt–mTOR inhibitors not only possess antitumor effects but can also alter tumor vasculature and oxygenation to improve the response to radiation and chemotherapy. In many cases, these changes are related to the downregulation of HIF-1α and VEGF. These effects are particularly well-documented from experiments with tissue culture and xenograft tumor systems (reviewed by Schuurbiers et al. [35]).

- In experiments with human prostate cancer cells, HIF-1 transcription was shown to be blocked in the absence of Akt or PI3K and stimulated by constitutively active Akt or dominant-negative PTEN [36].
- In breast cancer cell lines, interruption of the PI3K–Akt pathway by the selective PI3K inhibitor, LY294002, inhibited HIF-1α induction—resulting in a reduction of VEGF expression by 50% [37].

Interconnection between hypoxia, angiogenesis, and radiosensitivity was shown in lung carcinoma xenografts in which AKT signaling was inhibited by the protease inhibitor nelfinavir resulting in a decrease of both VEGF and HIF-1α expression leading to both a reduction of angiogenesis and a decrease in hypoxia in response to radiation [38]:

- NVP-BEZ235, a dual inhibitor of PI3K and mTOR, has been shown to improve tumor oxygenation and vascular structure over a prolonged period. In experiments designed to distinguish effects on the vasculature from those on tumor cells, drug administration coincident with radiation enhanced the delay in tumor growth without changing tumor oxygenation, establishing that radiosensitization is a component of the response. However, the enhanced growth delay was substantially greater after induction of vascular normalization, meaning that this treatment enhanced the tumoral radioresponse. Importantly, changes in vascular morphology persisted throughout the entire course of the experiment. The findings indicated that targeting the PI3K–mTOR pathway could modulate the tumor microenvironment to induce a prolonged normalization of blood vessels. The combination of NVPBEZ235 with irradiation resulted in substantial therapeutic gain [39].

- In a study with prostate cancer cells, BEZ235 rapidly inhibited PI3K and mTOR signaling in a dose-dependent manner and limited tumor cell proliferation and clonogenic survival in both cell lines independently of PTEN status. *In vivo*, BEZ235 pretreatment enhanced the efficacy of radiation therapy on PC3 xenograft tumors in mice without inducing intestinal radiotoxicity [40] (Figure 11.4).

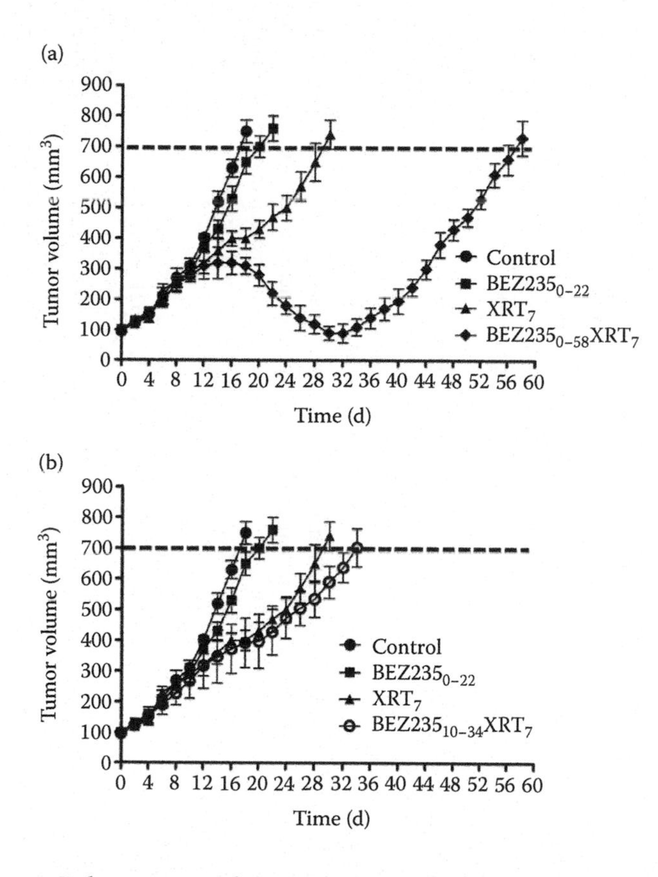

FIGURE 11.4 Enhancement of tumor radiosensitivity but not intestinal radiotoxicity by BEZ235. (a) PC3 tumor volume after 4 Gy/fraction ± 50 mg/kg BEZ235 2 h prior to irradiation for three consecutive days (mean ± SEM, $n \geqslant 6$). (b) Morbidity (% mice with tumors > 500 mm^3, $n \geqslant 12$). (From Potiron, V. et al., *Radiother Oncol* 2013;106(1):138–146. With permission.)

TABLE 11.4 Effect of Inhibitors of the PI3K–Akt–mTOR Pathway *In vitro* and *In vivo*

Model	Treatment	Effects	Conclusion	Reference
LC lines A549, H460	API-59CJ-OH (API) Targets AKT pathway (1–5 mmol/L)	Phosphorylation of DNA-PKcs Repair of DNA-dsb γH2AX foci. Apoptosis	↓ Clonogenic survival. Inhibition of DNA-PKcs-dependent DNA-dsb repair. No enhancement of radiation-induced apoptosis	[41]
TC Human sarcoma cell lines, dermal microvascular endothelial cells (HDMEC). Xenografts	Rapamycin	↓ Clonogenic assay (TC) ↑ GD, ↓ microvessel density (xenografts)	Rapamycin radiosensitizes sarcoma cells and tumor vasculature to radiation	[42]
Breast cancer cell lines MDA-MB-231, MCF-7	RAD001 (everolimus), derivative of rapamycin (10 nmol/L)	Induction of caspase-3 cleavage. An increase in G_2-M cell cycle arrest was seen in the combination treatment group	Prosurvival IR-induced Akt, mTOR signaling, attenuated by RAD001. Enhanced cytotoxicity of IR	[43]
Glioblastoma cells	Rapamycin (1–8 nM)	↓ Survivin by rapamycin + IR ↑ γH2AX ↑ p21 protein. enhanced G_1 arrest, accumulation of cells in G_0/subG_1, ↓ Survival	Survivin plays an important role in rapamycin-mediated apoptosis. Targeting survivin, an effective approach for radiosensitization	[44]

11.2.5 Evaluation of PI3K–Akt–mTOR Inhibitors in Preclinical Models

There have been many demonstrations of radiosensitization by compounds targeting the PI3K–Akt–mTOR pathway. Some of these are described in Section 11.2.4 as illustrative of specific modes of radiosensitization. Further examples are summarized in Table 11.4.

11.2.6 Clinical Applications of Inhibitors of the PI3K–Akt–mTOR Pathway

The PI3K pathway is initiated at the level of the RTKs discussed in the preceding chapter, and agents inhibiting the upstream RTKs are among the most established targeted therapies in oncology. This is particularly true for monoclonal antibodies (mAbs) directed against EGFR and HER2, both of which are RTKs that transduce signals at least in part through PI3K. Cetuximab and panitumumab both target the extracellular domain of EGFR, whereas trastuzumab inhibits ERbB2. Small-molecule tyrosine kinase inhibitors include gefitinib and erlotinib (against EGFR) and ErbB2 (lapatinib, which also targets EGFR). Several of these agents have been approved for use in specific cancers and there are very many others being investigated at various levels of clinical trials or are also working their way into clinical use. The application of some of these agents in conjunction with radiation is also being exploited at the clinical level.

The complexity of the PI3K–Akt–mTOR pathway and its pivotal position in cancer cell metabolism has spawned the development of a large number of compounds targeting different aspects of the pathway. These include dual mTOR-PI3K inhibitors, pure PI3K inhibitors, Akt inhibitors, and mTOR kinase inhibitors. Table 11.5 summarizes information about

TABLE 11.5 Clinical Trials Involving Agents or Treatment Strategies Inhibiting Components of the PI3K Pathway

% of Total	Class of Agent	Treatment Strategy	% of Total
27	Pan-PI3K inhibitor	Monotherapy	35
27	AKT Inhibitor	Combination with chemotherapy	15
19	PI3K–mTOR inhibitor	Combination with MEK inhibitor	14
10	mTORC1–mTORC2 inhibitor	Combination with chemotherapy and mAb	13
7	PI3Kδ inhibitor	Combination with inhibitor against TK other than MEK	10
5	Pan-PI3K or PI3K/mTOR inhibitor	Combination with mAb	6
4	PI3Kα inhibitor	Combination with hormonal therapy	5
1	PI3Kß inhibitor	Others	2

50 agents that are involved in recent and ongoing clinical trials as single agents or combined with chemotherapeutic or targeted molecular drugs [45]. Despite the promising results that have been achieved in preclinical models, there are currently no clinical trials involving PI3K pathway inhibitors and radiotherapy.

11.3 TARGETING Hsp90

11.3.1 Involvement of Hsp90 with Regulatory Proteins

Targeting a signaling molecule will affect radiosensitivity in a cell type-dependent manner. Overcoming cell type dependency requires either a multitarget approach or the targeting of a key molecule involved in the trafficking of a diverse set of regulating proteins. The molecular chaperone Hsp90 fulfills this requirement.

Heat shock protein 90 (Hsp90) is a ubiquitous molecular chaperone. It is one of the most abundant cytoplasmic proteins and plays a key role in regulating cellular response. Hsp90 modulates the degradation, folding, or transport of a diverse set of critical cellular regulatory proteins. Hsp90 client proteins including hormone or growth factor receptors, Raf-1, Akt, Src kinase, mutant p53, cyclin-dependent kinase 4, HIF1α, ErbB2, telomerase hTERT, survivin, and others are implicated in signaling pathways responsible for the progression of malignant cells and their resistance to therapeutics. Most Hsp90 clients, that is, those proteins that require its "chaperoning" activity for appropriate function, participate in some aspect of signal transduction including a wide variety of protein kinases, hormone receptors, and transcription factors. Hsp90 can also stabilize mutated proteins allowing them to maintain normal function despite genetic abnormalities. Although Hsp90 is an abundant protein in both normal and tumor cells, only the latter have a large pool of the "excited" Hsp90 form with high ATPase activity and greater susceptibility to inhibitors. A higher expression of Hsp90 in certain types of cancer is associated with a poor prognosis. Hsp90 is thus a unique molecular target enabling selective targeting of malignancies [46,47].

11.3.2 Inhibitors of Hsp90

Most of the known Hsp90 inhibitors are molecules blocking the ATP-binding site in the N-terminal domain of Hsp90 and thus perturbing ATP-dependent interactions with client proteins. The natural antibiotics geldanamycin (GA) and radicicol were previously identified as potent

Hsp90 inhibitors with antitumor activity but unacceptable hepatotoxicity. Geldanamycin is a benzoquinone ansamycin antibiotic that binds to the Hsp90 ATP binding site, rendering it inactive. Due to hepatotoxicity, it has a limited role in clinical studies. 17-Allylamino-17-demethoxygeldanamycin (17-AAG) is a second-generation derivative with a more favorable toxicity profile, although it is not water soluble. 17-(Dimethylaminoethylamino)-17-demethoxygeldanamycin (17-DMAG) is a third-generation water-soluble agent. These GA derivatives are less toxic Hsp90 inhibitors that retain anticancer potential [46,47]. The inhibitors have demonstrated effects in preclinical studies as either monotherapy or in combination with chemotherapy. Preclinical studies have also demonstrated radiosensitizing effects in cell lines derived from prostate, cervical, colon, lung, and pancreatic cancer in addition to HNSCC and glioma (reviewed in Zaidi et al. [48]).

11.3.3 Targeting Cancer Cells with Hsp90 Inhibitors

The affinity of binding of 17AAG to Hsp90 in cancer cells is approximately 100-fold higher than in normal cells [49]. This comes about because the Hsp90 of cancer cells is mainly sequestered into multiprotein complexes and has the high ATPase activity to ensure maturation and functioning of client proteins essential for cancer cells, whereas the major Hsp90 pool inside normal cells is in an uncomplexed, latent state with low ATPase activity [49]. Hsp90 chaperone dysfunction results in inactivation and proteasomal degradation of client proteins [46,47]; thus, Hsp90 inhibitors may impair signal transduction essential for cancer cell survival and proliferation.

11.3.4 Radiosensitization by Hsp90 Inhibitors

Hsp90 has been characterized as a determinant of tumor cell intrinsic radiosensitivity on the basis of results obtained with the Hsp90 inhibitors geldanamycin, radicicol, and the second-generation and third-generation inhibitors, which showed enhancement of radiosensitivity for a variety of human tumor cell lines. The potential of Hsp90 as a target for radiosensitization stems from the number of Hsp90 client proteins that are involved in radioresponse including erbB2, Akt, Raf, and Chk1. In addition, 17AAG and 17DMAG bind to Hsp90 in malignant cells with a higher affinity than in normal cells [49], suggesting that these drugs will enhance the IR-induced killing of tumor cells without aggravating damage to normal tissues.

11.3.4.1 Specific Mechanisms of Radiosensitization

11.3.4.1.1 Apoptosis The impairment of IR-responsive phosphorylation (activation) of Akt in Hsp90 inhibitor-treated cancer cells contributes to their radiosensitization through the disruption of PI3K/Akt, ErbB1/2, Raf-1, and extracellular signal-regulated kinase (ERK)-mediated signaling pathways, which aggravates an IR-triggered apoptotic scenario [50].

11.3.4.1.2 DNA Damage Repair Hsp90 inhibitors block DNA damage repair in irradiated cancer cells. 17AAG inhibits the homologous recombination repair of DNA DSBs in the radiosensitized cells of human prostate or lung carcinomas; although neither radiosensitization nor DSB repair inhibition were observed in normal fibroblasts treated in the same way. In human pancreatic carcinoma cells, 17DMAG inhibition of DSB repair occurred via an impairment of phosphorylation of the IR-responsive DNA-PKcs and a disruption of DNA-PKcs/ErbB1 interaction [51]. In non-small cell lung cancer cells, 17DMAG was synergistic with IR, reducing the activities of apurinic/apyrimidinic endonuclease and DNA polymerase-β (key enzymes in the base excision repair machinery).

11.3.4.1.3 Disruption of Checkpoint Control 17DMAG was shown to abrogate the activation of G_2- and S-phase cell cycle checkpoints that were associated with the suppression of IR-responsive activation of ATM. Abrogation of G_2 cell cycle checkpoint activation by 17DMAG was also associated with a reduction in radiation-induced activation of ATM, which resulted from reduced interaction between NBS1 and ATM. It seems that DSB repair inhibition and the abrogation of the G_2 checkpoint are independent events linked to different Hsp90 client proteins; however, for optimal radiosensitization, both the DSB repair inhibition and the abrogation of the G_2 checkpoint are required [51].

11.3.4.1.4 Targeting the Tumor Vasculature Aggressive tumors generate angiogenic growth factors stimulating the proliferation of EC and rendering them more radioresistant. Hsp90 inhibitors can act as suppressors of tumor-induced angiogenesis and radiosensitize tumor vasculature. Both 17AAG and 17DMAG exhibit antiangiogenic properties and disrupt HIF1α- and VEGF-involving proangiogenic pathways in cancer cells, suppressing the tumor vascularization [52,53].

It has been shown that clinically accessible (nanomolar) concentrations of 17AAG can radiosensitize human EC and abrogate the radioprotective effects of VEGF or basic fibroblast growth factor [30]. This radiosensitization of EC was associated with 17AAG-induced impairment of the Hsp90-dependent Akt activation in radioprotective PI3K/Akt signaling. Thus, the tumor-radiosensitizing effects attributed to 17AAG may not be strictly selective toward cancer cells, as has been suggested; however, such a "nonselectivity" would be highly beneficial in the treatment of cancer if the normal tissue target were the tumor vasculature [53].

11.3.5 Preclinical Studies

Approximately 20 studies describing the experimental demonstration of radiosensitization by Hsp90 inhibitors have been reviewed by Kabakov and coauthors [54]. The human tumor cell lines used included prostate carcinoma, glioma, cervical tumors, pancreatic carcinoma, lung carcinoma, non–small cell lung cancer, colon adenocarcinoma, head and neck squamous cell carcinoma, breast cancer bladder carcinoma, melanoma, and esophageal cancer.

Among the Hsp90 inhibitors investigated were Geldamycin, 17AAG, 17DMAG, Deguelin, LBH589, and BIIB021. All the Hsp90 inhibitors sensitized tumor cell lines and tumors growing *in vivo* to clinically relevant doses of sparsely ionizing radiation with end points including, enhanced cell death (apoptosis or mitotic catastrophe), cell cycle arrest, and repressed tumor growth. Mechanisms of radiosensitization were those described in Section 11.3.4. In some studies, the Hsp90 inhibitor-induced radiosensitization, although occurring in cancer cells, was not observed in normal cells. Some preclinical results of combined treatment with radiation and Hsp90 inhibitors are summarized in Table 11.6. A more recent publication described enhancement of radioresponse of xenograft NCI-H460 lung cancer cells by 17-AAG with or without celecoxib [55]. The enhancement ratio for 17-AAG + 5 × 2 Gy was 1.7 (Figure 11.5).

11.3.6 Clinical Results

There are currently no clinical trials open involving the combination of an Hsp90 inhibitor and radiotherapy. However, there are fifteen open cancer trials of 17AAG, including nine phase I and six phase II trials, with eight of those trials using 17AAG as a single agent and seven of those trials using 17AAG in combination with standard chemotherapy; there are also four active phase I trials using 17DMAG (www.clinicaltrials.gov).

TABLE 11.6 Hsp90 Inhibitor Treatments: Response in Cell Culture and Preclinical Models

Tumor Model	Treatment	Results and Conclusions	Reference
Glioma (U251), pancreatic cancer (MiaPaCa), prostate cancer DU145	2× Oral 17-DMAG 12 h, preradiation (5 Gy)	17-DMAG reduced the levels of three radiosensitivity-associated proteins: Raf-1, ErbB2, and Akt, with ErbB2 being the most susceptible. Corresponding concentrations of 17-DMAG enhanced radiosensitivity of each of the tumor cell lines (clonogenic assay). Sensitization attributed to a 17-DMAG-mediated abrogation of the G_2- and S-phase cell cycle checkpoints. 17-DMAG plus radiation to mice bearing tumor xenografts resulted in a greater than additive increase in tumor growth delay	[56]
Human cervical cancers, HeLa and SiHa	Clinically achievable concentrations of geldanamycin, 17-AAG, to tumor-bearing mice (HeLa) plus acute (12 Gy) or fractionated IR 16 h later	Akt1, extracellular signal-regulated kinase-1, Glut-1, HER-2/neu, Lyn, cAMP-dependent protein kinase, Raf-1, and VEGF expression downregulated in 17-AAG-treated cells. IR-induced cell death attributable to combination of apoptotic and nonapoptotic cell death. Radiosensitizing effects of 17-AAG were limited to transformed cells. Tumors radiosensitized to acute and fractionated radiation	[52]
HNSCC cell lines and xenografts	17-AAG, BIIB021 (synthetic and bioavailable Hsp90 inhibitor)	BIIB021 enhanced the *in vitro* radiosensitivity of HNSCCA cell lines with reduction in the expression of key radioresponsive proteins, increased apoptotic cells, and enhanced G_2 arrest. In xenografts studies, BIIB021 exhibited a strong antitumor effect, outperforming 17-AAG, either as a single agent or in combination with radiation, and improving the efficacy of radiation	[57]

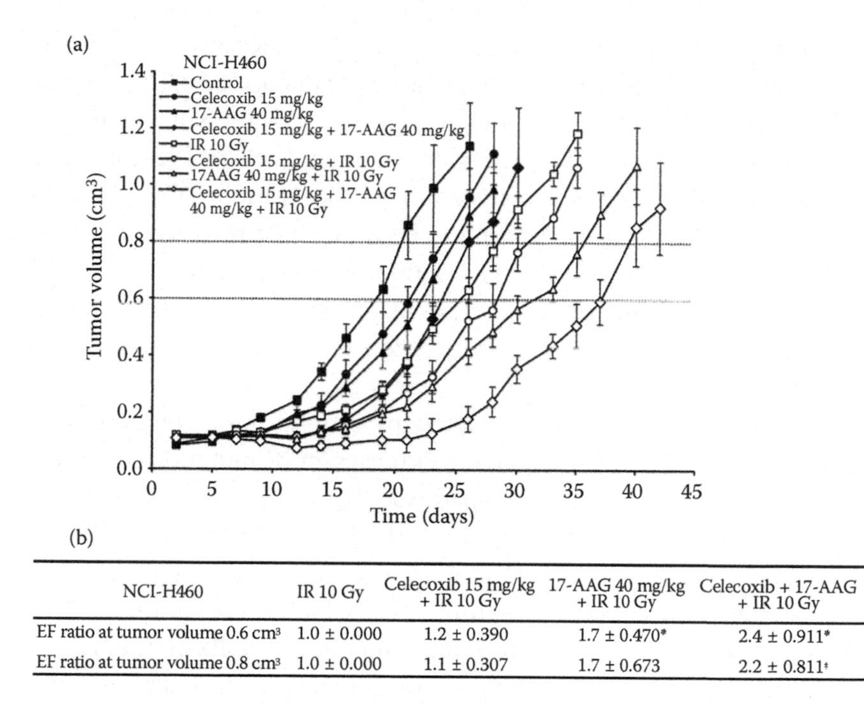

NCI-H460	IR 10 Gy	Celecoxib 15 mg/kg + IR 10 Gy	17-AAG 40 mg/kg + IR 10 Gy	Celecoxib + 17-AAG + IR 10 Gy
EF ratio at tumor volume 0.6 cm^3	1.0 ± 0.000	1.2 ± 0.390	1.7 ± 0.470*	2.4 ± 0.911*
EF ratio at tumor volume 0.8 cm^3	1.0 ± 0.000	1.1 ± 0.307	1.7 ± 0.673	2.2 ± 0.811†

FIGURE 11.5 (a) Combined treatment of 17-AAG and celecoxib effectively delayed tumor growth in BALB/C nude mice via enhancing radiosensitivity. Tumor-bearing mice were given i.p. with celecoxib (15 mg/kg), 17-AAG (40 mg/kg), or drug combination (celecoxib + 17-AAG) for seven consecutive days after 10 days postimplantation, with or without irradiation on tumor (2 Gy × 5) starting from the next day after drug administration. Control groups were given an equal volume of DMSO i.p. (Control, ■; 17-AAG, ▲; □ 10 Gy; △, 17-AAG + 10 Gy). (b) The enhancement factor (EF) ratio was determined at tumor volume 0.6 and 0.8 cm^3. Error bars ± SE. (From Kim, Y., and Pyo, H., *DNA Cell Biol* 2012;31(1):15–29. With permission.)

11.4 SUMMARY

This chapter deals with the downstream components of the signaling pathways that are initiated by growth factor receptors. Many of these molecules are mutated, abnormally expressed, or have alternative functions in neoplastic cells making them interesting as tumor-specific targets for radiosensitization. Ras is a proto-oncogene that, upon activation, can in turn activate a number of different signal transduction pathways. There is ample evidence from experimental systems that the expression of mutated Ras proteins is associated with a reduction in cellular radiosensitivity.

Raf1–MAPK and the PI3K–Akt are the best characterized RAS effector pathways. Activation of the MAPK pathway has been shown to mediate radioresistance under certain circumstances, but PI3K activity was found to be necessary for the radioresistance of Ras-transformed cells implicating PI3K–Akt as the critical effector pathway. The activation of Akt in turn, results in the stimulation of protein translation and cell proliferation, and the inhibition of apoptosis. Downstream of Akt is the mTOR, the activation of which leads to the phosphorylation of its downstream targets. Strategies aimed at exploiting Ras as a target for tumor radiosensitization have focused on the inhibition of farnesyltransferase and geranyl-geranyl transferases, which mediate the prenylation of the carboxy terminus of Ras proteins.

The PI3K pathway can be activated by amplification, by activating mutation of upstream receptor tyrosine kinases, or by mutations or downstream deletions. Inhibition in the pathway can similarly occur at several levels. Radiosensitization by inhibition of PI3K/Akt signaling has been shown to be brought about by a number of mechanisms including tumor cell apoptosis, destruction of the tumor vasculature, inhibition of DNA DSB repair, inhibition of HIF-1α induction, and normalization of tumor vasculature. Despite the promising results that have been achieved in preclinical models, there has been very limited interest in the clinical development of treatment combining PI3K pathway inhibitors and radiotherapy.

The protein kinase receptors and signal transduction intermediates described (the inhibition of which may cause radioenhancement) are the client proteins of the molecular chaperone Hsp90. The potential of Hsp90 as a target for radiosensitization stems from the number of Hsp90 client proteins that are involved in radioresponse including ErbB2, Raf-1, Akt, EGFR, and IGF-1R. Most of the known Hsp90 inhibitors are molecules blocking the ATP-binding site in the N-terminal domain of Hsp90 and perturbing ATP-dependent interactions with client proteins. The mechanisms of Hsp90 inhibition-induced radiosensitization of tumor cells are those described for the downregulation of the client proteins and include cell cycle arrest and enhanced apoptosis and DNA repair inhibition.

REFERENCES

1. Sklar, M. The ras oncogenes increase the intrinsic resistance of NIH 3T3 cells to ionizing radiation. *Science* 1988;239:645–647.
2. Bernhard, E., Stanbridge, E., Gupta, S. et al. Direct evidence for the contribution of activated N-ras and K-ras oncogenes to increased intrinsic radiation resistance in human tumor cell lines. *Cancer Res* 2000;60:6597–6600.

3. Kim, I., Bae, S., Fernandes, A. et al. Selective inhibition of Ras, phosphoinositide 3 kinase, and Akt isoforms increases the radiosensitivity of human carcinoma cell lines. *Cancer Res* 2005;65:7902–7910.
4. Russell, J., Lang, F., Huet, T. et al. Radiosensitization of human tumor cell lines induced by the adenovirus-mediated expression of an anti-Ras single-chain antibody fragment. *Cancer Res* 1999;59:5239–5244.
5. McKenna, W., Weiss, M., Endlich, B. et al. Synergistic effect of the v-myc oncogene with H-ras on radioresistance. *Cancer Res* 1990;50:97–102.
6. Pirollo, K., Hao, Z., Rait, A., Ho, C., and Chang, E. Evidence supporting a signal transduction pathway leading to the radiation-resistant phenotype in human tumor cells. *Biochem Biophys Res Commun* 1997;230:196–201.
7. Chakravarti, A., Chakladar, A., Delaney, M., Latham, D., and Loeffler, J. The epidermal growth factor receptor pathway mediates resistance to sequential administration of radiation and chemotherapy in primary human glioblastoma cells in a RAS-dependent manner. *Cancer Res* 2002;62:4307–4315.
8. Grana, T., Rusyn, E., Zhou, H., Sartor, C., and Cox, A. Ras mediates radioresistance through both phosphatidylinositol 3-kinase-dependent and Raf-dependent but mitogen-activated protein kinase/extracellular signal-regulated kinase kinase-independent signaling pathways. *Cancer Res* 2002; 62:4142–4150.
9. Kasid, U., Pfeifer, A., Brennan, T. et al. Effect of antisense c-raf-1 on tumorigenicity and radiation sensitivity of a human squamous carcinoma. *Science* 1989;243:1354–1356.
10. Kasid, U., Pirollo, K., Dritschilo, A., and Chang, E. Oncogenic basis of radiation resistance. *Adv Cancer Res* 1993;61:195–233.
11. Gupta, A., Bakanauskas, V., Cerniglia, G. et al. Ras regulation of radioresistance in cell culture. *Cancer Res* 2001;61:4278.
12. Rodriguez-Viciana, P., Warne, P., Khwaja, A. et al. Role of phosphoinositide 3-OH kinase in cell transformation and control of the actin cytoskeleton by Ras. *Cell* 1997;89:457–467.
13. Adjei, A. Farnesyltransferase inhibitors. *J Natl Cancer Inst* 2001;93:1062.
14. Delmas, C., Heliez, C., Cohen-Jonathan, E. et al. Farnesyltransferase inhibitor, R115777, reverses the resistance of human glioma cell lines to ionizing radiation. *Int J Cancer* 2002;100:43–48.
15. Brunner, T., Cengel, K., Hahn, S. et al. Pancreatic cancer cell radiation survival and prenyltransferase inhibition: The role of K-Ras. *Cancer Res* 2005;65:8433.
16. Lane, K., and Beese, L. Thematic review series: Lipid posttranslational modifications. Structural biology of protein farnesyltransferase and geranylgeranyltransferase type I. *Lipid Res* 2006;47:681–699.
17. Basso, A., Mirza, A., Liu, G., Long, B., Bishop, W., and Kirschmeie, P. The farnesyl transferase inhibitor (FTI) SCH66336 (lonafarnib) inhibits Rheb farnesylation and mTOR signaling. Role in FTI enhancement of taxane and tamoxifen anti-tumor activity. *J Biol Chem* 2005;280:31101–31108.

18. Bernhard, E., McKenna, W., Hamilton, A. et al. Inhibiting Ras prenylation increases the radiosensitivity of human tumor cell lines with activating mutations of ras oncogenes. *Cancer Res* 1998;58:1754–1761.
19. Cohen-Jonathan, E., Muschel, R., McKenna, G. et al. Farnesyltransferase inhibitors potentiate the antitumor effect of radiation on a human tumor xenograft expressing activated HRAS. *Radiat Res* 2000;154:125–132.
20. Berndt, N., Hamilton, A., and Sebti, S. Targeting protein prenylation for cancer therapy. *Nat Rev Cancer* 2011;11:775–791.
21. Martin, N., Brunner, T., Kiel, K.D. et al. A phase I trial of the dual farnesyltransferase and geranylgeranyltransferase inhibitor L-778,123 and radiotherapy for locally advanced pancreatic cancer. *Clin Cancer Res* 2004;10:5447–5454.
22. Moyal, E., Laprie, A., Delannes, M. et al. Phase I trial of tipifarnib (R115777) concurrent with radiotherapy in patients with glioblastoma multiforme. *Int J Radiat Oncol Biol Phys* 2007;68:1396–1401.
23. Larner, J., Jane, J., Laws, E., Packer, R., Myers, C., and Shaffrey, M. A phase I-II trial of lovastatin for anaplastic astrocytoma and glioblastoma multiforme. *Am J Clin Oncol Rep* 1998;21:579–583.
24. Hahn, S., Bernhard, E., Regine, W. et al. A phase I trial of the farnesyltransferase inhibitor L-778,123 and radiotherapy for locally advanced lung and head and neck cancer. *Clin Cancer Res* 2002;8:1065–1072.
25. Rengan, R., Cengel, K., and Hahn, S. Clinical target promiscuity: Lessons from ras molecular trials. *Cancer Metastasis Rev* 2008;27:403–414.
26. Faber, A., Li, D., Song, Y. et al. Differential induction of apoptosis in HER2 and EGFR addicted cancers following PI3K inhibition. *Proc Natl Acad Sci U S A* 2009;106:19503–19508.
27. Podsypanina, K., Ellenson, L., Nemes, A. et al. Mutation of Pten/Mmac1 in mice causes neoplasia in multiple organ systems. *Proc Natl Acad Sci U S A* 1999;96:1563–1568.
28. Samuels, Y., and Velculescu, V. Oncogenic mutations of PIK3CA in human cancers. *Cell Cycle* 2004;3:1221–1224.
29. Gupta, S., Ramjaun, A., and Haiko, P. Binding of ras to phosphoinositide 3-kinase p110α is required for ras-driven tumorigenesis in mice. *Cell* 2007;129:957–968.
30. Faivre, S., Kroemer, G., and Raymond, E. Current development of mTOR inhibitors as anticancer agents. *Nat Rev Drug Discov* 2006;5:671–688.
31. van der Heijden, M., and Bernards, R. Inhibition of the PI3K pathway: Hope we can believe in? *Clin Cancer Res* 2010;16(12):3094–3099.
32. Guertin, D., and Sabatini, D. Defining the role of mTOR in cancer. *Cancer Cell* 2007;12:9–22.
33. Toulany, M., Kasten-Pisula, U., Brammer, I. et al. Blockage of epidermal growth factor receptor-phosphatidylinositol 3-kinase-AKT signaling increases radiosensitivity of K-RAS mutated human tumor cells *in vitro* by affecting DNA repair. *Clin Cancer Res* 2006;12:4119–4126.
34. Shen, W., Balajee, A., Wang, J. et al. Essential role for nuclear PTEN in maintaining chromosomal integrity. *Cell* 2007;128:157–170.

35. Schuurbiers, O., Kaanders, J., van der Heijden, H., Dekhuijzen, R., Oyen, W., and Bussink, J. The PI3-K/AKT-pathway and radiation resistance mechanisms in non-small cell lung cancer. *J Thorac Oncol* 2009;4(6):761–767.
36. Zhong, H., Chiles, K., Feldser, D. et al. Modulation of hypoxia-inducible factor 1α expression by the epidermal growth factor/phosphatidylinositol 3-kinase/PTEN/AKT/FRAP pathway in human prostate cancer cells: implications for tumor angiogenesis and therapeutics. *Cancer Res* 2000;60: 1541–1545.
37. Blancher, C., Moore, J., Robertson, N., and Harris, A. Effects of ras and von Hippel-Lindau (VHL) gene mutations on hypoxia-inducible factor (HIF)-1alpha, HIF-2alpha, and vascular endothelial growth factor expression and their regulation by the phosphatidylinositol 3-kinase/Akt signaling pathway. *Cancer Res* 2001;61:7349–7355.
38. Pore, N., Gupta, A.K., Cerniglia, G.J. et al. Nelfinavir down-regulates hypoxia-inducible factor 1α and VEGF expression and increases tumor oxygenation: Implications for radiotherapy. *Cancer Res* 2006;66:9252–9259.
39. Fokas, E., Im, J., Hill, S. et al. Dual inhibition of the PI3K/mTOR pathway increases tumor radiosensitivity by normalizing tumor vasculature. *Cancer Res* 2012;72:239–248.
40. Potiron, V., Abderrhamani, R., Giang, E., Paris, F., and Supiot, S. Radiosensitization of prostate cancer cells by the dual PI3K/mTOR inhibitor BEZ235 under normoxic and hypoxic conditions. *Radiother Oncol* 2013;106(1):138–146.
41. Toulany, M., Kehlbach, R., Florczak, U. et al. Targeting of AKT1 enhances radiation toxicity of human tumor cells by inhibiting DNA-PKcs-dependent DNA double-strand break repair. *Mol Cancer Ther* 2008;7(7):1772–1781.
42. Murphy, J., Spalding, A., Somnay, Y., Markwart, S., Ray, M., and Hamstra, D. Inhibition of mTOR radiosensitizes soft tissue sarcoma and tumor vasculature. *Clin Cancer Res* 2009;15(2):589–596.
43. Albert, J., Kwang, K., Kim, W., Cao, C., and Lu, B. Targeting the Akt/mammalian target of rapamycin pathway for radiosensitization of breast cancer. *Mol Cancer Ther* 2006;5(5):1183–1189.
44. Anandharaj, A., Cinghu, S., and Park, W. Rapamycin-mediated mTOR inhibition attenuates survivin and sensitizes glioblastoma cells to radiation therapy. *Acta Biochim Biophys Sin* 2011;43:292–300.
45. Markman, B., Dienstmann, R., and Tabernero, J. Targeting the PI3K/Akt/mTOR pathway—beyond rapalogs. *Oncotarget* 2010;1(7):530–543.
46. Stravopodis, D., Margaritis, L., and Voutsinas, G. Drug mediated targeted disruption of multiple protein activities through functional inhibition of the Hsp90 chaperone complex. *Curr Med Chem* 2007;14:3122–3138.
47. Hwang, M., Moretti, L., and Lu, B. Hsp90 inhibitors: Multi-target antitumor effects and novel combinatorial therapeutic approaches in cancer therapy. *Curr Med Chem* 2009;16:3081–3092.
48. Zaidi, S., Huddart, R., and Harrington, K. Novel targeted radiosensitisers in cancer treatment. *Curr Drug Discov Technol* 2009;6:103–134.

49. Kamal, A., Thao, L., Sensintaffar, J., Boehm, Z., and Burrows, F. A high-affinity conformation of Hsp90 confers tumour selectivity on Hsp90 inhibitors. *Nature* 2003;425:407–410.
50. Mashida, H., Nakajima, S., Shikano, N. et al. Heat shock protein 90 inhibitor 17-allylamino-17-demethoxygeldanamycin potentiates the radiation response of tumor cells grown as monolayer cultures and spheroids by inducing apoptosis. *Cancer Sci* 2005;96:911–917.
51. Dote, H., Burgan, W., Camphausen, K., and Tofilon, J. Inhibition of Hsp90 compromises the DNA damage response to radiation. *Cancer Res* 2006;66:9211–9220.
52. Bisht, K., Bradbury, C., Mattson, D. et al. Geldanamycin and 17-allylamino-17-demethoxygeldanamycin potentiate the *in vitro* and *in vivo* radiation response of cervical tumor cells via the heat shock protein 90-mediated intracellular signaling and cytotoxicity. *Cancer Res* 2003;63:8984–8995.
53. Kim, W., Oh, S., Woo, J., Hong, W., and Lee, H. Targeting heat shock protein 90 overrides the resistance of lung cancer cells by blocking radiation-induced stabilization of hypoxia-inducible factor-1α. *Cancer Res* 2009;69:1624–1632.
54. Kabakov, A., Kudryavtsev, V., and Gabai, V. Hsp90 inhibitors as promising agents for radiotherapy. *J Mol Med* 2010;88:241–247.
55. Kim, Y., and Pyo, H. Cooperative enhancement of radiosensitivity after combined treatment of 17-(allylamino)-17 demethoxygeldanamycin and celecoxib in human lung and colon cancer cell lines. *DNA Cell Biol* 2012; 31(1):15–29.
56. Bull, E., Dote, H., Brady, K. et al. Enhanced tumor cell radiosensitivity and abrogation of G_2 and S phase arrest by the Hsp90 inhibitor 17-(dimethylaminoethylamino)-17-demethoxygeldanamycin. *Clin Cancer Res* 2004;10:8077–8084.
57. Yin, X., Zhang, H., Lundgren, K., Wilson, L., Burrows, F., and Shores, C. BIIB021, a novel Hsp90 inhibitor, sensitizes head and neck squamous cell carcinoma to radiotherapy. *Int J Cancer* 2010;126:1216–1225.

CHAPTER 12

Radiosensitization by Targeting the Tumor Microenvironment

12.1 THE TUMOR MICROENVIRONMENT

The microenvironment of malignant solid tumors differs markedly from that of normal tissues, particularly with respect to the pH, the distribution and availability of nutrients, and the oxygen concentration. Because the tumor microenvironment is a unique feature, it can be selectively targeted for cancer therapy with the understanding that it is not stable and will be changed by treatment.

12.2 TUMOR VASCULATURE AND ANGIOGENESIS

During fetal development, the formation and growth of new blood vessels from preexisting vasculature leads to the formation of a stable vasculature characteristic of healthy adults. In a normal adult, the vasculature is composed of endothelial cells (ECs), mural cells (pericytes), and basement membrane. ECs are the innermost layer of vascular walls and are in direct contact with pericytes along the length of the vessels while the basement membrane is a uniform and thin layer that covers almost the entire length of the ECs.

Exceptions to this stable condition occur during the cyclical growth of vessels in the ovarian corpus luteum or during pregnancy. Pathological conditions during which angiogenesis occurs are tissue repair during wound healing and during the growth of a tumor. Initiation of angiogenesis requires a shift in the balance between activators and suppressors. Important activators of angiogenisis include vascular endothelial growth factor-A (VEGF-A), matrix metalloproteinases (MMPs), placenta growth factor (PlGF), fibroblast growth factor (FGF), and hepatocyte growth factor (HGF). Endogenous inhibitors of angiogenesis include thrombospondins (THSBs), endostatin, angiostatin, and cytokines such as interleukin-12.

12.2.1 The Tumor Vasculature

The tumor vasculature has a number of distinct features compared with normal vessels:

- Tumor vessels lack the hierarchy of arterioles, capillaries, and venules.
- Tumor vasculature is disorganized and tortuous.
- Vessels in tumors are leakier than their counterparts in normal tissues because tumor-associated ECs (TAECS) are not in close contact with pericytes and are loosely connected to basement membrane.
- In some circumstances, tumor cells may line blood vessels via vasculogenic mimicry; compared with TAECs, which differ from ECs isolated from nontumor tissues in being enriched in transcripts indicative of their distinct molecular properties including PlGF, CD137, CD276, and CD109.

12.3 TUMOR CHARACTERISTICS EXPLOITABLE FOR RADIOSENSITIZATION: HYPOXIA

In 1955, Thomlinson and Gray [1] reported that partial oxygen pressure (pO_2) in malignant solid tumors can vary from well-oxygenated to oxygen levels low enough to be described as hypoxia, that is, $pO_2 \leq 2.5$ mm Hg. It was subsequently reported that the hypoxic fraction can be as high as 25% in malignant tumors, whereas in normal tissue, there is no region where pO_2 values are lower than 12.5 mm Hg. Tumor hypoxia is a subject of huge interest in radiation oncology because it is associated with tumor radioresistance, recurrence, and poor prognosis after radiation therapy.

12.3.1 Chronic Hypoxia

Cancer cells are characterized by accelerated proliferation, evasion of growth suppressors, immortality, and deregulated cellular energetics, and the vasculature of a malignant tumor is functionally and structurally defective [2]. Consequently, an imbalance exists in malignant solid tumors between oxygen supply and oxygen consumption, and some parts of the tumor will be severely compromised with respect to oxygenation and become hypoxic. The proliferation of tumor cells is dependent on the supply of oxygen and nutrients, and whereas a tumor blood vessel will be surrounded by actively proliferating cancer cells (the normoxic region), cancer cells approximately 100 μm or more away from tumor blood vessels will die and the region will become necrotic. Between these two distinct regions, there are chronically hypoxic regions in which cancer cells have access to sufficient oxygen from blood vessels to survive but not enough to actively proliferate [3,4].

12.3.2 Acute Hypoxia

Acute hypoxia was first reported in 1979 [4]. Structurally and functionally anomalous tumor vasculature causes the transient opening and closing of blood vessels. This leads to changes in the blood flow rate and fluctuations in perfusion and ultimately causes the generation of transient hypoxia even close to tumor blood vessels. It is believed that at least 20% of cancer cells in solid tumors will experience periods of acute hypoxia [5].

12.3.3 Hypoxia-Induced Factor 1

In addition to radiochemical mechanisms, hypoxia increases tumor radioresistance at the tissue level through biological mechanisms; particularly the important role played by the transcription factor, hypoxia-inducible factor 1 (HIF-1) [6]. HIF-1 is a transcription factor that induces the expression of a large array of proteins, most of which act to promote tumor cell proliferation in the specialized environment of a solid tumor. Of particular significance in the context of angiogenesis are a group of growth factors, particularly VEGF, which promote the growth of endothelial cells and protects them against the toxic effects of radiation. The role of HIF-1 in tumor response to radiation and the targeting of HIF-1 for radiosensitization will be dicussed in the second half of this chapter.

12.4 VASCULAR TARGETED THERAPIES

The pioneering work of Folkman [7] demonstrated that angiogenesis is crucial for the growth and survival of tumor cells, and these findings highlighted the importance of targeting the development of tumor vasculature and function as a therapeutic strategy.

12.4.1 Antiangiogenic Agents

Antiangiogenic agents (AAs) act to prevent the growth of new blood vessels in tumors. VEGF, a ligand with a central role in signaling pathways controlling tumor blood vessel development and survival, is the most important target of such agents. The binding of VEGF ligands activates receptor tyrosine kinases, designated VEGFR1, VEGFR2, and VEGFR3, which in turn activate a network of distinct downstream signaling pathways. However, the effects of VEGF receptor (VEGFR) signaling are not specific for the vasculature because VEGF also plays a role in many other processes. Angiogenesis is shown schematically in Figure 12.1 and antiangiogenic agents with radiosensitizing potential are listed in Table 12.1 [8,9].

TABLE 12.1 Antiangiogenic Agents with Radiosensitizing Potential

			Clinical Trials	
Category	**Agent**	**Preclinical**	**All**	**+RT**
Endogenous angiogenic inhibitors	Angiostatin,	Mouse lung Ca xenograft GBM	Yes	No
	Endostatin	HNSCC	Yes	Yes
Anti-VEGF antibody	Bevacizumab	Mouse lung Ca HNSCC, GBM, colon cancer xenografts	Yes	Yes
Anti-VEGFR-2 antibody	DC101	GBM, Breast xenografts	No	No
	Semaxanib (SU5416)	Mouse GBM	Yes	Yes
Anti-VEGFR TKIs	Vatalanib (PTK787)	Colon adenoca. xenograft	Yes	Yes
	Vandetanib (ZD6474)	CaLu-6NSCLC, LC49 colorectal cancer	Yes	Yes
	Cediranib (AZD2171)	NSCLC, colorectal xenografts	Yes	Yes
Multitarget TKI	Sunitinib (SU11248)	Mouse lung Ca	Yes	Yes
	SU6668	Mouse lung Ca	No	No
Others	TNP-470	HNSCC, GBM xenografts, mouse lung	Yes	Yes
	Thalidomide	HNSCC xenografts	Yes	Yes

Source: Compiled from data summarized by Yoshimura, M. et al., *Biomed Res Int*, 2013:685308, 2013; Ciric, E. and Sersa, G., *Radiol Oncol*, 44(2):66–78, 2010; and from http://www.clinicaltrials.gov.

Note: GBM, glioblastoma; HNSCC, head and neck squamous cell carcinoma; NSCLC, non-small cell lung cancer.

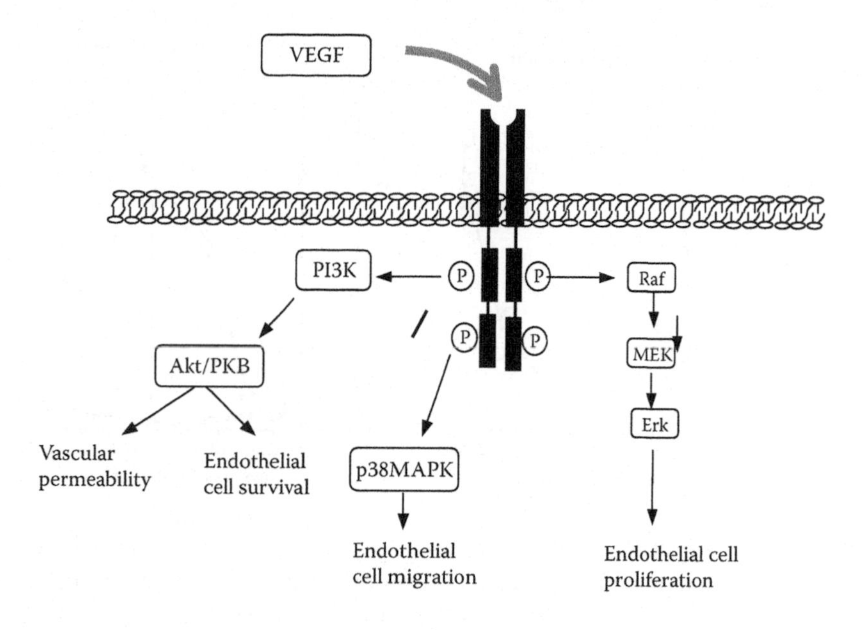

FIGURE 12.1 In response to hypoxia, VEGF is secreted by tumor cells. Binding of VEGF to its receptor VEGFR (located on endothelial cells) results in the dimerization and activation of VEGFR. Activation of VEGFR in turn results in activation of three receptor tyrosine kinases (TKs) which initiate several distinct downstream signaling pathways. Antiangiogenic agents target the process at different levels, that of the ligand VEGF (bevacizumab) the receptor VEGFR (DC101) and the VEGFR TK (Vandetanib).

12.4.1.1 Endogenous Angiostatic Agents

Synergistic antitumor effects, in combination with radiation therapy, have been reported for the endogenous angiogenesis inhibitors angiostatin and endostatin. In one study, antitumor interaction between radiation and angiostatin was shown for four tumors (Lewis lung mouse tumor and PC-3, SQ-20B, and D54 xenografts) at clinically relevant radiation doses. Combined treatment was cytotoxic to endothelial cells, but not to tumor cells [10].

12.4.1.2 Antibodies Targeting VEGF and VEGFR

Blocking VEGF with a neutralizing antibody has been shown to inhibit the growth of primary tumors and metastases [11], and to enhance the antitumor effects of radiation in preclinical studies [12]. Bevacizumab is a humanized monoclonal antibody that acts by binding and neutralizing

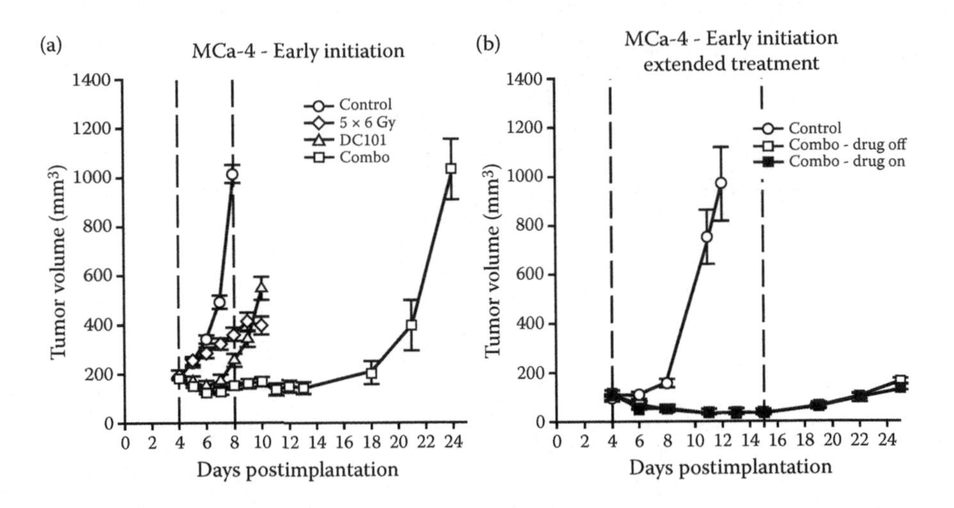

FIGURE 12.2 Effect of DC101 and radiation on the growth of mammary tumors. Effect of single and combined treatments on tumor growth inhibition (mean ± SE; 10 tumors/group). ○, controls; ◇, radiotherapy alone; △, DC101 alone; □, combined radiotherapy and DC101. Vertical dashed lines, treatment beginning and end times. (a) MCa-4 tumors, early initiation treatment (begun at volumes ~190 mm^3). (b) MCa-4 tumors, early initiation extended treatment (begun at volumes ~100 mm^3 and continued for 2 weeks). (From Fenton, B. et al., *Cancer Research* 64(16):5712–5719, 2004. With permission.)

VEGF. In combination with cytotoxic chemotherapy, bevacizumab showed a significant improvement in the survival of patients with advanced colorectal or lung cancer, and is currently approved for use in combination with cytotoxic chemotherapy in those diseases.

DC101 is an anti-VEGFR2 antibody that was reported to reduce the radiation dose necessary for tumor control in non-small cell lung cancer, U87 xenografts, and two human breast xenografts [13]. A synergistic effect was observed when irradiation was given several days after the administration of DC101 (Figure 12.2) [14].

12.4.1.3 Small-Molecule Tyrosine Kinase Inhibitors

Small-molecule tyrosine kinase inhibitors (TKIs) present another class of antiangiogenic agents. They act by preventing the activation of growth factor receptors, thus inhibiting downstream signaling pathways. In some cases, they are simultaneously active against receptors for different growth factors.

Clinically relevant TKIs are listed in Table 12.1 and include

- Vatalanib (PTK787/ZK222584). Radiation treatment combined with vatalanib delayed tumor growth in colon tumor xenografts [15].
- Vandetanib (ZD6474). Treatment with the combination of vandetanib and radiation led to significant enhancement of antiangiogenic, antivascular, and antitumor effects in an orthotopic model of lung cancer [16].
- Cediranib (AZD2171). A potent VEGFR TKI that has been reported to radiosensitize tumor xenografts [17].
- Sunitinib. A multikinase inhibitor targeting VEGFRs, platelet-derived growth factor receptor (PDGFR), and c-kit, and has shown significant efficacy in clinical trials for renal cancer [8].
- SU11657 inhibits VEGF, PDGF, and C-kit. It was shown to enhance the effect of radiochemotherapy on A431 tumors. Simultaneous inhibition of three targets was more effective than blockade of each target singly [18].

12.4.1.4 Other Agents

12.4.1.4.1 Thalidomide An orally administered drug that inhibits angiogenesis and has been recognized to have several antitumor and antimetastatic mechanisms. The Radiation Therapy Oncology Group (RTOG) conducted a phase III study to compare whole-brain radiation therapy (WBRT) with WBRT combined with thalidomide for patients with brain metastases; however, thalidomide with radiation therapy provided no survival benefit [19].

12.4.1.4.2 TNP-470 The antiangiogenic compound TNP-470 is an inhibitor of endothelial cells. The mechanism of action is not fully known, but the compound is believed to act by inhibiting methionyl aminopeptidase-2, an intracellular enzyme related to protein myristoylation. The effects of TNP-470 are mediated by the resulting inhibition of membrane proteins, such as nitric oxide synthase, which are translocated to the cell surface membrane by myristoylation [20].

The combination of radiotherapy and TNP-470 has been investigated in murine mammary and lung carcinomas. TNP-470 administered before IR did not induce tumor hypoxia, whereas TNP-470 given during fractionated radiotherapy decreased radiocurability [21]. This study suggested that scheduling radiation therapy and antiangiogenic therapy could influence the therapeutic outcome. In another study using the human glioblastoma line U87 grown in the flank or intracranially in the nude mouse, pretreatment with TNP-470 significantly enhanced the growth-retarding effect of ionizing IR, while preventing radiation-induced microvascular damage [20].

12.4.2 Vascular Targeted Therapies: Mode of Action

12.4.2.1 Vascular Normalization

As previously described, tumor angiogenesis is characterized by tortuous, irregular, and immature vessels as well as heterogenous density. Poor coverage with pericytes leads to a marked increase in vessel leakiness and high interstitial pressure in the tumor. As a consequence, blood flow in the tumor is insufficient to supply enough oxygen and nutrients even in well-vascularized areas in the tumor. Investigations pioneered by Jain [2] indicate that antiangiogenics may act through vascular normalization, resulting in improved tumor oxygenation. This apparently paradoxical model proposes that antiangiogenic agents cause the destruction of immature microvessels and the regularization of the vasculature by the recruitment of pericytes. In effect, causing the vascular system of the tumor to become more similar to that of normal tissue.

12.4.2.2 Sequence of Radiation Therapy and Antiangiogenic Therapy

If antiangiogenic agents improve tumor oxygenation by vascular normalization, it would be assumed that the optimal timing of radiation would be after antiangiogenic therapy. This was supported by a study in which bevacizumab was combined with anginex, an antiangiogenic peptide, and radiation therapy. It was found that significantly increased tumor oxygenation occurred in the 4 days after the start of drug treatment, and that radiation given during this period extended tumor growth delay [22].

The results of other experiments were more ambiguous. In one case, upregulation of HIF-1 was used as an indicator of hypoxia. If the radiation was delivered 24 h after bevacizumab treatment, when HIF-1 activity was not upregulated, enhanced antitumor effects were observed. However, by 72 h after bevacizumab treatment, when HIF-1 activity was upregulated, antitumor effects of combined treatment were lower than when radiation was delivered alone [23]. These findings suggest that the timing and duration for the optimal window for combining radiation with antiangiogenic agents seems to be tumor-dependent and host-dependent.

In fact, a vascular normalization window is not demonstrable for all antiangiogenic agents. ZD6474, an inhibitor of VEGFR and EGFR, was most effective when it was administered 30 min after radiation therapy as compared with concomitant administration or radiation alone, whereas PTK787, a VEGFR2 inhibitor, was also most effective when administered after fractionated irradiation, but not before or during radiation [24].

Upregulation of VEGF expression therapy can protect tumor endothelial cells from apoptosis due to radiation therapy. Both HIF-1 inhibitors, YC-1 and a neutralizing antibody against VEGF, dramatically induced apoptosis of endothelial cells, reduced microvessel density after radiation therapy, and delayed tumor growth. Similarly, cndostatin downregulates VEGF after radiation therapy and induces apoptosis, reducing the proliferation of endothelial cells after radiation therapy and significantly delaying tumor growth [25]. These antiapoptotic effects on endothelial cells are independent of the vascular normalization window and are yet another factor determining the optimal timing of the combination of antiangiogenic therapy and radiation (Figure 12.3).

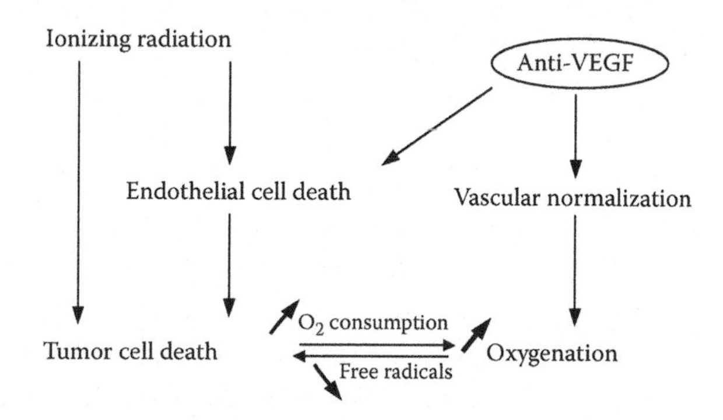

FIGURE 12.3 Interactions between inhibitors of angiogenesis and radiotherapy. The synergistic effects of the combination could be explained by a transient normalization of the tumor vasculature and by an indirect effect on tumor microenvironment.

12.4.2.3 The Role of Endothelial Cells

Apoptosis of endothelial cells is mediated by rapid generation of sphingolipid ceramide through the hydrolysis of cell membrane sphingomyelin by the acid sphingomyelinase (ASMase) enzyme. It has been shown that in response to a single high dose of radiation, an acute wave of ceramide-mediated endothelial apoptosis, initiated by ASMase, regulates tumor stem cell response. In cultured endothelium, VEGF prevents the radiation-induced ASMase activation occurring within minutes of radiation exposure and consequently represses apoptosis, whereas anti-VEGFR2 acts conversely, enhancing ceramide generation and apoptosis. Antiangiogenic agents derepressed radiation-induced ASMase activation only if delivered immediately before a single dose radiotherapy [26].

Combining high doses of radiation and antiangiogenic radiosensitization could have practical clinical applications. Recent evidence suggests that single-dose or hypofractionated radiotherapy, may provide therapeutic benefit to some human tumors, including those resistant to conventional fractionated radiotherapy schemes [27]. High local control rates have been observed for stereotactic body radiotherapy (SBRT) or stereotactic radiosurgery (SRS) suggesting that vascular damage may play an important role in the response to these modes of therapy in the clinic. Because of the difficulty often associated with the delivery of very high, single doses, there is a need for radiation sensitizers that would enable the use of more feasible lower dose per fraction protocols while maintaining the improvement in outcome provided by the higher exposures of single-dose or hypofractionated schemes.

12.4.3 Clinical Trials of Combined Radiation and Antiangiogenic Agents

The agents most widely explored in clinical trials are AAs targeting VEGF and its receptors. Of many that have been explored in clinical trials, three have been approved for clinical use; two small-molecule TKIs in monotherapy (sorafenib and sunitinib) for metastatic renal and hepatocellular carcinoma and an anti-VEGF monoclonal antibody (bevacizumab) in combination with chemotherapy for metastatic colorectal cancer, NSCLC, and breast cancer. None of these agents are currently approved in combination with radiation therapy. However, several phase I and II clinical trials have been concluded and the results are summarized in Table 12.2 [28,29], while ongoing clinical trials with bevacizumab and with small-molecule RTKIs

TABLE 12.2 Summary of Results of Clinical Trials Published 2005 to 2011

Drug	Tumor	N, Phase	Sequence	Chemo	Radiation	Results, Conclusions
Bevacizumab	Rectal adenocarcinoma	1. N = 11 (ph1) 2. N = 32 (ph2) 3. N = 11 (ph1) 4. N = 22 (ph2)	Neo + concom Neo + concom Concom Concom	1. 5 FU 2. 5 FU 3. Capecitabine/ oxaloplatin 4. Capecitabine	1–3 50.4 Gy/28 f 4 4.51 Gy/15 f	1. Feasible 2. Feasible 3. Feasible 4. Recommendation for phase 2 with amended dose of bevacizumab
	GBM GBM GBM GBM + AG	1. N = 51 (ph2) 2. N = 70 (ph2) 3. N = 15 (prospective) 4. Prospective	Concomitant + maintenance	Temozo-lamide	60 Gy 60 Gy/30 f 59.4 Gy/33 f 30 Gy/6 f (stereotactic)	1. Higher PFS with bevacizumab, similar OS 2. Improvement in PFS with bevacizumab 3. Feasible, but needs close monitoring 4. Feasible
	Pancreas adenocarcinoma	1. N = 46 (ph1) 2. N = 82 (ph2)	Concomitant + maintenace	Capecitabine	50.4 Gy/28 f 50.4 Gy/28 f	1. 3/30 tumor-associated bleeding duodenal ulcers. 2. Duodenal invasion excluded. Survival similar to conventional chemoradiation. 35.4%. Grade 3 toxicities
	ENT, various	N-43 (ph1)	Concomitant	5-FU, hydroxyurea	66–72 G 33–36 f	Recommendation: Maintain dose of bevacizumab with decreased doses of chemotherapy

(*continued*)

TABLE 12.2 Summary of Results of Clinical Trials Published 2005 to 2011 (Continued)

Drug	Tumor	N, Phase	Sequence	Chemo	Radiation	Results, Conclusions
Sunitinib	Liver (HCC)	N = 23 retrospective	1 w. pre-RT, during RT 12 w. post-RT		52 Gy/15 f	Acceptable response rates and safety profiles
	CNS tumors: Various	N = 23 (ph1)	Concom		14 to 60 Gy/f, 1.8 to 3.5 Gy	Safe, phase 2 required
	Metastatic Disease, various	N = 21 (21)	1 week before, concom and maintenance		40–50 Gy 10 f	Recommendation for phase 2: Highest Sunitinib dose + 50 Gy

Source: Based on a review by Mazeron, R. et al., *Cancer Treat Rev*, 37:476–486, 2011. From Palayoor, S. et al., *Int J Cancer*, 123:2430–2437, 2008.

Note: AG, anaplasic glioma; Concom, concomitant; d, day; f, fraction; GBM, glioblastoma; Gy: gray; HCC, hepatocellular carcinoma; N, number of patients; OS, overall survival; PFS, progression-free survival.

TABLE 12.3 Summary of Ongoing or Recently Completed Clinical Trials Involving Bevacizumab Combined with Radiotherapy or Radiochemotherapy

		Phase				
Tumor	**Treatment**	**1**	**1/2**	**2**	**3**	**Chemotherapy**
Brain	RT	2				Temolozomide
	CRT	1		17	2	
Rectal and colorectal	CRT	2	2	15		Capecitabine, oxaliplatin, 5-FU erlotinib
Lung	RT	1				Paclutaxel, carboplatin,
	CRT	1	3	8		cisplatin, etoposide, irinotecan
Pancreas	CRT	2		9		Capecitabine, oxaliplatin, 5-FU erlotinib, Tarceva gemcitabine
Head and neck	CRT	3		4		5-FU erlotinib, cisplatin, Paclitaxel, Docetaxel, Hydroxyurea, Cetuximab
Cervical cancer	RT			1		Cisplatin, carboplatin
	CRT			2		
Sarcoma	RT			2		
Prostate	CRT			2		Bicalutamide, Goserelin
Esophagus	CRT			3		Irinotecan, cisplatin

Source: http://www.clinicaltrials.gov.
Note: CRT, chemoradiotherapy; RT, radiotherapy.

are shown in Tables 12.3 and 12.4. Many of the trials have shown a promising antitumor response. However, increased toxicities have been observed in some studies, especially when the VEGF inhibitor is combined with chemoradiotherapy [30].

12.5 ANTIVASCULAR THERAPY

In contrast to antiangiogenic therapy, which is directed against the formation of neovasculature, antivascular therapy targets existing vasculature and has the potential to disrupt tumor and normal vessels.

12.5.1 Antivascular Agents

Vascular-disruptive agents (VDAs) cause a rapid shutdown of perfusion in the established tumor vasculature, leading to tumor cell ischemia and secondary tumor cell death. These agents have the potential to destroy existing tumor masses and may therefore be particularly suitable for treating large tumors, which are typically resistant to conventional therapies. In practice, many agents, such as those that target VEGF, will have both

TABLE 12.4 Summary of Ongoing or Recently Completed Clinical Trials Involving VEGF RTK Inhibitors or Multikinase Inhibitors Combined with Radiotherapy or Radiochemotherapy

			Phase				
Merge	**Tumor**	**Treatment**	**1**	**1/2**	**2**	**3**	**Chemotherapy**
Sunitib (SU11246)	HNSCC	RT	1				
		CRT	1				Cetuximab
	Sarcoma	RT	2	1			
	Breast	CRT			1		Capecitabine
	Esophagus	CRT				1	Irinotecan, cisplatin
	Prostate	RT	1				
	Glioblastoma	CRT					
Vandetanib (ZD6474)	HNSCC	CRT			1		Cisplatin
	Brain	RT	2				
		CRT	1		2		Temolozide
	NSCLC	RT		1			
	NSCLC Brain mets	CRT	2				Cisplatin
SU5416	Breast	CRT	1				Doxorubicin
	Prostate	RT	1				
	Sarcoma	CRT		2			Doxorubicin, Cyclophosphamide Ifosfamide
Cediranib (AZD2171)	GBM	CRT		1	1		Temolozomide
	Rectal ADC	RT		1			
	Brain mets (primary NSCLC)	RT	1				

Source: http://www.clinicaltrials.gov.
Note: ADC, adenocarcinoma; CRT, chemoradiotherapy; GBM, glioblastoma; mets, metastases; NSCLC, non-small cell lung cancer; RT, radiotherapy.

antivascular and antiangiogenic effects. Specific VDAs fall into two major classes of VDAs; the ligand-directed VDAs and the small-molecule VDAs.

12.5.1.1 Biological or Ligand-Directed VDAs

These drugs work by using antibodies, peptides, or growth factors that selectively bind to the endothelium. Coagulation or endothelial cell death is then achieved by coupling the vascular-targeting moiety with a toxin (e.g., ricin) or a procoagulant [31]. Alternatively, angiogenic endothelial cells can be targeted by specific peptide sequences that selectively bind to proliferating endothelial cells but not to quiescent cells [32].

12.5.1.2 Small-Molecule VDAs

These drugs, which work by inducing vascular collapse leading to extensive necrosis in tumors, are at a much more advanced stage of clinical development than ligand-based therapies. Small-molecule VDAs include flavonoids and tubulin-depolymerizing/binding agents. Flavone acetic acid and its derivatives, particularly 5,6-dimethyl-xanthenone-4-acetic acid (DMXAA), have a complex mechanism of action but are believed to work by inducing the release of vasoactive agents and cytokines, such as tumor necrosis factor-α (TNF-α), which leads to hemorrhagic necrosis [33]. The tubulin-binding agents (e.g., combretastatin A-4 disodium phosphate) are believed to work by selective disruption of the cytoskeleton in proliferating endothelial cells in tumors. The subsequent change in endothelial cell shape leads to vessel blockage, thrombus formation, rapid reduction in tumor blood flow, and secondary tumor necrosis [34].

12.5.2 Combined Treatment: Radiation and VDAs

Single-agent phase I studies have yielded modest results. This has been attributed to the presence of a viable rim of tumor cells at the tumor periphery after VDA treatment, as shown by preclinical studies. It has been suggested that increased blood flow in the adjacent normal tissue, together with probable rapid upregulation of angiogenic factors, such as VEGF, directly facilitates growth and expansion of the remaining rim of viable tumor cells [35]. Because these cells are well oxygenated, they present an excellent target for conventional cytotoxic therapies.

A rational approach to combining VDAs with radiation is to use the VDA drug to reduce or eliminate the radioresistant poorly oxygenated subpopulation of tumor cells and for radiation to kill the remaining well oxygenated peripheral cells. A number of preclinical studies performed on

TABLE 12.5 Preclinical and Clinical Trials Combining Vascular Disrupting Agents and Radiotherapy

Vascular Disrupting Agent	Preclinical Models	Clinical Trials	
		All	**+RT**
Tumor necrosis factor	Mammary ca. xenograft	Yes	Yes
Flavone acetic acid DMXAA	Mouse mammary Ca, mouse fibrosarcoma, Rat sarcoma, Mammary Ca xenograft	No	No
Combrestatin A-4 disodium phosphate	Rat sarcoma, mouse mammary carcinoma, Kaposi's sarcoma Rhabdomyosarcoma	Yes	Yes
ZD6126	Mouse mammary carcinoma, NSCLC xenograft, Glioblastoma xenograft Rat sarcoma	Yes	No
MN-029	Rat sarcoma	No	No

Source: Compiled from data summarized by Ciric, E. and Sersa, G., *Radiol Oncol*, 44(2):66–78, 2010 and http://www.clinicaltrials.gov.

rodent tumor models have reported enhanced tumor-killing when VDAs were given in combination with radiotherapy (Table 12.5).

The importance of scheduling was apparent from the results of one of the cited studies. No improvement in local control of murine CH3 tumors was seen when combretastatin A-4 disodium phosphate was given 60 min before radiation compared with improved results when the drug was given concurrently or after radiotherapy. A possible explanation is that the vascular shutdown induced by the VDAs may have rendered some tumor cells hypoxic at the time of irradiation, and that these cells later reoxygenated and survived. It is likely that blood flow needs to be reestablished in the remaining viable tissue to obtain maximum radiosensitization of the tumor. The greatest enhancement of the radiation response in fractionated dose regimens may be achieved when VDA is administered within a few hours after radiation and, under such conditions, antitumor effects may be greater than additive [36].

12.5.3 Clinical Application of VDAs

Vascular-disruptive agents are in a less advanced stage of clinical development, with only a few early trials concluded, mainly evaluating VDAs in monotherapy or chemotherapy combinations. Currently, the most widely explored VDA in clinical trials is combretastatin A-4 disodium phosphate, which has already been studied in several phase I trials evaluating dosage schedules and toxicity, and has recently entered phase II trials in combination with chemotherapy, radiation, and radioisotopes [37,38].

12.6 TARGETING HIF-1

An important role in the metabolism and treatment response of solid tumors is played by HIF-1, a transcription factor that induces the expression of a large array of proteins—most of which act to promote tumor cell proliferation in the specialized environment of a solid tumor [38].

12.6.1 Regulation of HIF-1 Activity

HIF-1 is a heterodimer composed of an α-subunit (HIF-1α) and a β-subunit (HIF-1β). The hypoxia-dependent activity of HIF-1 is regulated at multiple levels: translational initiation; degradation/stabilization, and upregulation of transactivation activity (Figure 12.4).

Under normoxic conditions HIF-α proteins are marked for degradation by hydroxylation by prolyl hydroxylases (PHD) and ubiquitination is mediated by the von Hippel–Lindau (VHL) protein. Degradation occurs in the proteasome. Under hypoxic conditions, PHD is inactive and HIF-α protein accumulates and translocates into the nucleus where it interacts

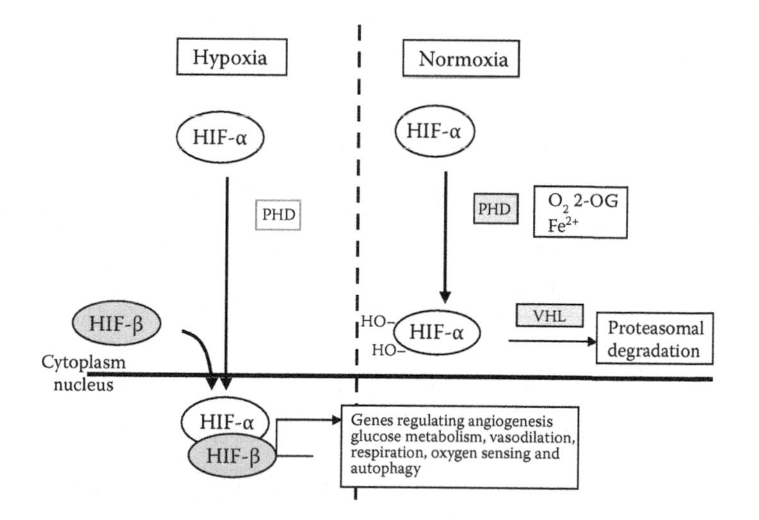

FIGURE 12.4 Under normoxia HIF-1α is hydroxylated by HIF-propyl hydroxylase then degraded along with VHL. Hypoxia inhibits this hydroxylation, thus stabilizing HIF-1α the concentration of which increases. HIF-1 is translocated into the nucleus where it dimerizes with HIF-1β and activates hypoxia-responsive genes including VEGF and PDGF.

with HIF-β to bind to specific DNA sequences. The resultant HIF-1 binds to its cognate transcriptional enhancer sequence, the hypoxia-responsive element (HRE), and induces the expression of an array of target genes regulating angiogenesis, glucose metabolism, vasodilation, respiration, oxygen sensing, and autophagy.

HIF-1α is activated in response to oxygen shortage, and is stabilized under normoxic conditions in cancer cells by transcriptional and post-transcriptional regulation in the following ways (Figure 12.5):

- After growth factor signaling, the transcription rate of HIF-1α gene is increased by a process involving activation of protein kinase C and NF-κB.
- Activation of the PI3K/Akt/mTOR/p70S6K pathway by reactive oxygen species increases the translational rate of HIF-1α protein [39].
- Upon receptor stimulation, tyrosine kinase transactivation promotes HIF-1 complex activation.
- HIF-1α can also be stabilized in various cell types by inflammatory cytokines, transforming growth factor-β, lipopolysaccharides, and by stem cell factors.

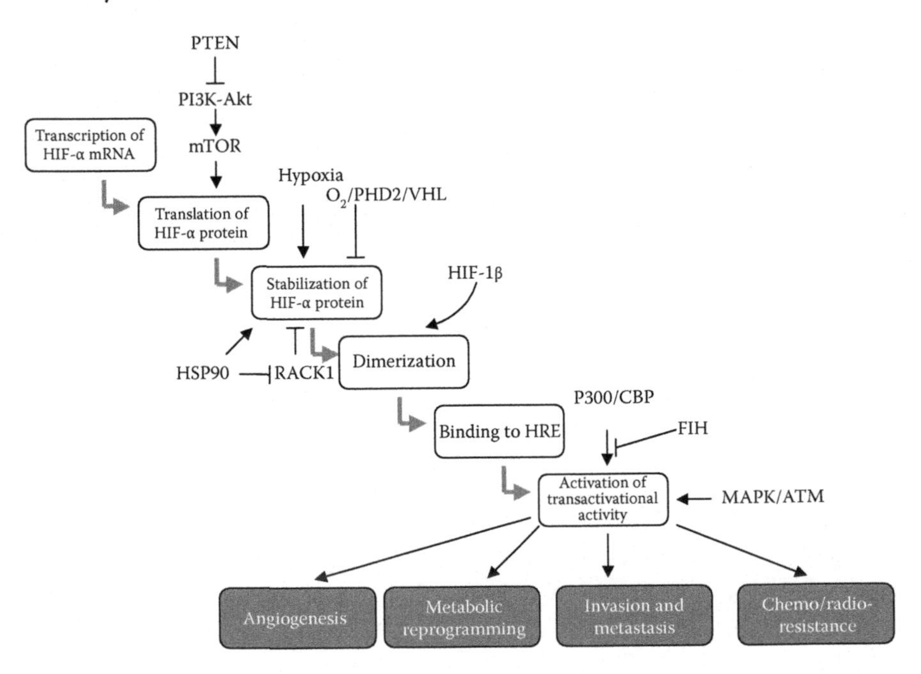

FIGURE 12.5 Stages in the activation of HIF-1.

12.6.2 Regulation of Radioresistance by HIF-1

Experimental observations of tumors in mice showed that radiation stimulated a twofold increase in HIF-1 protein and activity levels and a twofold increase in the expression of downstream cytokines, especially VEGF and bFGE, although irradiating tumor cells *in vitro* did not produce this effect. In irradiated tumors, it was shown that extensive HIF-1-activated regions existed that were also well-oxygenated 48 h after treatment, whereas in nontreated tumors, HIF-1 colocalized tightly with hypoxia. This suggests an apparent paradox, that RT activation of HIF-1 is associated with increased tumor oxygenation.

In fact, radiation does not activate HIF-1 signaling *in vitro* but radiation *in vivo* indirectly causes reoxygenation, which stimulates post-RT HIF signaling. This is not attributable to increased HIF-1 protein that, in fact, decreases within minutes of reoxygenation but rather by translational regulation of the HIF-signaling pathway. Dual mechanisms initiated by radiation simultaneously lead to HIF-1 activation; first, by reoxygenation stabilization of HIF-1 through free radical intermediates, and second, reoxygenation-mediated depolymerization of hypoxia-induced translational suppressors known as stress granules [40].

Immediately after irradiation, the predominant surviving cells are perinecrotic cells, which are hypoxic at the time of radiation and whose survival can be explained on the basis of the radiochemical oxygen effect. Subsequently, radiation-induced activation of HIF-1 is responsible for the translocation of the surviving cells toward tumor blood vessels. These cells secrete VEGF and other cytokines counteracting the destruction of tumor vasculature by radiation-induced apoptosis and facilitating recurrence and growth.

12.6.3 Targeting HIF-1

Basic and clinical researches have confirmed that the expression level of HIF-α, as well as absolute low pO_2, correlates with a poor prognosis and incidences of both tumor recurrence and distant tumor metastasis after radiation therapy [41]. As a countermeasure, each of the multiple steps responsible for the activation of HIF-1 has been exploited as a therapeutic target (Figure 12.6).

A number of drugs targeting successive stages of HIF-1 activation have been described:

- The stabilization of HIF-α protein is the most influential step in HIF-1 activity and is a similarly important target for treatment. YC-1, which was primarily synthesized with the aim of activating soluble

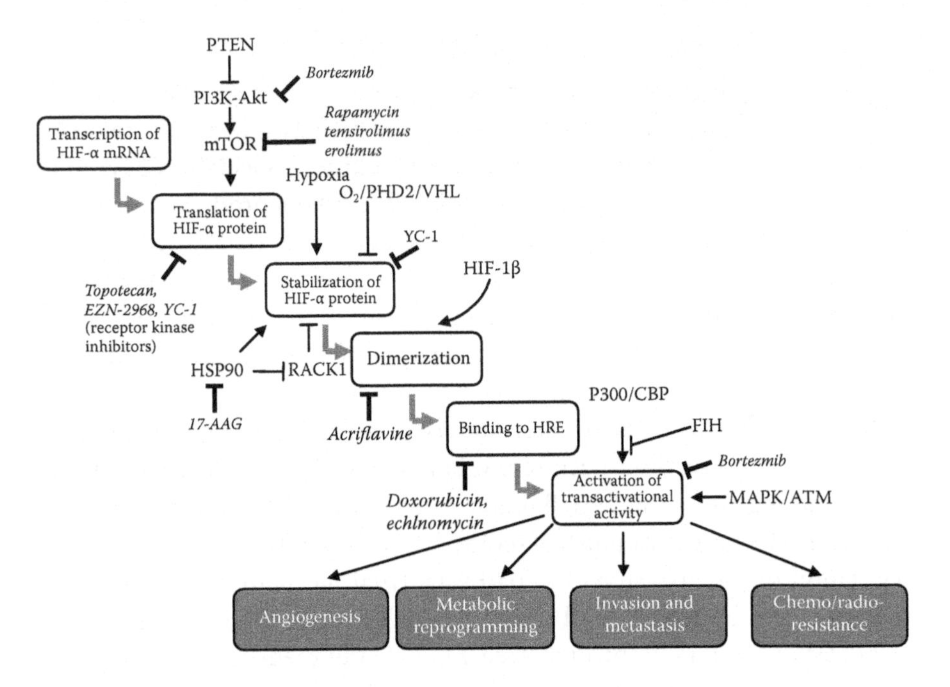

FIGURE 12.6 Inhibition of HIF-1 activation.

guanylate cyclase and inhibiting platelet aggregation, was reported to suppress the expression of HIF-1 target genes through the suppression of HIF-α accumulation and to significantly increase the antitumor efficacy of radiation therapy significantly [40,42]. HIF-1 degradation is promoted by an inhibitor of HSP90, 17-allylamino-17 demethoxygeldanamycin (17-AAG), which facilitates the RACK1-dependent ubiquitination of HIF-α, resulting in its degradation through the proteasome. Similarly, antioxidant reagents such as ascorbate and *N*-acetyl cysteine (NAC), promote the degradation of HIF-α protein by reducing Fe^{3+} to Fe^{2+}, which functions as a cofactor in the PHDs-VHL dependent degradation of HIF-α protein.

- Inhibiting the dimerization of HIF-1α with HIF-1β, which is required for HIF-1 DNA-binding and transcriptional activity. Acriflavine has been identified as an inhibitor of dimerization by directly binding to HIF-α [43]. Acriflavine treatment inhibited intratumoral expression of angiogenic cytokines, mobilization of angiogenic cells into peripheral blood, and tumor vascularization, resulting in the arrest of tumor growth [43] in an experimental model.

- An indirect approach is to inhibit the function of signaling pathways, which upregulate the expression of HIF-α, such as the PI3K–Akt–mTOR and Ras signaling pathways [44]. An mTOR inhibitor, RAD-001 reduced the level of HIF-α protein and its downstream gene products in a mouse model of prostate cancer with high oncogenic Akt activity [45]. Other mTOR inhibitors, such as rapamycin, temsirolimus (CCI-779) and everolimus (RAD-001), also showed the same effect [46]. In addition, it was reported that doxorubicin and echinomycin suppress the function of HIF-1 by inhibiting HIF-1 binding to HRE [43].

12.6.4 Radioenhancement by HIF Inhibitors

Direct and indirect inhibition of HIF-1 at various levels has been shown to be cytotoxic to tumor cells *in vitro* and to cause slowing or arrest of tumor growth *in vivo* by combined effects of cytotoxic, antiangiogenic and antivascular effects. *In vitro* increase in radiosensitivity as a result of HIF-1 inhibition has been demonstrated in tissue culture to be in some but not all cases to be selective for hypoxic cells. In tumors the results are more complex and not explicable solely on the basis of additive toxicity of HIF inhibition and RT. It is notable that the sequence of HIF-1 inhibitor and RT is crucial to the success in enhancing the therapeutic effect of RT. Inhibition of HIF-1 with unsuitable timing suppresses rather than enhances the effect of radiation therapy because its antiangiogenic effect increases the radioresistant hypoxic fraction in solid tumors [42]. Accumulated evidence indicates that the suppression of the postirradiation upregulation of HIF-1 activity is important for the best therapeutic benefit [40,42]. Some of the results obtained with tissue culture systems and preclinical tumor models are summarized in Table 12.6 [28,42,47–50].

12.6.5 Small-Molecule Compounds with Clinical Potential Reported to Inhibit HIF-1

The clinical application of HIF-1 inhibitors is in the very early stages of development. Although there are a number of potential methods for the inhibition of HIF-1 and the HIF-1 proteins, for preclinical investigation and clinical use the development of small-molecule inhibitors with good availability and pharmacological characteristics is highly important. Several of such drugs are under development these been observed to suppress HIF-1 or the HIF-proteins in preclinical studies. Notably all of them

TABLE 12.6 Targeting HIF-1 for Radiosensitization: Studies with Preclinical Models

Cell Lines or Tumor	Treatment, Strategy	Effect of Drug	Radiosensitization	Ref
C6 glioma, HN5 UMSCCa10 SC, Panc-1 pancreatic adenocarcinoma	PX-478, Orally available HIF-1 inhibitor. x-rays	Reduces HIF-1 protein levels & signaling.	Radiosensitizes hypoxic cells (clonogenic assay). Radiosensitization of xenografts. Is not simply explained by additive cell kill. PX-478 prevents postradiation HIF-1 signaling, and abrogates downstream stromal adaptation.	[47]
PC3 and DU 145 prostate carcinoma	PX-478 treatment for 20 h followed by x-rays	PX-478 decreased HIF-1α, protein and was cytotoxic in both cell lines with enhanced toxicity under hypoxia for DU-145. Induced S/G_2-M arrest in PC3. For both cell lines caused phosphorylation of H2AX.	For both cell lines radiosensitivity was enhanced under normoxic and hypoxic conditions with greater sensitization for PC3 cells. Prolongation of γH2AX expression in the irradiated cells. Conclusion: PX-478 acts as a radiosensitizer in cell lines under oxic/hypoxic conditions. Mechanisms are unclear.	[28]
AMC-HN3 SCC of the larynx	Cultured in hypoxia ± $CoCl_2$. HIF-1 inhibitor YC-1. x-rays	Hypoxia and $CoCl_2$ induced nuclear accumulation of HIF-1α, protein. Sub-G_1 fraction was increased in YC-1-treated hypoxic cells after irradiation.	Radiation > cytotoxicity oxic cells than in $CoCl_2$ cells. YC-1 in $CoCl_2$ cells inhibited HIF-α, expression enhanced radiosensitivity reducing SF to that of oxic cells.	[48]

The human cervical epithelial adenocarcinoma (HeLa)	YC-1 followed by RT or RT followed by YC-1	Hypoxic cells reoxygenated 6 h postirradiation, leading decrease in HIF-1 activity.	YC-RT: YC-1-mediated increase in tumor hypoxia suppresses the effect of radiation therapy RT-YC-1: Suppressed postirradiation increase in HIF-1 activity, delayed tumor growth. Indicates importance of treatment scheduling.	[42]
Human lung adenocarcinoma cell line, A549	10 μM YC-1 for 2 h; hypoxic conditions or air for 1 h before RT 2, 5, or 10 Gy	In hypoxia levels of HIF-1α > than in air. HIF-1α expression in hypoxic cells suppressed by YC-1 to the same level as that seen in cells exposed in air without YC-1.	In hypoxia no significant difference, at any dose between the SFs of cells treated ± YC-1 (clonogenic survival). OER at SF = 0.1 was 2.7 and 2.6 in the presence and the absence of YC-1, respectively.	[49]
Human gastric adenocarcinoma MKN45 & MKN28	1-(3-C-ethynyl-b-D-ribo-pentofuranosyl)cytosine (TAS106, ECyd) RT	In MNK45 xenografts TAS106 (0.5 mg/kg) suppressed HIF-1α expression and reduced the area of the hypoxic region in the tumor.	Reduced HIF-1α gene expression was associated with enhanced x-ray-induced apoptosis in hypoxic cells. In xenografts TAS106 + 2 Gy of x-rays enhanced apoptosis in the hypoxic region. Results suggest TAS106 acts as a radiosensitizer through the inhibition of HIF-1α expression.	[50]

Note: $CoCl_2$, cobalt chloride; OER, oxygen enhancement ratio; SC, squamous cells; SF, surviving fraction.

have been shown to sensitize cells in tissue culture and or xenografted tumors to radiation.

- PX-478 is an orally active small-molecule inhibitor of HIF-α, The mechanisms involved have yet to be fully elucidated. The inhibitory effect of this agent is independent of the tumor suppressor genes VHL and p53 and may be related to derangements in glucose uptake and metabolism due to inhibition of glucose transporter-1 (Glut-1). PX-478 has excellent activity against established human tumor xenografts, providing tumor regressions with prolonged growth delays which correlate positively with HIF-1 levels. PX-478 is a highly water soluble molecule, with good i.v., i.p. and p.o. antitumor activity. It is rapidly absorbed following oral and i.p. administration and has a good pharmacological profile. A phase 1 clinical trial has been conducted with PX-478 in which the drug is being administered orally to patients with lymphoma [28,47].
- YC-1 [3-(5′-hydroxymethyl-2′-furyl)-1-benzyl indazole] is a synthetic compound with a variety of pharmacologic actions including the inhibition of HIF-α. It has been widely used as a pharmacologic tool for investigating the physiologic and pathologic roles of HIF-1. YC-1 has been shown to inhibit the growth of several xenograft tumors in nude mice due to anti-HIF action and has also been shown to be a radiosensitizer. Mechanistically, it has been found that YC-1 accelerates HIF-1α degradation by targeting its amino acids 720 to 780 region or by inhibiting Mdm2 and that it inhibits the *de novo* synthesis of HIF-1α by inactivating the phosphatidylinositol 3-kinase/Akt/mammalian target of rapamycin pathway. However, the precise mechanism that underlies HIF-α downregulation by YC-1 remains uncertain [49,50].
- TAS106, (ECyd) (1-(3-C-ethynyl-β-D-ribo-pentofuranosyl)cytosine) is a ribonucleoside anticancer drug first synthesized in 1995. When TAS106 (ECyd) is incorporated into tumor cells, it is rapidly phosphorylated to ECyd 5′-triphosphate, which has the potential to inhibit RNA synthesis because of its inhibition of RNA polymerase, leading to cell death. Uridine/cytidine kinase (UCK) is a key enzyme for the first phosphorylation of TAS106 to ECyd 5′-monophosphate, generating cytotoxicity. The high activity of UCK in tumor cells relative to normal cells gives TAS106 an advantage in specificity for tumor therapy.

TAS106 treatment was shown to suppress the expression of HIF-1α proteins *in vitro* and *in vivo* as a result of suppression of HIF-1α itself. TAS106 was also shown to inhibit the mRNA expression of HIF-1α in MKN45 cells under hypoxia in a manner similar to that of actinomycin D, a transcription inhibitor suggesting that TAS106 downregulated HIF-1α expression transcriptionally [51]. Two ongoing or recently completed clinical trials have been reported.

12.7 SUMMARY

The microenvironment of malignant solid tumors differs markedly from that of normal tissues, particularly with respect to pH, the distribution and availability of nutrients, and of oxygen concentrations. Compared with normal vessels, tumor vasculature is disorganized and tortuous and lacks the hierarchy of arterioles, capillaries, and venules. Tumor blood vessels are leakier than their normal tissue counterparts, pericytes are only loosely connected to basement membrane, and in some circumstances tumor cells may line blood vessels. A consequence of the functionally and structurally defective blood supply of malignant tumors is imbalance between oxygen supply and oxygen consumption and some parts of the tumor become chronically hypoxic. In addition, functionally anomalous tumor vasculature causes the transient opening and closing of blood vessels which generates areas of acute hypoxia.

Targeting the development and function of the tumor vasculature is evolving as a therapeutic strategy. Vascular Targeted Therapies can be divided into antiangiogenic agents (AAs), which target the growth of new blood vessels in tumors and VDAs, which target established vasculature. Most antiangiogenic agents work by targeting VEGF, VEGFR and downstream signaling pathways. The most important example being bevacizumab (Avastin), a humanized monoclonal antibody that binds and neutralizes VEGF. Bevacizumab has been approved for clinical use in combination with chemotherapy for metastatic colorectal cancer, NSCLC and breast cancer. Bevacizumab is not currently approved in combination with radiation therapy but a number of clinical trials have been concluded whereas many others are ongoing.

Another class of antiangiogenic agents, small-molecule TKIs act by preventing activation of growth factor receptors, thus inhibiting downstream signaling pathways. Some of these have the advantage of being active against receptors for different growth factors. Sunitinib, for example, targets VEGFRs, PDGFR, and c-kit, and has shown significant efficacy in clinical trials for renal cancer. A number of TKIs in combination with radiotherapy are being investigated in phase I and II clinical trials.

It is is hypothesized that antiangiogenics target the nonfunctional and redundant vessels in tumors resulting in reduction in interstitial pressure and improved tumor oxygenation. In effect, the vascular system of the tumor becomes more like that of normal tissue (vascular normalization) and improved oxygenation is associated with increased radiosensitivity. In some but not all cases the sequence and timing of delivery of radiation and antiangiogenic is an important factor because radiation should be delivered during the vascular normalization window when tumor oxygenation is optimal.

Vascular-disruptive agents cause a rapid shutdown of perfusion in the established tumor vasculature, leading to tumor cell ischemia and secondary tumor cell death. These agents have the potential to destroy existing tumor masses and may be therefore particularly suitable for treating large tumors, which are typically resistant to conventional therapies. The only VDA to get to the clinical trial stage is combrestatin A-4 disodium phosphate, which has been evaluated in several Phase I trials and has recently entered Phase II trials in combination with chemotherapy, radiation and radioisotopes.

The hypoxic environment of solid tumors is conducive to the upregulation of HIF-1. HIF-1α is not only activated in response to oxygen shortage, but is also stabilized under normoxic conditions in cancer cells by transcriptional and posttranscriptional regulation. Basic and clinical studies have confirmed that the expression level of HIF-α, as well as absolute low pO_2, correlates with a poor prognosis and incidences of both tumor recurrence and distant tumor metastasis after radiotherapy. HIF-1 can be targeted by inhibitors of the multiple steps responsible for its activation. Small-molecule compounds with clinical potential reported to inhibit HIF-1 are under investigation and radiosensitization has been demonstrated in some cases.

REFERENCES

1. Thomlinson, R., and Gray, L. The histological structure of some human lung cancers and the possible implications for radiotherapy. *Br J Cancer* 1955;9:539–549.
2. Jain, R. Molecular regulation of vessel maturation. *Nat Med* 2003;9(6): 685–693.
3. Kizaka-Kondoh, S., Inoue, M., Harada, H., and Hiraoka, M. Tumor hypoxia: A target for selective cancer therapy. *Cancer Sci* 2003;94(12):1021–1028.

4. Brown, J. Evidence for acutely hypoxic cells in mouse tumours, and a possible mechanism of reoxygenation. *Br J Radiol* 1979;52(620):650–656.
5. Dewhirst, M. Relationships between cycling hypoxia, HIF-1, angiogenesis and oxidative stress. *Radiat Res* 2009;172(6):653–665.
6. Harada, H., and Hiraoka, M. Hypoxia-inducible factor 1 in tumor radioresistance. *Curr Signal Transd Ther* 2010;5(3):188–196.
7. Folkman, J. Tumor angiogenesis: Therapeutic implications. *New Engl J Med* 1971;285(21):1182–1186.
8. Yoshimura, M., Itasaka, S., Harada, H., and Hiraoka, M. Microenvironment and radiation therapy. *BioMed Res Int* 2013;2013:685308, 13 pp.
9. Ciric, E., and Sersa, G. Radiotherapy in combination with vascular-targeted therapies. *Radiol Oncol* 2010;44(2):66–78.
10. Murata, R., Tsujitani, M., and Horsman, M. Enhanced local tumour control after single or fractionated radiation treatment using the hypoxic cell radiosensitizer doranidazole. *Radiother Oncol* 2008;87:331–338.
11. Hoang, T., Huang, S., Armstrong, E., Eickhoff, J., and Harari, P. Enhancement of radiation response with bevacizumab. *J Exp Clin Cancer Res* 2012; 31:37.
12. Gorski, D., Beckett, M., Jaskowiak, N. et al. Blockade of the vascular endothelial growth factor stress response increases the antitumor effects of ionizing radiation. *Cancer Res* 1999;59(14):3374–3378.
13. Kozin, S., Boucher, Y., Hicklin, D., Bohlen, P., Jain, R., and Suit, H. Vascular endothelial growth factor receptor-2-blocking antibody potentiates radiation-induced long-term control of human tumor xenografts. *Cancer Res* 2001;61(1):39–44.
14. Fenton, B., Paoni, S., and Ding, I. Pathophysiological effects of vascular endothelial growth factor receptor-2-blocking antibody plus fractionated radiotherapy on murine mammary tumors. *Cancer Res* 2004;64(16): 5712–5719.
15. Hess, C., Vuong, V., Hegyi, I. et al. Effect of VEGF receptor inhibitor PTK787/ZK222548 combined with ionizing radiation on endothelial cells and tumour growth. *Br J Cancer* 2001;85:2010–2016.
16. Shibuya, K., Komaki, R., Shintani, T. et al. Targeted therapy against VEGFR and EGFR with ZD6474 enhances the therapeutic efficacy of irradiation in an orthotopic model of human non-small-cell lung cancer. *Int J Radiat Oncol Biol Phys* 2007;69(5):1534–1543.
17. Cao, C., Albert, J., Geng, L. et al. Vascular endothelial growth factor tyrosine kinase inhibitor AZD2171 and fractionated radiotherapy in mouse models of lung cancer. *Cancer Res* 2006;66(23):11409–11415.
18. Huber, P., Bischof, M., Jenne, J. et al. Trimodal cancer treatment: Beneficial effects of combined antiangiogenesis, radiation, and chemotherapy. *Cancer Res* 2005;65(9):3643–3655.
19. Chang, S., Lamborn, K., Malec, M. et al. Phase II study of temozolomide and thalidomide with radiation therapy for newly diagnosed glioblastoma multiforme. *Int J Radiat Oncol Biol Phys* 2004;60(2):353–357.

20. Lund, E.L., Bastholm, L., and Kristjansen, P. Therapeutic synergy of TNP-470 and ionizing radiation: Effects on tumor growth, vessel morphology, and angiogenesis in human glioblastoma multiforme xenografts. *Clin Cancer Res* 2000;6:971–978.
21. Murata, R., Nishimura, Y., and Hiraoka, M. An antiangiogenic agent (TNP-470) inhibited reoxygenation during fractionated radiotherapy of murine mammary carcinoma. *Int J Radiat Oncol Biol Phys* 1997;37(5):1107–1113.
22. Dings, R., Loren, M., Heun, H. et al. Scheduling of radiation with angiogenesis inhibitors anginex and avastin improves therapeutic outcome via vessel normalization. *Clin Cancer Res* 2007;13(11):3395–3402.
23. Ou, G., Itasaka, S., and Zeng, L. Usefulness of HIF-1 imaging for determining optimal timing of combining bevacizumab and radiotherapy. *Int J Radiat Oncol Biol Phys* 2009;75(2):463–467.
24. Zips, D., Hessel, F., Krause, M. et al. Impact of adjuvant inhibition of vascular endothelial growth factor receptor tyrosine kinases on tumor growth delay and local tumor control after fractionated irradiation in human squamous cell carcinomas in nude mice. *Int J Radiat Oncol Biol Phys* 2005;61:908–914.
25. Itasaka, S., Komaki, R., Herbst, R. et al. Endostatin improves radioresponse and blocks tumor revascularization after radiation therapy for A431 xenografts in mice. *Int J Radiat Oncol Biol Phys* 2007;67(3):870–878.
26. Truman, J., García-Barros, M., Kaag, M. et al. Endothelial membrane remodeling is obligate for anti-angiogenic radiosensitization during tumor radiosurgery. *PLoS One* 2010;5(8):e12310.
27. Yamada, Y., Bilsky, M., Lovelock, D. et al. High-dose, single-fraction image-guided intensity-modulated radiotherapy for metastatic spinal lesions. *Int J Radiat Oncol Biol Phys* 2008;71:484–490.
28. Palayoor, S., Mitchell, J., Cerna, D., Degraff, W., John-Aryankalayil, M., and Coleman, C. PX-478, an inhibitor of hypoxia-inducible factor-1alpha, enhances radiosensitivity of prostate carcinoma cells. *Int J Cancer* 2008;123:2430–2437.
29. Mazeron, R., Anderson, B., Supiot, S., Paris, F., and Deutsch, E. Current state of knowledge regarding the use of antiangiogenic agents with radiation therapy. *Cancer Treat Rev* 2011;37:476–486.
30. Ishikawa, H., Sakurai, H., Hasegawa, M. et al. Expression of hypoxic-inducible factor 1alpha predicts metastasis-free survival after radiation therapy alone in stage IIIB cervical squamous cell carcinoma. *Int J Radiat Oncol Biol Phys* 2004;60(2):513–521.
31. Thorpe, P. Vascular targeting agents as cancer therapeutics. *Clin Cancer Res* 2004;10:415–427.
32. Arap, W., Pasqualini, R., and Ruoslahti, E. Cancer treatment by targeted drug delivery to tumor vasculature in a mouse model. *Science* 1998;279:377–380.
33. Baguley, B.C. Antivascular therapy of cancer: DMXAA. *Lancet Oncol* 2003; 4:141–148.
34. Tozer, G., Prise, V., Wilson, J. et al. Mechanisms associated with tumor vascular shut-down induced by combretastatin A-4 phosphate: Intravital microscopy and measurement of vascular permeability. *Cancer Res* 2001;61:6413–6422.

35. Chaplin, D., and Hill, S. The development of combretastatin A4 phosphate as a vascular targeting agent. *Int J Radiat Oncol Biol Phys* 2002;54:1491–1496.
36. Siemann, D., and Horsman, M. Targeting the tumor vasculature: A strategy to improve radiation therapy. *Expert Rev Anticancer Ther* 2004;4:321–327.
37. Citrin, D., and Camphausen, K., eds. *Advancement of Antiangiogenic and Vascular Disrupting Agents Combined with Radiation*. New York: Springer Science, 2008.
38. Moeller, B., and Dewhirst, M. HIF-1 and tumour radiosensitivity. *Br J Cancer* 2006;95(1):1–5.
39. Semenza, G. Regulation of cancer cell metabolism by hypoxia-inducible factor 1. *Semin Cancer Biol* 2009;19(1):12–16.
40. Moeller, B., Cao, Y., Li, C., and Dewhirst, M. Radiation activates HIF-1 to regulate vascular radiosensitivity in tumors: Role of reoxygenation, free radicals, and stress granules. *Cancer Cell* 2004;5(5):429–441.
41. Semenza, G. Defining the role of hypoxia-inducible factor 1 in cancer biology and therapeutics. *Oncogene* 2010;29(5):625–634.
42. Harada, H., Itasaka, S., Zhu, Y. et al. Treatment regimen determines whether an HIF-1 inhibitor enhances or inhibits the effect of radiation therapy. *Br J Cancer* 2009;100(5):747–757.
43. Lee, K., Zhang, H., Qian, D., Rey, S., Liu, J., and Semenza, G. Acriflavine inhibits HIF-1 dimerization, tumor growth, and vascularization. *Proc Natl Acad Sci U S A* 2009;106(42):17910–17915.
44. Zundel, W., Schindler, C., Haas-Kogan, D. et al. Loss of PTEN facilitates HIF-1-mediated gene expression. *Genes Dev* 2000;14(4):391–396.
45. Majumder, P., Febbo, P., Bikoff, R. et al. mTOR inhibition reverses Akt-dependent prostate intraepithelial neoplasia through regulation of apoptotic and HIF-1-dependent pathways. *Nat Med* 2004;10(6):594–601.
46. Wysocki, P. mTOR in renal cell cancer: Modulator of tumor biology and therapeutic target. *Expert Rev Mol Diagn* 2009;9(3):231–241.
47. Schwartz, D., Powis, G., Thitai-Kumar, A. et al. The selective hypoxia inducible factor-1 inhibitor PX-478 provides *in vivo* radiosensitization through tumor stromal effects. *Mol Cancer Ther* 2009;8:947–958.
48. Moon, S., Chang, H., Roh, J. et al. Using YC-1 to overcome the radioresistance of hypoxic cancer cells. *Oral Oncol* 2009;45:915–919.
49. Oike, T., Suzuki, Y., Al-Jahdari, W. et al. Suppression of HIF-1α expression and radiation resistance in acute hypoxic conditions. *Exp Ther Med* 2012;3(1):141–145.
50. Yasui, H., Ogura, A., Asanuma, T. et al. Inhibition of HIF-1α by the anticancer drug TAS106 enhances X-ray-induced apoptosis *in vitro* and *in vivo*. *Br J Cancer* 2008;99:1442–1452.
51. Yasui, H., Inanami, O., Asanuma, T. et al. Treatment combining X-irradiation and a ribonucleoside anticancer drug, TAS106, effectively suppresses the growth of tumor cells transplanted in mice. *Int J Radiat Oncol Biol Phys* 2007;68:218–228.

CHAPTER 13

Phytochemicals

Chemopreventive, Radiosensitizing, Radioprotective

Phytochemicals are nonnutritive plant chemicals that have protective or disease-preventive properties. Plants produce these chemicals to protect themselves, but research has shown that they can also protect humans against disease. There are more than a thousand known phytochemicals. Some of the important groups are alkaloids (caffeine), carotenoids, flavenoids (polyphenols), isoflavones (phytoestrogens, genistein), organosulfides (glutathione; GSH), stylbenes (resveratrol), and phenolic acids (curcumin, ellagic acid). Recent investigations at the molecular and cellular level have elaborated some of the characteristics of phytochemicals that enable these compounds to attack a number of targets and fulfill diverse functions including, under the right circumstances, acting as radiosensitizers.

13.1 MECHANISM OF INTERACTION BETWEEN PHYTOCHEMICALS AND PROTEINS

A large number of phytochemicals have been identified as chemopreventive agents and recently as radiosensitizers or radioprotectors (Figure 13.1). However, in terms of identifying mechanisms at the chemical and biomolecular level, the bulk of the work has been done with one compound, curcumin, which has been investigated at all levels from chemistry to clinical trials. The mechanisms described in the following were largely

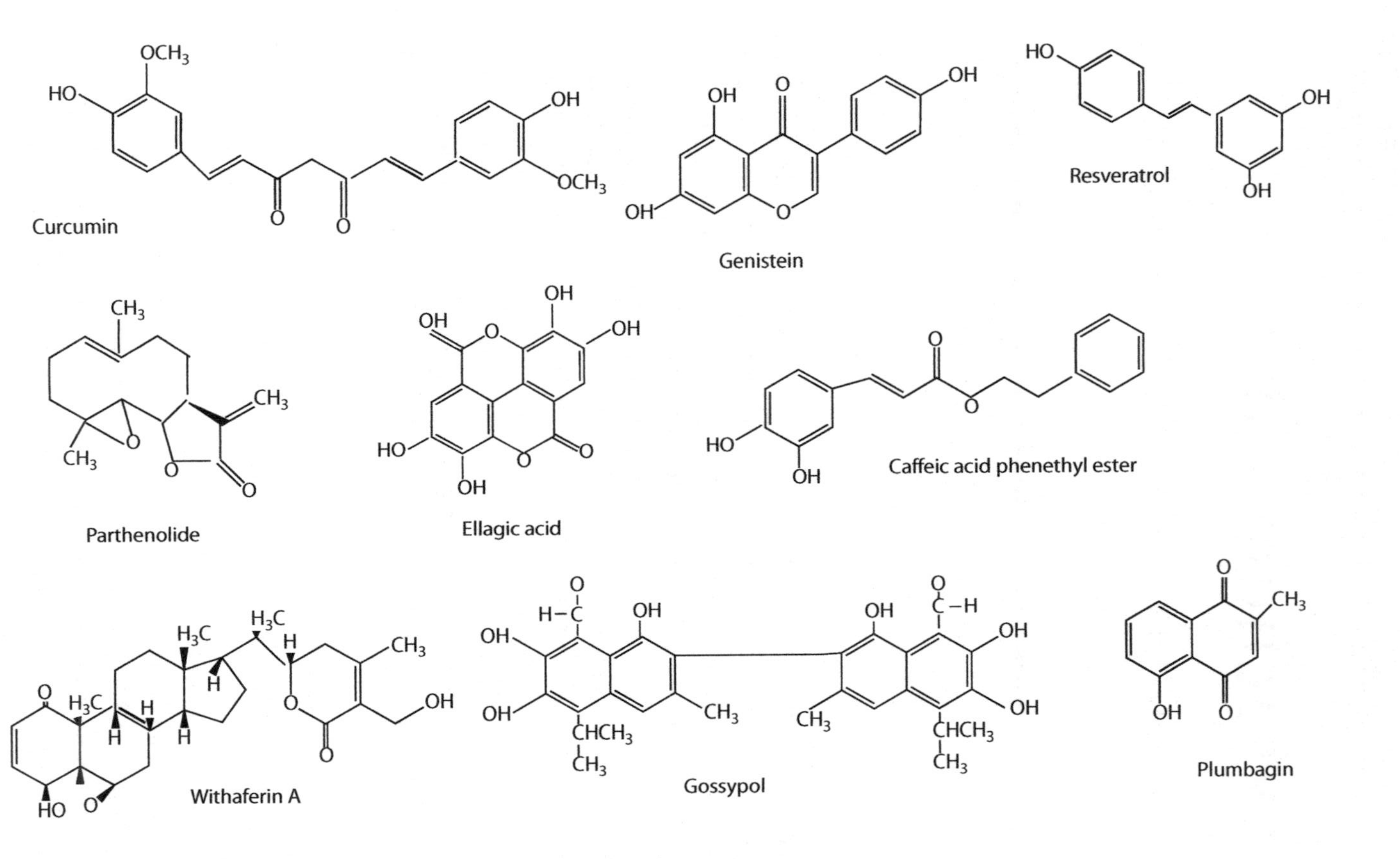

FIGURE 13.1 Structure of selected plant phytochemicals with radiosensitizing activity.

elaborated using curcumin, but the same rationale is applicable to some other phytochemicals.

13.1.1 Direct Interaction of Curcumin with Target Protein Molecules

(This section is based on reviews by Fang et al. [1] and Gupta et al. [2], and the articles referenced therein.)

The ability of curcumin to bind directly to diverse proteins with high affinity stems from its molecular structure and functionality. Curcumin is a diferuloyl methane molecule [1,7-bis (4-hydroxy-3-methoxyphenol)-1,6-heptadiene-3,5-dione] containing two ferulic acid residues joined by a methylene bridge. Curcumin has two hydrophobic phenyl domains that are connected by a flexible linker (Figure 13.2), and molecular docking studies have shown that curcumin can adopt many different conformations suitable for maximizing hydrophobic contacts with the protein to which it becomes bound. For example, the phenyl rings of curcumin can participate in π–π van der Waals interactions with aromatic amino acid side chains. Within curcumin's generally hydrophobic structure, the phenolic and carbonyl functional groups, which are located on the ends and in the center of the molecule, can participate in hydrogen bonding with a target macromolecule. This structure provides a strong and directed electrostatic interaction to increase favorable free energies of association. Because of its β-diketone moiety, curcumin undergoes keto-enol tautomerism and exists entirely in the enol form both in solution and in solid phase, and this keto-enol tautomerization provides curcumin with additional

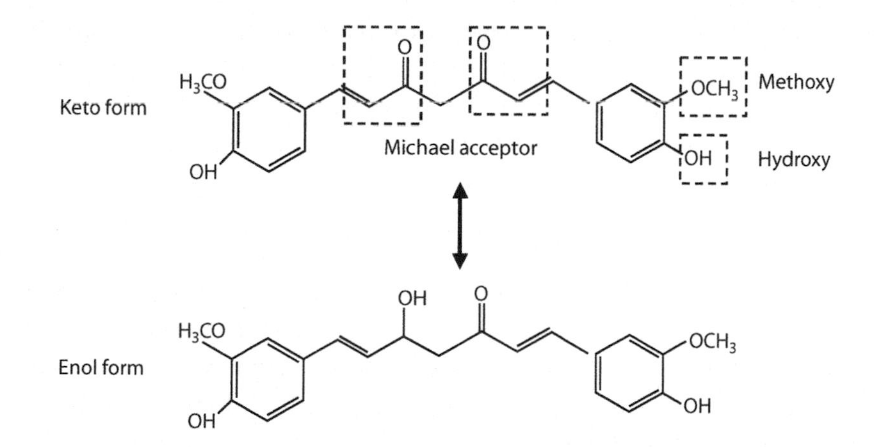

FIGURE 13.2 The molecular structures of curcumin known to interact directly with various proteins.

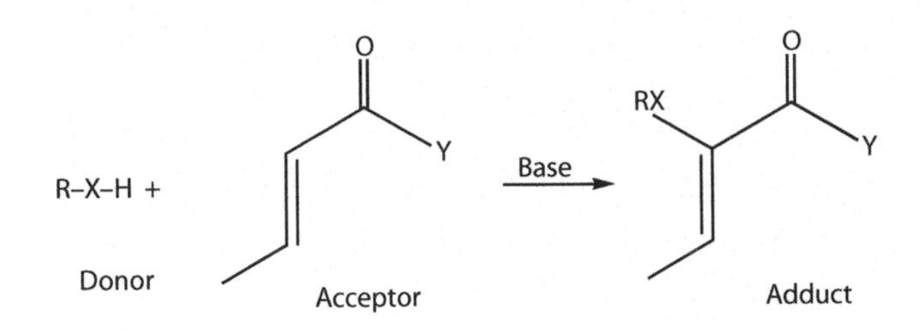

FIGURE 13.3 Schematic representation of Michael addition reaction. Nucleophiles (e.g., anions of OH, NH, SH, and CH acids) are good donors. Unsaturated carbonyl compounds (e.g., conjugated aldehydes, ketones, and esters) are good acceptors.

chemical functionality. The predominant enol form allows the midsection of the molecule to both donate and accept hydrogen bonds. The enol form also makes an ideal chelator of positively charged metals, which are often found in the active sites of target proteins. Finally, the keto-enol tautomerization allows curcumin to act as a Michael acceptor to nucleophilic attack, and curcumin has been found to bind covalently to nucleophilic cysteine sulfhydryls and the selenocysteine Se^- moiety. The Michael reaction is illustrated by a simple chemical equation (Figure 13.3) showing the addition of an exemplary nucleophile to a conjugated π-electron system, typically an α,β-unsaturated ketone or ester. The significant feature of this reaction is reversibility, which defines its biological importance. The combination of hydrophobic interactions, including π–π interactions, extensive hydrogen bonding, metal chelation, and covalent bonding covering such a large surface area gives curcumin many possible mechanisms to interact with target proteins. A number of the direct targets of curcumin, which may be significant in terms of radiosensitization, are listed in Table 13.1. This is by no means a complete list but covers many of those that are most significant.

13.1.2 Indirect Effects of Curcumin on Biological Systems

In addition to acting by binding directly to numerous proteins, curcumin has a diverse range of indirect molecular targets that mediate downstream reactions. These include transcription factors, enzymes, inflammatory mediators, protein kinases, drug resistance proteins, cell-cycle regulatory proteins, adhesion molecules, growth factors, receptors, cell-survival proteins, chemokines, and chemokine receptors. Given this range of direct and indirect targets, it can be envisioned how the pleiotropic activity of

TABLE 13.1 Selected Proteins That Are Directly Targeted by Curcumin

Target	Function	Mechanism	Radiosensitization
TxnRd1	Thioredoxin system is a ROS scavenger maintaining low molecular weight antioxidant levels in cells	C (curcumin) binds covalently to the highly nucleophilic selenocysteine residue in the COOH terminus of TxnRd1	Shown in SCC cell lines that inhibitory activity of Con TxnRd1 is required for curcumin-mediated radiosensitization. Indirect effects of targeting TrxRD1 has implications for radiation response
TNF-α	Proinflammatory. Effects mediated through production of ROS. Mediates both proapoptotic and antiapoptotic signaling	C binds to residues on the receptor-binding sites of TNF-α. Direct interaction with TNF-α by noncovalent and covalent binding	Upregulation TNF-α protein by radiation, which leads to an increase in NF-κB activity and induction of Bcl-2 protein inhibited by curcumin
COX-2	Mediates prostaglandin production. Involved in growth, metastatic spread, and resistance to therapy	C inhibited the enzyme activity by direct binding through residues in the active site. Inhibited PGE_2 production	Radiosensitization by Cox-2 inhibition seen in murine and human tumor models. Mechanisms may involve the elimination of PGs as protective molecules
NF-κB	Transcription factor. Regulator of inflammatory response and genes involved in cell survival and tumor progression	C directly inhibits NF-κB by (*i*) interfering with IκBα degradation; (*ii*) binding to p50 subunit of the NF-κB complex	Radiation selectively induces NF-κB in a number of tumor models. Inhibition of NF-κB enhances radiosensitivity
Bcl-2	Antiapoptotic protein overexpressed in solid and hematopoietic tumors, enhances survival	Curcumin, abrogated Bcl-2 activity by direct interaction with cavity 2 of Bcl-2 protein through multiple amino acids	Preclinical experiments show inhibition Bcl-2 sensitizes tumor cells to chemotherapy and radiotherapy

(*continued*)

TABLE 13.1 Selected Proteins That Are Directly Targeted by Curcumin (Continued)

Target	Function	Mechanism	Radiosensitization
DNA polymerase λ—	Eukaryotic polymerase involved in DNA repair	C derivative bound selectively to the N-terminal domain of pol λ	Radiosensitization by inhibition of DNA repair
ErbB2—(HER2/neu)	Transmembrane tyrosine kinase. Overexpression increases metastatic potential and resistance to anticancer agents	C binds to the kinase domain of ErbB2. Michael acceptor functionality of C required for both covalent association of C with ErbB2 and C-mediated ErbB2 depletion	Radiation activates ErbB2 leading to increased prosurvival signaling. Radiosensitization results from ErbB2 blockade
GSH	Tripeptide. Antioxidant protects against oxidative damage to important cellular components caused by ROS	C binds directly to glutathione forming glutathionylated products. Accelerated in the presence of GSTP1-1	GSH depletion by related phytochemical CAPE resulted in radiosensitization
HDACs	Deacetylate histone proteins. HDACs play a crucial role in epigenetic regulation of gene expression	C complexed with HDAC8. Makes close hydrophobic contacts with the active site residues	HDACs modulate chromatin structure and facilitate DNA repair. HDACs regulate the activity of a number of important nonhistone proteins; HDAC inhibitors have been shown to act as radiosensitizers in a number of preclinical studies (Chen et al. [3])

curcumin could be achieved. In fact, the situation is even more confused because there are a number of molecules or processes that can be both directly and indirectly targeted. Thus, the proapoptotic protein Bcl-2 is inhibited directly by interaction with curcumin (Table 13.1) and indirectly when the transcription faction nuclear factor-κB (NF-κB) is targeted by curcumin.

13.2 INTRACELLULAR SYSTEMS AND CONSTITUENT PROTEINS DIRECTLY TARGETED BY CURCUMIN AND OTHER RADIOSENSITIZING PHYTOCHEMICALS

The targets of curcumin and other phytochemicals that are implicated in radiosensitization effects can be grouped, for convenience, under certain general headings. In fact, as is true in molecular biology, these effects generally cannot be rigidly compartmentalized because there is overlap and crosstalk between the systems involved. An effort will be made here to present information in as systematic a fashion as possible.

13.2.1 Regulation of Oxidative Stress

Growing evidence suggests that cancer cells produce higher basal levels of reactive oxygen species (ROS) than normal cells. Under this persistent intrinsic oxidative stress, cancer cells develop an enhanced endogenous antioxidant capacity, which makes them more resistant to exogenous oxidants. One mode of action for phytochemicals acting as radiosensitizers is to target these radiation-induced antioxidant mechanisms.

13.2.1.1 Thioredoxin Reductase 1 and the Regulation of ROS

The upregulation of the antioxidant enzyme thioredoxin reductase 1 (TxnRd1) is observed in multiple primary human malignancies, and its loss has been associated with a reversal of tumor phenotype and a decrease in tumorigenicity. These observations suggest that malignant cells could be sensitized to oxidants, including radiation, by inhibition of this important antioxidant protein.

13.2.1.2 Trx–TrxR–NADPH System

Thioredoxin reductases (TxnRds) are a family of NADPH-dependent flavoproteins with a penultimate selenocysteine residue at the carboxy terminus. These enzymes exhibit a broad substrate specificity, which is due to the accessibility of the C-terminal redox-active site when reduced. TxnRds are ubiquitous with defined roles in diverse redox-regulated cellular

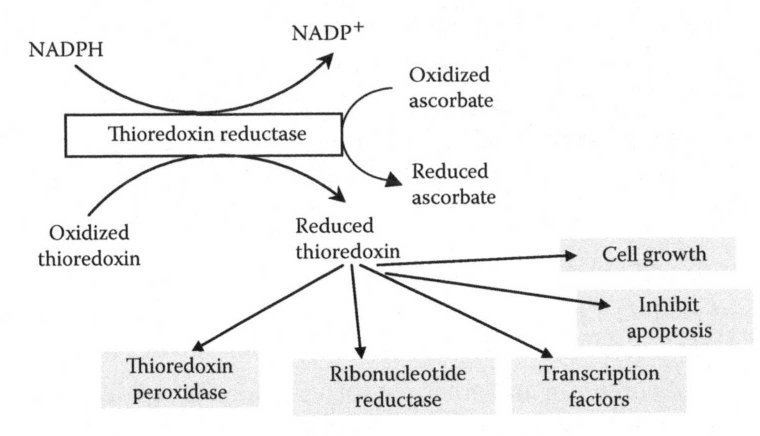

FIGURE 13.4 NADPH–Trx–TrxR system.

functions, including transcription, DNA damage recognition and repair, proliferation, apoptosis, and neo-angiogenesis.

TxnRds act in homodimeric form to catalyze NADPH-dependent reduction of thioredoxin and small molecular weight oxidants including ROS (Figure 13.4). TxnRd1 and TxnRd2 (the cytosolic/nuclear and mitochondrial form, respectively) are two ubiquitously expressed isoforms of this enzyme family. In response to oxidative stress, TxnRds sustain signaling pathways that regulate the transcription of genes to protect the cell from oxidative damage [4]. Cytosolic TxnRd1 expression is often upregulated in human cancers, where it is associated with aggressive tumor growth and poor prognosis. TxnRd1 has been shown to confer protection against the lethal effects of radiation in tumor cells, and agents that selectively target TxnRd1 have shown promising results as anticancer drugs in preclinical and clinical studies when used alone or when combined with radiation [4].

13.2.1.3 Interaction of TxnRd1 with Curcumin

Thioredoxin reductase 1 is among a number of identified molecular targets of curcumin. Curcumin-induced inhibition of TxnRd1 is dependent on the Michael acceptor function by which curcumin binds covalently to the highly nucleophilic selenocysteine residue in the COOH terminus of TxnRd1. Tetrahydrocurcumin, which lacks the α,β-unsaturated ketone moieties, does not inhibit TxnRd1 activity and does not radiosensitize tumor cells.

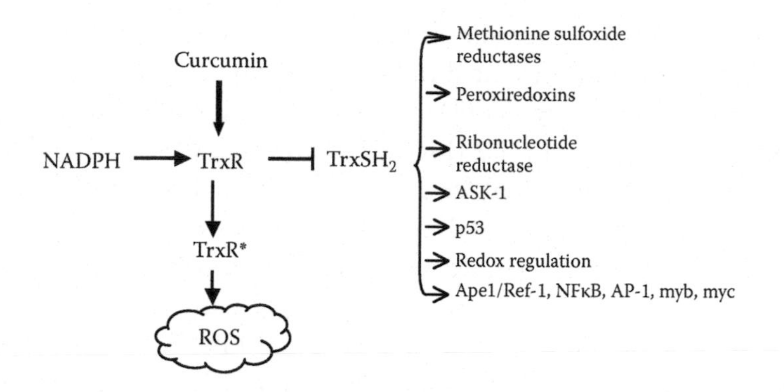

FIGURE 13.5 Cascade effects after modification of TrxR by curcumin. *Abbreviations*: AP-1, activator protein 1; TrxR*, modified TrxR.

It has been proposed that the preferential radiosensitization of tumor cells by curcumin is due, in part, to the targeting of the elevated expression of TxnRd1. The involvement of TxnRd1 in curcumin-mediated radiosensitization seems to be twofold. First, curcumin covalently binds to TxnRd1, irreversibly inhibiting its ability to reduce thioredoxin, which in turn is required for many of the pleiotropic antioxidant effects associated with TxnRd1 [5]. Second, modification of the selenocysteine residue of TxnRd1 converts the protein from an essential antioxidant enzyme to a protein with NADPH 1 oxidase activity, thereby elevating oxidative stress. Both effects would enhance the oxidative burden on the cell, leading to greater clonogenic cell death among other effects. Normal cells that express very low levels of TxnRd1 are not radiosensitized by curcumin (Figure 13.5).

Curcumin only inhibits the reduced form of TrxR. In the reduced form generated by NADPH, the active site residues Cys496 and Sec497 are present in the form of free –SH/–Se groups and are exposed at the surface of the enzyme making them easily attacked. This is in contrast to the oxidized enzyme where there is a nonreactive Cys496–Sec497 selenenylsulfide bridge [6].

13.2.1.4 Preclinical Experimental Demonstration of the Importance of TxrRd1 Knockdown by Curcumin in Radiosensitization

Experiments with FaDu and HeLa cells demonstrated that radiosensitization of reductase-1 [7]. In both cell lines, stable knockdown of TxnRd1 nearly abolished curcumin-mediated radiosensitization, indicating that inhibitory activity of curcumin of the antioxidant enzyme TxnRd1 was required for curcumin-mediated radiosensitization. TxnRd1 knockdown cells showed decreased radiation-induced ROS and sustained extracellular signal-regulated kinase 1/2 (ERK1/2) activation, which had been previously shown to be required for curcumin-mediated radiosensitization. Conversely, overexpressing catalytically active TxnRd1 in HEK293 cells, with low basal levels of TxnRd1, increased their sensitivity to curcumin alone and to the combination of curcumin and ionizing radiation. This work was extended by using two HNSCC cell lines, in which TxnRd1 is highly expressed in radioresistant HPV$^-$ HNSCC cell lines but not in radiosensitive HPV$^+$ HNSCC cells. Pretreatment with curcumin significantly increased the radiation sensitivity of the HPV$^-$ cells while having no significant effect on the radiation response in HPV$^+$ cell lines. In other experiments, the effect of dietary curcumin in an HPV$^-$ HNSCC xenograft was determined. Dietary curcumin and ionizing radiation resulted in a significant inhibition of tumor growth with a concomitant increase in animal survival (Figure 13.6) [8].

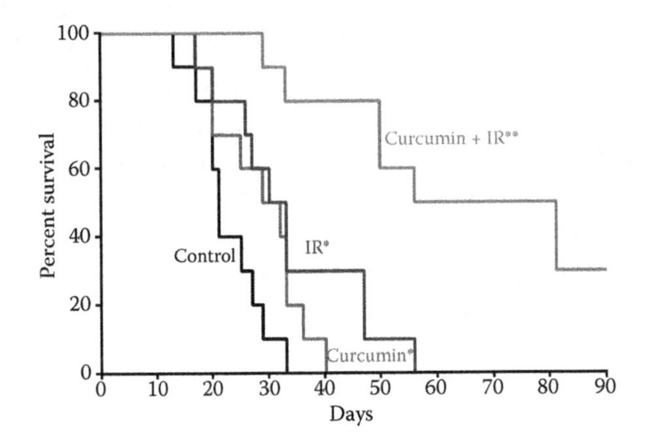

FIGURE 13.6 Effect of combined curcumin and radiation on animal survival. There was a statistical difference in mean survival between curcumin alone or radiation alone-treated groups when compared with untreated animals ($*p = 0.05$). The difference in mean survival between the combined treatment groups and all other groups was highly significant ($**p = 0.0001$). Three animals from the combined treatment group survived beyond the last data point collected on day 90. (From Tuttle, S. et al., *Cancer Biology Therapy* 13(7):575–584, 2012. With permission.)

13.2.1.5 Indirect Targeting TrxRd1 with Implications for Radiation Response

TrxR levels in tumor cell lines are often 10 times higher than those in normal tissues constituting up to 0.5% of total soluble proteins in mammary adenocarcinoma cell lines. This high level of TrxR fulfills a number of functions in the cancer cell.

13.2.1.5.1 Protection against Apoptosis A specific function of reduced thioredoxin in the prevention of apoptosis is through binding of ASK-1, a mitogen-activated protein kinase kinase kinase playing an essential role in apoptosis, which is activated by many stress-related and cytokine-related stimuli. Saitoh et al. [9] found that reduced Trx, but not oxidized Trx, bound directly to the N terminus of ASK-1 and inhibited ASK-1 kinase activity as well as the ASK-1-dependent apoptosis. When TrxR is inactivated by curcumin, it is unable to reduce oxidized Trx, and because oxidized Trx does not bind to ASK-1, the signaling cascade is allowed to proceed, ultimately inducing apoptosis.

13.2.1.5.2 DNA Damage Repair An important electron acceptor for reduced thioredoxin is ribonucleotide reductase, the S phase enzyme essential for synthesis of deoxyribonucleotides for DNA synthesis. Trx/TxnRd1 acts as an electron donor to the nuclear form of ribonucleotide reductase and this activity is suppressed by inhibition of TxnRd. Dysregulation of ribonucleotide reductase by lack of electrons from Trx/TxnRd1 leads, in turn, to catastrophic events in the S phase cells of tumor origin which, unable to complete DNA replication, will go into apoptosis. The ribonucleotide reductase subunit R2, which is induced by p53 [10], is essential for DNA repair.

Another DNA repair-related consequence of curcumin treatment involves the Trx-dependent upregulation transcription factor ApeI/Ref-1, which is both the apurinic/apyrimidinic endonuclease responsible for base excision repair and a transcription factor regulator together with Trx. There is evidence that only the reduced form of ApeI/Ref-1 is required for its DNA repair activity. Inactivation of this system will lead to failure to reconstitute abasic sites, stalling the replication fork and leading to loss of coding information.

13.2.1.6 Other Direct Targets of Curcumin with Potential Involvement in Radiosensitization

13.2.1.6.1 Survivin A member of the mammalian IAP family that inhibits apoptosis by blocking caspase 9 activation. It is overexpressed in tumors and absent from most normal tissue. High levels of survivin have been associated with tumor resistance to chemotherapy and radiotherapy. The antiapoptotic action of survivin is blocked when the protein is targeted by curcumin and it has been shown experimentally that inhibition of survivin plus radiotherapy results in significantly decreased cell survival.

13.2.1.6.2 Bcl-2 A member of the Bcl-2 family of proteins regulating both proapoptotic and antiapoptotic signaling in cells. Bcl-2 itself is an antiapoptotic protein that is overexpressed in a number of solid and hematopoietic tumors, and enhances cellular survival [11], contributing to the resistance to chemotherapy and radiotherapy. Curcumin interacts directly with cavity 2 of the Bcl-2 protein through multiple amino acids, abrogating Bcl-2 activity and thus enhancing apoptosis. Results of several investigations have shown that inhibiting Bcl-2 by a variety of means can potentiate the effect of radiotherapy.

13.2.1.6.3 EGFR Expression Epidermal growth factor receptor (EGFR) is a transmembrane glycoprotein with intrinsic tyrosine kinase activity. Upon binding with EGF or transforming growth factor-β, EGFR regulates a signaling cascade that, in turn, regulates cell growth and proliferation. Radiosensitization associated with targeting of EGFR has been described in Chapter 10.

13.2.1.6.4 Proteasome These are important targets of curcumin. Both carbonyl carbons of the β-diketone moiety of curcumin are highly susceptible to a nucleophilic attack by the hydroxyl group of the amino-terminal threonine (Thrl) of the β5 chymotrypsin-like (CT-like) subunit of the proteasome. Curcumin also hydrogen bonded with Ser96 in the β5-subunit. The direct binding of curcumin to β5-subunit was concomitant with inhibition of CT-like activity in several experimental models. The inhibition of proteasome activity by curcumin leads to the accumulation of ubiquitinated proteins and several proteasome target proteins and a subsequent induction of apoptosis. Radiosensitization associated with targeting of the proteasome has been described in Chapter 8.

13.2.1.6.5 ErbB2 (HER2/neu) A transmembrane tyrosine kinase whose overexpression has been shown to increase metastatic potential and resistance to anticancer agents. Therefore, therapeutic strategies that downregulate the level of ErbB2 protein or its activity (or both) could be potential treatments for ErbB2-overexpressing cancers. By binding to the kinase domain of ErbB2, curcumin has been shown to induce its ubiquitination and depletion. Further studies indicated that curcumin's Michael acceptor functionality was required for both covalent association of curcumin with ErbB2 and curcumin-mediated ErbB2 depletion. Radiosensitization associated with targeting of ErbB2 was also described in Chapter 10.

13.2.1.6.6 Glutathione GSH is a tripeptide and antioxidant with potential to prevent oxidative damage to important cellular components caused by ROS. Curcumin binds directly to glutathione, forming glutathionylated products that include monoglutathionyl and diglutathionyl adducts of curcumin. The initial rate of GSH-mediated consumption of curcumin was accelerated in the presence of glutathione-S-transferase (GSTP1-1).

13.2.1.6.7 Histone Acetyltransferase Histone acetylation plays an essential role in the epigenetic regulation of gene expression and is carried out by a group of enzymes called histone acetyltransferases (HATs), such as p300/CBP. It has been found that curcumin can specifically inhibit the activity of the p300/CBP family of HAT proteins by direct binding. It is proposed that the α,β unsaturated carbonyl groups in the curcumin side chain function as the Michael reaction sites and are required for its HAT-inhibitory activity. Concomitant with these observations, curcumin promoted proteasome-dependent degradation of p300. In addition to inducing p300 degradation, curcumin inhibited the acetyltransferase activity of purified p300 [12].

13.2.1.6.8 Histone Deacetylases The acetyltransferase activity of HAT is counterbalanced by a group of enzymes called histone deacetylases (HDAC), which deacetylate histone proteins. HDAC, in association with HAT, plays a crucial role in the epigenetic regulation of gene expression. Curcumin is a potent inhibitor of HDAC (IC_{50}, 115 μM). Molecular docking revealed that curcumin complexed with HDAC8 and adopted a stable binding pose extended toward the entrance cavity of the enzyme with close hydrophobic contacts with active site residues. The radiosensitizing effects of HAT and HDAC inhibition are discussed in Chapter 9.

13.3 TARGETING PROINFLAMMATORY SIGNALING PATHWAYS FOR TUMOR RADIOSENSITIZATION

Radiation-induced inflammatory signaling cascades promote tumor cell survival but, when unchecked, have detrimental effects on normal tissues.

The inflammatory signaling pathway is both inducible and may also have a degree of selectivity for the tumor cell and targeting this radioresistance pathway could radiosensitize tumor cells preferentially without sensitizing normal cells. Therefore, proteins that are transiently activated by radiation and play a central role in mediating proinflammatory signaling are important targets for radiosensitization.

13.3.1 NF-κB

NF-κB is constitutively activated in diverse solid malignancies, but not their normal counterparts. Radiation-induced NF-κB levels in cancer cells are transient, peaking within a few minutes of exposure to radiation and returning to baseline levels after a few hours. Nevertheless, this transiently activated NF-κB can provoke multiple radioresistance signals that attenuate the lethal effects of radiation. Not surprisingly, inhibition of NF-κB is perceived to be an exciting strategy to enhance tumor radiosensitivity.

13.3.1.1 Signaling through NF-κB

The transcription factor NF-κB is a family of closely related protein dimers that regulate inducible gene expression in various physiological settings [12]. The mammalian NF-κB family consists of five related proteins: p65 (RelA), RelB, c-Rel, p50/p105 (NF-κB1), and p52 (NF-κB2), which all share an amino-terminal REL homology domain (RHD) [13]. Dimers of NF-κB proteins bind to common sequence motifs in DNA (the κB sites) in promoters or enhancers of target genes, the transcription of which is regulated through the recruitment of transcriptional coactivators and transcriptional corepressors.

As described in Chapter 11, NF-κB is sequestered in the cytoplasm in association with its inhibitor IκBα in the form of an inactive complex. Stimulation of cells by inflammatory cytokines, UV light, or ROS results in the activation of IκB kinases (IKKα, β, and γ) and phosphorylation of IκBα on Ser32 and Ser36. The phosphorylated form of IκBα is ubiquitinated and degraded through the ubiquitin–proteasome system (UPS); unmasking the nuclear localization sequence on NF-κB, which then translocates into the nucleus, where it binds to the consensus sequence in the promoters of its target genes (Figure 13.7).

13.3.1.2 Inhibition of the NF-κB Signaling Cascade

The effective downregulation of NF-κB-mediated activity in target tissues can be achieved by inhibition of the key steps of the NF-κB pathway, by

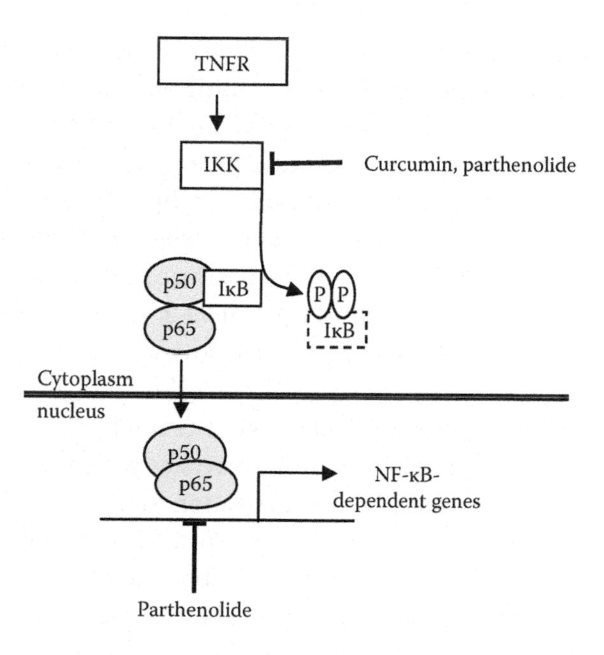

FIGURE 13.7 Targeting the NF-κB pathway. In the canonical NF-κB pathway, dimers such as p50/65 are maintained in the cytoplasm by interaction with inhibitor IκB molecules, mainly IκB. Binding of a ligand to a cell surface receptor, for example, TNF receptor (TNFR), recruits adaptors to cytoplasmic domain of the receptor, which in turn attracts an IKK complex. Activated IKK phosphorylates IκB at two serine residues, which leads to its ubiquitination and degradation by the proteasome. Released NF-κB is further covalently modified, enters the nucleus where it binds to DNA, and recruits chromatin remodeling enzymes to turn on target genes.

targeting the upstream components of NF-κB signaling, or by the pharmacological inhibition of the key components of the effector response.

13.3.1.2.1 Broad-Spectrum NF-κB Inhibition Plant phytochemicals are effective for tumor radiosensitizers through NF-κB inhibition. This approach renders tumor cells more radiosensitive by concurrently suppressing radiation-induced prosurvival signaling at multiple levels. Curcumin, for example, sensitizes by suppressing radiation-induced NF-κB activation directly (IKK activation, IkBa degradation), by inhibition of upstream activators of this pathway (Akt) [14], and by suppressing the NF-κB-regulated antiapoptotic, proliferative, angiogenic, invasive, and proinflammatory gene products [14]. Curcumin has also been shown to sensitize cancer cells to radiation by suppressing radiation-induced tumor necrosis factor-α (TNF-α) or TNF superfamily genes.

Radiosensitization by the soy isoflavone genistein is mediated by suppression of radiation-induced NF-κB leading to altered expression of regulatory cell cycle proteins such as cyclin B or p21WAF1/Cip1, and promoting G_2-M arrest [15]. The sesquiterpene lactone parthenolide sensitizes human hybrid CGL1 cells to radiation by inhibiting NF-κB and enhancing apoptosis [16]. In prostate cancer cells, parthenolide mediates radiosensitization by inhibiting radiation-induced NF-κB activation and the expression of its downstream target *SOD-2*, the gene encoding the antiapoptotic and antioxidant enzyme manganese superoxide dismutase.

13.3.1.2.2 Inhibition of Upstream Mediators by Curcumin

13.3.1.2.2.1 PI3K/Akt/PKB IKK activation acts as a converging point for NF-κB activation. However, NF-κB activation signals can be received from growth factor receptors and other prosurvival proteins upstream of IKK. A number of factors activate IKK, but the phosphatidylinositol 3-kinase (PI3K)/Akt/PKB pathway is probably the most important target with respect to tumor radiosensitization. The PI3K/Akt pathway is constitutively activated in various cancers and plays a critical role in promoting cell growth, and in TNF-α-mediated NF-κB activation. In addition, the PI3K/Akt pathway forms an important link in NF-κB-mediated expression of COX-2 and the radiation-induced NF-κB-mediated expression of MMP-9 [17], whereas inhibition of the PI3K/Akt pathway could attenuate NF-κB-regulated inflammatory gene expression [18]. Taken altogether, these reports suggest that Akt/PKB plays a crucial role in NF-κB activation through various stimuli—one of which is radiation.

13.3.1.2.2.2 Tumor Necrosis Factor α TNF-α is an essential component of the immune system, it is produced by several types of cells, but especially by macrophages. It has proinflammatory activities and contributes to a variety of autoimmune diseases. Curcumin binds directly to the receptor-binding site of TNF-α by covalent and noncovalent interactions and sensitizes cancer cells to radiation by suppressing radiation-induced TNF-α [19] or TNF superfamily genes. Inhibition of expression of both endogenous TNF-α and radiation-induced TNF-α protein is concomitant with inhibition of TNF-α-induced NF-κB activity.

13.3.1.2.2.3 Proteasome Inhibition The UPS is the major protein degradation machinery for intracellular proteins, which regulates protein turnover in a number of cell signaling pathways including cell proliferation, DNA repair, apoptosis, and the immune response. UPS functioning is crucial in regulating the NF-κB pathway and in fact, the proteasome inhibitor Velcade 1 (bortezomib or PS-341) has been shown to increase radiation-induced apoptosis and augment radiosensitivity in colorectal cancer cells *in vitro* and *in vivo* [20] through NF-κB inhibition. Curcumin potently inhibits the chymotrypsin-like (CT-like) activity of the 20S and 26S proteasome. In a preclinical study, inhibition of proteasome activity by curcumin in human colon cancer HCT-116 and SW480 cell lines was shown to lead to the accumulation of ubiquitinated proteins and several proteasome target proteins, and the subsequent induction of apoptosis.

13.3.1.3 Radiosensitization by Direct or Indirect Inhibition of the NF-κB Cascade: Preclinical Studies

- Human neuroblastoma cells. SK-N-MC cells were exposed to either 2 Gy alone, or pretreated with curcumin or NF-κB inhibitor peptide SN50 before irradiation. Curcumin or SN50 pretreatment significantly suppressed the radiation-induced NF-κB and enhanced the radiation-induced inhibition of cell survival. Microarray analysis revealed that curcumin enhanced the radiation-induced activation of caspases, other proapoptotic and death effector molecules, and inhibited antiapoptotic/survival molecules. In addition, curcumin markedly suppressed the radiation-induced TNF superfamily genes. In effect, curcumin acted as a potent radiosensitizer by overcoming the effects of radiation-induced NF-κB-mediated prosurvival gene expression [21].
- The effect of curcumin on radiation response of human glioma cell lines was investigated. Curcumin reduced cell survival in a p53- and caspase-independent manner, an effect that correlated with the inhibition of AP-1 and NF-κB signaling pathways through the prevention of constitutive JNK and Akt activation. Curcumin-mediated radiosensitization and chemosensitization of glioma cells correlated with reduced expression of bcl-2 and IAP family members as well as DNA repair enzymes (MGMT, DNA-PK, Ku70, Ku80, and ERCC-1) [22].
- In HCT 116 xenografts in nude mice, curcumin significantly enhanced the efficacy of fractionated radiation therapy as indicated

by prolongation of the time to tumor regrowth and reduction of the Ki-67 proliferation index. Curcumin suppressed NF-κB activity and the expression of NF-κB-regulated radiation-inducible gene products mediating radiation resistance including cyclin D1, c-myc, Bcl-2, Bcl-xL, cellular inhibitor of apoptosis protein-1, cyclooxygenase-2, matrix metalloproteinase-9, and vascular endothelial growth factor. The last effect was associated with decreased microvessel density. Overall, the results suggested that curcumin potentiated the antitumor effects of radiation therapy in colorectal cancer by suppressing NF-κB and NF-κB-regulated gene products, leading to inhibition of proliferation and angiogenesis [23].

- Curcumin was observed to inhibit proliferation and the post-irradiation clonogenic survival in colorectal cancer cell lines HCT116 (wild-type p53 and mutant K-ras), HT29 (mutant p53 and wild-type K-ras), and SW620. Transient radiation-induced NF-κB was suppressed by curcumin via inhibition of radiation-induced phosphorylation and degradation of IκBα, inhibition of IKK activity, and inhibition of Akt phosphorylation. Curcumin also suppressed NF-κB-regulated gene products (Bcl-2, Bcl-x_L, IAP-2, COX-2, cyclin D1, and vascular endothelial growth factor). The results suggest that transient radiation-inducible NF-κB activation provides a prosurvival response to radiation that may account for the development of radioresistance. The effect of curcumin was to block this pathway by inhibition of radiation-induced Akt phosphorylation and the subsequent inhibition of IKK activation [24].
- Curcumin was shown to downregulate MDM2 expression in cells with either wild-type or nonfunctional p53 with the effect acting at the transcriptional level. MDM2 transcription is regulated by the PI3K/mammalian target of rapamycin (mTOR)/erythroblastosis virus transcription factor 2 (ETS2) pathway, which is modulated by curcumin. Inhibition of MDM2 expression seems to be important for the anticancer, chemosensitization, and radiosensitization effects in human cancer models *in vitro* and *in vivo*, independent of p53 [24]. Curcumin inhibited the growth of PC3 xenografts and enhanced the antitumor effects of gemcitabine and radiation. In these tumors, curcumin reduced the expression of MDM2 (Figure 13.8) [25].

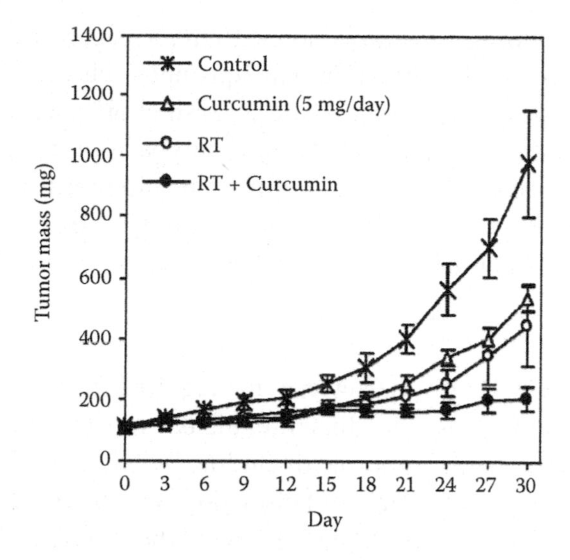

FIGURE 13.8 *In vivo* antitumor activity of curcumin administered alone or in combination with radiation to nude mice bearing PC3 xenografts. (From Li, M. et al., *Cancer Research* 67(5):1988–1996, 2007. With permission.)

13.3.2 STATs

The Jak–STAT signaling pathway is an evolutionarily conserved pathway essential for cytokine receptor signaling, which plays an essential role in regulating the immune response. The STAT proteins are a family of latent cytoplasmic transcription factors involved in cytokine, hormone, and growth factor signal transduction. The STAT proteins are activated by tyrosine phosphorylation, which acts as a molecular switch altering their conformation to allow specific binding to DNA and alter gene expression. Tyrosine phosphorylation can be induced by a variety of factors including Jak (Janus kinase) tyrosine kinases, cytokine receptors, G protein-coupled receptors, and growth factor receptors (such as EGFR and PDGF).

The biological effects of STATs include promotion of cell survival through increased expression of antiapoptotic proteins such as Bcl-2 and Bcl-XL. Persistent activation of STAT3 mediates tumor-promoting inflammation and promotes pro-oncogenic inflammatory pathways, including NF-κB and IL-6–GP130–Jak pathways. STAT-mediated signaling is involved in cross-talk with the NF-κB pathway at multiple levels. The role of STAT3 in radioresistance has been established on the basis of the results from several experimental studies (reviewed by Deorukhkar and Krishnan [26]). Stable transfection with shRNA against STAT3 results in enhanced radiosensitivity of human squamous cell carcinoma (A431) cells. Based on proteomic profiles of radioresistant prostate cancer cells, it has been shown that the radioresistant phenotype of the tumor cells with enhanced cell survival, proliferation, invasion, and motility is associated with multiple mechanisms including significant input from the radiation-induced activation of the Jak–STAT pathway.

13.3.3 COX-2

Cyclooxygenase (COX) is the key enzyme required for the conversion of arachidonic acid to prostaglandins. It exhibits two isoforms (COX-1 and COX-2), the latter being inducible (including by NF-κB), which plays an important role in cancer-related inflammation. Selective COX-2 inhibitors have been tested for their therapeutic efficacy when combined with radiation or chemotherapy and a number of radiosensitizing mechanisms have been identified. These include the arrest of cells in the radiosensitive G_2-M phase of the cell cycle, inhibition of DNA repair pathways, and NF-κB suppression.

The radiosensitizing action of COX-2 inhibitors might also be attributable to the inhibition of PGE_2 production. The COX enzymes are the rate-limiting enzymes for the conversion of arachidonic acid to prostaglandins. The major metabolite of COX-2 is PGE_2, which is reported to inhibit apoptosis and to act as a radioprotector. Curcumin has been shown to bind directly to COX-2 and to inhibit PGE_2 production. Increased tumor radioresponse as a result of COX-2 inhibition has been seen in murine tumor models and in human glioma xenografts. The mechanisms may involve the elimination of prostaglandins as protective molecules that are upregulated in response to chemotherapy and radiotherapy.

13.4 PHYTOCHEMICALS THAT HAVE BEEN SHOWN TO ACT AS RADIOSENSITIZERS

The preceding discussion of mechanisms by which phytochemicals can act as radiosensitizers centered on the polyphenol curcumin as the most extensively studied and widely used phytochemical. In fact, a number of phytochemicals used as traditional remedies and in cooking have been shown to be effective radiosensitizers in preclinical systems and to act by a variety of mechanisms similar to the plieotropic, multitasking model compound curcumin. The following phytochemicals have been shown to act as radiosensitizers of cancer cells.

13.4.1 Curcumin

Curcumin, the yellow polyphenol compound isolated from the rhizome of the plant *Curcuma longa* has been studied extensively for the past five decades for its chemopreventive potential and anticancer efficacy. The mechanisms of action of curcumin, as exemplified in preclinical studies, have been described previously.

13.4.2 Genistein and Soy Isoflavones

Genistein, the bioactive isoflavone of soybeans has been found to inhibit cell growth of tumor cell lines of various histologies. A number of publications have detailed the mechanisms an extent of radiosensitization by genistein.

- The effectiveness of combined treatment of radiation with genistein was investigated in HeLa cells. Inhibition of cell growth in response to combined treatment was significantly higher than that seen when the cells were treated with radiation or genistein alone. After radiation (4 Gy) combined with genistein, the apoptotic index was significantly increased and the cells were arrested in the G_2-M phase. Survivin mRNA expression was increased after radiation alone, whereas it was significantly decreased after combined treatment [27].
- Genistein was found to radiosensitize prostate cancer *in vitro* and *in vivo*. Pretreatment of tumor cells with genistein potentiated radiation-induced killing *in vitro* and in an orthotopic *in vivo* model; however, pure genistein alone was found to promote increased lymph node metastasis. This result prompted a switch from genistein alone to a mixture of soy isoflavones (genistein, daidzein, and glycitein), which inhibited cell survival and potentiated radiation cell-killing in PC-3 tumor cells *in vitro*. Increased cell-killing correlated with inhibition of antiapoptotic molecules Bcl-xL and survivin, upregulation of proapoptotic Bax molecules, and PARP cleavage; all suggesting activation of apoptotic pathways. *In vivo*, using the PC-3 orthotopic metastatic mouse model, combined treatment with soy isoflavones and radiation enhanced control of primary tumor growth and of

metastases, without an increase in metastasis to the lymph nodes observed when genistein was used alone [28].

- In other experiments with prostate cancer cells treated with soy isoflavones and radiation, downregulation of NF-κB and of APE1/Ref-1 was observed. APE1/Ref-1 acts as a redox activator of transcription factors, including NF-κB and HIF-1α. Radiation-induced phosphorylation of upstream molecules Src and STAT3 leads to the induction of HIF-1α. This can be prevented by pretreatment with isoflavones, which inhibit Src/STAT3/HIF-1α activation by radiation and the subsequent nuclear translocation of HIF-1α [28].
- In APE1/Ref-1 cDNA transfected cells, radiation-mediated increase in HIF-1α and NF-κB activities was enhanced, but this effect was inhibited by preradiation treatment with soy. Results of the transfection experiments indicate that APE1/Ref-1 inhibition by isoflavones can impair the radiation-induced transcription activity of NF-κB and HIF-1α. Together, these findings suggest that the increased responsiveness to radiation mediated by soy isoflavones could be due to pleiotropic effects of isoflavones blocking radiation-induced cell survival pathways including Src/STAT3/HIF-1α, APE1/Ref-1, and NF-κB [29].

13.4.3 Parthenolide

Parthenolide is a sesquiterpene lactone from the plant *Tanacetum parthenium*, which has been used orally or as an infusion for the treatment of multiple ailments. The molecule contains an α-methylene-γ-lactone ring and an epoxide moiety, which interact with nucleophilic sites of biologically important macromolecules. As a consequence, parthenolide has multiple targets and can instigate a range of *in vitro* and *in vivo* effects. Mechanistic studies done on various cancer models suggest that inhibition of NF-κB activity either via interaction with IKK or directly with the p65 subunit of NF-κB, is an important mode of action [16]. In addition, parthenolide is known to inhibit STAT and MAP kinase pathways and to regulate p53 activity by modulating MDM2 and HDAC1 levels [30]. Parthenolide reduces the cellular level of GSH in cancer cells, resulting in ROS accumulation and apoptosis [31]. In common with some other phytochemical radiosensitizers, parthenolide selectively targets cancer cells. Parthenolide has been shown to act as a radiosensitizer in a variety of cell lines:

- Parthenolide combined with radiation, increased the ROS levels and reduced the intracellular thiol levels in PC-3 cells, but decreased radiation-induced oxidative stress in normal human prostate epithelial cells (PrEC) cells [32].
- Parthenolide induced nonapoptotic cell death, inhibited proliferation, and increased the population doubling time from 23 to 49 h in PC-3 cells. It also inhibited radiation-induced NF-κB binding activity and enhanced the x-ray sensitivity of the (p53 null) PC-3 cells, leading to downregulation of its downstream target SOD2. Parthenolide also activates NADPH oxidase selectively in PC-3 cells, leading to a decrease in the level of reduced thioredoxin and of transcription factor, FOXO3a inhibiting MnSOD and catalase, which are downstream targets of FOXO3a [33].
- Differential susceptibility to the effect of parthenolide were observed in two radioresistant cancer cell lines, DU145 and PC-3, with DU145 cells showing higher sensitivity. Parthenolide inhibited NF-κB and activated the PI3K/AKT pathway in both cell lines but activated AKT in DU145 cells was maintained at a lower level than in PC3 cells due to the presence of functional PTEN, which suppresses the effect of activated phospho-AKT [34].
- Radiosensitization by parthenolide was investigated in PC-3 prostate cancer cells, a p53 null cell line with constitutively activated NF-κB. Parthenolide induced nonapoptotic cell death, inhibited PC-3 proliferation, and increased the population doubling time from 23 to 49 h. Parthenolide also inhibited constitutive and radiation-induced NF-κB binding activity and enhanced radiosensitivity [dose modification factor (DMF), 1.7]. Split-dose studies using 2 and 4 Gy fractions demonstrated that parthenolide completely inhibited split-dose repair in PC-3 cells. It was concluded that inhibition of NF-κB activity by parthenolide was responsible for radiosensitization and inhibition of split-dose repair [32].

13.4.4 Resveratrol

Resveratrol is a polyphenol isolated from the skin of grapes that has been shown to significantly alter the cellular physiology of tumor cells, as well as blocking the processes of tumor initiation and progression. A number of reports have described radiosensitization by resveratrol attributable to a variety of causes.

- Irradiation of cervical cancer cell lines, HeLa and SiHa, after pretreatment with resveratrol resulted in enhanced tumor cell killing by radiation, in a dose-dependent manner [35].
- Resveratrol differentially sensitized K-562 and HeLa cells. Pretreatment with resveratrol was shown to inhibit cell division and to induce arrest of treated cells in S phase [36].
- In another study of NCI-H838 cells, pretreatment with resveratrol significantly enhanced cell killing by radiation, with an SER of up to 2.2. Radiation activated NF-κB, and this effect was reversed by resveratrol pretreatment, which also brought about S phase arrest and enhanced radiosensitivity [37].
- Hormone-independent radioresistant DU145 prostate cancer cells were sensitized to radiation by pretreatment with resveratrol, which acted synergistically with radiation to inhibit cell survival *in vitro*. In this study, resveratrol was found to potentiate radiation-induced ceramide accumulation by promoting its *de novo* biosynthesis [38].
- Cells of an atypical teratoid/rhabdoid tumor, which express CD133, a marker for cancer stem-like cells, were radiosensitized by resveratol. Resveratrol alone inhibited STAT-3 and NF-κB-dependent transcription, and resulted in suppression of Bcl-xL expression, while activating the ATM–Chk2–p53 pathway. The effect of combined treatment was to significantly enhance the radiosensitivity and radiation-mediated apoptosis in these cells [39].

13.4.5 Plumbagin

Plumbagin, a naturally occurring naphthaquinone, derived from the plant *Plumbago zeylanica*, has been shown to chronically activate ERK1/2 and inhibit AKT activity in cancer cells [40], and to have free radical-generating properties. Treatment with plumbagin with or without radiation has been shown to be effective in inducing apoptosis.

- In a cervical cell line, a low dose of radiation (2 Gy) in combination with plumbagin induced apoptosis more effectively than did a higher dose of radiation (10 Gy) alone. Combined treatment induced a five-fold increase in the activation of caspase 3 in C33A cells and expression of apoptotic regulatory molecules Bcl-2, Bax, and survivin was modulated by plumbagin in combination with radiation.
- Plumbagin administered i.p. to mice with Ehrlich carcinoma cells growing as an ascites, caused a significant increase in the proportion of S phase and G_2-M cells with a corresponding decrease in the G_1 phase, whereas treatment with radiation only led to classic G_2 block. Combined treatment with plumbagin and radiation caused a pronounced G_2-M arrest together with an increase in the percentage of cells in S phase [41]. Drug or radiation alone significantly increased micronuclei induction at various posttreatment times and the combination of the two enhanced this effect additively.

13.4.6 Withaferin A

Withaferin A (WA), is a therapeutic plant withanolide isolated from the roots of *Withania sominifera*. WA has been shown to act as a radiosensitizer at lower doses and to be toxic to cancer cells as sole agent at higher doses.

- In an *in vitro* study, treatment of Chinese hamster ovarian cells with nontoxic doses of WA before irradiation enhanced cell killing, with dose-related SER.
- In Swiss albino mice inoculated intraperitoneally with Ehrlich ascites cells, WA treatment potentiated radiation-induced decrease in tumor growth [41].
- Similar effects were observed in mouse melanoma *in vivo* in which injection of WA, followed by 30 Gy of local gamma radiation, enhanced tumor response [42]. Treatment was most effective when the drug was injected i.p. 1 h before irradiation. In this case, inhibition of DNA repair was suggested as a mechanism of radiosensitization because WA has been shown to enhance radiosensitivity by interfering with homologous recombination, a major pathway of double-strand breaks (DSB) repair.

13.4.7 Caffeic Acid Phenethyl Ester

- Caffeic acid phenethyl ester (CAPE), an active component of honeybee propolis, has been reported to act as an antioxidant, anti-inflammatory, antiviral, immunostimulatory, and antimetastatic agent. CAPE has been shown to inhibit NF-κB by preventing the translocation of p65 subunit of NF-κB to the nucleus without affecting the TNF-α-induced IκBa degradation [43].

- Radiation sensitization by CAPE was evaluated in human lung cancer A549 cells and in normal lung fibroblast WI-38 cells [3]. A small but significant difference in radiation survival between cells treated with or without CAPE was seen, and radiosensitization was attributed to glutathione depletion occurring shortly after treatment.

- Daoy medulloblastoma cells pretreated with CAPE showed marked decrease in viability after radiation. At the molecular level, CAPE treatment of Daoy cells was found to effectively decrease levels of glutathione reductase and increase glutathione peroxidase [44]. Radiation-activation was reversed by CAPE pretreatment and CAPE was also shown to enhance radiation-induced apoptosis.

13.4.8 Ellagic Acid

Ellagic acid (EA) is a polyphenolic compound widely distributed in fruits and nuts. It is chemopreventive, inhibiting certain types of carcinogen-induced cancers. EA reduces viability in various cancer cell lines, and induces G_0-G_1 phase arrest of the cell cycle and apoptosis [45]. One of the mechanisms of induction of G_0-G_1 arrest is an increase of p53 and p21, and a decrease of CDK2 gene expression. Combined treatment of tumor with EA and radiation enhanced oxidative stress and cytotoxicity in tumor cells [46], whereas normal cells were protected against radiation damage.

EA-generated ROS in tumor cells were significantly increased when cells were treated with EA in combination with gamma radiation, and the decrease in mitochondrial potential and the loss of cell viability were significantly greater in tumor cells from mice treated with EA and radiation than in those treated with either agent alone. Antioxidant enzymes including SOD, catalase, GSH-Px, and glutathione reductase decreased in tumor cells after treatment with EA and radiation *in vivo* [45], and HeLa cells treated with EA and gamma radiation showed increased superoxide generation, upregulated p53 protein expression, and decreased levels of antioxidant enzymes.

13.4.9 (–)-Gossypol

A natural polyphenol product from cottonseed, (–)-gossypol has been identified as a potent small-molecule inhibitor of both Bcl-2 and Bcl-xL.

- In one study, (–)-gossypol enhanced radiation-induced apoptosis and growth inhibition in human prostate cancer PC-3 cells, which express a high level of Bcl-2/Bcl-xL proteins. Using a PC-3 xenograft model, it was that shown oral (–)-gossypol significantly enhanced the antitumor activity of radiation, leading to tumor regression in combination therapy. Furthermore, (–)-Gossypol plus radiation significantly inhibited tumor angiogenesis [47].
- Another study investigated radiosensitization by (–)-gossypol using a number of end points and a panel of human tumor cell lines. Cell cycle distribution, apoptosis, DNA DSB, and clonogenic cell survival were evaluated in A549, FaDu, H1299, MCF7, and Du145 cells. In A549 cells, gossypol strongly affected proliferation but caused only a modest arrest in G_1 phase, and no increase in the fraction of apoptotic cells or the number of additional DSB. Additional DSB were seen only in FaDu cells, where gossypol was extremely toxic. Preirradiation and incubation with (–)-gossypol resulted in varying levels of radiosensitization: most pronounced for FaDu and Du145 cells, less so for A549 cells, and almost no effect for H1299 and MCF7 cells. Sensitization did not result from increased apoptosis, but primarily from reduced DSB repair capacity, which could not be ascribed to changes in the level of repair proteins relevant for nonhomologous end-joining (Ku70, Ku80, and DNAPKcs) nor to changes in the level of higher phosphorylated inositols, which in fact were found to be enhanced by (–)-gossypol [48].

From these two studies, it seems that (–)-gossypol radiosensitizes as a result of increased apoptosis or because of reduced DNA repair capacity for unknown reasons. The origins of these effects and whether they are interrelated was not investigated.

13.5 RADIOPROTECTION BY PHYTOCHEMICALS: TARGETING THE PROINFLAMMATORY RESPONSE TO REDUCE THE SIDE EFFECTS OF RADIATION

(This section is based on reviews by Jagetia [49], Weiss and Landauer [50], and Lee et al. [51] and the articles referenced therein.)

Radiation is known to induce multiple biological responses at the cell and tissue level via the early activation of cytokine cascades. There is documented evidence of the elevation of proinflammatory cytokines including IL-6, IL-10, and TNFR1 being overexpressed in patients undergoing radiotherapy. In addition to mediating symptom burden acutely during a course of RT, many of the late sequelae of radiation are also attributable to deranged inflammatory cytokine signaling. Late side effects, which occur after long latent periods of months or years, include radiation-induced fibrosis, atrophy, and vascular damage. The early phases of fibrogenesis after irradiation can be seen as a wound-healing response characterized by an almost immediate upregulation of proinflammatory cytokines such as TNF-α, IL-1, and IL-6 and many growth factors in the irradiated tissue. Chemokines are released, which recruit inflammatory cells from the surrounding tissue into the irradiated area. In the entire process, TGF-β acts as a key fibrogenic cytokine. Inflammation seems to be an inherent component of radiation-induced chronic effects, which would suggest that inhibition of inflammation, including the suppression of both acute inflammatory and chronic fibrogenic cytokine gene expression, is likely to reduce normal tissue toxicity.

Curcumin is a nontoxic, anti-inflammatory compound. Curcumin and its related compounds demonstrate anti-inflammatory activity, both *in vitro* and *in vivo*, by inhibiting inflammatory cytokine production and curcumin has been shown to modulate arachidonic acid metabolism. In addition, curcumin not only blocks inflammatory cytokine and chemokine production, but also stimulates normal cell proliferation.

The mechanisms by which curcumin affects multiple biochemical and inflammatory conditions seems to be cell-specific and stimulus-specific, and to exert effects via transcription factors (NF-κB, COX-2) and through redox homeostasis. Curcumin has direct antioxidant activity (15), and it is a potent inhibitor of prostaglandin synthesis. It inhibits the activation of both AP-1 and AP-1 binding, and blocks JNK activation. Other anti-inflammatory properties of curcumin may come from its ability to act as a potent inhibitor of NF-κB activation, as has already been described.

The results of a recent experimental study exemplified the action of curcumin as an anti-inflammatory and radioprotectant. In this study, radiation reduced the severity of early skin reactions in C3H/HeN mice. Preradiation or postradiation treatment with curcumin provided similar levels of acute cutaneous radioprotection, whereas protection in the chronic phase of radiation skin toxicity was only observed after posttreatment administration of curcumin. The acute inflammatory cytokine mRNA expression levels in skin tissue most affected by radiation were IL-1β, IL-6, IL-1Ra, TNF-α, and lymphotoxin-β. The effects in muscle tissue were less pronounced than the effects in skin; the acute inflammatory cytokine mRNA expression levels in muscle tissue most affected by radiation included IL-1β, IL-18, IL-1RA, and lymphotoxin-β. The reduced muscle response is consistent with the lesser degree and longer latency of muscle fibrosis compared with cutaneous tissue fibrosis. The addition of curcumin, even after a high single radiation dose (50 Gy), reduced cytokine levels to close to nonirradiated values. Individual variability in acute and fibrogenic cytokine expression corresponded with the individual cutaneous toxicity experienced by the animal.

Radioprotection by curcumin has been demonstrated in a number of experimental studies:

- Data from several animal studies has confirmed that curcumin has a strong radioprotective function. The mouse bone marrow micronucleus test was used assess the protective role of three dietary agents, including curcumin, in mice exposed to gamma radiation. The data from this study indicated that oral administration of curcumin 2 h before or immediately after exposing the animals to whole-body irradiation significantly reduced the frequency of micronucleated polychromatic erythrocytes [52].
- Similar protective effects were noticed in mice pretreated with curcumin before irradiation, where curcumin reduced the number of bone marrow cells with chromosomal damage [53]. Such radioprotective effects of curcumin were still noticeable when much higher radiation doses (up to 10 Gy) were used in rats in which curcumin pretreatment significantly reduced the number of micronucleated cells and inhibited superoxide dismutase activity with a concomitant increase in catalase activity in liver tissues [54].

- In an *in vivo* study, curcumin treatment protected against radiation-induced oral mucositis in animals exposed to localized irradiation of their tongues [55]. The dose–effect curve for the incidence of this lesion in animals treated with a formulation containing curcumin was significantly shifted to the right and ED_{50} value was higher by a DMF of 1.24.
- In another study, it was found that curcumin-treated mice developed less cutaneous toxicity after radiation, and this was associated with a reduction in mRNA expression levels of interleukins, TNF-α, lymphotoxin-β, and TGF-β1 in irradiated skin and muscle tissues [56].

A method that might possibly increase the bioavailability of curcumin would be to combine it with alkaline foods. Using this stratagem, and based on the anticytokine activity and the long history of curcumin consumption without side effects, it seems that curcumin is an attractive and potentially effective nontoxic drug for the beneficial modulation of cutaneous toxicity resulting from radiation exposure, particularly during the treatment for breast cancer, head and neck cancers, and sarcomas.

13.6 CLINICAL APPLICATIONS OF PHYTOCHEMICALS

In the first part of this chapter, an effort has been made to describe the myriad molecular roles of curcumin and other phytochemicals, which make these compounds effective radiosensitizers. As is the case with many of the targeted molecular radiosensitizers described in preceding chapters, there is voluminous literature describing *in vitro* experiments elucidating molecular pathways, a smaller body of work describing *in vivo* studies with preclinical models, not surprisingly, because these studies are often more difficult, more time-consuming, and more expensive. Clinical trials are at the apex of this pyramid and are by far the most difficult to initiate and accomplish of all the aspects of therapeutic development. In addition, at this level of evaluation, the use of a particularly promising agent as a radiosensitizer is the last therapeutic application to be considered, and so the results in this area are particularly sparse.

Curcumin, which is the phytochemical most frequently considered in a therapeutic context, is the major active component of turmeric, a yellow compound derived from the plant *C. longa*. It has been used for centuries in traditional medicines and has also long been part of the daily diet in Asian countries, without any indication of toxicity. Extensive research over the past 30 years has indicated that this molecule has therapeutic potential against a wide range of diseases, including cancer and many inflammatory diseases. Numerous lines of evidence indicate that curcumin is highly pleiotropic with anti-inflammatory, hypoglycemic, antioxidant, wound healing, and antimicrobial activities and possesses chemosensitization, chemotherapeutic, and radiosensitization activities as well. In terms of the radiation response, curcumin has been shown in experimental systems to radiosensitize cancer cells and to be protective of normal cells. This combination of characteristics would seem to make curcumin ideal for use in combination with radiotherapy.

Nevertheless, even after half a century of research, naturally occurring compounds (including the most investigated ones such as curcumin and epigallo-catechin gallate [EGCG] [extract of green tea]) have yet to find a place as accepted prophylactic treatments. Among the reasons for the limited success in the clinical setting is the poor bioavailability and rapid rate of metabolism. Pharmacokinetic studies on curcumin have shown that the systemic bioavailability of oral curcumin is low in humans, limiting its use as a radiosensitizer or radioprotector. The quest to utilize traditional natural compounds for their chemotherapeutic and chemopreventive potential in the clinical setting has motivated drug delivery scientists to devise advanced drug delivery systems such as nanoparticles (NPs), liposomes, microemulsions, and implants. Although research on most of these delivery systems has shown the potential for enhanced bioavailability, two important aspects would need attention: (*i*) the occurrence of rapid drug metabolism (e.g., for curcumin), which may be mitigated by the application of combination therapies such as the inclusion of enzyme inhibitors like piperine; and (*ii*) the need for frequent parenteral dosing to maintain effective therapeutic concentrations in the blood.

13.6.1 Methods Proposed to Improve Phytochemical Availability

13.6.1.1 Adjuvants

Rapidly occurring glucuronidation in the liver reduces the activity of curcumin, an effect that can be avoided if glucuronidation is inhibited by piperine, a known inhibitor of hepatic and intestinal glucuronidation. Inhibition of glucuronidation by piperine is potentially an important stratagem for increasing the bioavailability of curcumin. Some other agents have shown synergistic effects when used in combination with curcumin in *in vitro* studies. A combination of curcumin and genistein was superior to the individual effects of either curcumin or genistein, and eugenol and terpeniol have been shown to enhance skin absorption of curcumin.

13.6.1.2 Delivery Systems

13.6.1.2.1 Solid Lipid Nanoparticles Nanoparticle-based delivery systems are suitable for highly hydrophobic agents, such as curcumin, circumventing the pitfalls of poor aqueous solubility. NPs as drug delivery vehicles enable passive targeting of tumors and other inflamed tissues due to increased vascular leakiness that results from the increased production of cytokines and angiogenesis cascades at these sites. Solid lipid nanoparticles are spherical lipid NPs with a high specific surface area that can be easily modified for rapid internalization by cancer cells, and stealth properties that lessen uptake by the reticuloendothelial system. They are highly versatile drug delivery systems able to cross the blood–brain barrier for a variety of compounds and less toxic as compared with polymeric NPs.

13.6.1.2.2 Liposomes The spherical bilayer vesicles with an aqueous interior formed by the self-association behavior of amphiphilic phospholipids with cholesterol molecules. These lipid-based particulate carriers can significantly enhance the solubility of poorly water-soluble chemopreventives. The lipophilic nature of many phytochemicals, including curcumin and resveratrol, make them suitable candidates for liposomal drug delivery.

13.6.1.2.3 Microemulsions/Microencapsulation Microemulsions are one of the most widely used drug delivery systems capable of providing high drug entrapment efficiency with long-term stability of hydrophobic molecules. They have a dynamic microstructure that results spontaneously from mixing lipophilic and hydrophilic excipients in presence of suitable surfactants. This microstructure results in high drug solubilization capacity along with free and fast drug diffusion that, coupled with high lipophilicity, endow them with a high potential for delivering lipophilic compounds like curcumin across lipophilic cell membranes and through skin.

13.6.1.2.4 Solid Polymeric Implants Polymeric implantable drug delivery systems have shown great potential for systemic delivery of various therapeutic agents. The implants, consisting of drug homogeneously entrapped in a polymeric matrix, achieve sustained localized delivery coupled with bioavailability by slowly releasing the encapsulated drug at the site of implantation. The slow release kinetics of implants extends drug release over a prolonged period suitable for the delivery of a sensitizer during fractionated radiation treatment.

13.6.1.3 Derivatives and Analogues

The chemical structure of curcumin plays a pivotal role in its biological activity and researchers hope to improve the biological activity of curcumin by structural modifications. A second strategy to improve the biological activity of curcumin is to chelate it with metals. The presence of two phenolic groups and one active methylene group in a curcumin molecule makes it an excellent ligand for any chelation. Several metal chelates of curcumin are reported to possess biological activity over that of free curcumin.

13.6.2 Clinical Trials Involving Curcumin

Approximately 40 ongoing or recently completed clinical trials involving curcumin or other phytochemicals are listed in the database (www.clinicaltrials.gov). Of these, the majority are concerned with chemoprevention. One large group is focused on testing various procedures with respect to the mode of delivery, pharmacokinetics, and bioavailability of curcumin. A smaller group is concerned with the use of curcumin in actual cancer treatment in conjunction with chemotherapy (in all cases) and involving chemoradiation (in two cases). One of these is a phase II trial of capecitabine, curcumin, and radiotherapy in the treatment of rectal cancer, while the other is a phase II study of curcumin versus placebo for chemotherapy-treated breast cancer patients undergoing radiotherapy. Another group of trials are investigating curcumin for the protection or amelioration of treatment-related symptoms. One trial targets radiation-induced dermatitis, while another trial targets radiation-induced mucositis. Finally, a smaller group of trials is concerned with cancer patients but does not target the actual disease but instead aims to control cancer-related symptoms.

13.7 SUMMARY

Many of the phytochemicals identified as chemopreventive agents have been shown more recently to act as radiosensitizers or radioprotectors. In terms of identifying the mechanisms of action at the chemical and biomolecular level, most studies have focused on one compound, curcumin, which has been shown to bind directly and with high affinity to a wide and diverse range of proteins. A number of direct targets of curcumin are important in terms of radiosensitization but in addition to interacting directly with target proteins, the effects of curcumin are manifest by the response of a diverse range of indirect molecular targets that are downstream of the direct target. These include transcription factors, enzymes, inflammatory mediators, protein kinases, drug resistance proteins, cell-cycle regulatory proteins, adhesion molecules, growth factors, receptors, cell survival proteins, chemokines, and chemokine receptors.

An important group of enzymes that are directly targeted by curcumin are those involved in radiation-induced antioxidant mechanisms. The preferential radiosensitization of some tumor cells by curcumin has been shown to be due to the targeting of the elevated expression of thioredoxin reductase (TxnRd1) in these cells. The involvement of TxnRd1 in curcumin-mediated radiosensitization seems to be twofold. First, curcumin covalently binds to TxnRd1, irreversibly inhibiting its ability to reduce thioredoxin; and second, modification of the selenocysteine residue of TxnRd1 converts the protein from an essential antioxidant enzyme to a protein with NADPH1 oxidase activity, thereby elevating oxidative stress. Both effects enhance the oxidative burden on the cell culminating as cell death among other effects. There are indirect consequences of targeting TxnRd1, which have implications for radiation response, including protection against apoptosis and enhancement of DNA repair and these effects will also be counteracted by curcumin. Other direct targets of curcumin with potential involvement in radiosensitization are survivin and Bcl2.

Another major approach to tumor radiosensitization by curcumin is the targeting of radiation-induced proinflammatory signaling pathways. The inflammatory signaling pathway is both inducible and may also have a degree of selectivity for the tumor cell and targeting this radioresistance pathway could radiosensitize tumor cells preferentially without sensitizing normal cells. One component of this response, the transcription factor NF-κB in cancer cells, is transient; peaking in tumor cells within a few minutes of radiation exposure and returning to baseline levels after a few hours. Nevertheless, this transiently activated NF-κB can provoke multiple radioresistance signals, which attenuate the lethal effects of radiation.

Curcumin sensitizes cancer cells to radiation by a multifaceted approach to suppression of radiation-induced NF-κB activation. This is achieved by inhibition of NF-κB activation directly (IKK activation, IkBα degradation), inhibition of upstream activators of this pathway (Akt), and finally, by suppressing the NF-κB-regulated antiapoptotic, proliferative, angiogenic, invasive, and proinflammatory gene products. Curcumin also sensitizes cancer cells to radiation by suppressing radiation-induced TNF-α or TNF superfamily genes.

Curcumin has been shown to be an effective radioprotector experimentally and in some clinical settings. This action of curcumin is directly related to its role as a nontoxic, anti-inflammatory compound. Radiation is known to induce multiple biological responses at the cell and tissue level via the early activation of cytokine cascades, and there is convincing evidence that many of the early and late sequelae of radiation are attributable to deranged inflammatory cytokine signaling. Radioprotection by curcumin affects multiple biochemical and inflammatory conditions and seems to be cell-specific and stimulus-specific. Effects are exerted through transcription factors (NF-κB, COX-2), redox homeostasis, and inhibition of prostaglandin synthesis.

Despite the fact that, at the chemical and molecular level, plant phytochemicals display many desirable characteristics, their application in cancer therapeutics has thus far had limited success even after many years of research. The most important reasons for this are their poor bioavailability and rapid rate of metabolism. A number of strategies have been proposed to overcome this, including adjuvants which, when administered with the phytochemical, inhibit metabolism or facilitate absorption, delivery agents like nanoparticles, or improving the liposome structure of the molecule to improve biological activity.

REFERENCES

1. Fang, J., Lu, J., and Holmgren, A. Thioredoxin reductase is irreversibly modified by curcumin. A novel mechanism for its anticancer activity. *J Biol Chem* 2005;280(26):25284–25290.
2. Gupta, S., Prasada, S., Kima, J. et al. Multitargeting by curcumin as revealed by molecular interaction studies. *Nat Prod Rep* 2011;28(12):1937–1955.
3. Chen, M., Wu, C., Chen, Y., Keng, P., and Chen, W. Cell killing and radiosensitization by caffeic acid phenethyl ester (CAPE) in lung cancer cells. *J Radiat Res* 2004;45:253–260.
4. Karimpour, S., Lou, J., Lin, L. et al. Thioredoxin reductase regulates AP-1 activity as well as thioredoxin nuclear localization via active cysteines in response to ionizing radiation. *Oncogene* 2002;21:6317–6327.
5. Mustacich, D., and Powis, G. Thioredoxin reductase. *Biochem J* 2000;346:1–8.
6. Syng-Ai, C., Kumari, A., and Khar, A. Effect of curcumin on normal and tumor cells: Role of glutathione and bcl-2. *Mol Cancer Ther* 2004;3:1101–1108.

7. Jawadi, P., Hertan, L., Kosoff, R. et al. Thioredoxin reductase-1 mediates curcumin-induced radiosensitization of squamous carcinoma cells. *Cancer Res* 2010;70:1941–1950.
8. Tuttle, S., Hertan, L., Daurio, N. et al. The chemopreventive and clinically used agent curcumin sensitizes HPV⁻ but not HPV⁺ HNSCC to ionizing radiation, in vitro and in a mouse orthotopic model. *Cancer Biol Ther* 2012;13(7):575–584.
9. Saitoh, M., Nishitoh, H., Fujii, M. et al. Mammalian thioredoxin is a direct inhibitor of apoptosis signal-regulating kinase (ASK) 1. *EMBO J* 1998;17(9):2596–2606.
10. Tanaka, H., Arakawa, H., Yamaguchi, T. et al. A ribonucleotide reductase gene involved in a p53-dependent cell-cycle checkpoint for DNA damage. *Nature* 2000;404(6773):42–49.
11. Kaufmann, S., and Vaux, D. Alterations in the apoptotic machinery and their potential role in anticancer drug resistance. *Oncogene* 2003;22:7414–7430.
12. Marcu, M., Jung, Y., Lee, S. et al. Curcumin is an inhibitor of p300 histone acetylatransferase. *Med Chem* 2006;2:169–174.
13. Ghosh, S., and Hayden, M. New regulators of NF-κB in inflammation. *Nat Rev Immunol* 2008;8:837–848.
14. Sandur, S., Deorukhkar, A., Pandey, M. et al. Curcumin modulates the radiosensitivity of colorectal cancer cells by suppressing constitutive and inducible NF-κB activity. *Int J Radiat Oncol Biol Phys* 2009;75:534–542.
15. Raffoul, J., Wang, Y., Kucuk, O., Forman, J., Sarkar, F., and Hillman, G. Genistein inhibits radiation-induced activation of NF-κB in prostate cancer cells promoting apoptosis and G2/M cell cycle arrest. *BMC Cancer* 2006;6:107.
16. Mendonca, M., Chin-Sinex, H., Gomez-Millan, J. et al. Parthenolide sensitizes cells to X-ray-induced cell killing through inhibition of NF-kappaB and split-dose repair. *Radiat Res* 2007;168:689–697.
17. Cheng, J., Chou, C., Kuo, M., and Hsieh, C. Radiation-enhanced hepatocellular carcinoma cell invasion with MMP-9 expression through PI3K/Akt/NF-κB signal transduction pathway. *Oncogene* 2006;25:7009–7018.
18. Kim, J., Lee, G., Cho, Y., Kim, C., Han, S., and Lee, H. Desmethylanhydroicaritin inhibits NF-κB-regulated inflammatory gene expression by modulating the redox-sensitive PI3K/PTEN/Akt pathway. *Eur J Pharmacol* 2009;602:422–431.
19. Karin, M. The IkB kinase—a bridge between inflammation and cancer. *Cell Res* 2008;18:334–342.
20. Russo, S., Tepper, J., Baldwin, A. et al. Enhancement of radiosensitivity by proteasome inhibition: Implications for a role of NF-κB. *Int J Radiat Oncol Biol Phys* 2001;50:183–193.
21. Aravindan, N., Madhusoodhanan, R., Ahmad, S., Johnson, D., and Herman, T. Curcumin inhibits NFkB mediated radioprotection and modulate apoptosis related genes in human neuroblastoma cells. *Cancer Biol Ther* 2008;7:569–576.

22. Dhandapani, K., Mahesh, V., and Brann, D.W. Curcumin suppresses growth and chemoresistance of human glioblastoma cells via AP-1 and NF-κB transcription factors. *J Neurochem* 2007;102:522–538.
23. Kunnumakkara, A., Diagaradjane, P., Guha, S. et al. Curcumin sensitizes human colorectal cancer xenografts in nude mice to γ-radiation by targeting nuclear factor-kB-regulated gene products. *Clin Cancer Res* 2008;14:2128–2136.
24. Ozes, O., Mayo, L., Gustin, J., Pfeffer, S., Pfeffer, L., and Donner, D. NF-κB activation by tumour necrosis factor requires the Akt serine–threonine kinase. *Nature* 1999;401:82–85.
25. Li, M., Zhang, Z., Hill, D., Wang, H., and Zhang, R. Curcumin, a dietary component, has anticancer, chemosensitization, and radiosensitization effects by down-regulating the MDM2 oncogene through the PI3K/mTOR/ETS2 pathway. *Cancer Res* 2007;67(5):1988–1996.
26. Deorukhkar, A., and Krishnan, S. Targeting inflammatory pathways for tumor radiosensitization. *Biochem Pharmacol* 2010;80:1904–1914.
27. Zhang, B., Liu, J., Pan, J., Han, S., Yin, X., and Wang, B. Combined treatment of ionizing radiation with genistein on cervical cancer HeLa cells. *J Pharmacol Sci* 2006;102:129–135.
28. Wang, B., Wang, J., Lu, J., Kao, T., and Chen, B. Antiproliferation effect and mechanism of prostate cancer cell lines as affected by isoflavones from soybean cake. *J Agric Food Chem* 2009;57:2221–2232.
29. Singh-Gupta, V., Zhang, H., Banerjee, S., Kong, D., Raffoul, J., and Sarkar, F. Radiation-induced HIF-1alpha cell survival pathway is inhibited by soy isoflavones in prostate cancer cells. *Int J Cancer* 2009;124:1675–1684.
30. Rocha, S., Martin, A., Meek, D., and Perkins, D. p53 represses cyclin D1 transcription through down regulation of Bcl-3 and inducing increased association of the p52 NF-kappaB subunit with histone deacetylase. *Mol Cell Biol* 2003;23:4713–4727.
31. Won, Y., Ong, C., and Shen, H. Parthenolide sensitizes ultraviolet (UV)-B induced apoptosis via protein kinase C-dependent pathways. *Carcinogenesis* 2005;26:2149–2156.
32. Watson, C., Miller, D., Chin-Sinex, H. et al. Suppression of NF-kappaB activity by parthenolide induces X-ray sensitivity through inhibition of split-dose repair in TP53 null prostate cancer cells. *Radiat Res* 2009;17(4):389–396.
33. Sun, Y., Clair, D.S., Xu, Y., Crooks, P., and Clair, W.S. NADPH oxidase-dependent redox signaling pathway mediates the selective radiosensitization effect of parthenolide in prostate cancer cells. *Cancer Res* 2010;70:2880–2890.
34. Sun, Y., Clair, D.S., Fang, F., Warren, G., Rangnekar, V., and Crooks, P. The radiosensitization effect of parthenolide in prostate cancer cells is mediated by nuclear factor-kappa B inhibition and enhanced by the presence of PTEN. *Mol Cancer Ther* 2007;6:2477–2486.
35. Zoberi, I., Bradbury, C., Curry, H., Bisht, K., Goswami, P., and Roti, J. Radiosensitizing and anti-proliferative effects of resveratrol in two human cervical tumor cell lines. *Cancer Lett* 2002;25:165–173.

36. Baatout, S., Derradji, H., Jacquet, P., Ooms, D., Michaux, A., and Mergeay, M. Enhanced radiation-induced apoptosis of cancer cell lines after treatment with resveratrol. *Int J Mol Med* 2004;13:895–902.
37. Liao, H., Kuo, C., Yang, Y., Lin, C., Tai, H., and Chen, Y. Resveratrol enhances radiosensitivity of human non-small cell lung cancer NCI-H838 cells accompanied by inhibition of NF-kappa B activation. *J Radiat Res* 2005;46:387–893.
38. Scarlatti, F., Sala, G., Ricci, C., Maioli, C., Milani, F., and Minella, M. Resveratrol sensitization of DU145 prostate cancer cells to ionizing radiation is associated to ceramide increase. *Cancer Lett* 2000;253:124–130.
39. Kao, C., Huang, P., Tsai, P. et al. Resveratrol-induced apoptosis and increased radiosensitivity in CD133-positive cells derived from atypical teratoid/rhabdoid tumor. *Int J Radiat Oncol Biol Phys* 2009;74:219–228.
40. Yang, S., Chang, S., Wen, H., Chen, C., Liao, J., and Chang, C. Plumbagin activates ERK1/2 and Akt via superoxide, Src and PI3-kinase in 3T3-L1 cells. *Eur J Pharmacol* 2003;25:21–28.
41. Devi, P., Sharada, A., and Solomon, F. In vivo growth inhibitory and radiosensitizing effects of withaferin A on mouse Ehrlich ascites carcinoma. *Cancer Lett* 1995;16:189–193.
42. Sharada, A., Solomon, F., Devi, P., Udupa, N., and Srinivasan, K. Antitumor and radiosensitizing effects of withaferin A on mouse Ehrlich ascites carcinoma in vivo. *Acta Oncol* 1996;35:95–100.
43. Ang, E., Pavlos, N., Chai, L., Qi, M., Cheng, T., and Steer, J. Caffeic acid phenethyl ester, an active component of honeybee propolis attenuates osteoclastogenesis and bone resorption via the suppression of RANKL-induced NF-kappaB and NFAT activity. *J Cell Physiol* 2009;221:642–649.
44. Lin, Y., Chiu, J., Tseng, W., Wong, T., Chiou, S., and Yen, S. Antiproliferation and radiosensitization of caffeic acid phenethyl ester on human medulloblastoma cells. *Cancer Chemother Pharmacol* 2006;57:525–532.
45. Hayeshi, R., Mutingwende, I., Mavengere, W., Masiyanise, V., and Mukanganyama, S. The inhibition of human glutathione S-transferases activity by plant polyphenolic compounds ellagic acid and curcumin. *Food Chem Toxicol* 2007;45:286–295.
46. Varadkar, P., Dubey, P., Krishna, M., and Verma, N. Modulation of radiation-induced protein kinase C activity by phenolics. *J Radiol Prot* 2001;21:361–370.
47. Xu, L., Yang, D., Wang, S. et al. (−)-Gossypol enhances response to radiation therapy and results in tumor regression of human prostate cancer. *Mol Cancer Ther* 2005;4:197–205.
48. Kasten-Pisula, U., Windhorst, S., Dahm-Daphi, J., Mayr, G., and Dikomey, E. Radiosensitization of tumour cell lines by the polyphenol Gossypol results from depressed double-strand break repair and not from enhanced apoptosis. *Radiother Oncol Rep* 2007;83:296–303.
49. Jagetia, G. Radioprotection and radiosensitization by curcumin. *Adv Exp Med Biol* 2007;595:301–320.
50. Weiss, J., and Landauer, M. Protection against ionizing radiation by antioxidant nutrients and phytochemicals. *Toxicology* 2003;189:1–20.

51. Lee, K., Bode, A., and Dong, Z. Molecular targets of phytochemicals for cancer prevention. *Nat Rev Cancer* 2011;11:2011–2018.
52. Abraham, S., Sarma, L., and Kesavan, P. Protective effects of chlorogenic acid, curcumin and beta-carotene against gamma-radiation-induced in vivo chromosomal damage. *Mutat Res* 1993;303:109–112.
53. Thresiamma, K., George, J., and Kuttan, R. Protective effect of curcumin, ellagic acid and bixin on radiation induced genotoxicity. *J Exp Clin Cancer Res* 1998;17:431–434.
54. Thresiamma, K., George, J., and Kuttan, R. Protective effect of curcumin, ellagic acid and bixin on radiation induced toxicity. *Indian J Exp Biol* 1996;34:845–847.
55. Rezvani, M., and Ross, G. Modification of radiation-induced acute oral mucositis in the rat. *Int J Radiat Biol* 2004;80:177–182.
56. Okunieff, P., Xu, J., Hu, D. et al. Curcumin protects against radiation-induced acute and chronic cutaneous toxicity in mice and decreases mRNA expression of inflammatory and fibrogenic cytokines. *Int J Radiat Oncol Biol Phys* 2006;65:890–898.

CHAPTER 14

Delivery Methods for Radioenhancing Drugs

14.1 DELIVERY OF RADIOSENSITIZING DRUGS

In several earlier chapters in this book, references have been made to methods other than the standard perfusion techniques aimed at facilitating the delivery of radioenhancing drugs in conjunction with fractionated radiation. In the preceding chapter, for instance, methods proposed to improve the delivery of phytochemicals are described including solid lipid nanoparticles, liposomes, and solid polymeric implants. Nanoparticles, including liposomes that have a prolonged circulation time and facility for enhanced tumor targeting, have been mentioned in several other chapters as an option for drug delivery. In this chapter, nanoparticles and a totally different approach, implantable devices for intratumoral drug delivery, will be considered in greater detail.

14.1.1 Nanoparticles

Nanoparticles are usually considered to be particles that are less than 100 nm in diameter. The interest surrounding these particles for drug delivery in the treatment of cancer and other diseases is largely related to the enhanced permeability and retention (EPR) effect. This is the property by which certain sizes of molecules (typically liposomes, nanoparticles, and macromolecular drugs) tend to accumulate in tumor tissue much more than they do in normal tissues. The explanation for this is that the newly formed tumor vessels are usually abnormal in form and architecture. They consist of poorly aligned defective endothelial cells with wide fenestrations, lacking a smooth muscle layer, or innervation with a wider lumen, and impaired functional receptors for angiotensin II. Furthermore, tumor tissues usually lack effective lymphatic drainage.

14.1.2 Gold Nanoparticles

Gold nanoparticles (GNPs) have already been discussed in Chapter 3. The GNP itself is both the sensitizer and the delivery vehicle, in effect, the messenger and the message. There are a number of reasons why nanoparticles, and gold nanoparticles in particular, are an ideal material for radiosensitization:

- High biocompatibility, gold is very inert.
- Gold nanoparticles can enhance the effect of radiation throughout a large tumor volume thus eliminating the need for delivery to all tumor cells.
- Nanoparticles have a low rate of systemic clearance. This allows the sensitizing material time to be absorbed into the tumor tissue.
- Nanoparticles have better permeation into tumor tissue. This, along with lower clearance rate, results in the EPR effect.
- Gold atoms can be specifically delivered to the tumor tissue by attaching targeting moieties such as antibodies. A nanoparticle of 10 to 15 nm in size contains 50,000 to 75,000 atoms within it, resulting in a much higher efficiency of delivery.
- The gold nanoparticles can be varied in size or shapes (such as spheres cube, rods, cones, or other three-dimensional structures) to optimize delivery to the tumor.
- Pharmacokinetics are readily performed with gold nanoparticles because they are easy to image and quantify.

Surface coating using polymeric material has led to better regulation of the pharmacokinetic and targeting properties of the gold nanoparticles. The gold nanoparticle itself provides a large number of ligand-binding sites directly proportional in number to the size of the nanoparticle. The chemistry for ligand attachment is relatively easy, and the surface properties allow for the binding of multiple ligands to the same nanoparticle. The possibility of attaching multiple ligands allows for the attachment of polymeric materials such as polyethylene glycol (PEG) along with other targeting moieties. PEG has been shown in multiple studies to reduce the uptake of nanoparticulate formulations by the reticuloendothelial system, further prolonging the retention of gold nanoparticles within the circulatory system.

As described in Chapter 3, the efficacy of the metal-based formulations depends on the energy of the radiation along with the type, amount, and location of material within the tissue. Better targeting and pharmacokinetic profiling of the nanoparticles will generate much more efficient therapy with reduced adverse effects to surrounding healthy tissue. Electrons from the gold particles and its effects were directly proportional to the number of particles in the proximity of the DNA.

14.1.2.1 Coated Gold Nanoparticles

Studies by a number of groups using PEG-coated gold nanoparticles showed the increased therapeutic efficacy of the formulation for radiosensitization [1]. Various sized nanoparticles were studied and, in each case, a concentration-dependent increase in the efficiency of cancer cell killing was observed. This increased efficacy was attributed to the EPR benefits of the PEG coating rather than its effects on energy redistribution. Results of studies to determine the toxicological effects of PEG-coated gold nanoparticles in healthy tissue suggest that there is no enhanced toxicity associated with these coated nanoparticles [2], although effects may be size-dependent or concentration-dependent or vary with the route of administration.

Combination of the gold nanoparticles with other radiosensitizers either by coadministration or by conjugation have been utilized to increase cytotoxicity and improve targeting.

- In one study, GNPs were used as a carrier for the delivery of ss-lapachone, a novel anticancer agent with cytotoxic and radiosensitizing capabilities. The combination showed significantly enhanced activity whereas the introduction of an anti-EGFR antibody as a targeting moiety further enhanced the effect [3].
- Human epidermal growth factor receptor-2, ErbB2 (HER-2)-targeted gold nanoparticles have been synthesized by conjugating trastuzumab (Herceptin) to 30 nm gold nanoparticles. Herceptin acts as

both a targeting moiety as well as a monotherapeutic agent. These conjugated nanoparticles increased the cytotoxic effects of radiation by 3.3-fold as compared with radiation alone, whereas nontargeted nanoparticles showed only 1.7-fold increase in efficiency [4].

14.1.3 Polymeric Nanoparticles

Polymeric nanoparticles have been formulated using various chemotherapeutic agents either alone or in combination to act as radiosensitizers. There are a number of reports in the literature of radioenhancement by drugs delivered by nanoparticles.

- In one study, polyglycolic-lactic acid (PGLA) nanoparticles were used to deliver two drugs, paclitaxel and etanidazole. Both the individual drugs and the combination enhanced the susceptibility of the cells to radiation. In particular, the prolonged release of the drug from the formulation allowed radiosensitization of radioresistant hypoxic cells. The effect of the combination was greater than that of either drug individually [5].
- Genexol-PM, a clinically approved formulation of paclitaxel, was studied as a radiosensitizer using non-small cell lung cancer mouse xenograft models. Again, this formulation was found to be both a better radiosensitizer than the formulation drug (with effective concentration half of that of the free drug) as well as a safer therapeutic with much reduced exposure of the drug to the healthy lung tissue [6].
- A nanomiceller composite formulation of doxorubicin displayed significantly enhanced radiation sensitization capability toward multicellular spheroids of the A549 lung cancer cell line [7]. The formulation cells treated showed significantly higher radiotoxicity as compared with cells treated with drug alone.
- Biodegradable lipid polymer nanoparticles have also been made using docetaxel as the entrapped drug and targeted to cancer tissue using folate. The targeted nanoparticles showed better radiosensitizing properties compared with drug alone or unmodified nanoparticles. The studies also showed that the radiosensitizing effects using nanoparticulate formulations depend significantly on the time gap between the dosing of the formulation and the radiation [8].

- An important target for the reversal of radiation resistance is the epidermal growth factor receptor (EGFR; see Chapter 10). EGFR is overexpressed in many types of cancers and anti-EGFR treatments has been shown to increase the therapeutic activity of radiotherapy. PLGA nanoparticle-encapsulated antisense EGFR oligonucleotides were combined with radiotherapy in the treatment of SCCVII squamous cells. The results indicated that antisense EGFR nanoparticles enhanced radiosensitivity by inhibition of EGFR-mediated mechanisms of radioresistance [9].
- Curcumin (Chapter 13) has been shown to have both radiosensitizing and radioprotective effects depending on the cell types and concentrations. Curcumin is also known to act on multiple essential pathways in cancer responsible for radiation resistance. A targeted PLGA nanoparticle formulation of curcumin was developed and tested for its chemotherapeutic and radioresistant effects on cisplatin-resistant ovarian cancer cell lines [10,11].

14.2 INTRATUMORAL SUSTAINED DRUG RELEASE DEVICES

The toxicities of drugs used in chemotherapy protocols are frequently the dose-limiting factor, and the problems of normal tissue toxicity are compounded by combined drug use or mixed modality treatments. In many cases, this precludes the use of optimal drug doses required for tumor control or the use of multiple drug doses during fractionated radiation treatment over a prolonged period. If, however, the drug can be localized at the disease site and released slowly over time, then systemic drug toxicities could be decreased while simultaneously increasing drug concentrations at the tumor site. These considerations support a role for a sustained release intratumoral drug delivery system in mixed modality therapy.

14.2.1 Polymeric Slow-Release Systems

Various systems have been used experimentally to deliver anticancer drugs to the tumor site in a sustained, slow-release manner, and some of these have been utilized to deliver radiosensitizing drugs. One example is the delivery of cis-DDP by OPLA-Pt, an open-cell polylactic acid polymer. Implantation of this device adjacent to a murine mammary tumor produced a greater than additive response, in terms of growth delay, when combined with radiation [11]. The same device, used for the treatment of spontaneous tumors in dogs, alone or in combination with radiation, gave an improved response over conventional treatment [12].

14.2.2 Biodegradable Polymeric Systems

In these devices, the drug is typically trapped physically within the polymer matrix and is released as the polymer degrades in response to its local environment. The degradation rate of the polymer device can be changed by using different polymers (or different combinations of copolymers), thereby effectively controlling the rate of drug release. An additional advantage is that because the copolymer is biodegradable, surgical removal of the spent delivery device is not required.

A class of polymers, which have been used with some success to localize drugs to treatment sites, are the polyanhydrides. Bis(*p*-carboxyphenoxy) propane/sebacic acid (CPP/SA) is a polyanhydride-based polymer that has been extensively studied and found to be biocompatible with non-toxic degradation products. In experimental studies, this copolymer has been used as a vehicle for direct intratumoral delivery of taxol [13] and of buthionine sulfoximine (BSO) to potentiate the effects of 4-hydroperoxy-cyclophosphamide [14]; all studies being done in a rat model of malignant glioma. Clinically, phase I to III trials have been conducted using CPP/SA (20:80) for the delivery of BCNU intracranially in patients with malignant glioma [15].

One group has used the CPP/SA copolymer in a series of investigations exploring the use of this vehicle for the intratumoral delivery of several different types of chemosensitizing and radiosensitizing drugs including cis-DDP, 5-fluorouracil, etanidazole, tirapazamine, and bromodeoxyuridine.

14.2.2.1 Cisplatinum

For the RIF-1 mouse fibrosarcoma intratumoral implant of the PCPP/SA polymer containing cis-DDP, it was found that the drug was continuously released over 8 to 10 days as the polymer degraded, and that the drug was localized at the tumor site and maintained at high concentration relative to other organs such as the kidney [16]. Cis-DDP delivered intratumorally was more effective than when the drug was delivered systemically by intraperitoneal injection in potentiating the effect of radiation. The most effective treatment protocol was cis-DDP delivered by polymer implanted 2 days before an acute dose of radiation (dose-modifying factor [DMF] = 2.2). Multifraction treatments in which polymer implantation was on the same day as the commencement of treatment also showed potentiation of the effect, with the DMF for fractionated treatment remaining relatively constant (1.5–1.9) for 5, 8, and 12 fractions (see Chapter 6, Figure 6.4).

14.2.2.2 5-Flurouracil

For intratumoral implant of 5-FU, there was a dose-dependent relationship between the amount of 5-FU administered by polymer implant and the duration of tumor growth delay (TGD). For the highest concentrations of 5-FU implanted (3.5 mg), the TGD was significantly greater than that produced by intraperitoneal injection of 5-FU (100 mg/kg; $p = 0.0045$) but was not significantly different from the TGD when the drug was delivered by osmotic pump. Combined treatment with external beam irradiation and intratumoral 5-FU gave a striking increase in TGD compared with either treatment used singly. Similarly fractionated radiation combined with intratumoral 5-FU produced an increase in TGD, although the effect was not as great as that seen for acute radiation [17].

14.2.2.3 Nitroimidazole Hypoxic Sensitizers

The nitroimidazoles are a class of radiosensitizer that are tumor-specific by virtue of the fact that they target hypoxic cells and, given an effective delivery system, can potentiate the effect of radiation. Etanidazole, a nitroimidazole hypoxic cell radiosensitizer, has been extensively studied experimentally and in clinical trials in conjunction with fractionated radiation. The results were disappointing, partly due to systemic toxicities that developed when the drug was administered over a long treatment schedule [18]. Interestingly, intratumoral injections of nitroimidazoles have been found to be effective as radiosensitizers and promising results have been obtained using this approach in the treatment of advanced disease of the oral cavity [19], bladder tumors [20], and carcinoma of the cervix [21].

For subcutaneous Rif-1 tumors, etanidazole/polymer implant had little effect on the response of the tumor to radiation. Fractionated radiation combined with polymer implant gave a longer TGD than that for radiation alone, but the difference was not statistically significant. In contrast, for intramuscular RIF-1 tumors, implantation of etanidazole/polymer potentiated the effect of radiation treatment to a significant extent. The effect was greatest for an acute dose of 16.7 Gy, where the TGD increased from 11 days for radiation alone to 19 days for radiation plus polymer implant. Treatment of intramuscular tumors with fractionated radiation (5 × 6 Gy) increased the TGD from 18.8 days for radiation alone to 23.4

days when polymer/etanidazole was implanted at the start of fractionated treatment [22]. The intramuscular tumor was shown by EF5 labeling to have a hypoxic fraction, although this was not detectable in the subcutaneous tumor.

14.2.2.4 Tirapazamine

The response to combined treatment with tirapazamine/polymer and acute radiation did not differ from the response to radiation alone for subcutaneous tumors. For fractionated treatment, there was a small increase in TGD when radiation was combined with polymer implant, which was not quite statistically significant. In contrast, the response of intramuscular tumors was affected by polymer/tirapazamine implant to a much greater extent. The greatest increase in TGD (from 11 to 18 days) was seen when acute radiation was combined with polymer implant. When fractionated radiation was combined with polymer implant, the TGD was increased from 18.8 to 23 days [22]. It was hypothesized that for both etanidazole and tirapazamine, the lack of effect in subcutaneous tumors was attributable to the smaller size of the hypoxic fraction in this tumor model, and this was confirmed using the hypoxia marker EF5.

14.2.2.5 Radiosensitization by BrdUrd

14.2.2.5.1 Subcutaneous Tumor Model RIF-1 tumors were treated with radiation after implantation of polymer/BrdUrd, and the time required for the tumor to regrow was measured. Implantation of polymer alone or polymer loaded with 20% or 30% BrdUrd, without irradiation, did not affect the rate of tumor growth. For treatment with single radiation doses, no radiosensitization was seen when BrdUrd was implanted 1 day before radiation. When the interval between implant and radiation exposure was increased to 3 days, there was a small increase in TGD compared with polymer implant, which was not statistically significant. Tumors treated with 5 or 10 dose fractions were radiosensitized only if the polymer was implanted 3 days before the first radiation dose. Even for prolonged radiation treatments, it seems to be necessary, if radiosensitization is to occur, to have a period of preradiation during which BrdUrd incorporation can take place. The most effective treatment was 10 radiation fractions combined with BrdUrd implant 3 days before the first fraction; in this case, the period of growth delay for mice in which the tumor eventually regrew was almost 80 days compared with 45 days for radiation treatment alone. In the polymer-implanted 10-fraction group, two out of eight mice were cured of the tumor (no regrowth seen for 120 days), whereas no cure was seen for the radiation-only group (Chapter 4, Figure 4.6) [23].

14.2.2.5.2 Intracranial Model The glioma model chosen, the rat C6 astrocytoma, proved to be extremely refractory to radiosensitization by BrdUrd *in vitro* and the situation was not greatly improved by incorporation of 5-FU or of methotrexate (MTX) into the treatment scheme. Although BrdUrd + PALA was the most effective radiosensitizing combination, it was nevertheless only modestly effective. At the level of cell killing achieved with 10 Gy, the addition of BrdUrd + PALA had a dose-modifying effect of 1.25. In view of these findings, the results of the *in vivo* study were unexpected. Treatment was highly effective, 40% of the rats treated with 10 Gy + BrdUrd were still alive at 180 days after tumor implant whereas for BrdUrd + PALA, the survival was 83% [22,24]. Possibly, the tissue culture model is not representative of the *de novo* and salvage pathway nucleotide synthesis occurring *in vivo* whereas another factor may be that a much higher intratumoral concentration of PALA, which is itself cytotoxic, could be achieved by polymer implant than was used *in vitro* (Chapter 4, Figure 4.7).

Similar results were reported by Williams et al. [25], who observed radiosensitization by halogenated pyrimidines based on tumor control in an animal model as an end point. In that study, drug delivery was also by an intratumoral biodegradable polymer implant and radiosensitization was observed in terms of increase in TGD.

14.2.2.6 Intratumoral Sustained Drug Release Combined with Low Dose Rate Radiation

This mode of radiation therapy is highly appropriate for combination with an intratumoral drug delivery system because one of the constraints of the latter system is that it is only suitable for the treatment of solid tumors that are accessible for implant. There are similar limitations on brachytherapy, and sites that are appropriate for interstitial radiation treatment are similarly accessible to interstitial chemotherapy by radiosensitizer or other drug.

In one experiment, a consistent dose–response relationship was shown between tumor growth delay and total dose. In combined treatments, tumors were first implanted with two polymer rods that had been formulated to deliver either 5-FU or cis-DDP. For irradiation by 125I implant, the activity of the sources and the duration of the implant were adjusted in such a way that the nominal dose delivered was always approximately 20 Gy. When 125I implant irradiation was combined with 5-FU, greater than additive responses were seen for both short (30 h) and long (96 h) 125I treatment times. In contrast, short-duration (30 h) 125I irradiation combined with cis-DDP was the least effective treatment, giving a less than additive combined response whereas cis-DDP combined with 96 h exposure to 125I was the most effective treatment. With the exception of 30 h 125I exposure combined with cis-DDP, the DMF was greater than 1.0 for all treatments, indicating that the combined effects of the treatments was greater than additive. The most effective combination was cis-DDP combined with 96 h exposure to 125I implant. It was conjectured that the inverse dose rate effect seen when cis-DDP is combined with low dose rate radiation is related to the inhibition of repair of radiation damage by cis-DDP [26].

14.3 SUMMARY

The results of radiation–drug interaction are affected both by the mode of radiation delivery and by the form and rate at which the drug is delivered. Some modes of drug delivery, other than the standard perfusion procedures, are particularly appropriate to combining radioenhancing drugs with fractionated radiation. One example is nanoparticles including liposomes, which have a prolonged circulation time and facility for enhanced tumor targeting. Gold nanoparticles are particularly effective and, in this case, the GNP itself is both the sensitizer and the delivery vehicle. There are a number of reasons why nanoparticles and gold nanoparticles in particular are ideal as sensitizers including biocompatibility, low rate of systemic clearance, good permeation in the tumor, and the fact that they can be easily decorated with targeting moieties such as antibodies. Combination of the gold nanoparticles with other radiosensitizers either by coadministration or by conjugation has also been utilized to increase cytotoxicity and improve targeting. Studies of PEG-coated gold nanoparticles showed increased therapeutic efficacy of the formulation for radiosensitization.

Polymeric nanoparticles have been formulated using various chemotherapeutic agents either alone or in combination to serve as radiosensitizers. PGLA nanoparticles have proven particularly effective and have been used to deliver paclitaxel, etanidazole, curcumin, and antisense EGFR oligonucleotides. Biodegradable lipid polymer nanoparticles have also been made using docetaxel.

The other class of delivery systems described are slow-release drug delivery devices, which are implanted in the tumor. These come in various forms; one of the most thoroughly investigated are biodegradable polymeric systems in which the drug is trapped physically within the polymer matrix and is released as the polymer degrades in response to its local environment. The degradation rate of the polymer device can be changed by using different polymers (or different combinations of copolymers), thereby effectively controlling the rate of drug release. An additional advantage is that because the copolymer is biodegradable, surgical removal of the spent delivery device is not required. One of the most successful of these is a copolymer of a polyanhydride with sebacic acid, CPP/SA. This device has been used for the intratumoral delivery of chemosensitizing and radiosensitizing drug, including cisplatinum, 5-fluorouracil, etanidazole, tirapazamine, and bromodeoxyuridine. In all cases, when the implant was combined with acute or fractionated radiation, radioenhancement was demonstrated.

REFERENCES

1. Liu, C., Wang, C., and Chen, S. Enhancement of cell radiation sensitivity by pegylated gold nanoparticles. *Phys Med Biol* 2010;55:931–945.
2. Cho, W., Kim, S., Han, B., Son, W., and Jeong, J. Comparison of gene expression profiles in mice liver following intravenous injection of 4 and 100 nm-sized PEG-coated gold nanoparticles. *Toxicol Lett* 2009;191: 96–102.
3. Jeong, S., Park, S., Yoon, S. et al. Systemic delivery and preclinical evaluation of Au nanoparticle containing beta-lapachone for radiosensitization. *J Control Release* 2009;139:239–245.
4. Chattopadhyay, N., Cai, Z., Kwon, Y., Lechtman, E., Pignol, J., and Reilly, R. Molecularly targeted gold nanoparticles enhance the radiation response of breast cancer cells and tumor xenografts to X-radiation. *Breast Cancer Res Treat* 2013;137:81–91.
5. Jin, C., Bai, L., Wu, H., Tian, F., and Guo, G. Radiosensitization of paclitaxel, etanidazole and paclitaxel+etanidazole nanoparticles on hypoxic human tumor cells in vitro. *Biomaterials* 2007;28:3724–3730.
6. Werner, M., Cummings, N., Sethi, M. et al. Preclinical evaluation of Genexol-PM, a nanoparticle formulation of paclitaxel, as a novel radiosensitizer for the treatment of non-small cell lung cancer. *Int J Radiat Oncol Biol Phys* 2013;86:463–468.
7. Xu, W., Han, M., Dong, Q. et al. Doxorubicin-mediated radiosensitivity in multicellular spheroids from a lung cancer cell line is enhanced by composite micelle encapsulation. *Int J Nanomed* 2012;7:2661–2671.
8. Werner, M., Copp, J., Karve, S. et al. Folate-targeted polymeric nanoparticle formulation of docetaxel is an effective molecularly targeted radiosensitizer with efficacy dependent on the timing of radiotherapy. *ACS Nano* 2011;5:8990–8998.

9. Ping, Y., Jian, Z., Yi, Z. et al. Inhibition of the EGFR with nanoparticles encapsulating antisense oligonucleotides of the EGFR enhances radiosensitivity in SCCVII cells. *Med Oncol* 2010;27:715–721.
10. Veeraraghavan, J., Natarajan, M., Lagisetty, P. et al. Impact of curcumin, raspberry extract, and neem leaf extract on rel protein-regulated cell death/radiosensitization in pancreatic cancer cells. *Pancreas* 2011;40:1107–1119.
11. Douple, E., Xi, X.-F., Yang, L., and Brekke, J. Potentiation of radiotherapy by cisplatin released via biodegradable implants (OPLA-Pt) in a murine solid tumor. Radiation Research Society 40th Annual Meeting, Salt Lake City, 1992.
12. Straw, R., Withrow, S., Douple, E. et al. Effects of cis-diaminedichloroplatinum (II) released from D,L-polylactic acid implanted adjacent to cortical allografts in dogs. *J Orthop Res* 1993;12:871–877.
13. Walter, K., Kahan, M., Gur, A. et al. Interstitial taxol delivered from a biodegradable polymer implant against experimental malignant glioma. *Cancer Res* 1994;54:2207–2212.
14. Sipos, E., Witham, T., Ratan, R., Baraban, J., and Brem, H. Biothionine sulfoximine potentiates the antitumor effect of 4-hydroxycyclophosphamide when delivered locally in the intracranial 9L-glioma model. *Proc Am Assoc Cancer Res* 1995;36:308.
15. Brem, H., Piantdosi, S., Burger, P. et al. Placebo-controlled trial of safety and efficacy of intra-operative controlled delivery by biodegradable polymer of chemotherapy for recurrent gliomas. *Lancet* 1995;345:1008–1012.
16. Yapp, D., Lloyd, D., Zhu, J., and Lehnert, S. Tumor treatment by sustained-intratumoral release of cisplatin: Effects of drug alone and combined with radiation. *Int J Radiat Oncol Biol Phys* 1997;39:497–504.
17. Berrada, M., Yang, Z., and Lehnert, S. Tumor treatment by sustained intratumoral release of 5-fluorouracil: Effects of drug alone and in combined treatments. *Int J Radiat Oncol Biol Phys* 2002;54:1550–1557.
18. Coleman, C., Noll, L., and Riese, N. Final report of the phase I trial of continuous infusion etanidazole (SR-2508); a Radiation Therapy Oncology Group Study. *Int J Radiat Oncol Biol Phys* 1992;22:577–580.
19. Saunders, M., and Dische, S. Clinical results of hypoxic cell radiosensitization from hyperbaric oxygen to accelerated radiotherapy, carbogen and nicotinamide. *Br J Cancer* 1996;74(Suppl. XXVII):S271–S278.
20. Awwad, H.K., Abdel Moneim, H., Abdel Baki, H., Omar, S., and Farag, H. The topical use of misonidazole. 13th International Cancer Congress, New York, 1983, 303–306.
21. Garcia-Angulo, A., ed. *Hyperthermic Oncology*. London/Philadelphia: Taylor and Francis, 1988.
22. Yapp, D., Lloyd, D., Zhu, J., and Lehnert, S. Radiosensitization of a mouse tumor model by sustained intra-tumoral release of etanidazole and tirapazamine using a biodgradable polymer implant device. *Radiother Oncol* 1999;53:77–84.
23. Doiron, A., Yapp, D., Olivares, M., Zhu, J., and Lehnert, S. Tumor radiosensitization by sustained intratumoral release of bromodeoxyuridine. *Cancer Res* 1999;59:3677–3681.

24. Li, Y., Owusu, A., and Lehnert, S. Treatment of intracranial rat glioma model with implant of radiosensitizer and biomodulator drug combined with external beam radiotherapy. *Int J Radiat Oncol Biol Phys* 2004;58:519–527.
25. Williams, S., Pettaway, C., Song, R., Papandreou, C., Logothetis, C., and McConkey, D. Differential effects of the proteasome inhibitor bortezomib on apoptosis and angiogenesis in human prostate tumor xenografts. *Mol Cancer Ther* 2003;2:835–843.
26. Berrada, M., Yang, Z., and Lehnert, S. Sensitization to radiation from an implanted 125I source by sustained intratumoral release of chemotherapeutic drugs. *Radiat Res* 2004;162:64–70.

Index

Page numbers followed by f and t indicate figures and tables, respectively.

C

D

E

F

H

J

K

O

P

Q

R